The Strangest Man

The Strangest Man

The Hidden Life of Paul Dirac, Quantum Genius

GRAHAM FARMELO

faber and faber

First published in 2009
by Faber and Faber Limited
3 Queen Square London WC1N 3AU

Typeset by Faber and Faber Limited
Printed and bound in England by MPG Books Ltd, Bodmin, Cornwall

A CIP record for this book
is available from the British Library

ISBN 978-0-571-22278-0

2 4 6 8 10 9 7 5 3 1

To my mother and the memory of my late father

Contents

Prologue 1

The Strangest Man 7

Abbreviations in Notes 439
Notes 441
Bibliography 495
List of Plates 507
Acknowledgements 509
Index 515

[T]he amount of eccentricity in a society has generally been proportional to the amount of genius, mental vigour, and moral courage which it contained. That so few now dare to be eccentric, marks the chief danger of the time.

JOHN STUART MILL, *On Liberty*, 1869

We are nothing without the work of others our predecessors, others our teachers, others our contemporaries. Even when, in the measure of our inadequacy and our fullness, new insight and new order are created, we are still nothing without others. Yet we are more.

J. ROBERT OPPENHEIMER, Reith Lecture, 20 December 1953

Prologue

[A] good deal of unkindness and selfishness on the part of parents towards children is not generally followed by ill consequences to the parents themselves. They may cast a gloom over their children's lives for many years.
SAMUEL BUTLER, *The Way of All Flesh*, 1903

All it took was a single glass of orange juice laced with hydrochloric acid. A few minutes later, it was clear that his digestive problems were due to a chronic deficiency of stomach acid. For months, he had been admitted to hospital every few weeks to be fed vitamins intravenously, but the doctors had no idea why his digestion was so poor. Now, following the orange-juice experiment, a laboratory test on the chemical contents of his stomach confirmed the conclusion that his stomach contained far too little acid. The simple prescription of a pill to be taken after every meal ended almost eight decades of digestive problems. As a result, Kurt Hofer, the friend who suggested the experiment and made the correct diagnosis, became the reluctant health guru to Paul Dirac, one of the most revered – and strangest – figures in the history of science.

Hofer and Dirac both worked at Florida State University but otherwise appeared to have little in common. Hofer – just over forty years of age – was a top-drawer cell biologist, a spirited raconteur who told all comers of his early family life among Austrian mountain farmers and his moment of cinematic glory as a well-paid extra in *The Sound of Music*. Hofer's eyes glittered when he told his stories, his thickly accented voice swooped and surged for emphasis, his hands chopped and shaped the air as if it were dough. Even in this lively company, Dirac was unresponsive, speaking only when he had a pressing question to ask or, less often, a comment to make. One of his favourite phrases was: 'There are always more people who prefer to speak than to listen.'[1]

Dirac was one of the pre-eminent pioneers of quantum mechanics, the modern theory of atoms, molecules and their constituents. Arguably the most revolutionary scientific breakthrough of the

I

twentieth century, quantum mechanics uprooted centuries-old prejudices about the nature of reality and what can, in principle, be
known for certain about the universe. The theory also proved to be
of enormous utility: it underpins the whole of modern microelectronics and has answered many basic questions that had long defied
straightforward answers, such as why electricity flows easily through
wire but not through wood. Yet Dirac's eyes glazed over during talk
of the practical and philosophical consequences of quantum physics:
he was concerned only with the search for the fundamental laws that
describe the longest strands in the universe's fabric. Convinced that
these laws must be mathematically beautiful, he once – uncharacteristically – hazarded the unverifiable conjecture that 'God is a mathematician of a very high order.'[2]

The ambitions of Kurt Hofer were more modest than Dirac's.
Hofer had made his name in cancer and radiation research by carefully carrying out experiments and then trying to find theories to
explain the results. This was the conventional, bottom-up technique
of the English naturalist Charles Darwin, who saw his mind 'as a
machine for grinding general laws out of large collections of facts'.[3]
Dirac, a classic example of a top-down thinker, took the opposite
approach, viewing his mind as a device for conjuring laws that
explained experimental observations. In one of his greatest achievements, Dirac used this method to arrange what had seemed an
unlikely marriage – between quantum mechanics and Einstein's theory
of relativity – in the form of an exquisitely beautiful equation to
describe the electron. Soon afterwards, with no experimental clues to
prompt him, he used his equation to predict the existence of antimatter, previously unknown particles with the same mass as the corresponding particles of matter but with the opposite charge. The
success of this prediction is, by wide agreement, one of most outstanding triumphs of theoretical physics. Today, according to the cosmologists' standard theory of the early universe – supported by a
wealth of observational evidence – antimatter made up half the material generated at the beginning of the Big Bang; from this perspective,
Dirac was the first person to glimpse the other half of the early universe, entirely through the power of reason.

Hofer liked to compare Dirac with Darwin: both English, both
uncomfortable in the public eye, both responsible for changing the
way scientists think about the universe. A decade before, Hofer was

amazed when he heard that Dirac was to move from one of the world's leading physics departments, at the University of Cambridge in England, to take up a position at Florida State University, whose physics department was ranked only eighty-third in the USA. When the possibility of his appointment was first mooted, there were murmurings among the professors that it was unwise to offer a post to an old man. The objections ended only after the Head of Department declared at a faculty meeting: 'To have Dirac here would be like the English faculty recruiting Shakespeare.'[4]

Around 1978, Hofer and his wife Ridy began to pay visits to the Diracs on most Friday afternoons, to wind down for a couple of hours after the week's work. The Hofers set off from their home near the campus in Tallahassee at about 4.30 p.m. and took the two-minute walk to 223 Chapel Drive, where the Diracs lived in a modest, single-storey house, a few paces from the quiet residential street. At the front of the house was a flat, English-style lawn, planted with a few shrubs and a Pindo palm tree. The Hofers were always welcomed warmly by Dirac's smartly dressed wife Manci, who laughed and joked as she dispensed sherry, nuts and the latest faculty gossip. Dirac was painfully spare and round-shouldered, dressed casually in an open-necked shirt and an old pair of trousers, content to sit and listen to the conversation around him, pausing occasionally to sip his glass of water or ginger ale. The chatter ranged widely from family matters to local politics at the university, and from the earnest utterances of Mrs Thatcher on the steps of Downing Street to the most recent sermon from Jimmy Carter in the White House garden. Although Dirac was benign and receptive during these conversations, he was so reserved that Hofer often found himself trying to elicit a response from him – a nod or a shake of the head, a few words, anything to make the conversation less one-sided. Just occasionally, Dirac would be moved to contribute a few words about one of his private enthusiasms – Chopin's waltzes, Mickey Mouse and any television special featuring Cher, the brassy chanteuse.

During the first two years or so of these visits, Dirac showed no sign of wanting to talk about himself or of having any deep feelings, so Hofer was ill prepared when, one Friday evening in the spring of 1980, Dirac's vacuum-packed emotions burst into the open. 'I remember it well. It was pretty much like all my other visits except that I was alone,' Hofer says. 'My wife decided not to come as she was tired, heavily pregnant with our first child.' At the beginning of the visit, Dirac behaved normally and looked alert and ready to absorb the conversations around him. After the customary pleasantries, the Diracs took Hofer by surprise when they ushered him through the formal front room – where they always talked during their Friday chats – to the less formal family room at the rear of the house, adjoining the' kitchen and overlooking the garden. The Diracs' pre-war taste was reflected in the decor of this room, dominated by the wood of the floorboards, the panelling on all four walls, and the huge 1920s sideboard covered with framed photographs of Dirac in his prime. A mock-Baroque chandelier hung from the ceiling and, on most of the walls, there were paintings with no trace of modernity.

As usual, Manci and Hofer chatted convivially while the frail Dirac sat motionless in his favourite old chair, occasionally looking through the glass sliding doors to the garden. For the first half an hour or so of the conversation, he was, as usual, mute but came vibrantly to life when Manci happened to mention his distant French ancestors. Dirac corrected one of Manci's historical facts and began to speak about his family origins and his childhood in Bristol, talking fluently in his quiet, clear voice. Like a well-rehearsed actor, he spoke confidently, in carefully articulated sentences, without pausing or correcting himself. 'I was startled – for some reason, he had decided to take me into his confidence,' Hofer says. 'I'd never seen him talk so eloquently in private.'

Dirac described his roots in the rural villages of Bordeaux, in western France, and how his family migrated to the Swiss canton of Valais at the end of the eighteenth century. It was in Monthey, one of the region's industrial towns, that his father was born. As soon as Dirac began to talk about his father, he became agitated, and he turned away from his wife and Hofer, adjusting his pose so that he was staring straight into the fireplace. Hofer was now looking directly at the profile of the top half of Dirac's body: his hunched shoulders, his high forehead, his straight and upward-pointing nose, his white smudge of a moustache. The air conditioning and television were switched off, so the room was silent except for the occasional rumblings of traffic, the barking of neighbourhood dogs, the rattling of the lid on the simmering casserole in the kitchen. After spelling out his ancestry with the precision of a genealogist, Dirac reached the part of his story where his father arrived in Bristol, married Dirac's mother and started a family. His language remained simple and direct, but, as he began to talk about his childhood, his voice tightened. Hofer, watching Dirac's silhouette sharpen with the fading of the early evening light, was transfixed.

'I never knew love or affection when I was a child,' Dirac said, the normally neutral tone of his voice perceptibly tinged with sorrow. One of his main regrets was that he, his brother and younger sister had no social life but spent most of their time indoors: 'we never had any visitors'. The family was dominated, Dirac recalled, by his father, a tyrant who bullied his wife, day in, day out, and insisted that their three children speak to him in his native French, never in English. At mealtimes, the family split into two: his mother and siblings would eat in the kitchen and speak in English, while Dirac sat in the dining room with his father, speaking only in French. This made every meal an ordeal for Dirac: he had no talent for languages, and his father was an unforgiving teacher. Whenever Dirac made a slip – a mispronunciation, a wrongly gendered noun, a botched subjunctive – his father made it a rule to refuse his next request. This caused the young Dirac terrible distress. Even at that time, he had digestive problems and often felt sick when he was eating, but his father would refuse him permission to leave the table if he had made a linguistic error. Dirac would then have no option but to sit still and vomit. This did not happen just occasionally, but over and over again, for years.

Hofer was aghast, scarcely able to believe his ears. 'I felt extremely embarrassed, like I was witnessing a friend pouring out his most terrible secrets to his psychiatrist,' he recalls. 'Here he was, a man famous for equability and his almost pathological reticence, openly talking of the demons that had haunted him for nearly seventy years. And he was as angry as if these awful events had happened yesterday.'

Manci barely stirred, except once to bring food and alcohol and to slow down the preparations for dinner. She knew that on the very rare occasions her husband chose to tell his story, it was best to keep well out of his way and to let him get it off his chest. As the evening turned colder, she brought him a blanket and draped it over his legs, covering him from his lap down to his ankles. Hofer braced himself as Dirac resumed and explained why he was so quiet, so ill at ease with normal conversation: 'Since I found that I couldn't express myself in French, it was better for me to stay silent.'

Dirac then moved on to talk about other members of his family: 'I was not the only one to suffer,' he said, still agitated. For thirty-seven years, his mother was locked in a disastrous marriage to a man who treated her like a doormat. But it was Dirac's brother who felt the brunt of their father's insensitivity: 'It was a tragedy. My father bullied him and frustrated his ambitions at every turn.' In what appeared to be a change of tack, Dirac mentioned that his father always appreciated the importance of a good education and that he was respected by his colleagues as a conscientious, hard worker. But this was only a brief respite. Seconds later, Dirac was struggling to control his rage when he spelt out the conclusion he eventually reached about the extent of his debt to his father: 'I owe him absolutely nothing.' That final rasp made Hofer flinch; he could not help but grimace. Dirac hardly ever spoke an unkind word about anyone, but here he was, denouncing his own father with a vehemence most people reserve for the cruellest abusers.

Dirac stopped talking abruptly, just after nightfall. His monologue had lasted over two hours. Hofer knew that any words from him would be inappropriate, so he said his subdued goodbyes and walked home, numb and drained. Soon to be a father himself, he reflected on his own youth as part of a close and loving family: 'I simply could not conceive of any childhood as dreadful as Dirac's.'[5] Time tends to embellish, distort and even create childhood memories: could it be that Dirac – usually as literal-minded as a computer – was exaggerating? Hofer could not help asking himself, over and again: 'Why was Paul so bitter, so obsessed with his father?'

Later that night, after talking with his wife Ridy about Dirac's account of his young life, Hofer made up his mind to find out more about it. 'I thought he might open up again during our later get-togethers.' But Dirac never mentioned the subject again.

One

English home life to-day is neither honorable, virtuous, wholesome, sweet, clean, nor in any creditable way distinctively English. It is in many respects conspicuously the reverse [. . .].
GEORGE BERNARD SHAW, Preface to *Getting Married*, 1908

As Kurt Hofer had seen, the elderly Paul Dirac was fixated on his father Charles. But most of Dirac's acquaintances knew nothing of this: at home, he allowed no photographs of his father to be displayed, and he kept his father's papers locked in his desk. Dirac examined them from time to time and talked with distant relatives about his father's origins, apparently still trying to understand the man he believed had blighted his life.[1]

Dirac knew that his father had endured a childhood no less miserable than his own. By the time Charles Dirac was twenty, in 1888, he had done three stints of national service in the Swiss army, dropped out of university in Geneva and left home, without telling his family where he was heading.[2] He became an itinerant teacher of modern languages – the subject he had studied at university – and held posts in Zurich, Munich and Paris, before he fetched up two years later in London. English was one language that he did not speak well, so it is not clear why he chose to live in Britain; perhaps it was because it was the world's wealthiest economy, with plenty of teaching jobs at relatively high salaries.

Six years later, Charles Dirac had acquired a sheaf of complimentary references. One, written by the headmaster of a school in Stafford, stated that Monsieur Dirac 'is possessed of very great patience combined with firmness [. . .] I believe he is much liked both by his colleagues and pupils.' His employer in Paris had praised 'his capacity to analyze and generalize, which enabled him to point out my mistakes and help me to ascertain scientifically why they were mistakes'. Charles settled in Bristol, a city famous for the high quality of its schools, and he became Head of Modern Languages at the rapidly expanding Merchant Venturers' School on 8 September 1896, contracted to teach thirty-four hours a week for an annual salary of one hundred and eighty pounds.[3] He stood out among the teachers because of his conscientiousness, his thick Swiss-French accent and his appearance: a short, stocky, slow-moving man with a drooping moustache, a receding hairline and a face dominated by a huge forehead.

Mellowest of British industrial cities, Bristol was known for the friendliness of its people, its mild and wet climate and the hilly roads that wend their way down to the moorings on the river Avon, eight miles from the coast. Bristol was then a thriving manufacturing centre, producing Fry's chocolate, Wills's cigarettes, Douglas motorcycles and many other commodities. Together, these industries had eclipsed the declining trade in shipping, which had been the city's main source of wealth for centuries, some of it based on the slave trade.[4] Most of the city's wealthiest maritime figures were members of the Merchant Venturers' Society, a secretive group of industrialists with a strong philanthropic tradition. It was the generosity of the Society that had made possible the founding of Charles's school together with the high standard of its workshop and laboratory facilities.[5]

During a visit to the Central Library a few months after his arrival in Bristol, Charles met Florence Holten, the guileless nineteen-year-old librarian who would become his wife. Though no beauty, she was attractive and possessed features that she would later pass on to her most famous child: her oval face was framed by dark, curly hair, and a firm nose darted out from between her dark eyes. Born into a family of Cornish Methodists, she was brought up to believe that Sunday should be a day of rest, that gambling was sinful and that the theatre was decadent and best avoided.[6] She had been named after the nurse Florence Nightingale, whom her father Richard met during the Crimean War, where he served as a young soldier before becoming a seaman.[7] He was often away for months at a time, leaving behind his wife and six children, of whom Flo was second eldest.[8]

Flo Holten and Charles Dirac were an odd couple. She was twelve years younger than him, a daydreamer uninterested in pursuing a career, whereas Charles was strong-minded and industrious, devoted to his job. The couple had been raised in different, scarcely compatible religions. She was from a family of devout Methodists and so had been raised to frown on alcohol, whereas Charles had been brought up in a Roman Catholic home and liked a glass of wine with his meals. Catholicism had been the cause of riots in Bristol and other

English cities, so Charles may at first have kept his religious beliefs to himself. If he did disclose them, his relationship with the young Flo would have raised eyebrows in her circle.[9]

Despite the possible sectarian tensions, by August 1897 Charles and Flo were engaged, though Flo was feeling sore. Charles had chosen to 'break the spell' of their relationship to visit his mother Walla, a dressmaker in Geneva, leaving his fiancée to sulk in Bristol's incessant rain. His father had died the year before. He had been a highly strung junior schoolteacher and later a stationmaster at Monthey station in south-west Switzerland but was dismissed for repeatedly being drunk on duty, leaving him plenty of time to pursue his interest in writing romantic poetry.[10] The Swiss stretch of the Rhône valley had been home to the Dirac family since the eighteenth century, when – according to family lore – they moved from the Bordeaux area in western France. The names of many of the towns in this region and its vicinity end in -ac, such as Cognac, Cadillac and the little-known village, about ten kilometres south of the Angoulême, called Dirac.[11] Charles believed his family had originated there, but there is no evidence for this among the family records, now stored in the town hall of Saint Maurice (near Monthey), where the colourful Dirac coat of arms – featuring a red leopard with a three-leaf clover in its right paw, below three downward-pointing pine cones – is one of many painted on the walls.[12]

Uneven postal delays caused Charles's letters from Switzerland to arrive out of order, infuriating Flo, who wished that 'letters went by electricity like tram cars'; a century would elapse before long-distance lovers benefited from the type of communication she was vaguely envisioning – electronic mail.[13] Lonely and disconsolate, she repeatedly read Charles's notes and, when her family was not looking over her shoulder, replied with newsy letters of how they could not resist teasing her about her pining for 'my own boy'. Struggling to put her longing into words, she sent him a poem full of ardour; in return, he sent a posy of Alpine flowers which she hung round his photograph.

Almost two years later, Flo and Charles were married 'according to the rites and ceremonies of the Wesleyan Methodists' in Portland Street Chapel, one of the oldest and grandest of Bristol's Methodist churches. The couple moved into Charles's residence in 42 Cotham Road – probably in rented rooms – a short walk from Flo's family home in Bishopston, in the north of the city. Following custom and practice, Flo stopped doing paid work and stayed at home to do the housework and read about the first skirmishes of Britain's latest imperial venture, the Boer War in South Africa. Soon, she had other things on her mind: the Diracs' first son Felix was born on the first Easter Sunday of the new century.[14] Nine months later, the country mourned the passing of an era when Queen Victoria, having reigned for an unprecedented sixty-three years, died in the arms of her grandson, Kaiser Wilhelm II. Soon after a period of national grief, mitigated only by relief at the ending of the war, the family prepared for a new beginning of its own. In July 1902, they moved into a slot in one of the new terraces on Monk Road, to a roomier, two-storey home that Charles named after his native town of Monthey. The Diracs would soon need extra space as Flo was again pregnant, with only a few weeks to go before the birth.[15]

On Friday, 8 August 1902, Bristol's eyes were on London, where King Edward VII was to be crowned on the following day. Thousands took the train from Bristol to the capital to see the coronation procession, but the celebrations were a sideshow in the Dirac household. On that Friday morning, Flo gave birth at home to a healthy six-pound boy, Paul Adrien Maurice Dirac. He was, as his mother later recalled, a 'rather small', brown-eyed baby, who slept contentedly for hours in his pram in the patch of the front garden.[16] His mother worried that he ate less food than most children, but the family doctor reassured her that Paul 'was OK, perfectly proportioned'.[17] His parents nick-named him 'Tiny'.

When Felix and Paul were young, they resembled each other, each a quiet, round-faced cherub with a thick bonnet of black, curly hair. Flo dressed them stylishly in thick woollen waistcoats topped with stiff, white-lace Eton collars that reached out to their shoulders, like the wings of a huge butterfly. From family letters and Flo's later testimony, it appears that the boys were close and liked to be with their father, whose top priority was to encourage them to learn. With the virtual absence of visitors and opportunities to mix outside their immediate family, Paul and Felix probably did not appreciate they were being brought up in a singularly unusual environment, a hothouse of private education overseen by a father who would speak to them only in French and a mother who would talk only in English.

According to one witness, the young Paul Dirac believed that men and women spoke different languages.[18]

But Paul and Felix were let off the leash occasionally. Their mother sometimes took them to the Bristol Downs so that they could play on the vast expanse of grassy parkland stretching from the cliffs of the Avon Gorge to the edges of the city's suburbs.[19] From their favourite spot on the Downs, the Dirac boys had an excellent view of the Clifton Suspension Bridge, one of the most famous creations of Isambard Kingdom Brunel, the charismatic engineer who also left Bristol with its Floating Harbour and Temple Meads railway station, two of the city's finest monuments.

In the summer, the family would take a bus trip to the beach at nearby Portishead, where the boys learned to swim. Like most families of their modest means, the Diracs rarely took vacations, but, in 1905, they went to Geneva to visit Charles's mother, who had an apartment a stone's throw from the lake and ten minutes' stroll from the railway station.[20] The brothers spent hours by the lakeside statue of the philosopher Jean-Jacques Rousseau, playing together and watching the artificial geyser shoot its jet of water ninety metres towards the sky. When the seventy-year-old Dirac told this story, one of his earliest memories, he liked to point out that his first trip to Switzerland took place at the same time as Einstein was having his most successful spurt of creativity in Berne, only a short train journey from Geneva. That year, Einstein wrote four papers that changed the way people think about space, time, energy, light and matter, laying the foundations of quantum theory and relativity. Twenty-three years later, Dirac would be the first to combine the theories successfully.

There exist two vivid snapshots of life in the Dirac household in the summer of 1907, shortly before Paul started school, a year after the birth of his sister Betty. The first is the correspondence between Charles Dirac and his family when he was in Trinity College, Cambridge, attending the International Esperanto Congress. Earlier in the year, Charles had qualified to teach the language, which he championed in Bristol for the rest of his life.[21] When Charles was away, his family showered him with loving notes. Flo's affectionate gusto was almost as intense as it had been in the heat of their passion, ten years before. Up to her ears in the chaos of having to look after the three children – taking them for walks, feeding the pet mice,

cooking Paul his favourite jam tarts – she had the undivided attention of her boys: 'It is very quiet without you, the boys are sticking to
me for a change.' She assured her husband that his family at home
'all had a nice dinner, mutton, peas, junkets [a sweet dessert]'. The
boys missed Charles terribly, Flo told him, just as she did: 'I shall
miss you in the bye-bye [i.e. bed] tonight.'[22] Flo enclosed in her letters to Charles notes from Felix and from Paul, who wrote in stick-
letter capitals of the welfare of the mice and, most importantly, his
love for him: 'Tiny hopes Daddie has not forgotten little Tiny' and 'I
love you very much. Come home soon to your own Tiny Dirac
xxxxx.' Charles replied with a postcard, written mainly in English
but with a little French, promising to bring home some Esperanto
chocolate and concluding, 'I would not go out if I did not have to.'

Nothing in this loving correspondence bears any sign of the terrible home life that Dirac described to Kurt Hofer. Charles's use of
English words appears to be inconsistent with the French-only linguistic regime that Paul claimed his father practised, and his father's
tone bears no sign of the heartlessness that Paul remembered.

It is clear that Charles was as keen as any other father to keep a
photographic record of his children. At about this time, he purchased a camera – probably one of the fashionable Kodak box
Brownies – to take pictures of his children, many of them showing
Felix, Paul and Betty reading avidly. Charles also wanted a portrait
of his family to be taken by a professional and for the result to be
printed on postcards for family and friends. The photograph, the
only surviving image of the entire family, was taken on 3 September
and gives us the second impression of the Diracs in 1907.[23] Flo
looks demure and serious, her long hair tied up at the back, baby
Betty on her lap. Felix is leaning towards her, smiling broadly and
looking directly into the camera like Paul, whose left arm rests on
his father's right leg, apparently seeking reassurance. Charles leans
forward to the camera, eagerly, his alert eyes shining. He steals the
picture.

This photograph of a happy family is subverted by Dirac's later
memories of trauma and unhappiness. In one stinging memory, his
parents bawled at each other in the kitchen while he and his siblings
stood in the garden, frightened and uncomprehending. He once
remarked in an interview that his parents 'usually ate separately',
though twenty years later friends wrote that he told them he 'never'

saw his parents have a meal together – apparently a rare example of his being caught exaggerating.[24] The rift between his parents was, according to Dirac, responsible for his dining-table ordeals. Three times every day, the tinkling of cutlery, the clatter of saucepans on the gas stove, the waft of cooking smells through the house presaged the ritual that he loathed. In none of the surviving accounts of the dining arrangements did he explain why he alone sat with his father, while his brother and sister ate with their mother in the kitchen. The only partial explanation that Dirac ever gave was that he could not sit in the kitchen because there were insufficient chairs.[25] But this says nothing about the mystery of why Charles singled out him, not Felix or Betty, for special treatment.

The dining ritual was particularly harrowing on winter mornings, Dirac remembered. He would sit at the table with his father in the silent room, warmed by the burning coal in the fireplace and lit by a few oil lamps. Charles would be dressed in his three-piece suit, ready to cycle to the Merchant Venturers' School, always anxious not to be late for Assembly. His wife, scrambling and disorganised in the kitchen, made his anxieties worse by serving breakfast – usually large portions of piping-hot porridge – much too late for comfort. While he was waiting for his breakfast, Charles gave his first French lesson of the day to his younger son. Quite apart from Dirac's hatred of these arrangements, he grew to dislike eating mainly because his parents insisted, even when his appetite had been sated and he felt sick, that he must eat every morsel of food on his plate.[26]

For the young Dirac, this was normality. In his early thirties, he wrote to a close friend of the sourness of his home life: 'I did not know of anyone who liked someone else – I thought it did not happen outside novels.'[27] In another letter, he wrote: 'I found it to be the best policy as a child [. . .] to make my happiness depend only on myself and not on other people.'[28] According to Dirac, his best defence against the unpleasantness and hostility he perceived all around him was to retreat into the bunker of his imagination.

Dirac first experienced the company of children outside his family shortly after his fifth birthday, when he started at the small and intimate Bishop Road Junior School.[29] This was his first opportunity to socialise, to get a sense of other children's lives, of other domestic customs and practices. But he apparently made no attempt to talk to other children: he remained silent and continued to live in his own private world.

The school was round the corner from his home, so close that he could hear its bell ringing at the start of the day. Despite the daily hurry of the breakfast routine, he and his brother always arrived on time.[30] Dirac's class typically consisted of about fifty children crammed into a room about twenty-five feet square, the pupils sitting in rows of identical wooden desks, learning in an atmosphere that was, by today's standards, extremely disciplined and competitive.[31] At the end of their time at school, children had to compete for scholarships that would help to pay for their senior education. Success meant that the child's parents would have to pay little or nothing; failure often meant that the child would be sent out to work.

Paul and Felix were recognisably brothers, but Felix had a rounder face, was a few inches taller and was more heavily built.[32] He was placid and well behaved, though given to lapses of concentration, as his headmaster pointed out when he wrote across his school report: 'The boy appears to me to be a perpetual dreamer. He must wake up!' Felix appears to have taken the advice, as he soon improved and did well in most subjects, especially drawing.[33]

From Dirac's later descriptions of his early life, we might expect him to have been an unhappy child, but there are no signs of this in the extant descriptions of him at the time. Twenty-seven years later, when his mother wrote a short poem about him for her own amusement, she described him as 'a cheerful little schoolboy', and added that he was 'contented' and 'happy'.[34] In official reports written when he was eight, teachers at Bishop Road do not comment on his demeanour, saying only that he was 'well behaved', 'an intelligent boy' and 'a very steady worker'. But there are indications that Dirac was not performing to his potential. A few teachers allude to this, most notably the Headmaster, who, on seeing that Dirac had only just managed to be ranked in the top third of the class, wrote on his report in November 1910, 'I expected to find you higher.'[35]

Among the boys Dirac did not get to know at Bishop Road School was Cary Grant, then known as Archie Leach and living in poverty about half a mile from Monk Road. In the classrooms and playground of the Bishop Road School, Dirac acquired the distinctively warm Bristol accent, which sounds slightly hickish to other native English speakers, evocative of farmers in the south-west of the country. Like other young natives of Bristol, Dirac and Grant added an L to the pronunciation of most words that end in the letter A, a practice that is now dying out, though many English people still recognise Bristol as the only city in Britain to be able to turn ideas into ideals, areas into aerials.[36] Cary Grant shed this accent when he emigrated to the United States, but Dirac kept it all his life. He spoke with a gentle intonation and an unassuming directness that would surprise the many people who expected him to talk like the plummy-voiced English intellectual of popular caricature.

Like his brother, Dirac's ranking in the class gradually improved. He was good though not exceptional at arithmetic, and he did well in most subjects that did not involve his meagre practical skills. Soon after his eighth birthday, his teacher described him as 'An intelligent boy, but must try hard with his hand-work', drawing attention to his poor marks for handwriting (45 per cent) and drawing (48 per cent). His disappointed teacher commented that he should have done better than thirteenth in the class. Two years later, Dirac was consistently at or near the top of his class, his overall grade occasionally lowered by his relatively weak performance in history and brush-work.[37] At home, he pursued his extra-curricular hobby of astronomy, standing in his back garden at night to check the positions of the visible planets and constellations and, occasionally, to follow the track of a meteor hurtling across the sky.[38]

The school did not teach science but did give classes in freehand drawing and also technical drawing, a subject that provided Dirac with one of the foundations for his unique way of thinking about science. His mother later drew attention to his 'most beautiful hands', suggesting that his long and bony fingers equipped him well to be an artist.[39] Technical drawing, used by engineers to render three-dimensional objects on a flat piece of paper, is now taught at very few English junior schools, and rarely at senior level. Yet, in the early twentieth century, it was a compulsory subject for half the pupils: for a few lessons each week, the class would split into two: the girls studied needlework, while the boys were taught technical drawing. In these classes, Dirac learned to make idealised visualisations of various manufactured products by showing them from three orthogonal points of view, making no allowance for the distortions of perspective.[40]

Britain was among the slowest of the wealthier European countries to introduce technical drawing into its schools and did so only in the wake of the Great Exhibition in 1851. Although the Exhibition was a great popular success, the most perceptive of its 6.2 million visitors saw evidence that mass technical education in Britain would have to improve substantially if the country were to retain its economic hegemony against growing competition from the USA and Germany. The Government agreed, enabling the Great Exhibition's prime mover Sir Henry 'King' Cole to change the technical curriculum of English schools so that boys were taught technical drawing and given an appreciation of the beauty of manufactured objects as well as natural forms.[41] There was, however, a backlash to this practical notion of beauty in the form of the Aesthetic Movement, which flourished in England from the mid-1850s. The movement's leader in France was the flamboyant poet and critic Théophile Gautier, a weight-lifting habitué of the Louvre's Greek galleries.[42] His phrase 'Art for art's sake' became the motto of the English aesthetes, including Oscar Wilde, who shared Gautier's belief that formal, aesthetic beauty is the sole purpose of a work of art. This view would later be distantly echoed in Dirac's philosophy of science.

Sir Henry Cole's reforms endured: the guidelines set out by him and his associates were being used in Bishop Road School when Dirac began his formal schooling. In 1909, the educationist F. H. Hayward summarised the prevailing philosophy that underlies the contemporary teaching of art: 'drawing aims at truth of conception and expression, love of beauty, facility in invention, and training in dexterity [. . .] nature study and science lessons cannot proceed far without it.'[43] Hayward urged that students should practise their drawing skills by trying to represent accurately both natural and manufactured objects, including flowers, insects, tables, garden sheds and penknives. In autumn 1912, Dirac was asked to draw a penknife, and he did it competently enough – like all his other drawings, it includes not a line of embellishment.[44]

The school took pains to teach its pupils how to write legibly, according to textbook rules that Dirac and his brother apparently studied closely.[45] They developed a similar style of handwriting – consistent with the rules set out in the books they studied – neat, easy to read and virtually devoid of flourishes, except for the unusual forming of D, with a characteristic curl at the top left. Dirac did not change this calligraphy one iota for the rest of his life.

In the early summer of 1911, school inspectors noted that 'the boys who are particularly bright and responsive are being carefully trained in habits of self-reliance and industry.' Nearly three years later, when Dirac was in his final year at the school, the inspectors visited Bishop Road again and wrote warmly of this 'progressive' school and the practical education it offered: 'a keen, vigorous and thoughtful head [teacher]. Staff [are] earnest, painstaking [. . .] Drawing is well taught and handwork is resourceful, the boys make a number of useful models and are allowed considerable freedom in their choice while the work is so taken as to train them in habits of self reliance, observation and careful calculation and measurement.'[46]

Bishop Road School wanted to give its pupils the skills they needed to get good jobs. But, for Dirac, the most important consequence of this practical approach was that it helped to shape his thinking about how the universe works. As he was sitting at his desk in his tiny Bristol classroom, producing an image of a simple wooden object, he had to think geometrically about the relationships between the points and lines that lie in a flat plane. In his mathematics classes, he also learnt about this type of Euclidean geometry, named after the ancient Greek mathematician who reputedly discovered it. So, Dirac studied geometry using both visual images and abstract mathematical symbols. Within a decade, he would transfer this geometric approach from concrete technological applications to the abstractions of theoretical physics – from an idealised, visual representation of a wooden fountain-pen stand to an idealised, mathematical description of the atom.

Later in life, Dirac would say that he never had a childhood. He knew nothing of the rites of passage of most other young boys – long weekend afternoons spent stealing eggs from birds' nests, scrumping from nearby orchards, dashing out in front of trams. In many ways, as a child he seems to have behaved much as Newton had done. 'A sober, silent, thinking lad [. . .] never was known scarce to play with the boys abroad' was how one of Newton's friends described him: the description applies equally well to Dirac as an infant.[47]

Dirac was not interested in sport, with the exception of ice-skating, which he learned with Betty and Felix at the nearby Coliseum rink, the talk of Bristol when it opened in 1910.[48] Decades later, his mother recalled that he would sit quietly, reading books that he had placed neatly around him and learning long poems that he would recite to his family.[49] She shed some light on his sheltered childhood when she spoke to reporters in 1933: '[his father's] motto has always been to work, work, work, and if the boy had showed any other tendencies, then they would have been stifled. But that was not necessary. The boy was not interested in anything else.'[50] There is little doubt that Charles Dirac impressed his sedulous work ethic on his younger son, who later wrote admiringly of his father's conscientiousness:

One day while cycling [to school, my father fell off his bike], trying to avoid a child who ran out in front of him, and broke his arm. He was very conscientious, so he continued to the school and continued with his teaching, in spite of the broken arm. Eventually, the head master found out about it and sent him home, and told him not to come back until he was better.[51]

Paul was also aware that his father was exceptionally careful with money. In April 1913, Charles took the biggest financial decision of his life by purchasing a more expensive and more spacious home. The family moved from the cramped terrace of Monk Road to a neat semi-detached residence a few minutes' walk away in a slightly more salubrious part of Bristol, at 6 Julius Road. The Diracs now had a home befitting Charles's status in the community, with separate rooms for their two boys so that Dirac now had a place to escape, a private place where he could work alone. The family still kept themselves to themselves, inviting no visitors into their home, apart from Flo's family, her guests – all female – at a monthly afternoon tea party and the steady stream of pupils who took private language lessons from her husband.[52]

Like many parents, Charles entered all his children for scholarship exams.[53] When Felix was nine years old, he failed one of these exams, leading his father to demand an explanation from his teachers; Betty also failed the exam a few years later. Paul had no such problems: he passed every scholarship exam with flying colours and, thus, unlike Felix and Betty, ensured he was educated at minimal expense to his parents.

Dirac could see new technology making its imprint on Bristol. The city centre was a patchwork of centuries-old buildings and brand-new ones, many of them emblazoned with advertisements for new services and products.[54] Open-topped motor cars vied for space on the roads with horse-drawn carriages, bone-shaking bicycles and the trams that made their jerky way round the city. When a programme of road construction began, in the early years of the century, cars began to dominate the city. In late 1910, Dirac had witnessed the beginnings of the Bristol aviation industry, one of the first and largest in Britain. The leading figure in this new Bristol industry was the local entrepreneur Sir George White, who founded the British and Colonial Aeroplane Company and supervised the building of some of the earliest aircraft in a tram shed in Filton, a few miles north of the Diracs' home. Long afterwards, Dirac told his children that he would rush out into the back garden to see aeroplanes precariously taking off from the new airfield less than a mile away.[55] It seems that he wanted to find out more about this new technology: among the papers he kept from his youth were details of a programme at a local technical college, beginning in December 1917: 'Ten Educational Lectures on Aeronautics'.[56]

Dirac and his brother stood out among the boys in Bishopston as they both spoke good French even before they started school. According to one report, local boys would stop the Dirac brothers on the streets and ask them to speak a few sentences of French.[57] This knowledge of French was also obvious to the students at their next school, where the language was taught by the school's most feared disciplinarian – their father.

Two

In the world of commerce,
In the crafts and arts,
Sons of her are honour'd
Nobly bear their parts;
While in sports and pastimes
They have made a name,
Train'd to wield the willow,
Learn'd to 'play the game'.
　　　Verse of the Merchant Venturers' School song[1]

On 4 August 1914, when Dirac was preparing to start at senior school, he heard that Britain was at war – the first conflict to involve every industrialised country in Europe. 'The European War', which would claim more British lives than any other, was to be the backdrop to the whole of his secondary education at the Merchant Venturers' School.

Like most other British cities in the UK, Bristol quickly prepared for the war, the urgency of the preparations heightened by the statement by the Boer War hero Lord Kitchener that the conflict would be decided by Britain's last million men. On the last day of August, in his capacity as Secretary of State for War, Kitchener sent a telegram to the Bristol Citizens' Recruiting Committee asking them to form a battalion of 'better class young men', and within a fortnight some 500 professional men had volunteered for the 'Twelfth Gloucesters', part of 'Kitchener's Army'.[2] Within a few weeks, the focus of the city's industries had changed from making money to supplying the military with everything from boots and clothes to cars and aircraft. Even the Coliseum ice-rink was commandeered as a site to assemble warplanes.

The first casualty lists were published barely a month after the declaration of war. The Bristol newspapers reported that the Allies had contained the initial German onslaught and that the battle lines had hardened to form a series of linked fortifications that stretched from the Franco-Belgian border on the coast right through to the Franco-

Swiss border, close to where Charles Dirac had been brought up. After Parliament passed the Aliens Registration Act, Bristol was one of the UK cities to be declared a 'prohibited area'. Charles had to register with the authorities as a foreigner, although he was hardly a threat to British security. By the time his elder son arrived at the all-boys Merchant Venturers' Secondary School, Charles had spent almost a third of his forty-eight years as its Head of French, doing more than any other teacher to extend the school's reputation for excellence beyond its established forte of technical subjects to modern languages.

It took Charles about fifteen minutes to cycle from his home to the school in Unity Street, in the heart of the city. The building was round the corner from the Hippodrome, Bristol's newest and swankiest music hall, where the young Cary Grant secured his first job, as a trainee electrician helping to operate the lighting rigs – soon after Paul started at the school. The school's Edwardian-Gothic building had been opened in April 1909, after the previous school on the site had burnt down. Everyone in the vicinity of the new school heard the clatter and rumblings from the basement workshops. The vibrations were so violent that the school's near-neighbour, Harvey's wine merchants, complained of the incessant disturbance to their cellars.[3]

The behaviour of Charles Dirac, whose pupils nicknamed him 'Dedder', emerges clearly in the testimonies of several of his fellow teachers and his students obtained by the Oxford University physicist Dick Dalitz in the mid-1980s. One of Dirac's fellow students, Leslie Phillips, gave a sense of the reputation of Monsieur Dirac:

He was *the* disciplinarian in the school, precise, unwinking, with a meticulous, unyielding system of correction and punishments. His registers, in which he recorded all that went on in the class were neat and cabalastic; no scholar could possibly understand their significance. Later, as a senior, I began to realize the humanity and kindness of the man, the twinkle in the eyes. But to us in the junior school, he was a scourge and a terror.[4]

Dedder was well known for his old-fashioned, strictly methodical approach to teaching and for springing random tests on his students, so that they always had to be prepared. If he caught them cheating in these tests or in their homework, he punished them with four half-hour periods of detention on Saturday afternoons. 'You never wrote this. Saturday at four for cribbing,' he told Cyril Hebblethwaite, later

Lord Mayor of Bristol. Most teachers routinely meted out corporal punishment by whacking errant boys across their backsides with a slipper or cane with an enthusiasm that bordered on the sadistic. But there is no record that Charles was fond of this form of chastisement, either at school or at home.

It is easy to imagine Monsieur Dirac's terrified pupils looking at Paul and Felix and wondering, probably out loud, 'What's he like at home?' Their father's strict classroom regime did, however, bring the benefit of a supply of comics that he had confiscated and brought home for his children.[5] The young Dirac read these cheap 'penny dreadfuls', black-and-white comics full of slapstick cartoons, juvenile jokes, detective stories, sensational tales of soldierly adventure and even the occasional topical reference to the build-up of the German military.[6] This one concession to popular culture in the Dirac home gave the young Paul an enduring taste for comics and cartoons.

The boys' mother also inflicted her share of pain on them by keeping their hair in tight curls and making them wear knickerbockers long after they were fashionable. They wore short breeches and garters so tight that, when they were removed, they each left an angry red line around the boys' legs. Dirac long remembered the taunts of his fellow pupils for being what nowadays would be damned as 'uncool'.[7] Such was his induction into that most characteristic of English anxieties, embarrassment.

Like all parents at that time, Charles and Flo worried that their children would catch tuberculosis, then responsible for one in every eight deaths in the UK.[8] It was particularly brutal in culling adult males: it accounted for more than one death in three among men aged fifteen to forty-four. The Dirac children were all born during the first decade of a government-funded anti-tuberculosis campaign that urged all citizens to get out into the open air, to take plenty of outdoor exercise and thus to get plenty of fresh air into their lungs. This philosophy may have encouraged Charles to decline to pay for his sons' tram fares to and from school and therefore to force them to walk there and back twice a day (they had lunch at home). Paul later resented what he believed was his father's meanness, though it probably led him to acquire a taste for taking long walks, soon to become one of his obsessions.[9]

*

It took only weeks for Dirac to establish himself as a stellar pupil at the Merchant Venturers' School. Except for history and German, he shone at every academic subject and so was usually ranked as the top student of his class.[10] The curriculum was wholly practical, with no room for music nor – to Dirac's relief – Latin and Greek. Instead, the school focused on subjects that would equip its boys to take up a trade, including English, mathematics, science (though not biology), some geography and history. What made the education at this school special was the high quality of the teaching of technical skills such as bricklaying, plasterwork, shoemaking, metalwork and technical drawing. For the previous fifty years, government inspectors had praised the school for giving one of the best technical educations available to any child in the country.[11]

In the school's laboratories, Dirac learned how to fashion pieces of metal into simple products, how to operate a lathe, how to cut and saw, how to turn a screw thread. Away from the clatter of machinery, the puddles of oil and the coils of swarf, he learned more of the art of technical drawing. These lessons built on the introductory classes at Bishop Road and showed Dirac how to produce plans for more complicated objects, developing his ability to visualise them from different angles. In his 'geometric drawing' classes, Dirac considered cylinders and cones, and he learned how to see in his mind's eye what happens when they are sliced at different angles and then viewed from various perspectives. He was also taught to think geometrically about objects that are not static but moving, and he learned how to draw the path of, for example, a point on the outside of a perfect circle as it rolls along a straight line, like a speck of dust on the outside of a wheel rolling along a road. To students who first encounter these shapes – curved, symmetrical and often intricate – they are a source of delight. If, as is likely, Dirac wondered how to describe these curves mathematically, his technical-drawing teachers would probably have been unable to enlighten him as they were usually former craftsmen with little or no mathematical expertise.

Although Dirac focused intensely on his college work, he was well aware of the scale of the war. All day long, convoys of trucks passed through Bristol with their supplies for the soldiers at the front, and huge guns were towed through the streets, shaking nearby buildings. At night, the streetlamps were extinguished to make the city a difficult target for the expected convoys of German airships, although they never arrived. The city's rapidly expanding aviation industry was on a war footing, so the threat of aerial bombing was clear to Dirac, who passed a busy aircraft factory every time he walked to and from school.[12]

Unreliable news of the conflict trickled back from the battlefronts through newspapers and by word of mouth. The Government's censorship policy prevented journalists from reporting on the full extent of the carnage, but readers could form a broad picture of the conflict and its ramifications. In February 1916, the Germans began their campaign to try to wear down the French Army at Verdun, and in July the British Army attacked on the Somme. Casualty figures soared, although the battle lines changed only slowly. In April 1917, the Germans introduced unrestricted U-boat warfare, aiming to cut supplies of food and other resources to the UK and thereby to force the enemy to the conference table. This brought the United States into the war, and Bristol celebrated by giving its schoolchildren a half-day holiday on 4 July, Independence Day.[13] Meanwhile, Russia was in turmoil, with the fall of the monarchy in February followed nine months later by Lenin's Bolshevik revolution.

Every day, the Dirac family read about these events in the local and national newspapers. The inside pages of the *Bristol Evening News* showed head shots of uniformed teenage soldiers, with a few lines that listed their regiment, when they fell and whom they left behind. Despite the depressing regularity of these reports, the recruitment campaigners maintained a constant flow of army volunteers, many of them younger than the minimum legal age of eighteen. Some of the boys shipped out to the killing fields were only a year older than Dirac. The nearest he came to military service was a brief stint in the Cadet Corps in 1917, but around him there was plenty of evidence of the experiences of less fortunate young men. He would certainly have seen legions of wounded and maimed soldiers hobbling around the city, having returned from France for treatment.[14]

But the war was a boon for Dirac's education.[15] The exodus of the school's older boys depleted the higher classes and enabled Dirac and other bright children to fill the gaps and therefore make quick progress. He excelled at science, including chemistry, which he studied in a silence that he broke on one occasion, a fellow student later remembered, when the teacher made an error, which

24

Dirac gently corrected.[16] In the foul-smelling laboratories, Dirac learned how to investigate systematically how chemicals behave and learned that all matter is made of atoms. The famous Cambridge scientist Sir Ernest Rutherford gave an idea of the smallness of atoms by pointing out that if everyone in the world spent twelve hours a day placing individual atoms into a thimble, a century would elapse before it was filled.[17] Although no one knew what atoms were made of or how they were built, chemists treated them as if they were as palpable as stones. Dirac learned how to interpret the reactions he saw in the laboratory test tubes simply as rearrangements of the chemicals' constituent atoms – his first glimpse of the idea that the way matter behaves can be understood by studying its most basic constituents.[18]

In his physics lessons, he saw how the material world could be studied by concentrating, for example, on heat, light and sound.[19] But the mind of young Dirac was now venturing far beyond the school curriculum. He was beginning to realise that underneath all the messy phenomena he was studying were fundamental questions that needed to be addressed. While the other boys in his class were struggling to get their homework done on time, Dirac was sitting at home, reflecting for hours on the nature of space and time.[20] It occurred to him that 'perhaps there was some connection between space and time, and that we ought to consider them from a general four-dimensional point of view'.[21] He appears to have shared much the same opinion as the Time Traveller in the 1895 novel *The Time Machine* by H. G. Wells, whose science-fiction novels he read: 'There is no difference between Time and any of the three dimensions of Space except that our consciousness moves along it.'[22] Such an opinion had wide currency at the end of the nineteenth century, and Dirac may have read the Traveller's words when he was a child.[23] In any case, the young Dirac was mulling over the nature of space and time before he had even heard of Einstein's theory of relativity.

Dirac's teacher, Arthur Pickering, gave up on teaching him with the rest of the boys and sent him to the school library with a book list. Pickering once set the prodigy a set of tough calculations to keep him busy at home that evening, only to hear from Dirac on his way home that afternoon that he had already done them.[24] And Pickering opened up another new vista to Dirac when he suggested that he look beyond simple geometry to the theories of the German mathematician Bernhard Riemann, who had proposed that the angles of a triangle do not always add up to exactly 180 degrees.[25] Just a few years later, Dirac would hear how Riemann's geometric ideas – superficially without relevance to science – could shed new light on gravity.

Charles Dirac understood as well as anyone that his younger son had an exceptionally fine mind coupled with formidable powers of concentration. By imposing a rigorous educational regime at home, Charles had produced a workaholic son in his own image, as he presumably intended. What Charles did not apparently appreciate as acutely as other people was Paul's odd behaviour. The young Dirac's fellow students certainly regarded him as strange. In testimonies given sixty years later, several of them described him as a very quiet boy; two accounts speak of 'a slim, tall, un-English-looking boy in knickerbockers with curly hair', and 'a serious-minded, somewhat lonely boy [who] haunted the library'.[26] Even at that time, he had a monomaniacal focus on science and mathematics. Games did not appeal to him and, when he was obliged to play, his participation seems to have been superfluous: one of his fellow schoolboys later remembered that Dirac's style of holding a cricket bat was 'peculiarly inept'. As an old man, Dirac attributed his dislike of team games to his having to play soccer and cricket with the older and bigger boys on the Merchant Venturers' playing fields.[27]

His appreciation of literature was also extremely limited. He never understood the appeal of poetry, though he did read novels written to appeal to young boys, including adventure stories and tales of great battles, scrutinising each text with the care of a literary critic.[28] As a nine-year-old, Bishop Road School had awarded him a prize of Daniel Defoe's *Robinson Crusoe*, a novel that always strikes a chord with those who are happy to be away from the crowd – almost, but not quite, alone.[29]

It was the mathematics and science lessons that did most to shape Dirac's way of thinking. Decades later, when his history teacher Edith Williams renewed contact with him, she told him that, when he was a student in her class, she 'always felt you were thinking in another medium of form and figures'.[30] By every account of Dirac's behaviour in his mid-teens, he had the same personality characteristics as today's pasty-faced technophiles who prefer using the latest software and gadgets to mixing with other people and who are happiest sitting alone at their computer screens. From a modern perspective, the young Dirac was an Edwardian geek.

At the Merchant Venturers' School, the class sizes shrunk and the range of lessons narrowed. When Dirac began at the school in September 1914, there were thirty-seven boys in his class; by the time he left in July 1918, four months before the end of the war, there were eleven. At the Speech Day, July 1918, he received a prize – as he had done every year – and heard the Headmaster announce that ninety-six boys had been killed and fifty-six wounded in the year 1916–17.[31] For the rest of his life, he would remember these litanies of death.

Nor was there any respite at home from the gloom. In Dirac's eyes, when his father returned home from school, his persona changed from the school's fair-minded and respected disciplinarian to bullying tyrant. He still imposed his linguistic regime at the dinner table, where wartime shortages and rationing had made Flo's meals simpler and less abundant. By the beginning of 1918, there were long, morale-sapping queues for bread, margarine, fruit and meat. The price of a chicken rose to a guinea, a week's wages for a manual labourer.[32] The shortages encouraged many families, including the Diracs, to cultivate fruit and vegetables, and it was mainly for this reason that Paul Dirac took up gardening, though the hobby would also have given him another reason to escape the atmosphere inside the house.[33]

Another source of unhappiness in the Dirac family was that Charles and Flo each had a favourite child: Paul was his mother's, Betty her father's, with Felix left out in the cold.[34] As a student, Felix had done almost as well as his younger brother at Bishop Road, but the gap between their abilities at senior school became so wide that it began to cause serious friction between them. The two brothers no longer walked around together but were continually bickering. In his later life, Dirac was uncharacteristically forthright about the reason for the rift: 'having a younger brother who was brighter than he was must have depressed him quite a lot'.[35] This is a telling remark. Dirac was never socially sensitive and, as an old man, was exceptionally modest and given to understatement, so he was probably making light of how painful Felix found the experience of being academically outclassed by his younger brother.

As he came to the end of his studies at the school, Felix had set his heart on becoming a medical doctor. His father, however, had other ideas: he wanted Felix to study engineering. This subject was popular among young people, just as Bernard Shaw had foreseen in his novel *The Irrational Knot*: a new class of engineer-inventors would go 'like a steam roller' through the effete boobies of the aristocracy.[36] The future appeared to be in the hands of H. G. Wells's 'scientific samurai'. It certainly seemed sensible for Felix to use his practical skills to take a course that would virtually guarantee him employment. As Charles probably realised, for Felix to train to be a doctor would entail six expensive years of training, with little prospect of the costs being off-set by Felix winning one of the scarce scholarships to medical school. Felix tried to stand firm, but Charles forced him to climb down, doing more harm to their relationship than he probably realised.[37]

The cheapest and most convenient place for Felix to study was at the university's Faculty of Engineering, housed in the Merchant Venturers' Technical College, which shared the same premises and facilities as the Merchant Venturers' School.[38] Probably with a good deal of resentment, Felix began his course in mechanical engineering there in September 1916, his studies funded by a City of Bristol University Scholarship.[39]

Paul never considered studying anything other than a technical subject.[40] He could have taken his pick from dozens of science courses, and seriously considered taking a degree in mathematics, but decided against it after he learned that the likely outcome would be a career in teaching, a prospect that held no appeal for him.[41] In the end, in the absence of a strong preference of his own, he decided to follow his brother – and, apparently, their father's advice – by studying engineering at the Merchant Venturers' College, supported by a generous scholarship.[42]

In September 1918, Felix was preparing to begin the final year of his engineering course, which he had been finding hard going – throughout, he had languished near the bottom of his class. At the same time, Paul, aged only sixteen, was about to join the ranks of the engineering students – two years younger than the other students in his class. Felix must have known that others were comparing his talent with his brother's and that he would not emerge well from the comparison.

Three

A report by the Bristol Advisory Committee, working in conjunction with the Employment Exchange, issued early in 1916, threw light on the effect of the war on the labour of young people in the preceding year. It stated that boys were almost generally fired by the ambition to become engineers [. . .]

GEORGE STONE and CHARLES WELLS (eds), *Bristol and the Great War*, 1920

On the overcast morning of Monday, 11 November 1918, Dirac set off from his home as usual to walk to the Merchant Venturers' College. It was the beginning of his seventh week at the college, and appeared to be like any other day. But when he arrived, he found that all lectures had been cancelled. He soon heard the reason: suddenly and unexpectedly, the war had ended.

By midday, the centre of Bristol had become the site of a vast, anarchic carnival. During a day of noisy jubilation not seen before in living memory, English reserve was abandoned. Church bells rang out, businesses shut down, everyone felt licensed to drape themselves in the national flag, to march the streets, to bash empty biscuit tins and dustbin lids and anything that would make a lot of noise.[1] All over the city, Union Jacks hung from windows, lamp posts and from the hundreds of trams and motor vehicles that had been commandeered for the day without demur from the police. Among the groups of marchers repeatedly singing 'Rule Britannia' was a group of American soldiers on the way to war, each of them holding a corner of the Union Jack. Nearby, a group of grammar-school students carried an effigy of the Kaiser, once a resident of Bristol.[2] Dirac's fellow Merchant Venturers' students caroused around the city, singing the song they had composed for the occasion. Dirac long remembered the chorus they sang at the top of their voices: 'We are the boys who make no noise,' followed even more loudly by 'Oo-ah, oo-ah-ah.'[3]

The Prime Minister David Lloyd George spoke that day in the House of Commons of the curious mixture of regret and optimism in the country after 'the cruellest and most terrible War that has ever scourged mankind. I hope we may say that thus, this fateful morning, came an end to all wars.' Fate, however, had yet more cruelty in store: the Spanish Flu pandemic that broke out towards the end of the conflict cost even more lives than the war. To try to slow the spread of the virus, Bristol's schools had been closed, leaving thousands of children wanting to spend the afternoons laughing at new film comedians such as Fatty Arbuckle, but they were thwarted by the closing of the cinemas during school hours by the local Council's Malvolios.[4]

The novelist and poet Robert Graves remarked perceptively that before August 1914, the country was divided into the governing and governed; afterwards, although there were still two classes, they had changed into 'the Fighting Forces [. . .] and the Rest, including the Government'.[5] The new divisions were clear at the Merchant Venturers' College after the war: Dirac saw young men returning from the battlefront suddenly outnumber the original intake of students, whose closest brush with the enemy had been through reading newspaper reports. The soldiers had returned to a brief welcome, but they had to settle down quickly to normal life, encumbered by disfigurement and by shell shock and other psychological damage. These men, most of them still in uniform, brought a new grittiness and pragmatism to the lecture rooms. Dirac later observed: 'the new students had a more mature outlook on life, and in the Engineering Faculty they were especially eager to learn results of practical importance and [they] did not have much patience with theory.'[6]

The returning soldiers were among the thousands who flocked to that year's Christmas treat in Bristol: the opportunity to see and take a tour around the inside of a captured German submarine U86. It was moored in the docks, the Union Jack flag fluttering on one of its masts above the German naval ensign. Everyone knew the significance of the display: the tank, the machine gun, the aircraft, radio and poison gas had all played their part in the war, but none had seemed more menacing than the submarine. Now this most feared weapon was impotently on show, like a dead shark.

Engineering was evidently not the subject best suited to the talents of the young Dirac. The course at the Merchant Venturers' College was more practical than theoretical and therefore exposed his limited manual skills while not making the most of his mathematical gifts.[7] True to form, Dirac strode ahead in mathematics and was 'a student who got all the answers exactly right, but who had not the faintest idea of how to deal with apparatus'.[8] Not only was he maladroit, his mind was on other things: he spent much of his time in the physics library, reflecting on the fundamentals of science.[9] With no money and nothing else to do during the day, Dirac would walk down from his home in Julius Road to the college and work in the libraries six days a week.[10] He did, however, make his first friend among the other thirty-one students in the class: Charlie Wiltshire, another solitary young man with a mathematical bent.

They were taught mathematics by Edmund Boulton, nicknamed 'Bandy', as his gait gave the impression that he had just dismounted a mare. Not a strong academic, Bandy showed his class how to tackle textbook mathematical problems in orthodox ways, only for Dirac repeatedly to proffer simpler and more elegant solutions. Soon Dirac and Wiltshire were segregated so that they could work at a pace that would not shame everyone else. Poor Wiltshire may have felt better if he had stayed behind, as he found the task of keeping up with his friend's mathematical progress 'utterly hopeless'. Within a year, they had completed the mathematical content of their degree, but Wiltshire was permanently scarred. Over thirty years later, he wrote that the experience of trying to stay abreast of Dirac had left him with a 'pronounced inferiority complex'.[11]

Mathematics was only a small part of Dirac's curriculum: he spent most of the time fumbling in the laboratories with Wiltshire or trying to stay alert during lectures. Unlike most students, he did not like to be spoon-fed and preferred to learn in private, ideally alone in the library, where he would flit back and forth between passages in books and journals, making his own links and associations. One course of lectures that did keep Dirac on his toes was given by the hard-driving head of the electrical-engineering department, David Robertson, a theoretically minded engineer who had been confined to a wheelchair after contracting polio.[12] Dirac admired Robertson for arranging his life methodically and for the way he used clever labour-saving initiatives to help overcome his disability. It was difficult for Robertson to deliver standard chalk-and-blackboard presentations, so he used a precursor of digital presentation software: a continuous series of lantern slides lit – none too reliably – by a flickering carbon arc lamp.[13] Robertson rushed through his commentary, giving no quarter to the intellectual limitations of his audience or to their need to write legible notes. Dirac's favourable opinion of him was not shared by the great majority of his students, who were left trailing in frustration and despair.[14]

Robertson ensured that the electrical-engineering course was built on solid theoretical foundations. Dirac and his colleagues specialised in electrical engineering only in their final year, after they had been given a grounding in physics, chemistry, technical drawing and other types of engineering – civil, mechanical and automotive. No one could reasonably accuse the course of being out of touch with business: Dirac was taught the elements of management, contract law, patents, bookkeeping and accountancy. He even learnt about income tax.[15]

The course was based in the engineering laboratories. Dirac spent many hours every week there, working with Wiltshire, learning about the mechanical structures and machinery that underpinned industry, including bridges, pulleys, pumps, internal combustion engines, hydraulic cranes and steam turbines. He measured the strength of materials by stretching them until they snapped and by observing how much they bent under stress. The course on electrical engineering was extremely thorough, and Dirac learned about the subject from its roots – simple experiments in electricity and magnetism – through to the minutiae of the design and operation of the latest hardware of the electricity-supply industry. H. G. Wells could not have asked for a more thorough training for a future leader in his technocratic utopia.

The university Engineering Society organised trips to local factories, partly to give the students a sense of the noise and grime in which most of them would soon be working. A posed photograph taken on one of these trips in March 1919 shows the physical appearance of Dirac and his fellow students, all of them male. Each of them is wearing a tie, a hat and an overcoat, several of them have a stick, and a few are still in military uniform. The sixteen-year-old Dirac is standing at the front, hands in his pockets, looking blankly at the camera with a hint of adolescent rebelliousness. It is the first of many photographs of him as a young man to show confidence and resolve shining out of his eyes.[16]

*

Six Julius Road was a cold and unloving refuge to Dirac, but for many local people he seemed to be part of an admirable home. The reputation of Charles Dirac was still on the rise: he had become one of the 'Big Four' housemasters at the Merchant Venturers' School, and his private language classes were thriving at home. A few minutes after the beginning of each tutorial, in the small study overlooking the front garden, Flo knocked on the door to bring Charles and his student a pot of tea and a plate of biscuits – part of the attentive service students took for granted at that address. She spent most of her time running the house but liked to while away afternoons reading romantic novels and the poetry of Robert Browning, Robert Burns and Rudyard Kipling. In an exercise book, she wrote out some of her favourite verse and a collection of aphorisms that indicated her penchant for the Victorian virtues: 'Control, give, sympathise: these things must be learnt and practised: self-control, charity and sympathy.'[17]

The Diracs' daughter Betty was as timid as her brothers. Most such girls of her generation began a menial job straight after leaving junior school, but Charles and Flo wanted her to continue her education at the nearby Redlands Girls' School, where she studied without special enthusiasm or achievement. It was convenient for her father to accompany her to school after 1919, when his school relocated to Cotham Lawn Road, ten minutes' walk from the Diracs' home. The move was unpopular with its teachers, though it was made palatable for Charles by a sweetener – promotion to the more lucrative post of Associate University Lecturer. His colleagues in the staffroom respected him as one of the most effective teachers in Bristol, though many regarded him as odd. He did nothing to shed this reputation when he told one of them that he had been trepanned: presumably a surgeon had drilled a tiny hole into his head, intending to let out evil spirits.[18]

To some of Charles's fellow teachers, there was a whiff of fraudulence about him: they found out that the letters B. ès. L. (*Baccalauréat-ès-Lettres*) that he almost always put after his name signified only that the University of Geneva had pronounced him able to embark on higher education. He had spent only a year at the university, as an *auditeur*, taking notes but not a degree. One of his colleagues later chuckled as he recounted the minor staffroom scandal involving Charles: as he was not eligible to wear the full academic dress, he bought a gown and asked his wife to make him a hood in red, white and blue. She knew nothing of the deception and only found out about it several years later.[19]

In the spring of 1919, for reasons that are not clear, Charles Dirac sought British nationality for the first time. He wrote urgently to the Swiss authorities, saying that after teaching in the UK for thirty years, 'professional reasons' made it essential that he renounce his Swiss nationality.[20] When he submitted his application to the British authorities, he said he wanted the right to vote after the government had withdrawn it, following the recent amendment to the Aliens Registration Act, which also denied Flo – as the wife of a 'foreign national' – the right to vote in future general elections (she had voted for the first time six months before, in common with other British women over thirty years of age). Perhaps, too, he wanted his daughter and elder son to be eligible for the scholarships available only to British citizens? Whatever his motivation, Charles swore allegiance to George V in front of a justice of the peace in Bristol on 22 October 1919.[21] On that day, his children also became Britons, having previously been classed as Swiss, a status that, according to Betty's later recollections, caused her to be teased in the playground for being 'one of those Europeans'.[22] Paul Dirac was no longer a foreigner, but, to many British eyes, he would always have the air of one.

In the early summer of 1919, when Paul's first-year results confirmed his potential as a top-flight student, Felix became the first person in his extended family to be awarded a degree, though only with third-class honours. The disparity between the brothers' academic talents had never been so stark, so it is probably no coincidence that the relationship between them became seriously troubled at about this time. In the pained and elliptical comments Dirac made later about Felix, he remarked they would often 'get into a row', though he gives no details of the arguments.[23] One possibility is that they were seeded by Felix's jealousy and sense of inferiority, nourished by Paul's lack of empathy with his brother and by his inability to muster tactful words that were sorely needed to preserve Felix's sense of self-worth. Among his colleagues in his later career, Paul Dirac was famous for not understanding the feelings of others and for his lack of tact. It is unlikely that he was any different when he was a young man.

After Felix had taken his degree, he left home and moved two hundred miles away to Rugby, which was rapidly changing from one of the East Midlands' sleepy market towns into a booming centre of the new electrical technology. Felix took a three-year student apprenticeship at the British Thomson-Houston Company, on a starting wage of a pound a week, giving him a measure of financial independence. Meanwhile, his penniless brother continued to study engineering – while moonlighting in physics – at the Merchant Venturers' College. As he had already chomped his way through the mathematics part of the course, he seemed destined to spend the remaining two years of his engineering degree fumbling his way through his laboratory exercises and listening to his lecturers drone their way through the syllabus. Having cast around for a challenge, he amused himself in the library by hunting down the longest German words in the technical dictionaries (hyphens barred) and reading about the subject that most interested him, physics.[24] His scientific imagination was ripe for a challenge, and, a few weeks after he began his second year at university, it arrived.

No event in Dirac's working life ever affected him as deeply as the moment when relativity 'burst upon the world, with a tremendous impact', as he remembered nearly sixty years later.[25] Einstein became a media figure on Friday, 7 November 1919, when *The Times* in London published what appeared to be just another post-war edition, including the news that the King supported the proposal of an Australian journalist for two minutes' commemorative silence on the anniversary of Armistice Day. On page 12, the sixth column featured a 900-word article that most readers probably passed over, unless the headline, 'Revolution in Science', captured their attention. Yet this was a momentous piece of journalism, and it helped to propel Einstein from relative obscurity in Berlin to international celebrity; soon, his moustachioed face and frizzled mane of black hair were familiar to newspaper readers all over the world. The unsigned article reported the apparent verification of a theory by Einstein that 'would completely revolutionize the accepted fundamental physics' and thereby overturn the ideas of Isaac Newton that had held sway for over two centuries.[26] The observations were made by two teams of British astronomers who had found that the deflection by the Sun of distant starlight during the recent solar eclipse was consistent with Einstein's theory but not Newton's. When he was an old man, Dirac remembered this as a time of special excitement: 'Suddenly Einstein was on everyone's lips [. . .] [E]veryone was sick and tired of the war. Everyone wanted to forget it. And then relativity came along as a wonderful idea leading to a new domain of thought.'[27]

Dirac, Charlie Wiltshire and their fellow students were fascinated by Einstein's new theory and tried to find out what the fuss was about. This was not an easy task. Their teachers, like most academics in the UK, were no more knowledgeable than their students about this alleged scientific revolution. Apart from occasional articles in scientific journals such as *Nature*, the primary sources of knowledge about the new theory of relativity were newspapers and magazines, whose editors gave commentators thousands of column inches to speculate – usually facetiously – about the new theory and its apparent defiance of common sense. On 20 January 1920, *Punch* featured an anti-Semitic poem that exemplified popular puzzlement with the theory that had originated behind the lines of the UK's bitter enemy:

> Euclid is gone, dethroned,
> By dominies disowned,
> And modern physicists, Judaeo-Teuton,
> Finding strange kinks in space,
> Swerves in light's arrowy race,
> Make havoc of the theories of Newton.

The pages of the newspapers and magazines were replete with advertisements for scores of half-baked accounts of Einstein's work churned out only months after the theory came to public attention.[28] At that time, there were no science journalists, so Dirac and his friend Wiltshire had to rely on popular articles written by scientists, notably Arthur Eddington, the Quaker astronomer and mathematician at the University of Cambridge and the only person in Britain to have mastered the theory. He had even got his hands dirty in one of the eclipse expeditions that produced crucial support for the theory.

In a stream of entertaining articles and books, Eddington deployed witty, down-to-earth analogies that made even the most complex abstract ideas accessible and arresting. His skill is exemplified in the account he gave in 1918 of Einstein's famous equation $E = mc^2$. Other authors could only crank out a dreary and barely comprehensible explanation of the equation's neat connection between the energy E equivalent to a mass m, and the speed of light in a vacuum (symbolised by the letter c). Eddington knew better. In his explanation, he used the equation to do a calculation that he knew would interest his readers: he worked out the total mass of the light that the Sun shines onto the Earth and then used the result to comment on the controversial question of whether to keep daylight-saving time:

the cost of light supplied by gas and electricity companies works out at something like £10,000,000 an ounce. This points the moral of Daylight Saving: the Sun showers down on us 160 tons of this valuable stuff every day; and yet we often neglect this free gift and prefer to pay £10,000,000 an ounce for [light of] a much inferior quality.[29]

Eddington and other writers fuelled Dirac's interest in understanding how the material universe works. But he spent most of his time studying for his engineering degree, struggling to concentrate in lectures, mastering the theoretical concepts, doing experiments and writing them up in immaculate accounts that feature scarcely a single crossing-out. To the modern eye, they almost look as if they had been printed by machine in a special typeface that successfully mimics ordinary human handwriting, with every repeated letter reproduced identically.[30]

Charlie Wiltshire was one of the very few people who glimpsed the human side of Dirac. To most people, he looked like a cold-hearted solipsist, uninterested in human contact, engaged only by mathematics, physics and engineering. Even in those repressed times, Dirac appeared to be exceptionally narrow-minded and inhibited.[31]

Soon after his eighteenth birthday, Dirac had to spend time away from his sheltered environment for the first time. He travelled to Rugby, where his brother Felix was one of the small army of young apprentices in the local factories, to spend the summer as a trainee engineer, and, perhaps, to see whether he was suited to factory work. By the end of his month-long stay, the answer was clear.

Dirac worked in the British Thomson-Houston electrical goods factory, located on a ninety-acre site next to the railway station. The factory dominated the town. It was said that everyone who lived in Rugby either worked there or knew someone who did. Certainly, everyone in the town was familiar with the saw-tooth profile of the factory's roofs, one of them bearing the sign 'Electrical Machinery'.

And everyone, wherever they stood, could see the smoke billowing from two chimneys that pointed to the sky like a pair of smouldering lances.

Dirac arrived in Rugby sporting a new wristwatch, a device that had a decade before been regarded as effeminate for men (and outré for women) but had become respectable after soldiers in the war had found them useful.[32] He lodged above a draper's shop on a street corner, precisely midway between the factory's two entrances, a few minutes' walk away. Dirac was one of about a hundred vacation students who provided menial labour, mainly in the relatively quiet testing laboratories well away from the turbine-construction area, when many of the workers were on holiday. It was a slow-news summer, enlivened only by the dramatic lockout of the Electrical Trades Union and by a local polo match in which one of the players was the Secretary of State for War, Winston Churchill.[33]

Flo regularly wrote to Paul, the first of several hundred letters that she sent him between then and her death. It seems that he kept all of them. These first letters were warm and newsy, telling him of Betty's new dog, how 'Daddy missed you when he had all the grass to cut' and of the new overcoat she was going to have done up for him ('I showed it to Pa & he wants it for himself'). Flo repeatedly complained that he was not telling the family enough about what he was doing. 'Do you ever come across Felix?' she asked.[34] The answer was that the two brothers did pass each other on the streets of Rugby, but they did not exchange a word.[35] Their relationship had deteriorated into a state of cold hostility; Paul apparently offered his brother the same expressionless stare that he gave almost everyone else. Either their mother did not know of her sons' falling out, it seems, or she was too blinkered to notice.

Dirac's employers in Rugby gave him the only poor report he would receive in his entire life. David Robertson later showed him the damning comments and disclosed that he was the only vacation student from Bristol ever to receive an unfavourable report. It judged Dirac to be 'a positive menace in the Electrical Test Department', to 'lack keenness' and to be 'slovenly', making it clear between the lines that Dirac would be unwise to seek a future on the factory floor.[36]

In late September 1920, Dirac returned to Bristol to prepare for his final undergraduate year, when he specialised in electrical engineering. His passion, however, was the theory of relativity. One of his frustrations was that he could not find an accessible technical account of the theory that would explain, step by step, how Einstein had developed his ideas. Of the academic disciplines that contributed the reams of piffle Dirac read about relativity, none was more prolific than philosophy. One commentator wrote: 'A philosopher who regards ignorance of a scientific theory as insufficient reason for not writing about it cannot be accused of complete lack of originality.'[37] The writer of those words was one of the most talented young philosophers working in Britain, Charlie Broad. Having originally wanted to be an engineer, he trained in both philosophy and science at Cambridge and acquired more expertise in relativity theory than the great majority of physicists, many of whom knew next to nothing about Einstein and his work. In the autumn of 1920, soon after Broad was appointed as the Professor of Philosophy at the University of Bristol, he gave a series of lectures for final-year science students on scientific thought, billed to include a description of Einstein's theory.[38] Dirac and several other engineering students sat in on these lectures, though few of them were sitting alongside Dirac to the end, as the going quickly became tough and the material had little to do with engineering. For Dirac, the course was a memorable experience, as it was for Broad, who wrote thirty years later in his autobiography:

there came to these lectures one whose shoe-laces I was not worthy to unloose. This was Dirac, then a very young student, whose budding genius had been recognized by the department of engineering and was in the process of being fostered by the department of mathematics.[39]

Broad was a wonderfully idiosyncratic lecturer. He always appeared with a carefully prepared script, and he read every sentence twice, except for the jokes, which he delivered three times. Although he spoke drearily, his content was compelling, jargon-free and spiked with witty references to Charles Dickens, Conan Doyle, Oscar Wilde and other literary figures. Trenchancy was one of his strongest suits. During a warning about the snake oil of most popular accounts of relativity, he counselled that 'popular expositions of the Theory are either definitely wrong, or so loosely expressed as to be dangerously misleading; and all pamphlets against it – even when issued by eminent Oxford tutors – are based on elementary misunderstandings.'[40]

Broad's treatment of relativity in his course was unconventional to the point of quirkiness. He taught Einstein's first theory and his more general version together, taking a unified approach and concentrating on the basic ideas rather than on the mathematics. Broad's aim was to make it clear that the theories give 'a radically new way of looking at Nature'.[41] The first of Einstein's theories is usually dubbed the 'special theory' because it deals only with observers who move in straight lines at constant speeds with respect to one another; for example, passengers on two trains moving smoothly on parallel tracks. Einstein based his theory on just two simple assumptions: first, that when each of the observers measures the speed of light in a vacuum, they will always find the same value, regardless of their speed; and, second, that measurements made by the observers will lead them to agree on all the laws of physics. Einstein's great insight was to see that if these assumptions were followed to their logical conclusion, a new understanding of space, time, energy and matter emerged.

A casualty of Einstein's theory was the widely accepted belief that the universe is pervaded by an ether, which Broad argued had become superfluous:

there was supposed to be a peculiar kind of matter, called Ether, that filled all Space. On these theories the Ether was supposed to produce all kinds of effects on ordinary matter, and it became a sort of family pet with certain physicists. As physics has advanced, less and less has been found for the Ether to do.[42]

Contrary to the theory, the existence of such a substance would imply that there is a uniquely privileged frame of reference, so relativity implies that the ether is an unnecessary assumption and may well not exist, unless experiments say otherwise. Einstein also noted that measurements of space and time are not, as almost everyone else thought, independent but are inextricably linked, leading to the idea of a unified space-time, a concept introduced by his former teacher Hermann Minkowski, a German mathematician. Finally, Einstein showed that an inevitable consequence of this new way of thinking was his equation $E = mc^2$, implying that the mass of a small coin is equivalent to the vast energy needed to run a city for days or indeed to raze it. An apocalyptic vision of this power had already been presented by H. G. Wells, shortly before the outbreak of the First World War, in his novel *The World Set Free*.

For most purposes, the predictions of Einstein's special theory were extremely similar to the corresponding ones made by Newton's theory. The two sets of predictions, however, were noticeably different at speeds approaching the speed of light in a vacuum: Einstein claimed that, under these conditions, his theory was more accurate, though it would be several decades before the superiority was convincingly demonstrated by experimenters. In the meantime, Einstein's reasoning made it possible to amend the description of anything given by Newton's theory and produce a 'relativistic' version – one that agreed with the principles of the special theory of relativity. Two years later, Dirac took up a new hobby, aiming to produce relativistic versions of Newtonian theories – an activity he pursued like an engineer upgrading tried-and-tested designs to ones that perform to a higher specification: 'There was a sort of general problem one could take, whenever one saw a bit of physics expressed in a non-relativistic form, to transcribe it to make it fit in with special relativity. It was rather like a game, which I indulged in at every opportunity.'[43]

Einstein's second theory of relativity applied to *all* observers, including ones who are accelerating; for example, observers who fall freely under the action of gravity. In this 'general theory of relativity', Einstein proposed a geometric picture of gravity, replacing Newton's concept that an apple and every other mass is subject to a force of gravity by a radically new way of describing the situation. According to Einstein, every mass exists in a curved space-time – roughly analogous to a curved sheet of rubber – and the motion of the mass at every point in space-time is determined by the curvature of space-time at that point. Because the theory is relativistic, information cannot be transmitted faster than light, and all energies contribute to mass (via $E = mc^2$) and therefore to gravity. It turns out that, in the Solar System, where almost all matter has comparatively low density and travels much more slowly than light, the predictions of Einstein's theory of gravity are in extremely good agreement with Newton's. But, in some situations, they can be distinguished, and one of the most straightforward ways of doing so involved measuring the bending of starlight by its gravitational attraction to the Sun during a solar eclipse: Einstein's theory predicted that this deflection would be twice Newton's value. This was the prediction that Eddington and his colleagues believed they had verified in their solar-eclipse experiments.

It was during one of the early lectures in Broad's course that Dirac had a revelation about the nature of space and time. Broad was talking about how to calculate the distance between two points. If they lie at the sharpest corners of a right-angled triangle, then every schoolchild knows that the distance between the points (the hypotenuse) is given by Pythagoras's Theorem: the square of this distance is equal to the *sum* of the squares of the lengths of the other two sides. In the space-time of the special theory of relativity, things are different: the square of the distance between two points in space-time is equal to the sum of the squares of the spatial lengths *minus* the square of the time. Dirac later recalled 'the tremendous impact' on him of Broad's writing down that minus sign.[44] This dash of chalk on Broad's blackboard told Dirac that his schoolboy ideas about space and time were wrong. He had assumed that the relationship between space and time could be described using the familiar Euclidean plane geometry, but if that had been true, every sign in the formula for the distance between two points would have been positive. Space and time must be related by a different kind of geometry. Pickering, Dirac's mathematics teacher at the Merchant Venturers' School, had already introduced him to the Riemannian geometry that Einstein had used to describe curved space-time. In this way of looking at space and time, the angles of a triangle may not add up to 180 degrees as they do in ordinary Euclidean space. In Einstein's general theory of relativity, matter and energy are linked with the space and time in which they exist: matter and energy determine how much space-time is curved, and the curvature of space-time dictates how matter and energy move. Thus, Einstein offered a new explanation of why the apple in the tree in Newton's garden fell: it was not the gravitational pull of the Earth that was responsible but the planet's curvature of space-time in the region of the apple.[45]

Inspired by Broad's lectures, and by Eddington's semi-popular book *Space, Time and Gravitation*, Dirac soon taught himself the special and general theories, another early sign of his special talent as a theoretician. The mathematical complexities of Einstein's general theory so terrified most physicists that they found excuses not to bother with it, whereas Dirac – an engineering undergraduate, not a registered student of physics – studied it voraciously. While other nineteen-year-olds were seeking beauty in the flesh, he sought it in equations.

*

Broad was sceptical of the contribution philosophy can make to advance the understanding of the natural world (he called it 'aimless wandering in a circle'), but his lectures persuaded Dirac that the subject was worth pursuing. One text he took out of the library was John Stuart Mill's A System of Logic, which the young Einstein had studied some fifteen years before.[46] Mill had been the nineteenth century's pre-eminent British philosopher, the most cogent voice of empiricism, the belief that human beings should ground every concept in verifiable experience.[47] His approach to ethics was largely utilitarian, believing that the ultimate good is one that brings the most happiness to the greatest number of people and that the rightness of any human action should be judged according to its contribution to public happiness. Mill was influenced by other empiricists, notably by his friend Auguste Comte, the French pioneer of the positivist belief that all true knowledge is scientific, including knowledge about 'sociology', a word that Comte coined. Mill had no time for the Kantian 'intuitionist' view that some truths are so exalted that they transcend experience: he dismissed as meaningless many unverifiable statements made by bishops, politicians and others he regarded as airy-fairy moralists. Mill's views and his feet-on-the-ground public spiritedness were enormously influential among Victorians and have become the essence of the liberal English consensus. He influenced Dirac, and many others, more than they knew.

A System of Logic, published in 1843, is a plain-spoken if laborious account of how empiricism can shape every aspect of human life.[48] The book features Mill's agenda for science, which assumes that there is an underlying 'uniformity of nature'. The aim of scientists should be to explain more and more observations in terms of fewer and fewer laws, every one of them grounded in experience and induced from it. For Mill, the agreement between an experimental measurement and a corresponding theoretical prediction does not imply that the theory is correct, as there may well be many other theories that give equally good agreement. He argued that scientists have the never-ending task of finding theories that are in ever-better agreement with empirical observations.

In a memoir he wrote in his seventies, Dirac said he gave 'a lot of thought' to philosophy, trying to understand what it could contribute to physics. He recalled that he read A System of Logic 'all through', which we can safely interpret to mean that he read and pondered almost every word of it, his usual practice.[49] Although he found it 'pretty dull', it introduced to him the important idea that the disparate scientific observations and theories he had learned about had an underlying unity. Furthermore, science should seek to describe this unity using the fewest possible laws of nature, each of them formulated in the simplest possible way. Although this probably influenced the thinking of the young Dirac, he concluded that philosophy was not an effective way of finding out what makes nature tick. Rather, as he put it in an interview in 1963, 'it's just a way of talking about discoveries which have already been made'.[50]

The best way of understanding nature's regularities, he was coming to believe, was through mathematics. Dirac's lecturers in the engineering classes had drummed into him that mathematical rigour is unimportant; mathematics is simply a tool to obtain useful answers that are correct or, at least, accurate enough for the purpose in hand. One exponent of this pragmatic approach to the mathematics of engineering was Oliver Heaviside, an acid-tongued recluse who had invented a battery of powerful techniques that made it easy to study the effects of passing pulses of electric current through electrical circuits. No one quite understood why these methods worked, but he didn't care: what mattered to him was that they gave correct results, with a speed more rigorous methods could not match and without generating inconsistencies with other parts of mathematics. Engineers prized Heaviside's methods for their usefulness, but mathematicians mocked them for their lack of rigour. Heaviside had no time for pedantry ('Shall I refuse my dinner because I do not understand digestion?'[51]) and rejected the attacks of his detested opponents. He even entitled his autobiography after them: *Wicked People I Have Known*.[52]

Dirac studied Heaviside's techniques and later remarked that there was 'some sort of magic' about them.[53] Another of the engineers' clever tricks that impressed Dirac concerned the calculation of the stresses exerted on materials; for example, by a gymnast balancing on a beam. Engineers routinely calculate these stresses using special diagrams that generate correct answers much more quickly than the mathematicians' rigorous techniques. In his classes, Dirac used this method to represent stresses in this way and saw its power; within a few years, he would use similar techniques in a different context, to understand atoms.[54]

One of the lessons he learned in his engineering classes was the value of approximate theories. In order to describe how something works, it is essential to take into account the quantities that do most to affect its behaviour and to single out the quantities unimportant enough to be ignored. David Robertson taught Dirac a lesson he later regarded as crucial: even approximate theories can have mathematical beauty. So, when Dirac studied electrical circuits, the stresses on revolving shafts in engines and the windings of the rotors in electric dynamos, he was aware that the underlying theories had, like Einstein's general theory of relativity, a mathematical beauty.

It was probably Dirac's reflections on Einstein's theory that first led him to believe that the goal of theoretical physicists should be to find equations that describe the natural world, but his studies of engineering were the source of a proviso: that the fundamental equations of Nature are only approximations.[55] It was the job of scientists to find ever-better approximations to the truth, which always lies tantalisingly beyond their reach.

Apart from the embarrassing report Dirac had been given in Rugby, his record during his degree was almost flawless: only once in three years did he fail to top his class in every subject (the spoilsport was the assessor of a Strength of Materials course who ranked him second).[56] But it was clear that his real talents were in theoretical subjects and mathematics. Early in 1921, within a few months of completing the degree, his father suggested that he set his sights on studying at Cambridge.[57] Early in February, Charles wrote to St John's College, almost certainly acting on the advice of Ronald Hassé, head of Bristol University's mathematics department and a member of Cambridge University's network of talent-spotters. Hassé was a graduate and research student of the college, notable as the first person in Cambridge to speak of Einstein's 'theory of relativity'.[58]

Charles enquired whether the college would let him have details of 'any open scholarship in mechanical science or mathematics' that his son could apply for.[59] The college responded swiftly and arranged for Dirac to make his trip to Cambridge in June 1921, to sit the college's entrance examination.[60] Dirac's application to the college, made when he had just turned nineteen, is the earliest extant example of his adult handwriting. It shows that he wrote with the precision and clarity of a calligrapher, each letter standing upright with some of the capitals decorated unobtrusively with a tiny curlicue.[61]

Dirac passed the entrance examination handsomely, winning an annual scholarship of £70, which was disappointingly short of the minimum of £200 a year that he needed to live in Cambridge.[62] Charles argued that it was 'out of the question' to give his son the additional money as he earned only £420 a year and had no other income, neglecting to mention his lucrative private tuition. Bristol council refused to help because Charles and Paul had become British citizens only two years before and were therefore ineligible for financial assistance. Disappointed, Charles later wrote to Cambridge asking to be kept informed if any other opportunities should arise for his son. He concluded, 'I am sorry to trouble you, but I believe the boy has an exceptionable [sic] head for mathematics and I am trying to do my best for him.'[63] When an official at St John's College offered tactfully to advise him further if he would provide more information about his family's finances, Charles did not reply.[64]

Although Paul's Cambridge application had stalled, by July he had a first-class honours degree in engineering, a qualification that he and his father hoped would all but guarantee him employment. However, his graduation coincided with the worst depression in the UK since the industrial revolution: unemployment soared to two million. To every job application, Dirac drew a blank. Thus, the most talented graduate Bristol had ever produced found himself unemployed. But this turned out to be a stroke of luck.

Four

Mathematics [. . .] does furnish the power for deliberate thought and accurate statement, and to speak the truth is one of the most social qualities a person can possess. Gossip, flattery, slander, deceit all spring from a slovenly mind that has not been trained in the power of truthful statement.

S. T. DUTTON, *Social Phases of Education in the School and the Home*, London, 1900

What might have happened to Dirac if he had got one of the jobs he applied for, perhaps in the burgeoning aviation industry? Might the loss to physics have been offset by a commensurate gain for aeronautics? That these are questions of virtual history is due to the mathematician Ronald Hassé, who deftly steered Dirac's career from engineering to science. Things could easily have worked out quite differently. In September 1921, when Dirac was at a loose end and looking for jobs, David Robertson suggested to Dirac that, rather than hang around doing nothing, he should do an electrical-engineering project.[1] Dirac dabbled in some experiments, but, after a few weeks, Hassé wooed him back to the lecture theatres in the mathematics department, having arranged for him to do a full mathematics degree free of charge and for him to skip the first year's work so he could complete it in two years.

Dirac's fellow mathematics students were struck by his punctuality. For the first lectures of the day, beginning at 9 a.m., he was always the first to arrive, silently occupying a seat in the front row and showing no interest whatever in his fellow students. He spoke only when spoken to and talked only in clipped, matter-of-fact sentences that bore no trace of emotion. One of the students later recalled that no one even knew the name of the 'tall, pallid youth' or showed much interest in him until the results of the Christmas examination results revealed that the new student 'P. A. M. Dirac' was top of the class.

Some of the students resolved to make some enquiries about their mysterious colleague. They were surprised to learn that although he was eighteen months younger than anyone else in the class, he already had a degree in engineering. One of his characteristics was that although he was preternaturally silent, he did stir if he spotted a serious scientific error. In one such incident, after a lecturer had filled two and a half blackboards with symbols and left almost all the students frantically scribbling as they tried to keep up with him, he realised that he had made a mistake. He stood back from the blackboard and turned to Dirac: 'I have gone wrong, can you spot it?' After Dirac identified the error and explained how to put it right, the lecturer thanked him and resumed his exposition.[2]

In Dirac's first year of his new course, he studied pure mathematics – the branch of mathematics pursued with no concern for its applications – and applied mathematics, employed to solve practical problems. One of his lecturers was Peter Fraser, a farmer's son from the Scottish Highlands, a bachelor who lived much of his life in a reverie and liked to tramp the countryside while contemplating the higher truths of mathematics. He did no original research and never wrote a research paper but channelled all his intellectual energy into his teaching. Dirac believed he was the best teacher he ever had.[3]

Shortly before 9 a.m. on Mondays, Wednesdays and Fridays, Dirac was in his seat, awaiting the next episode of Fraser's teaching of a special type of mathematics, known as projective geometry, largely a French invention derived from studies of perspective, shadows and engineering drawing. One of its founders was Gaspard Monge, a draughtsman and mathematician who much preferred to solve mathematical problems using geometric ideas rather than complicated algebra. In 1795, Monge founded the descriptive geometry that Dirac had used in the first technical drawings he made in Bishop Road School, representing objects in three orthogonal points of view. Jean-Victor Poncelet, an engineer in Napoleon's army, built on Monge's ideas to set out the principles of projective geometry when he was a prisoner in Russia in 1812. His ideas and their consequences were to become the mathematical love of Dirac's life.

When most students come across projective geometry, they find it an unusual branch of mathematics because it primarily taxes their powers of visualisation and does not feature complicated mathematical formulae. What matters in projective geometry is not the familiar concept of the distance between two points but the *relationships* between the points on different lines and on different planes. Dirac became intrigued by the techniques of projective geometry and by their ability to solve problems much more quickly than algebraic methods. For example, the techniques allow geometers to conjure theorems about lines from theorems about points, and vice versa – 'that appealed to me very much', Dirac stressed forty years later.[4] To him, an impressionable young mathematician, this was a powerful demonstration of the power of reasoning to probe the nature of space.[5]

Fraser also persuaded Dirac of the value of mathematical rigour – an uncompromising respect for logic, consistency and completeness – something he had, as an engineering student, been taught to wink at.[6] In Dirac's studies of applied mathematics, he learned how to describe electricity, magnetism and the flows of fluids using powerful equations that yielded neat solutions, all consistent with experimental observations. He also used Newton's laws of mechanics to study the contrived examples that inform the education of every applied mathematician: rigid ladders resting against walls, spheres rolling down inclined planes, and beads sliding around circular hoops.[7] Dirac filled several exercise books with his answers, most of them flawless. He did most of this work in his bedroom, his escape from the family he perceived to be unloving and a refuge from Betty's yapping dog. Betty was developing into an unambitious, self-deprecating young woman, in awe of her brother Paul's intelligence, content to while away hours doing nothing. Her father doted on her, as Bishopston local Norman Jones remembered sixty years later when he said that his main recollection of Charles Dirac was 'seeing him always carrying an umbrella, struggling up the hill [. . .] often with his daughter, of whom he was very fond'.[8]

Dirac saw Felix only occasionally, at weekends, when he returned from his lodgings in the Black Country of the Midlands, near Wolverhampton. The brothers were still not on speaking terms.

In the final year of his course, Dirac should have been given the choice of specialising in either pure or applied mathematics. He wanted to take the pure option but did not get his way. His fellow student on the honours mathematics degree programme, Beryl Dent – the strong-minded daughter of a headmaster – had the upper hand because she was paying for her tuition, unlike Dirac. She expressed a firm preference for studying applied mathematics, and her wishes carried the day, perhaps partly because it was easiest for the lecturers to teach the same courses to the two students. So, for the first time since he began senior school, Dirac had to work alongside a young woman, but his relations with her were strictly formal; they seldom spoke.[9]

Dirac spent the 1922–3 academic year with his head down, building on the applied mathematics that he had learned the year before. One bonus for him was that his course included a few lectures on the special theory of relativity, though he probably knew more about the subject than his lecturer.[10] By the time he had finished, he had acquired considerable expertise in Newtonian mechanics. Although he knew that Einstein had found fault with Newton's laws of mechanics, they worked extremely well for all real-world applications, so it made good sense to master them, as tens of thousands of other students – including Einstein himself – had done before.

During his mathematics degree, Dirac encountered the ideas of William Hamilton, the nineteenth-century Irish mathematician and amateur poet. He was a friend and correspondent of William Wordsworth, who served science well by helping to persuade Hamilton that he would do better to spend his time on mathematics rather than on poetry. Among his discoveries, Hamilton was most enamoured with his invention of quaternions, mathematical objects that behave peculiarly when they are multiplied together. If two ordinary numbers are multiplied, the same result emerges regardless of their order of multiplication (for example 6 × 9 has the same value as 9 × 6). Mathematicians say that such numbers 'commute'. But quaternions are different: if one quaternion is multiplied by a second, the result is *different* from the result obtained if the second is multiplied by the first. In modern language, quaternions are said to be 'non-commuting'.[11] Hamilton believed that quaternions have many practical applications, but the consensus was that they are mathematically interesting but scientifically infertile.

Dirac also heard about Hamilton's reformulation of Newton's laws of mechanics. Hamilton's approach largely dispensed with the idea of force and, in principle, enabled scientists to study any material thing – from a simple pendulum to cosmic matter in outer space – much more easily than was possible using Newton's methods. The key to Hamilton's technique was a special type of mathematical object that comprehensively describes the behaviour of the thing under study, the Hamiltonian, as it became known. Hamilton's methods became another of Dirac's fixations and were to become his favourite way of setting out the fundamental laws of physics.

The mathematics degree did not present a sufficient challenge to keep Dirac occupied, so Hassé encouraged him to take as many of the undergraduate physics courses as his timetable allowed. Once again, Dirac chose to study fundamental subjects which were not covered in his syllabus. In one course, he studied the electron, the particle discovered twenty-five years before in the Cavendish Laboratory in Cambridge by J. J. Thomson, a man equally adept at investigating nature theoretically and – despite his ham-fistedness – experimentally. Several of Thomson's colleagues thought he was joking when he argued that the electron was smaller than the atom and was a constituent of every atom; to many scientists, the idea that there could exist matter smaller than the atom was inconceivable. Yet he was proved right, and, by the time Dirac first became acquainted with the electron, textbooks routinely ascribed electric current to the flow of Thomson's electrons.

Dirac also attended lectures in atomic physics given by Arthur Tyndall, a kindly and articulate man with a keen eye for scientific talent. Tyndall introduced Dirac to what was to prove one of the central insights of twentieth-century physics: the idea that the laws of 'quantum theory', which describe nature on the smallest scale, are not the same as the scientific laws that describe everyday matter. Tyndall illustrated this by describing how the energy of light arrives not in continuous waves but in separate, tiny amounts called quanta. At first, this idea was not taken seriously, as virtually all scientists were convinced that light behaves as waves. Their faith rested on the unarguable success of the theory of light published several decades before by the Scottish physicist James Clerk Maxwell, the Cavendish Laboratory's first professor. According to this theory, checked by many experiments, the energy of light and all other types of electromagnetic radiation is delivered not in lumps but continuously, like water waves lashing against a harbour wall.

Quantum theory had been discovered – largely by accident – by Max Planck, the Berlin-based doyen of German physics. He happened on the idea of quanta when he was analysing the results of some apparently obscure desktop experiments that investigated the radiation bouncing around inside the reflecting walls of ovens at steady temperatures (the experiments aimed to help German industry improve the efficiency of lighting devices).[12] The quantum emerged stealthily from the darkness of those ovens through the ingenuity of Planck, who brilliantly guessed a formula for the variation in the intensity of the radiation with its wavelength, at every temperature setting of the oven. In the closing weeks of 1900, Planck found he could explain the formula for the 'blackbody radiation spectrum' only if he introduced a concept that seemed completely contrary to Maxwell's theory: the energy of light (and every other type of radiation) can be transferred to atoms *only* in quanta.

The conservative Planck did not view this quantisation as a revolutionary discovery about radiation but as 'a purely formal assumption' needed to make his calculations work. Einstein first recognised the true importance of the idea in 1905, when he took the concept of radiation quanta literally and demonstrated that the reasoning Planck had used to derive his black-body radiation spectrum formula was hopelessly flawed. The challenge was to do better than Planck by finding a logical derivation of the formula.

When Planck discovered the quantum of energy, he also realised that its size is directly determined by a new fundamental constant, which he denoted h and others dubbed Planck's constant. It figures in almost every equation of quantum theory, but nowhere in the previously successful theories of light and matter, retrospectively labelled 'classical theories'. The minuscule size of the constant means that the energy of a typical quantum of light is tiny; for example, a single quantum of visible light has only about a trillionth of the energy of the beat of a fly's wing.

In these lectures, Tyndall introduced Dirac to a new way of thinking about light, to new physics. But although Tyndall was admired for his clear presentations, quantum physics was then vague, provisional and messy, so it was impossible for him to present to Dirac the kind of tidy, well-reasoned course that he preferred, underpinned by clear principles and concise equations. This may explain why, if Dirac's later recollections are correct, his first course in quantum theory made virtually no impact on him. His main interest remained relativity.

Despite his earlier setback, Charles Dirac had not lost hope of sending Paul to Cambridge. Late in March, Ronald Hassé wrote to the applied mathematician Ebenezer Cunningham, one of the Fellows of St John's

College, reminding him of Dirac's failure to win a local scholarship that would have enabled him to take up the place that he had won two years earlier. Hassé pointed out that he was 'certain to get first class honours in June', and that he was 'an exceedingly good mathematician', interested mainly in 'general questions – relativity, quantum theory etc., rather than in particular details, and is, I think, very keen on the logical side of the subject'. Among his perceptive comments, Hassé did include some provisos about the young Dirac's character: 'He is a bit uncouth, and wants some sitting on hard, is rather a recluse, plays no games, is very badly off financially.' Those minor points aside, Hassé warmly recommended that the college should accept Dirac if he could find the funds to eke out a living.[13]

This time, Paul Dirac was successful. In August, after he heard that he had won a place at Cambridge, he asked to study relativity with Eddington's Quaker colleague Cunningham, who had introduced a muddled version of Einstein's special theory of relativity to the UK shortly before the Great War.[14] At that time, Cunningham and Eddington were streets ahead of the majority of their Cambridge colleagues, who dismissed Einstein's work, ignored it or denied its significance.[15] But Cunningham was not available: he had given up supervising graduate students after the war, when he had been pilloried as a conscientious objector, most woundingly by authorities who prevented him from working in schools on the grounds that he 'was not a fit person to teach children'.[16] The supervisor chosen for Dirac was another mathematical physicist, Ralph Fowler, a generous-spirited man with the build of Henry VIII and the voice of a drill sergeant. He was not a master of relativity but the foremost quantum theorist in the country and an expert in linking the way materials behave to the en-masse behaviour of their atoms. For Dirac, wanting above all to study relativity, this was not encouraging news.

Two scholarships – one of £70 per year from St John's College, the other from the Government's Department of Scientific and Industrial Research for £140 per year – were sufficient to fund Dirac's first year in Cambridge, provided he lived frugally, as was his wont.[17] The arrangements seemed to have fallen into place, but, in September, he received bitter news: the university required students to settle their bills at the beginning of term, but his government grant was going to arrive too late. He feared that he would again have to forgo his place, all for the sake of £5.

But his father came to the rescue by handing him the money he desperately needed to be sure of solvency in Cambridge. Dirac was touched. This was a crucial act of compassion, he later said, and it minded him to forgive his father for the browbeatings round the dinner table and all the other earlier miseries.[18] Charles Dirac did not seem so bad after all.

Five

[. . .] I could behold
The antechapel where the statue stood
Of Newton with his prism and silent face,
The marble index of a mind for ever
Voyaging through strange seas of Thought, alone.
WILLIAM WORDSWORTH, *The Prelude*,
Book III, 'Residence at Cambridge', 1805

Cambridge has never been the most welcoming place. Visitors who first arrive by rail are often surprised when they realise that the station is almost a mile from the town centre. This rebuffing nudge was quite intentional. Four decades before the station opened in 1845, the authorities had helped to fight off proposals to link the town to London with a canal, but pressure to make Cambridge part of the emerging railway network was irresistible. They did, however, ensure that the station was about twenty minutes' walk from the nearest college so that students would be less tempted to flit off to London and that outsiders would think twice about invading the town's privacy. In 1851, the Vice Chancellor of the university complained to the directors of the railway company that 'they had made arrangements for conveying foreigners and others to Cambridge at such fares as might be likely to tempt persons who, having no regard for Sunday themselves, would inflict their presence on the University on that day of rest'.[1]

As soon as Dirac – and every other new, luggage-laden student – emerged from the station, he had to trek to the city centre or join the queue for one of the few buses that took passengers to Senate House Hill. On Monday, 1 October 1923, when he walked through the stone portals of St John's College to register his arrival, he entered an unfamiliar world of tradition, camaraderie and privilege.[2] He would have been greeted by college porters – resplendent in their liveries and silk hats – each of them charged with keeping an eye on the students and with an obligation to report any errant behaviour. The college admitted only men, many of them in jodhpurs and flat caps and talking in voices that advertised their breeding. Dirac's social standing was given away by his cheap suit – purchased from the Bristol Co-Op – his gauche manners and, on the odd occasion when he spoke, his accent. There was also something out of the ordinary about his appearance. A small and well-tended black moustache lay above his snaggled top teeth, his wan face topped with a thatch of black curly hair and dominated by his assertively pointed nose. Not quite six feet tall and recognisably his father's son, Dirac had bright eyes, a large forehead that revealed a receding hairline and, already, the slightest of stoops.

The sense of tradition in the college is most powerfully expressed in its architecture. Some of it was four centuries old, its construction funded by the posthumous largesse of Henry VIII's bookish paternal grandmother, Lady Margaret Beaufort. The enduring presence of these buildings reminds students that their academic home will remain long after all but the most talented of them have been forgotten. Dirac arrived there with no great ambition, and he was unaware of his academic standing relative to his fellow science students, though he had already decided to do only the most challenging fundamental research. This tradition dates back to Galileo, the founder of modern physics, who took the first steps to cast what he called 'the book of nature' in the language of mathematics. He did this at the turn of the seventeenth century, almost a hundred years after the completion of the first buildings of the college. In this sense, St John's is older than physics.

College life reflected the origins of British academia. The earliest scholars had been monks, all wearing the same clothes, and all going about their contemplative lives within an agreed set of timetables and rules. In 1923, all the official students of the college and the rest of the university were male, each of them required to wear a gown and mortarboard in public. Any student who went into town incorrectly attired knew he ran the risk of being nabbed by one of the university's private policemen (proctors or 'progs') or their assistants ('bulldogs'), who roamed the streets after dusk.[3] A transgression of the dress code was punished by a fine of 6s 8d, no laughing matter for any young man keen to preserve his spare money, though not nearly as serious as the penalty for being caught with a woman in his room.[4]

The students were waited on hand and foot. By 6 a.m., the invariably female bed-makers ('bedders') were hanging around the stone staircases, ready to begin their morning's work. The gyps – man-servants – were available all day to clean, wash up and run errands for the students and for the Fellows (also known as 'dons'). Such service was not, however, available to young Dirac in his first year. He spent it in a cold and damp shoebox of a room in a four-storey Victorian house, a fifteen-minute walk from St John's, sharing with two other lodgers. At a cost of almost £15 a term, the landlady Miss Josephine Brown delivered coals and wood for their fires, supplied gas for the lamps that lit their musty little rooms, provided them with crockery and cleaned their boots. Like all the other landladies approved by the university, Miss Brown was obliged to keep a record of any failure of Dirac's to return home by 10 p.m. Always early to bed, he would not have given her any trouble.[5]

Dirac had his first experience of grand dining in Hall, where he took his meals.[6] The room is magnificently appointed, with an elab-orately decorated wooden ceiling, Gothic stained-glass windows and dark-wood panels hung with portraits of some of the college's most distinguished alumni, including William Wordsworth. The formali-ties began at 7.30 p.m. with the arrival of the procession of Fellows and other senior members of college at their long table, under the calm gaze of Lady Margaret, whose portrait in oils hung above them. The students were already seated in their gowns along the six rows of benches, either side of three long rows of tables, each of them set with crisp white linen tablecloths, the college coat of arms worked into the damask.

It was expected that every head should be dutifully cocked, every pair of hands solemnly crossed in silence as one of the students read the Latin grace from a tablet. The moment he finished, a hundred conversations surged to fill the hall.

The menus, written by hand in French, described the three courses in a style that would meet the approval of a Paris gourmet. The meal might begin with scalloped cod or lentil soup, move on to a main course of jugged hare or boiled tongue and end with gooseberry pie and cream or a plate of cheese with cress and radishes, or even sar-dines on toast.[7] Much of this rich food was wasted on Dirac, whose poor digestion made him favour more basic fare, which he ate slowly and in only modest quantities.

Dirac's fellow diners consisted mainly of the young men of the

Brideshead generation (in Evelyn Waugh's novel, Charles Rider and Sebastian Flyte were then beginning their final year over in Oxford). Most of them had been privately educated at schools such as Eton, Harrow and Rugby, where they had learned Latin and Greek and the art of discoursing easily about the fashionable topics of the day, such as T. S. Eliot's modernist poetry, or of passing supercilious judgement on Shaw's latest provocation. Dirac was ill equipped to join them.

Every night, alcohol circulated up and down the dinner table in Hall, loosening the students' tongues, freeing them to shout ever more loudly to make themselves heard over the din. Amid the cacophony, Dirac sat impassively, a teetotaller in the Methodist tradition, silently sipping water from his glass. He had left Bristol never having consumed a cup of tea or coffee, so his first sampling of these drinks was an event for him.[8] Neither much appealed to him, though he did have the occasional weak and milky tea, its caffeine dose scarcely exceeding homoeopathic levels. Decades later, he told one of his children that he drank coffee only to give himself courage before giving a presentation.[9]

Dirac's manner at the dinner table became the stuff of legend. He had no interest in small talk, and it was common for him to sit through several courses without saying a word or even acknowledging the students sitting next to him. Too diffident even to ask someone to pass the salt and pepper, he made no demands at all on his fellow diners and felt no obligation to maintain the momentum of any dialogue. Every opening conversational gambit would be met with silence or with a simple yes or no. According to one story still in circulation in St John's College, Dirac once responded to the comment 'It's a bit rainy, isn't it?' by walking to the window, returning to his seat, and then stating 'It is not now raining.'[10] Such behaviour quickly persuaded his colleagues that further questioning was both unwelcome and pointless. Yet he did prefer to eat in company and to hear intelligent people talking about serious matters, and it was by listening to such conversations that Dirac slowly learned about llife outside science.

He was fortunate to go up to Cambridge at this time. The colleges had just seen the departure of the last students in military uniform, which took precedence over academic dress until the students were officially demobilised.[11] Now that Britain was under no threat of another international conflict, this was an optimistic time, and the next generation of students was anxious to get back to academic work. Dirac was studying in the university's largest department, mathematics, famous for its high standards and its competitiveness. Among the students, the highest cachet was reserved for those who both excelled in their studies and who competed successfully in sport, which is why Hassé had thought it relevant to remark in his reference for Dirac that he 'played no games'. Most students took at least some part in the social life in Cambridge – chatting in the new coffee bars, singing in choirs, slipping out in the evening to the cinema or to see an ancient Greek play.[12] None of this interested Dirac. Even by the standards of the most ambitious swot, he was exceptionally focused on his work, though dedication is no guarantee of success, as thousands of students find out every year. He had been consistently top of the class in the academic backwater of Bristol, but he had no idea whether he would be able to compete with the best students in Cambridge. From the moment Dirac and his colleagues arrived, the dons were watching every one of them, always on the lookout for a student of truly exceptional calibre – in Cambridge parlance, 'a first-rate man'.[13]

It did not take long for the extent of Dirac's talent to become clear to his supervisor, Fowler, who took a brisk interest in his progress, giving him carefully chosen problems to tackle, constantly encouraging him to hone his mathematics. Students who brought Fowler a good piece of work were rewarded with his favourite exclamation, 'Splendid!', and, more often than not, a pat on the back. He was an inspirational presence in the department, but sometimes unpopular: by spending much of his time working at home or on trips to the Continental centres of physics, he often frustrated the students who yearned for the succour of his advice. But Dirac was not so dependent; he was content to be lightly supervised, to work alone and to generate many of his own projects. Soon, he realised that he had been lucky to have been allocated the most effective supervisor of theoretical physics in Cambridge.

Fowler's manner was unique in the mathematics department. The prevailing culture was intensely formal, and the academics – every one of them male and dressed like a banker – kept their heads down in their offices and college rooms. The use of first names was all but forbidden: even the friendliest of colleagues referred to each other by their surnames and, outside the common room, conversations rarely lasted longer than politeness deemed necessary. Opportunities for them to meet outside the college were minimal as there was no tradition of communal tea and coffee breaks and no programme of seminars. Nor was there any of the staff–student socialising now almost de rigueur in modern university life. Apart from Fowler's guidance, Dirac was left to his own devices. He soon settled into a private routine that would have rendered him invisible among the thousands of his fellow students. With no room of his own in the department, he worked on problems that Fowler set him, read recommended books and the latest journals and reviewed the notes he had made during the lectures. He relaxed only on Sundays. If the weather was fine, he set off in the morning for a few hours' walk, dressed in the suit he wore all week, his hands joined behind his back, both feet pointing outwards as he made his way around the countryside in his metronomic stride. One of his colleagues said he looked like 'the bridegroom in an Italian wedding photograph'.[14]

Dirac would put his calculations firmly at the back of his mind, aiming to clear his head so that he could approach his work fresh on Monday morning. Pausing only to eat his packed lunch, he looked every inch the city gent inspecting the local countryside: to the north, there was the winding valley of the river Great Ouse and to the east, the geometrical network of fenland drains and Tudor-style buildings with their Dutch gables.[15] He would return in time for dinner at St John's and then walk back to his digs through the foggy backstreets of Cambridge, most of them unlit. On Monday morning, he was ready for another six days' uninterrupted study.

Dirac's reserve did not prevent him from meeting many of the country's most famous scientists soon after he arrived. Among them was the man who had introduced him to the technicalities of relativity theory, Arthur Eddington. He was a young-looking forty-year-old, always neatly dressed in his three-piece suit, the knot of his dark tie poised just below the top button of his starched shirt. For someone so eminent, he was surprisingly lacking in confidence – he often sat with his arms crossed defensively, weighing his words carefully. His unique strength as a scientist lay in his hybrid skills as a mathematician and astronomer, giving him the ideal qualifications to play a leading role in tests of the general theory of relativity. He was one of the few scientists who could work on the experiments because, as a Quaker, he was registered as a conscientious objector. Unknown to most of his colleagues, Eddington had used his reputation to contrive the media hullabaloo that followed the announcement in November 1919 that the solar eclipse results supported the prediction of Einstein's theory rather than Newton's.[16]

Dirac attended his lectures and, like most people who first encountered him through his dazzling prose, was disappointed to find that he was an incoherent public speaker who had the habit of abandoning a sentence, as if losing interest, before moving on to the next one.[17] But Dirac admired Eddington's mathematical approach to science, which would become one of the most powerful influences on him. There was no love lost between Eddington and the other great figure of Cambridge science, the New Zealand-born Ernest Rutherford. The two men had sharply contrasting personalities and diametrically opposed approaches to physics. Whereas Eddington was introspective, mild-mannered and fond of mathematical abstraction, Rutherford was outgoing, down to earth, given to volcanic temper tantrums and dismissive of grandiose theorising. 'Don't let me catch anyone talking about the universe in my department,' he growled.[18]

Unlike Eddington, Rutherford did not look in the least like an intellectual.[19] By the time Dirac first felt his surprisingly limp handshake, Rutherford was a burly fifty-two-year-old, with a walrus moustache, staring blue eyes and given to filling his pipe with a tobacco so dry that it went off like a volcano when he lit it. Everyone knew when he was in a room as he spoke more loudly than anyone else. To the people who saw him waddling down Trumpington Street, he had the brash, confident air of a man who had done well out of life by running a chain of betting shops. But his appearance was deceptive: he was the most accomplished experimental scientist alive, as he was the first to confirm. His most famous discovery, the atomic nucleus, followed after he suggested to two of his students that they should investigate what happens when they fired subatomic particles at a thin piece of gold foil. After he heard that a few of the particles were deflected backwards, Rutherford imagined his way into the heart of the atom and concluded that the core of every atom is positively charged and occupies only a tiny fraction of its space, 'like a gnat in the Albert Hall', as he put it.[20] He first identified the existence of atomic nuclei in the summer of 1912, when he was working at the University of Manchester, eight years before he moved to Cambridge to become J. J. Thomson's successor as Director of the Cavendish Laboratory. Soon after he arrived there, he made one of his bold predictions about atomic nuclei by proposing that most of them are made not only of protons, each positively charged, but also of hitherto-unidentified particles with about the same mass but no electrical charge. Rutherford encouraged his colleagues to hunt for these 'neutrons', but their desultory experiments drew a blank.

The mid-1920s were not a productive time for Rutherford as he was no longer making ground-breaking discoveries but was devoting his prodigious energy to directing the Cavendish Laboratory, which he ruled like an absolute but benevolent monarch. The laboratory was tucked away in Free School Lane, a side street that was a few minutes' walk from the Department of Mathematics, but a world apart. Built in 1871, the Victorian Gothic façade of the laboratory was much the most impressive part of the building. After walking through the front door, visitors found themselves in a dingy corridor next to a hall half-filled with haphazardly parked bicycles. To the modern eye, the laboratories look like the kind of functional workshops Heath Robinson might have set up in his garage: bare brick walls and wooden floors, pedal-operated lathes, hand-operated vacuum pumps, glass-blowing equipment, sturdy benches covered with greasy tools and some pieces of equipment so primitive that they would be hard to sell from a junk shop. The authorities in Cambridge had worried whether an environment like this was worthy of a university for gentlemen, but they acknowledged that it had established itself as an exceptionally productive centre for physics research, and at only modest cost. In 1925, the total budget of the laboratory, including all salaries and equipment, was £9,628.[21]

Although Rutherford was disdainful of mathematical physicists – or pretended to be – he welcomed tame theorists who would do difficult calculations for him, such as his son-in-law and golfing partner Fowler, the only theorist to have his own office in the Cavendish. Visiting theoreticians had nowhere to sit except in the squalid, unheated library, a shabby tearoom that reeked of congealed milk and stale biscuits.[22] Many of the older theoreticians reciprocated Rutherford's disdain by having nothing to do with activities at the Cavendish, but some of the younger students accepted Rutherford's invitations to attend the laboratory's regular Wednesday afternoon seminars, preceded by tea – often poured by Lady Rutherford – and, sometimes, Chelsea buns.[23] At the Cavendish, Dirac came to know two of Rutherford's 'boys', who were to become his closest friends: the Englishman Patrick Blackett and Russian Peter Kapitza. Both had been trained as engineers, but their personalities were quite different, exemplifying the two extremes that Dirac liked most: shy introverts like himself (Blackett) and boisterous extroverts (Kapitza).[24] In their different ways, these two men would powerfully influence Dirac, drawing him out of his shell in his early years at Cambridge, keeping him at the hub of experimental activity, introducing him to dozens of new acquaintances he would not otherwise have made and to a field that had previously been of no interest to him: politics.

Blackett and Kapitza had recently turned up at the Cavendish, like jetsam thrown up by the war. Blackett had arrived first, in January 1919, when he was twenty-one years old and still in his navy uniform. He had been given a first-rate technical education at a naval college and, days after graduating, went to war, aged sixteen. On 31 May 1916, the first day of the battle of Jutland, the most violent naval conflict of the war, he was at one of the twin fifteen-inch turrets of HMS *Barham*, relentlessly bombarded by German warships too distant to see. By the end of the day, he was walking on the deck – the air thick with TNT fumes and disinfectant – among the charred corpses, some with their limbs blown off.[25]

Three weeks after arriving in the Cavendish, he resigned his commission and took a degree in natural sciences to prepare himself for a life in experimental physics. He cut a suave, romantic figure: six feet two inches tall, slim, handsome as a movie star, yet with the haunted demeanour of a midshipman who had seen his mates die in agony in front of his eyes. In the laboratory, he quickly proved to be an ingenious experimenter, with the scientific virtues of imagination and scepticism. One colleague noted that he was 'not easily convinced even by his own ideas'.[26]

In almost any other laboratory, Blackett would have stood out as the finest student of his generation. However, in that exceptional phase in the history of the Cavendish, he had plenty of competition, especially in the chunky form of Kapitza, who had earlier beaten Blackett to the scholarship for the university's best laboratory student, one of several small victories that helped to fuel Blackett's resentment of him. Kapitza had settled in the UK in 1921 looking – as one of his Trinity colleagues observed – 'like a tragic Russian prince', insecure and depressed after the deaths of four members of his close family within a few months at the end of 1919: scarlet fever took the life of his infant son, shortly before his father, wife and baby daughter fell victim to Spanish Flu.[27] In the summer of 1921, after braving an initial rejection, he persuaded Rutherford to take him on as a student in the Cavendish. Kapitza idolised Rutherford for his straightforwardness, his energy and his uncanny ability to ask nature the right questions to make it yield its deepest secrets. When Rutherford was out of earshot, Kapitza referred to him as 'the Crocodile', the young Russian's favourite creature: Kapitza collected poems about crocodiles and even welded a metal model of one to the radiator of his open-topped Lagonda.[28] Kapitza's name for his boss may have been an unconscious reference to the reptile that appeared prominently in books by the Soviet Union's most popular children's writer, Korney Chukovsky. Like most parents in Russia, Kapitza had probably read his children the famous stories of the crocodile who swallows people and dogs but who good-naturedly disgorges them unharmed. Chukovsky encouraged his readers to regard the crocodile with a mixture of fear and admiration, just as Kapitza saw Rutherford.[29]

By the time Dirac arrived in Cambridge, Kapitza was one of the town's most colourful characters. Although he did not speak any language well – even, it was said, his own – he loved to talk, words tumbling incessantly out of one side of his mouth. He chatted merrily in his high-pitched voice, delighting his colleagues with his card tricks and the amusing stories he told in 'Kapitzarene', a language that seemed to consist of Russian, French and English in roughly equal parts. He returned to the Soviet Union every year to see his family and to advise on the programme of industrialisation being pushed by Lenin's successor, Joseph Stalin. He was playing a dangerous game, as the economist John Maynard Keynes told his wife in October 1925 after Kapitza mentioned that he was planning to visit Russia to advise Trotsky on their country's electrification programme, having secured a firm promise that he could return to Cambridge: 'I believe that they will catch him sooner or later [. . .] he is a wild, disinterested, vain, and absolutely uncivilized creature, perfectly suited by nature to be a Bolshie.'[30]

Dirac had no such reservations. Near the end of his life, in a nostalgic account of his early days with Kapitza, Dirac wrote that he was immediately taken with his boldness and self-confidence.[31] They shared a passion for science and engineering, but much divided them: Kapitza delighted in chit-chat, whereas Dirac ignored it; Kapitza loved literature and theatre, whereas Dirac had little time for either; and Kapitza was sceptical of the abstractions of theoretical physics, which were meat and drink to Dirac.

On Kapitza's first day in the Cavendish, he was surprised by one of Rutherford's first instructions, forbidding him to spread Communist propaganda in the laboratory.[32] Kapitza worked sedulously at his bench but in his spare time never made any secret of his support of Lenin's politics and pleasure at the defenestration of Russia's land-owning aristocracy during the 1917 revolution. As he wrote later, although he never joined the Communist Party, he always supported its goals: 'I am in complete sympathy with the socialist reconstruction directed by the working class and with the broad international-ism of the Soviet Government under the guidance of the Communist Party.'[33]

In the early 1920s, the British Government was worrying about the stability of the country's institutions, concerned that Communists would infiltrate and subvert them.[34] It is hardly surprising that, only two years after he arrived in Cambridge, an anonymous informer had tipped off the Government's Security Service MI5 with a report 'to the effect that Kapitza is a Russian Bolshevist'.[35] In collaboration with the Metropolitan Police Special Branch, they kept him under surveillance, anxious that he did not suspect for a moment that he was being watched.

It was probably Kapitza who introduced Dirac to Soviet ideology, a subject that would later become a crucial ingredient of their friend-ship. In the mid- to late 1920s, such beliefs were not in vogue in Cambridge, as the great majority of students and dons were not seri-ously interested in politics.[36] The only prominent Marxist don was the economist Maurice Dobb, who, like Kapitza, was based at Trinity College. The tenor of political conversations in its senior common room was the soul of moderation, equilibrium being guar-anteed by moderates such as Rutherford and by a bevy of conserva-tives that included the poet and classicist A. E. Housman and Charlie Broad, who had moved to Cambridge and was living in the rooms once occupied by Newton.

Kapitza liked to compare himself to Dickens's Mr Pickwick, and it was an apposite comparison: each, with winning brio, had founded a club whose members had elected him to be their permanent president. In setting up the Kapitza Club in October 1922, he had shaken his postgraduate colleagues out of their lethargy and persuaded them to attend a weekly seminar on a topical subject in physics. The talks usually took place in Trinity College on Tuesday evenings, after a good dinner. The speakers, normally volunteers from the club's members, spoke with the aid only of a piece of chalk and a blackboard mounted on an easel and had to be prepared for a series of interruptions, mediated by Kapitza with the quick wit and élan of a modern-day game-show host.[37]

The rules of the club were that a student could become a member only by giving a talk and that his membership would be withdrawn if he missed a few meetings. Soon after Dirac's arrival in Cambridge, he started going to the club and joined the less frequent, more theoretically inclined $\nabla^2 V$ Club, named after a common symbol in mathematical physics. This club – the nearest the theoreticians came to having a seminar programme – was attended by dons as well as students, so its proceedings were more in keeping with the stiff ambience of the mathematics department. Rutherford attended them only rarely, scoffing that theorists 'play games with their symbols, but we in the Cavendish turn out the real facts of nature'.[38]

Despite all these new experiences, the postcards Dirac sent home did little more than confirm he was still alive:

Dear Father and Mother
I am coming home next Thursday. I expect I shall arrive by a late train.
  Love to all
  Paul[39]

All his postcards were like this. They each bore a sepia photograph of a Cambridge scene and about a dozen sterile words, consisting entirely of facts and brief summaries of the weather. His mother set the pace of the correspondence by writing to him almost weekly letters that continued until the middle of Dirac's career, giving her view of life in 6 Julius Road and her relationship with Charles. At this stage, the letters give no sign that the family was unusual: pooterish, chatty and steeped in maternal affection, they continually stress how much he was missed – an emotion that Dirac never reciprocated. Charles Dirac apparently did not write to him, though Flo went out of her way to underline that his father was 'very anxious' to know how he was getting on.[40]

Flo told her son how excited the family was by its new toy, a radio. The Diracs were in the first generation of families to buy a receiver, scarcely a year after they first became available in 1922. Their home did not yet have a mains supply of gas or electricity, so Charles had to walk down to the local tram station to charge up the radio's accumulator (its battery). It was worth the inconvenience: the new device livened up 6 Julius Road, replacing the day-long silence with a soundtrack of programmes from the new British Broadcasting Corporation, including talks, concerts and news. The Diracs would gather round the radio each night to hear the newsreader orate as if he were addressing a funeral. On 22 January 1924, they heard that Ramsay MacDonald had been appointed Britain's first Labour Prime Minister. The party that had begun as the creature of the trade unions was in Downing Street, its agenda and rhetoric moderate enough to avoid panicking the British public, always wary of rapid change.[41] Flo reported to Dirac that his father was 'pleased that the Labour government have got in at last. It is the best for teachers' salaries.'[42]

In Flo's letters, she hardly mentions Felix. In the spring of 1924, still based near Wolverhampton, he was earning a modest wage as a draughtsman and was cycling home to Bristol during his short vacations.[43] Stooped over his drawing board, his rimless eyeglasses perched on his nose, he spent his days making technical drawings for a manufacturer of heavy machinery and advising engineers in the workshops. A steady worker, he was admired for his politeness and reliability by his colleagues, who knew – as he must have done – that he could look forward to nothing more in his professional life than mediocrity. In private, he began to pursue interests that set him apart from his parents and brother: he became a Buddhist and dabbled in astrology, seeking help from a guru, the Revd. Sapasvee Anagami Inyom, based in south-west London. To judge from his communications to Felix, this counsellor was a theosophist, someone who sought knowledge of God through a mixture of Hindu and Buddhist teachings.[44] His letters – long on generalities, short on specifics – each began with a florid salvo ('Greetings in the Glorious Love, Joy and Peace in the Three Gems') and continued with pages of windy reassurance. By embarking on this spiritual path, Felix was abandoning both the Methodism of his mother's family and his father's Catholicism, and by following astrology he was perhaps cocking a snook at his brother, who, like every other scientist, will have dismissed the notion that local stars and planets influence human fortunes as fatuous.

Unlike his brother, Felix showed an interest in the opposite sex. He acquired a girlfriend, and the relationship became serious enough for his father to suggest that Felix and his girlfriend should visit the family home when Paul was present so that the whole family could meet her. He may well have been disappointed by his mother's rejection of the idea, and it appears that his brother was miffed. In the first public interview Paul gave about his family life, almost forty-five years later, he laughed when he quoted the words his mother used to veto the request – 'Oh no, she mustn't, she might go after Paul' – and, unusually, gave his description of the incident a dab of colour by commenting on his mother's protectiveness: 'I rather resented it.'[45] He said nothing about whether he would have accepted the invitation to meet the young woman but implied that – in this isolated case – his father behaved much more reasonably than his mother. Paul's account of her behaviour appears to be the only criticism he ever made of her in public or private, perhaps a sign of the anger she caused him by her possessiveness towards him and the insensitivity she showed to his brother. This is a rare example of his recalling empathy with his brother or anyone else.

After his arrival in Cambridge, Dirac realised that if he was to work on truly fundamental research, he had some catching up to do. The University of Bristol had given him an excellent technical training and a basic grounding in mathematics, but there were several gaps in his education. Among the most serious was his ignorance of the unified theory of electricity and magnetism set out fifty years before by James Clerk Maxwell. This theory, with Darwin's theory of evolution, was the most important scientific advance of the Victorian era and did for electricity and magnetism what Einstein's general theory of relativity would later do for gravity. Maxwell described electricity and magnetism in a handful of equations and used them to predict successfully that visible light consists of electromagnetic waves (or 'electromagnetic radiation'). Such light waves fall within the small range of wavelengths that human eyes can see. Electromagnetic waves with shorter wavelengths than visible light include ultraviolet radiation and X-rays; waves with longer wavelengths include infrared radiation and microwaves.

Dirac first learned about Maxwell's equations in lectures given by Ebenezer Cunningham, who found the precocious Bristol engineer-mathematician to be assertive and quick to ask questions about physics that he did not understand.[46] Maxwell's equations must have been thrilling to Dirac: in just a few lines of mathematics, they could explain the results of every experiment on electricity, magnetism and light that he had ever done in Bristol, and much else besides. When he heard about the equations, he saw why Einstein's light quanta had, until a few years before, been so widely ridiculed: the idea flatly contradicted the accepted Maxwellian view that light consisted of waves, not particles. However, nine months before Dirac arrived in Cambridge, news from Chicago suggested that Einstein might be right: the American experimenter Arthur Compton had found that, in some circumstances, electromagnetic radiation – including, pre-sumably, visible light – really can behave not as waves but as discrete particles.[47] He had scattered X-rays from free electrons and found that he could explain his measurements only if each scattering is due to a collision between two particles, like a pair of snooker balls strik-ing one another. This is just as Einstein had suggested – the radiation and the electrons were both behaving as particles – in contradiction to the wave picture. Many physicists refused to believe these results, but Dirac was one of the few who took them in his stride, unencum-bered by years of familiarity with the deceptive success of Maxwell's theory.

One of the scientists who dismissed the new photon picture of light as nonsense was the Danish theoretician Niels Bohr. He had made his name in 1913, when he built on Rutherford's suggestion that every atom contains a tiny nucleus. Rutherford's picture could not explain the experimental discovery that atoms emit and absorb light with certain definite wavelengths (each type of atom that gives out visible light, for example, emits only light with a particular set of colours). It is as if each atom has its own 'song', composed of light, not sound – instead of musical notes, each played with a characteristic loud-ness, every atom can give out light with its own set of colours, each colour with a characteristic brightness. Scientists had, somehow, to understand the composition of every atomic melody. Bohr came up with his idea soon after he heard that the colours of the light emitted by hydrogen – the simplest atom, containing only one electron – had an extremely simple pattern, first spotted in 1885 by Johannes Balmer, a Swiss schoolteacher. He happened on a simple but mysterious formula that accounted for the colours of the light given out by these atoms, a mathematical encapsulation of hydrogen's signature tune. Every other atom was more complicated and much harder to understand. Bohr's achievement was to take the cue from the hints in this pattern, to build a theory of the hydrogen atom and then to generalise it to every other kind of atom.

Bohr's atom had a positively charged nucleus, which has most of the atom's mass, orbited by negatively charged electrons which are tethered by the attractive force between the opposite charges. In much the same way, the planets are held in their orbits around the Sun by the attractive force of gravity. He imagined that the electron in a hydrogen atom could move around in its nucleus in only certain circular orbits – called by others 'Bohr orbits' – each of them associated with a particular value of energy, 'an energy level'. Each of these orbits had its own whole number, known as a quantum number: the orbit closest to the nucleus was labelled by the number one, the next orbit by the number two, the next orbit by three, and so on. Bohr's innovation was to imagine that the atom gives out light when it jumps (or, in other words, makes a transition) from one energy level to another of a lower energy, simultaneously emitting a quantum of radiation that has an energy equal to the difference between the energies of the two levels. Bohr was saying, in effect, that matter at the atomic level behaves very differently from everyday matter: if the apple that fell in Newton's garden were able to lose energy by descending down a set of allowed energy values, it would not have fallen smoothly but would have made its way jerkily to the ground, as if bumping its way down an energy staircase. But the energy values of the apple are so close together that their separation is negligible and the fruit appears to slide smoothly down the staircase. Only in the atomic domain are the differences between energy values significant enough for the transitions to be jerky.

Bohr's theory offered a simple understanding of Balmer's mysterious formula. In just a few lines of undemanding high-school algebra, any physicist could derive the formula using Bohr's assumptions, leaving the satisfying impression that the pattern of hydrogen's colours was comprehensible. Yet Bohr's theory was only a qualified success: according to the laws of electromagnetism, it was absurd. Maxwell's theory said that the orbiting electron would shine – continuously give out electromagnetic radiation – and thus gradually radiate its energy away. So it would not take long before the orbiting electron would spiral to its doom in the nucleus, with the result that the atom would not exist at all. The only way Bohr could counter this was to assert, by fiat, that orbiting electrons do not give off such radiation, that Maxwell's theory did not work on the subatomic scale.

With a remarkable sureness of intuition, Bohr extended his ideas to all other atoms. He suggested that each atom has energy levels and that this helped to explain why the different chemical elements behaved so differently – why, for example, argon is so inert but potassium is so reactive. Einstein admired the way Bohr's ideas explained Balmer's formula and the insights they gave into the differences between each type of atom, hinting at an understanding of the very foundations of chemistry. As Einstein remarked in his autobiographical notes, Bohr's theory exemplified 'the highest form of musicality in the sphere of thought'.[48]

But no one properly understood the relationship of Bohr's atom to the great theories of Newton and Maxwell. These theories came to be described as 'classical', to distinguish them from their quantum successors. A fundamental question was, how, precisely, does the theory of the very small merge into the theory of the comparatively large? To answer this, Bohr developed what he called the correspondence principle: the quantum description of a particle resembles the classical theory more and more closely as the particle's quantum number becomes larger. Similarly, if a particle vibrates rapidly and therefore has a very small quantum number, quantum theory must be used to describe it; classical theory will almost certainly fail.

This principle was too vague for Dirac: he preferred theoretical statements to be expressed in an equation with a single, lapidary meaning, not to be set out in words that philosophers could dispute. But he was fascinated by Bohr's theory of the atom. He had not heard of it in Bristol, so Fowler's lectures on the theory were an eye-opener. Dirac was impressed that Bohr had come up with the first tractable theory of what was going on inside atoms. Dirac spent long after-noons in the libraries studying his notes from Fowler's lectures and poring over the classic textbook *Atomic Structure and Spectral Lines*, by the Munich theoretician Arnold Sommerfeld. Required reading for every student of quantum theory, the book set out Bohr's picture of the atom and showed how it could be refined and improved. Sommerfeld gave a more detailed description in which the possible orbits of the electron are not circular (as Bohr had assumed) but elliptical, like the path of a planet round the Sun. He also improved on Bohr's work by describing the motion of the orbiting electron not using Newton's laws but using Einstein's special theory of relativity. The result of Sommerfeld's calculation was that the measured energy levels should differ slightly from the levels predicted by Bohr, a con-clusion supported by the most sensitive experiments. Bohr knew as well as everyone else in atomic physics that his theory was fatally flawed and therefore only provisional; what was unclear was whether the theory that succeeded it would be based on a few tweaks to Bohr's ideas or on a radically new approach.

At the same time as he was learning and applying Bohr's theory, Dirac was immersed in geometry, which he studied privately and at weekly tea parties held on Saturdays by the mathematician Henry Baker, a close friend of Hassé's. Now approaching his retirement, Baker was an intimidating man with the thick moustache which was, in those days, almost mandatory. His parties took place at four o'clock on Saturday afternoons in the Arts school, a grim Edwardian building only a short walk from the Cavendish. Apart from the porter and a few cleaners, the School was as lifeless as a museum at midnight until Dirac and fifteen or so other aspiring scholars of geometry arrived and knocked on the front door. Baker regarded these meetings as his opportunity to promote his love of geometry to his most able students. The subject needed him: for almost a century, it had been the most fashionable branch of mathematics in Britain, but its popularity was waning as fashion began to favour mathemat-ical analysis and the study of numbers.[49]

The parties – better described as after-hours classes for devotees – were friendly but tense with formality and protocol. Each gathering began promptly at 4.15 p.m., and, in the time-honoured way at English universities, could not begin until everyone had been served a cup of tea and a biscuit. The only students allowed to be late were the sportsmen – rowers, rugby players and athletes who would arrive red-faced and settle down hurriedly after depositing their knapsacks full of sweaty kit. Each week, Baker arranged in advance for one of the students to give a talk to the party before submitting to a grilling by the audience, most of them writing with one hand and smoking with the other. Baker was a spirited teacher, a no-nonsense mediator but a stern host – he had no compunction about berating any student whose attention showed the slightest sign of wandering. For several of the young men, the parties were a chore, but they were a highlight of Dirac's week: '[they] did much to stimulate my interest in the beauty of mathematics'. He learned that it was incumbent on mathematicians to express their ideas neatly and concisely: 'the all important thing there was to strive to express the relationships in beautiful form'.[50]

It was at one of these parties that Dirac gave his very first seminar, about projective geometry. From his fellow students and Baker, he also became acquainted with a branch of mathematics known as Grassmann algebra, named after a nineteenth-century German mathematician. This type of algebra resembled Hamilton's quaternions, as they are both non-commuting: one element multiplied by another gives a different result if the two are multiplied in a different order. Some applied mathematicians jeered that Grassmann's ideas were of little practical use, but such concerns did not trouble Baker. He warned his students to expect no public recognition for anything they achieved in pure mathematics, whereas 'if you discover a comet you can go and write a letter to "The Times" about it'.[51]

Baker was the type of don Cambridge academics called 'deeply civilised' – a subject specialist whose enthusiasms were grounded in high culture. One of his hobbies was the culture of ancient Greece, and he was fascinated by the Greeks' love of beauty, which he believed was as good a stimulus to a scientific life as any. This may be one reason why Dirac drew attention to the aesthetic appeal of Einstein's theory of gravity in a talk he gave at one of Baker's gatherings, having pointed out that its predecessor, Newton's law of gravity, 'is of no more interest – (beauty?) – to the pure mathematician than any other inverse power of distance'.[52] This is Dirac's first recorded mention of 'beauty'. In Bristol, he had been encouraged to take an aesthetic view of mathematics; now, in Cambridge, he had found again that the concept of beauty was in vogue. The popularity of the concept was at least partly due to the enduring success of *Principia ethica*, published in 1903 by the philosopher George Moore, one of Charlie Broad's colleagues in Trinity College. Writing with a refreshing absence of jargon, Moore made the incisive suggestion that 'the beautiful should be defined as that of which the admiring contemplation is good in itself'.[53] Soon the talk of intellectuals, *Principia ethica* was admired by Virginia Woolf and her colleagues in the Bloomsbury Group and declared by Maynard Keynes to be 'better than Plato'. Over a century before, Immanuel Kant had rendered the subject of beauty too complex and intimidating for most philosophers, but Moore made it accessible again in a way that commanded respect.[54] Although *Principia ethica* did not consider the aesthetics of science, Moore's common-sense approach to beauty probably influenced his scientific colleagues at Trinity, including Rutherford and the college's most eminent pure mathematician, G. H. Hardy: both often talked about the beauties of their subject. Kapitza, too, looked on experimental physics not as 'business', as it was to several of his colleagues, but as a kind of 'aesthetic enjoyment'.[55]

Although Dirac was not interested in philosophy, this fascination with the nature of beauty had powerful resonances for him. Like many theoreticians, he had been moved by the sheer sensual pleasure of working with Einstein's theories of relativity and Maxwell's theory. For him and his colleagues, the theories were just as beautiful as Mozart's *Jupiter Symphony*, a Rembrandt self-portrait or a Milton sonnet. The beauty of a fundamental theory in physics has several characteristics in common with a great work of art: fundamental simplicity, inevitability, power and grandeur. Like every great work of art, a beautiful theory in physics is always ambitious, never trifling. Einstein's general theory of relativity, for example, seeks to describe all matter in the universe, throughout all time, past and present. From a few clearly stated principles, Einstein had built a mathematical structure whose explanatory power would be ruined if any of its principles were changed. Abandoning his usual modesty, he described his theory as 'incomparably beautiful'.[56]

Dirac was extremely hard to read. Usually, he looked blank or wore a thin smile, whether he was making headway with one of his scientific problems or depressed by his lack of progress. He seemed to live in a world in which there was no need to emote, no need to share experiences – it was as if he believed he was put on Earth just to do science.

His belief that he was working solely for himself led to one of his rare spats with Fowler. Soon after Dirac began in Cambridge, Fowler gauged the ability of his new student by asking him to tackle a non-trivial but tractable problem: to find a theoretical description of the breaking up of the molecules of gas in a closed tube whose temperature gradually changes from one end to the other.[57] Some five months later, when Dirac found the solution, he wanted to file it away and forget it, a suggestion that dismayed Fowler: 'if you're not going to write your work up, you might as well shut up shop!'[58] Dirac succumbed and forced himself to learn the art of writing academic articles. Words did not come easily to him, but he gradually developed the style for which he was to become famous, a style characterised by directness, confident reasoning, powerful mathematics, and plain English. All his life, Dirac had the same attitude to the written word as his contemporary George Orwell: 'Good prose is like a window pane.'[59]

That first paper was a piece of academic throat-clearing, of little consequence and unrelated to the fundamental theories of physics that Dirac loved. In his next three papers, however, he was on the more congenial ground of relativity. In his first paper on the subject, he clarified a point in Eddington's mathematical textbook on Einstein's general theory of relativity, and in the next two applied the special version of the theory first to atoms jumping between energy levels and then to soups of atoms, electrons and radiation. It was not until the end of 1924 that he produced an outstanding piece of work, an exploration – using Bohr's atomic theory – of what happens to the energy levels of an atom when the forces acting on it change slowly. Although Dirac came to no startling conclusions, his paper attested to his mastery of Bohr's theory and of Hamilton's mathematical methods. Yet Dirac was starting to believe that such exercises were hollow. The more he thought about the Bohr theory, the more dissatisfied he was with its weaknesses. Others shared this dissatisfaction: physicists all over Europe feared that a logical theory of the atom might simply be beyond the human mind.

Six

My grief lies all within,
And these external manners of lament
Are merely shadows to the unseen grief
That swells with silence in the tortured soul.
WILLIAM SHAKESPEARE, *Richard II*,
Act IV, Scene 1

Towards the end of Dirac's graduate research, Ebenezer Cunningham described him as 'quite the most original student I have met in the subject of mathematical physics' and 'a natural researcher'.[1] By the time he returned to Bristol for Christmas in 1924, he had every reason to be pleased with himself: he had written five good papers – well above the average for even a strong graduate student – with little help from Fowler or any other senior colleague. He was certain to get his Ph.D. But Dirac knew that his work had so far involved mainly tidying up loose ends in other people's projects and that he had not done nearly enough to deserve a place with Bohr and Einstein at the forefront of theoretical physics. For the moment, Dirac was biding his time in the green room, awaiting inspiration, before he could step out on the international stage.

Throughout the preceding year, Dirac may have noticed that his mother's letters indicated her deepening unhappiness and that she was manoeuvring him into the position of a confidant. Early in the summer, she had complained of having little money of her own, a theme that was to become a leitmotif of her correspondence with him. Charles earned a respectable salary and supplemented it by giving private tuition but was always worried about money and had – like many a husband at that time – no compunction about giving his wife only enough to run the house. With virtually no money of her own and too proud to turn to her siblings, she was reduced to asking Paul for money: '[Pa] is grousing about the bills just now especially the grocer's, so I am wondering if you will be able to spare a few shillings a week next time you are home?'[2] Though Dirac does not appear to have responded in writing, it is reasonable to suppose that he was disturbed by it as he was living frugally on his grant and had no additional income from teaching. To give his mother money would reduce him to penury.

In June, he had moved out of his digs into one of the grandest buildings in the college, the neo-classical New Court, built in the early nineteenth century.[3] In his rooms in the west wing of the building, he had for the first time the benefits of being able to work in complete private, disturbed only by the cleaner and bed-maker. Many well-off students put their individual stamp on their own patch of the college by bringing their own furniture, oriental rugs, paintings and trinkets. Dirac's room was as bare as a jail cell, but the accommodation gave him all he needed: peace and quiet, regular meals and warmth. The only irritation for him was the regular ringing of the chapel bell: a few years later, he told a friend that it 'gets on my nerves sometimes' – so much so that 'I am a little afraid of [it]'.[4] But his mother knew that he was happier in Cambridge than he was in Bristol, and she feared that he would no longer be content in the modest and ill-kept family home now that he had gone up in the world. Shortly before he returned to Bristol for the Christmas vacation, she prepared his bedroom, beating the carpet and scrubbing the floor, 'the best I can do to such a shabby room'.[5]

Felix had settled in Birmingham, living in lodgings in the southwest of the city and working in the machine-testing laboratory of a factory. With no sign that his career was about to move up a gear, it may have been hard for him to hear his parents talk about the successes of his younger brother in Cambridge. Felix had good reason to be envious: he was still tethered to a stool in a drawing office, plying a trade that brought him little money and, it seems, little satisfaction. Still regretting that his father had refused to let him study medicine, Felix volunteered for the Ambulance Corps, evening work that gave him glimpses of the doctor's life he had longed for. He was sharing none of this with his brother – they lived separate lives, all fraternal affection spent.

Early in the cold and dreary January of 1925, Felix snapped. He left his job, though he took care to remain on good terms with his employer, the technical manager in the Testing Machine Department, who certified that he always found Felix 'to be obliging, courteous, and painstaking in his work'.[6] He stopped writing to his parents and sister and did not tell either them or his landlady what he had done or that he was living off his savings. He pretended still to be at work, leaving his digs in the morning and returning for his evening meal, sometimes attending classes at the nearby Midland Institute.

By the end of winter, his savings ran out. His landlady did not suspect that anything was wrong until the first Thursday evening in March, when he did not return for dinner.[7]

The chilly, overcast morning of 10 March began like any other term-time Tuesday for Paul Dirac. There was a hint of spring in the air. As usual, before beginning his day's work, he walked across the stone courts of St John's to the Porter's Lodge to see if there was any mail in his pigeonhole. He found a tiny envelope – small enough to fit in the palm of his hand – postmarked in Bristol late on the previous night, though it was not the weekly note from his mother. He opened the folded letter and saw that it was from his mother's sister Nell. She began uneasily, asking him to bear up for the news that she was about to convey because his 'parents are so greatly upset'. Felix was dead.[8]

His body had been discovered four days before under a holly bush on the edge of a field two miles south of the Shropshire village of Much Wenlock. Smartly dressed in a suit and bow tie, Felix had a spanner in one of his pockets and was still wearing his bicycle clips, though his cycle was nowhere to be seen. The people who found him assumed that he had killed himself by taking poison, as an empty glass bottle lay next to his corpse. He carried no identifying papers and left no final message; the only clue to his identity was the case of his glasses, which bore the name of an optician in Wolverhampton.[9]

Not so long ago, Dirac had loved his brother and looked up to him, shared the same bedroom and the same handed-down comics, ran with him on the Bristol Downs and followed him to university. They had been split by arguments, resentments and jealousies, all of them now rendered pathetically insignificant by grief. Now, the act of suicide had made reconciliation impossible.

Dirac's feelings about all this are not known, as there is no documentary evidence of his reactions. If he behaved according to type, he will have received the news with the calm of a statue and told no one in Cambridge about it, apart, perhaps, from Fowler. But it is possible to speculate on his emotions from the testimonies of the few close family members with whom he shared his pain decades later, if only for a few moments.[10] If we extrapolate the feelings he showed then back to 1925, it is reasonable to conclude that the passing of Felix left his brother with a tapeworm of anger, sadness and guilt gnawing inside him.

The news of Felix's death had been all over Bristol late on the Monday afternoon: the *Evening News* announced the death in a front-page article under the headline 'Dead in a Field'.[11] A report on the following day noted that Felix's death had caused 'a profoundly painful sensation in the city', hinting that the tragedy was all the more incomprehensible because the deceased was 'the son of one of the most respected gentlemen connected with education in this city'.[12] Charles and Flo did not read the report when it was published as they were in Shropshire to identify their son's body and attend the first stage of the inquest. Dirac had just received his aunt's letter and may have wondered why his parents had not wired him as soon as they heard the news. Did they really believe that he would not want to be among the first to hear of his brother's death? Four decades later, Dirac told friends that he was shocked by his parents' distress. The death of his brother was 'a turning point' for him: 'My parents were terribly distressed. I didn't know they cared so much. [. . .] I never knew that parents ought to care for their children, but from then on I knew.'[13]

If these and his other recollections of his early family life are accurate, they indicate the extent of his emotional detachment. He appears to have been unaware of many of the experiences that do most to shape the lives of children – the fondness of their parents, the importance of family rituals, the day-to-day entanglements of family life. Nor does he ever even allude to the possibility that the coldness of the Dirac household could have been due at least in part to his own insensitivity. These are among the strongest clues that he suffered from what amounted to a kind of emotional blindness.

From Dirac's portrayals of his father's cold-hearted tyranny and his mother's overweening maternalism, it would be natural to expect that the suicide of Felix would have hurt his mother much more than his father. But it was the other way round. Charles was poleaxed. This was no ordinary grief: his doctor advised him to rest for a year; his family feared for his sanity and even worried that he might take his own life.[14] Flo, by contrast, took it all in her stride, though she was distressed that she had misunderstood Felix and had not seen the disaster as it approached. In a memorial poem to him she wrote thirteen years later, she wrote, 'He had dropped the mask.'[15]

On a bitterly cold Sunday, two weeks after Charles and Flo first heard of their son's death, they attended a memorial service for him at a nearby church. When Flo returned home, she wrote to Dirac with a mother's firmness: 'Mind you meet Pa on Thursday & <u>stick to him all the time after the inquest</u>, there's a dear boy, & bring him home safely whatever he may hear.'[16] Dirac did as she requested: a few days later, he travelled to the enquiry, held within a mile of the hills where Felix had been found, a part of the country finely etched into the English imagination by Housman's bitter, nostalgic poetry. At the enquiry, Dirac and his heartbroken father sat next to each other when they listened to the coroner read his report. He began by noting that the body had been found on Friday 6 March. The corpse was of a man about twenty-five years old, five feet nine inches tall, with thin features, dark hair, a slight moustache and good teeth. Felix had taken his life, the coroner concluded, by 'taking cyanide of potassium whilst of unsound mind'.[17]

Witnessing Charles Dirac's grief taught his son a lesson: no matter how painful life might become, he would never commit suicide, because the price paid by his family would be too great.[18] Betty was no less affected: in her later life, she never spoke about the circumstances of Felix's suicide, though she once remarked to her children that he had been killed in a car accident.[19]

It appears that Dirac kept working to his usual routine. Fowler had gone on sabbatical in Copenhagen to work with Bohr, leaving Dirac in the care of the young astrophysicist Edward Milne. He set Dirac the task of investigating the processes going on at surfaces of stars such as the Sun, a problem that Dirac solved efficiently, though once again he did not come up with any eye-catching conclusions.[20] For several months, Dirac's productivity plummeted. He never explained why, but it is reasonable to speculate that he was slowed down by grief and, possibly, that he was turning his attention from tackling readily solvable problems to looking for a truly fundamental research problem. Dirac had yet to show that he had the ability to identify such a challenge, the hallmark of a great scientist. But it is clear that he was developing the talent: he returned to the unexplained question of understanding black-body radiation, which had first led Planck to the idea of energy quanta.

Dirac investigated a daring new idea first introduced by a twenty-six-year-old French student, Louis de Broglie, in his Ph.D. thesis. De Broglie used special relativity to argue with startling boldness and originality that every subatomic particle – including electrons – should have an associated wave of a nature yet to be understood.[21] Dirac was inured to thinking of the electron as a particle, for example, in orbit around an atomic nucleus, so de Broglie's notion of a wave-like electron seemed to be a mathematical fiction of no importance to physicists.[22] He carried out some initial calculations but put the work aside after concluding that he had done nothing worth publishing. Having sniffed the scent of an important problem, he had then lost it; but he would soon return.

In early May, almost two months after the death of Felix, Dirac was looking forward to the visit of Niels Bohr, widely regarded as the world's leading atomic scientist (he had won the Nobel Prize for physics two years before). Then approaching his fortieth birthday, he was an imposing figure: tall, charismatic and good-natured, with a huge head and a heavily built body that still bore traces of youthful athleticism.[23] His sprawling hands had once helped him to become a top Danish goalkeeper, narrowly missing selection for his country's soccer team in the 1908 Olympics. Those hands now spent much of the time relighting his pipe or cigarettes; like his fellow chain-smoker Rutherford, Bohr was a serial cadger of matches. The two men had worked together in Manchester for three months in the early summer of 1912, and Bohr had come to regard Rutherford as a 'fatherly presence'. It was an improbable friendship. Both were profound, intuitive thinkers and impatient with mathematical thinking, but their modes of expression were entirely different: Rutherford was a straight talker whose bluntness could make a navvy blush, whereas Bohr – an inveterate mumbler – was almost always polite and struggled to articulate the tortuous debate going on inside his head. His words were well worth hearing, however, and his audiences sat in silence, straining to hear his every word.[24]

Bohr gave his talk, 'Problems of Quantum Theory', on 13 May and spoke again at the Kapitza Club three days later. He underlined his view that the current atomic theory was only provisional and that a better-founded one was sorely needed. Bohr was also unhappy with the need to describe light sometimes as particles and at other times as waves. Shortly before, he had failed to resolve the dichotomy, and he was now gloomy about the state of quantum physics. Such confusion intimidates mediocre thinkers, but for the most able ones it signals an opportunity to make their name. One student who was bright enough, in Bohr's estimation, to solve the problems of quantum theory was the German prodigy Werner Heisenberg, based in Göttingen but soon to visit Cambridge.[25] He was very different to Dirac: widely cultured and with a fondness for conversation and patriotic songs which had been nurtured around campfires during his years in the German Youth Movement. Heisenberg would declare over a glass of beer that 'physics is fun', a phrase that would not have entered the heads of the serious men who had founded the subject eighty years before.[26]

On the cool Tuesday evening of 28 July, the sweet summer air calm and damp after a day of wind and light showers of rain, Heisenberg addressed the Kapitza Club, his first presentation in Cambridge. He expected to be met with the university's famous formality but, instead, found himself talking in a makeshift college room, with several members of his audience having to sit on the floor. It is not clear whether Dirac was awake throughout Heisenberg's seminar or even if he attended it.[27] Some of the physicists who attended vaguely remembered that Heisenberg spoke about the light emitted and absorbed by atoms and that he remarked in a coda that he had written an article about a new approach to atomic physics. Later, Heisenberg could be sure only that he did mention this article to his host Fowler, but no one in Cambridge – or even Heisenberg himself – appears to have realised that they had been part of history in the making.[28]

Dirac returned home for the summer break having secured funding for another three years' research from the Royal Commission for the Exhibition of 1851, which dispensed scholarships funded by the Exhibition's unexpected profits. Dirac's application had been recommended by Maynard Keynes and included encomia from Cunningham, Fowler and the physicist and astronomer James Jeans, who affirmed that Dirac had 'ability of the highest order in mathematical physics'.[29] Much was expected of the young Dirac, though he had published nothing of consequence since his brother's suicide.

Dirac probably had to fend off his mourning parents' requests for him to return to Bristol. His father had already tried to persuade him to apply for the post of Assistant Lecturer in Mathematics at the university, but there can never have been any question that Dirac would accept such a post – he was starting to become aware of his academic worth.[30] And he was still awaiting a challenge equal to his talent.

Early in September 1925, a postman walked up the steep path to the front door of 6 Julius Road and delivered an envelope that changed Dirac's life. The package, sent by Fowler, contained fifteen pages of the proofs of a paper sent to him by its author, Werner Heisenberg, who had made several corrections to it in his slanting handwriting.[31] This article, written in German, contained the first glimpse of a completely new approach to understanding atoms. Most supervisors would have kept the proofs to themselves, to get a head start on their fellow researchers. Fowler, however, sent the proofs to Dirac with a few words scribbled on the top right-hand corner of the front page: 'What do you think of this? I shall be glad to hear.'

The paper, technical and complex, would not have been easy reading for Dirac, whose training at the Merchant Venturers' had given him only a modest command of German. He could, however, see that this was not just another run-of-the-mill exercise in the mathematics of quantum theory. Bohr's theory featured quantities such as the position of the electron and the time it takes to orbit its nucleus, but Heisenberg believed that this was a mistake, as no experimenter would ever be able to measure them. He made this point when he summarised the aim of his theory in the article's introductory sentence: 'The present paper seeks to establish a basis for theoretical quantum mechanics founded exclusively upon relationships between quantities which in principle are observable.'[32] Heisenberg knew that it would be extremely difficult to come up with a complete atomic theory built along the lines he envisaged in a single flourish. That would have been too big a task. Instead, he attempted something simpler, by trying to set out a theory of an electron moving not in three dimensions of ordinary space but in just *one* dimension, that is, in a straight line. Such an electron exists only in the mind of the theoretical physicist, but if this prototype theory worked, then maybe it would be possible to extend it and produce a more realistic version of the theory, one that could be applied to atoms.

Heisenberg considered how classical theory describes his electron, moving back and forth, and how quantum theory might account for it, bearing in mind that the two theories must merge smoothly, according to the correspondence principle. The new theory looked completely different from its classical counterpart. For example, there is no mention in the quantum theory of single numbers to represent the electron's position; instead, position is replaced by numbers in a square array, an example of what mathematicians call a matrix. Each number in this array is a property of a *pair* of the electron's energy levels and represents the likelihood that the electron will jump between that pair of energy levels. So, each number can be deduced from observations of the light given out by the electron when it jumps between them. In this way, Heisenberg demonstrated how to build an entirely new atomic theory solely in terms of *measurable* quantities.

This picture looks bizarre to anyone coming to it for the first time. With astonishing boldness, Heisenberg had abandoned the assumption that electrons can be visualised in orbit around a nucleus – an assumption no one had previously thought to question – and replaced it by a purely mathematical description of the electron. Nor was this description easy to accept: for example, if it were to apply to ordinary matter, an object's precise location would not be measured with a ruler but would be given in terms of an array of numbers that give the chances of its making transitions to other energy states. This was no one's idea of common sense. In making an imaginative leap like this, Heisenberg was behaving rather like a painter who had switched from Vermeer's classically descriptive style to one based on the abstractions of Mondrian. But whereas painters can use abstraction simply as a technique for producing an attractive image that may or may not refer to real things, abstraction for physicists is a way of representing things en route to the most accurate possible account of material reality.

Dirac initially found Heisenberg's approach too complicated and artificial, so he put the paper aside, dismissing it as being 'of no interest'.[33] About ten days later, however, Dirac returned to it and was struck by a point that Heisenberg made in passing, almost halfway through the paper. Heisenberg wrote that some of the quantities in the theory have a peculiar property: if one quantity is multiplied by another, the result is sometimes different from the one obtained if the sequence of multiplication is reversed. This was exemplified by the quantities he used to represent position and momentum of a piece of matter (its mass multiplied by its velocity): position multiplied by momentum was, strangely, not the same as momentum multiplied by position. The sequence of multiplication appeared to be crucial. Heisenberg later remarked that he mentioned this point as an embarrassing aside, hoping that it would not put off the paper's reviewers and encourage them to think the theory was too far-fetched to be worth publishing. Far from being disconcerted, Dirac saw that these strange quantities were the key to a new approach to quantum physics. Several years later, his mother told an interviewer that Dirac was so excited that he broke his rule of saying nothing about his work to his parents and did his best to explain non-commutation. He did not try again.[34]

Unlike Heisenberg, who had never come across non-commuting quantities before, Dirac was well acquainted with them – from his studies of quaternions, from the Grassmann algebra he had heard about at Baker's tea parties, and from his extensive studies of projective geometry, which also features such relationships.[35] So, Dirac was not only comfortable with the appearance of such quantities in the theory, he was excited by them, although at first he did not understand their significance, nor did he know how to build on Heisenberg's ideas. What Dirac did notice was that Heisenberg had not constructed his theory to be consistent with special relativity so, true to form, Dirac played his favourite game of trying to produce a version of Heisenberg's theory that was consistent with relativity, but he soon gave up.[36] At the end of September, Dirac prepared to return to Cambridge, convinced that the non-commuting quantities in the theory were the key to the mystery. To make progress, he needed to find the lock – a way of interpreting these quantities, a way of linking them to experimentally observed reality.

One person who, unknown to Dirac, shared his excitement about the theory was Albert Einstein, who wrote to a friend: 'Heisenberg has laid a big quantum egg.'[37]

At the beginning of October, Dirac began his final year as a postgraduate student. With Fowler's encouragement, he set aside his books of intricate calculations based on the Bohr theory, well aware that – if Heisenberg's theory was right – those calculations were all but worthless.

It was during a Sunday walk in the countryside, soon after term began, that Dirac had his first great epiphany. Long afterwards, he could not recall the exact date, though he clearly remembered those first exciting hours of discovery.[38] He was, as usual, trying to forget about his work and let his mind wander in the tranquillity of the flat Cambridgeshire countryside. But on that day, the non-commuting quantities in Heisenberg's theory kept intruding into his conscious mind. The crucial point was that two of these quantities, say A and B, give different results according to the order in which they are multiplied: AB is different from BA. What is the significance of the difference AB – BA?

Out of the blue, it occurred to Dirac that he had come across a special mathematical construction, known as a Poisson bracket, that looked vaguely like AB – BA. He had only a faint visual recollection of the construction, but he knew that it was somehow related to the Hamiltonian method of describing motion. This was characteristic of Dirac, as he was much more comfortable with images than with algebraic symbols. He suspected that the bracket might provide the connection he was seeking between the new quantum theory and the classical theory of the atom – between the non-commuting quantities in Heisenberg's theory and the ordinary numerical quantities in classical theory. Fifty-two years later, he remembered, 'The idea first came in a flash, I suppose, and provided of course some excitement, and then of course came the reaction "No, this is probably wrong". [. . .] It was really a very disturbing situation, and it became imperative for me to brush up my knowledge of Poisson brackets.'

He hurried home to see if he could find anything about the Poisson bracket from his lecture notes and textbooks, but he drew a blank. So he had a problem:

There was just nothing I could do, because it was a Sunday evening then and the libraries were all closed. I just had to wait impatiently through that night without knowing whether this idea was any good or not, but still I think that my confidence grew during the course of the night. The next morning I hurried along to one of the libraries as soon as it was open [. . .].[39]

A few minutes after Dirac entered the library, he pulled from one of the shelves the tome that he knew would provide the answer to his question: *A Treatise on the Analytical Dynamics of Particles and Rigid Bodies* by the Edinburgh University mathematics professor

Edmund Whittaker. The index directed him first to page 299, where Whittaker set out the mathematical formula for the bracket. Sure enough, as Dirac had surmised, the Poisson bracket, which first appeared over a century before in the writings of French mathematician Siméon-Denis Poisson, had the form of two mathematical quantities multiplied together minus two related quantities multiplied together, the multiplication and minus signs making it appear similar to the expression AB – BA.[40] In one of his greatest insights, Dirac saw that he could weave an entire carpet from this thread – within a few weeks of uninterrupted work he had set out the mathematical basis of quantum theory in analogy to the classical theory. Like Heisenberg, he believed that mental pictures of the tiniest particles of matter were bound to be misleading. Such particles cannot be visualised, nor is it possible to describe them using quantities that behave like ordinary numbers, such as position, speed and momentum. The solution is to use abstract mathematical quantities that *correspond* to the familiar classical quantities: it was these relationships that Dirac pictured, not the particles that they described. Using the analogy with the Poisson bracket, together with the correspondence principle, Dirac found connections between the abstract mathematical quantities in his theory, including the crucial equation connecting the symbols associated with the position and momentum of a particle of matter:

position symbol × momentum symbol – momentum symbol × position symbol = h × (square root of –1)/(2 × π)

where h is Planck's constant and π is the ratio of the circumference to the diameter of every circle (its value is about 3.142). The square root of minus one – the number that, when multiplied by itself gives minus one – plays no role in everyday life but is common in mathematical physics. So there was nothing new on the right-hand side of the equation. The most mysterious part of the equation was on the left-hand side, especially for those unwise enough to think of the position and momentum symbols as anything other than abstractions: they are not numbers or measurable quantities but *symbols*, purely mathematical objects.

To all but mathematical physicists of the most austere disposition, Dirac's description looked remote from reality, but, in the right hands, it was possible to manipulate his abstract symbols to make concrete predictions. In Eddington's words, 'The fascinating point is that as the development process proceeds, actual numbers are *exuded* from the symbols.'[41] By this, Eddington meant that the underlying symbolic language yielded, after mathematical manipulation, numbers that experimenters could check. The value of the theory depended on whether these predictions agreed with the readings on counters, dials and detecting screens. If the theory did that successfully and was logically consistent, it must be judged a success, according to Dirac, no matter how peculiar it looked.

Fowler appreciated that his student had done something special. Dirac's theory, much more ambitious than Heisenberg's prototype description of the artificial case of an electron jiggling about in a straight line, sought to describe the behaviour of *all* quantum particles in *all* circumstances throughout *all* time. He knew, however, that the most important priority was to demonstrate that his theory could account for the most important general observations that experimenters had made about atoms. In a few lines of algebra, Dirac demonstrated that energy is conserved in his theory – as it is in the everyday world – and that when an atomic electron jumps from one energy level to another, it gives out a quantum of light whose energy is equal to the difference between the two levels. This indicated that the theory was able to reproduce Bohr's successes, without having to assume that electrons are in orbit, like planets round a star, doomed to cascade into the nucleus. For Dirac, it was meaningless to use such graphic images – quantum particles can be described only using the precise, rarefied language of symbolic mathematics.

Although Dirac had been inspired by Heisenberg's paper, the two men had sharply different approaches to their subject. Heisenberg proudly referred to his paper as 'the great saw', a tool to cut off the limb on which the old Bohr theory rested.[42] Dirac, on the other hand, sought to build a bridge between Newtonian mechanics and the new theory. His dream was that all the mathematics that Hamilton and others had used to recast Newton's theory of mechanics would have exact counterparts in the new theory. If Dirac was right, physicists would be able to use the infrastructure of 'classical mechanics' – the stuff of hundreds of textbooks – in the construction of the new theory, which had been named the year before by Heisenberg's senior colleague, Max Born: 'quantum mechanics'.

By early November, Dirac had written his paper and had given it an ambitious title that would catch the attention of even the most casual browser: 'The Fundamental Equations of Quantum Mechanics'. Fowler was delighted. Only a few months before, he had described his student's ability to 'push forward the mathematical development of his ideas' and to 'view old problems in a fresh and simpler way'.[43] Now he could alter the focus of his praise of Dirac from his potential to his achievement. Fowler's highest priority now was to ensure that the paper was published as quickly as printing schedules allowed; if one of Dirac's competitors managed to submit a similar paper before him, then, according to the unwritten rules of the scientific community, Dirac would be regarded as an 'also ran'. Like sport, science is supposed to be an activity in which the winner takes all. Fowler had recently been elected a Fellow of the UK's academy of science, the Royal Society, qualifying him to send manuscripts for publication in its proceedings in the confident expectation that they would be accepted without delay.

For most physicists in Cambridge, the discovery of quantum mechanics was a non-event. Apart from his discussions with Fowler, Dirac made no effort to draw his colleagues into the new revolution in physics that he knew was afoot. Word was beginning to spread, however, that he was a 'first-rate man' in the making, though his wispy, almost wordless presence gave no clue to the depth and subtlety of his thinking. It appears to have been at about this time that his colleagues invented a new unit for the smallest imaginable number of words that someone with the power of speech could utter in company – an average of one word an hour, 'a Dirac'. On the rare occasions when he was provoked into saying more than yes or no, he said precisely what he thought, apparently with no understanding of other people's feelings or the conventions of polite conversation.

During a meal in St John's Hall, he crushed a fellow student who was devoting his time to workaday problems in classical physics: 'You ought to tackle fundamental problems, not peripheral ones.'[44] This was Rutherford's credo, too, though his approach was more down to earth. Rutherford was wary of the theorists' effusions about their latest hieroglyphics until the results were useful to experimenters. Quantum mechanics had yet to do that. Most physicists found it implausible that nature could be so perverse as to favour a theory that required thirty pages of algebra to explain the simplest atom's energy levels, rather than Bohr's theory, which explained them in a few lines. For Rutherford and his boys, the real sensation that autumn was not the revelations about quantum mechanics but the discovery that electrons have spin. Made at the University of Leiden by two Dutchmen, this discovery took everyone by surprise. In terms of the Bohr picture of the atom, it was easy to envisage crudely what was going on: the orbiting electron is spinning, just as the Earth spins like a top around its north–south axis. Though soon to be taken for granted, many leading physicists thought the idea that the electron has spin was ridiculous.[45]

One of the postgraduate students who first heard in Cambridge that term about the discovery of spin was Robert Oppenheimer, a dapper, well-to-do American Jew just arrived from Harvard, then riddled with anti-Semitism. He was emotionally fragile, unsure of what he wanted to do with his life but outwardly confident and always keen to display the breadth and depth of his cultural interests. After Rutherford refused to accept him as a student, he spent a few unproductive weeks working with J. J. Thomson, then well over the hill. Oppenheimer disliked Cambridge life – the 'rather pallid science clubs', the 'vile' lectures, having to live in 'a miserable hole'. He saw fellow American students 'literally dying off under the rigors of disregard, climate, and Yorkshire pudding'.[46] By the end of his first term in Cambridge, Oppenheimer was judged by a close American friend to have 'a first class case of depression'.[47]

Dirac mentioned none of his new student acquaintances in his postcards home, and virtually nothing about his work. His frustrated parents had to wait six weeks for him even to confirm that his lodgings were comfortable. Flo, having seen her son ratchet up his work rate after tumbling to the importance of Heisenberg's first paper, began what was to become her ineffectual refrain: 'Don't work too hard; have some fun if it comes your way.' Dirac's father was still a broken man, suffering in the cold weather and – in his wife's words – shuffling around 'so slowly that he is like a block of ice'.[48]

One of Flo's favourite subjects was national and local politics, but that autumn she wrote little about them, probably because there was not much to write about: Britain was stable and quietly prospering. As the country entered the second half of the 1920s, it seemed at last to be coming to terms with its memories of the war, encouraged by the growing international consensus that disagreements should never again be resolved on the battlefield. This understanding was manifest in the hailed Treaty of Locarno, a non-aggression pact between France, Germany and Belgium, guaranteed by the two supposedly impartial powers, Italy and the UK. Some English schools celebrated by giving their pupils a day off when the treaty was signed in London on 1 December, the day the Royal Society published Dirac's first paper on quantum mechanics. Fowler had managed to cut the time between the submission of the paper and its publication from the usual three months to three weeks.

Word passed around the cognoscenti of quantum theory that a star had been born. Dirac's earlier work had gone largely unnoticed, but here was a paper that appeared to have been written by a mature mathematician and physicist.[49] One of those who had not heard of Dirac before his first work on the new theory was Heisenberg's boss in Göttingen, Max Born.[50] Though given to understatement rather than hyperbole, in his memoir he described his first reading of Dirac's early work on quantum mechanics as 'one of the greatest surprises of my life [. . .] the author appeared to be a youngster, yet everything was perfect in its way and admirable'.[51]

Heisenberg, too, was jolted by the paper. On 23 November, a few days after he received the proof copy Dirac sent him, Heisenberg replied in a two-page letter (in German) that began a fifty-year friendship.[52] He began graciously by telling Dirac that he had read his 'beautiful work with great interest', adding that 'There can be no doubt that all your results are correct, insofar as one believes in the new theory.' The discoverer of the new theory was unsure of whether he had hit on ideas of lasting value.

What followed must have made Dirac's heart sink: 'I hope you are not disturbed by the fact that part of your results have already been found here some time ago.' Born had independently found the relationship between the position and momentum symbols, a connection that Dirac probably thought he had been first to make. Also, Heisenberg's theory accounted for the Balmer formula for hydrogen atoms, according to a virtuoso calculation by Heisenberg's slightly older friend Wolfgang Pauli, an Austrian theoretician known for his brilliance, his unsparing intellectual aggression and for drinking a glass of wine too many in the nightspots of Hamburg. Heisenberg's note bore the disappointing message that other European theoreticians were on the same track and the deflating prospect that they would repeatedly beat him into print.

In the ten days following his first letter, Heisenberg wrote Dirac three more warm and complimentary notes, pointing out technical difficulties and minor errors in Dirac's first paper and seeking to clarify details. He concluded his letter of 1 December: 'Please do not take these questions that I write to you as criticisms of your wonderful work. I must now write an article on the state of the theory [. . .] and still wonder at the mathematical simplicity with which you have overcome this problem.'[53] Dirac knew that he was facing some of the toughest competition theoretical physics had to offer. Heisenberg was working in Göttingen not only with Born and his student Pascual Jordan but also in association with some of the world's leading mathematicians. The trio of Born, Heisenberg and Jordan were working in the Göttingen tradition of a close relationship between the theoretical physicists, mathematicians and experimenters, in sharp contrast with the virtual separation of the communities in Cambridge, where individuality was prized. So, in the undeclared contest to be the first to develop quantum mechanics into a complete theory, the combined might of the mathematicians and physicists in Göttingen was pitted against the loner Dirac. He knew that Heisenberg had given his German competitors a head start of two months.

It would take several years before quantum mechanics crystallised into a complete theory. During that time, it was a work in progress by about fifty physicists. In retrospect, they resembled a group of construction workers who had agreed on a common project – to build a new theory of the behaviour of matter – though not on how to accomplish it. In this case, the construction site was dispersed across north-western Europe, and virtually all the builders were male, under thirty, intensely competitive and craving the respect of their peers as well as the blessing of posterity. There was no official leader, so the workers were free to concentrate on any part of the project they liked. In this quasi-anarchy, some tasks were sure to be done by several people at the same time so, when useful results emerged, there would be quarrels about who most deserved the credit for them. All the workers had their favourite tools and their own preferred way of getting the job done. Some approached it philosophically, some mathematically and some with their eyes on what experiment could teach them. Some concentrated on the project's grand plans and others on its details. Most of them liked to collaborate and to bounce ideas off their colleagues, while a few others – notably Dirac – had no wish to be in anyone's team. It was rarely easy to see which of the new ideas were duds and which were gems, nor was it obvious whose approaches to the problem were the most promising. Not that any physicist felt bound by a need to take an entirely consistent approach; all that mattered was getting the job done, by whatever means were available. In the end, prizes for a new scientific theory tend to be awarded as they are in architecture for a new building – not to the people who talked most eloquently during the construction but to those who set out its vision and who did most to realise it.[54]

Dirac knew that he and his colleagues had taken only the first step towards the building of a complete theory of quantum mechanics. There was much to do.

Seven

A door like this has cracked open five or six times since we got up on our hind legs. It's the best possible time to be alive, when almost everything you knew is wrong.

TOM STOPPARD, *Arcadia*, 1993, Act 1, Scene 4

Einstein admired the new quantum mechanics, but he was suspicious of it. On Christmas Day 1925 in Berlin, he wrote to a close friend that it seemed implausible to him that something so simple as a number representing a quantum particle's position should have to be replaced by an array of numbers, 'a genuine witches' multiplication table'.[1] Seven weeks later, he was coming to the conclusion that the theory was wrong.[2]

Dirac had no such qualms – he was sure that Heisenberg had pointed the best way ahead. Yet although Dirac was working with Heisenberg's theory, their approaches to it were quite different: whereas Heisenberg thought the theory was revolutionary, for Dirac it was an extension of classical theory.[3] While Heisenberg and his Göttingen colleagues strove constantly to account for experimental results, Dirac's priority was to lay the theory's 'substrata', following a favourite term of Eddington's. Dirac was following Einstein in taking a top-down approach, beginning with mathematically precise formulations of fundamental principles and only afterwards using the theory to make predictions.

A few weeks after Christmas – the first the Dirac family had spent without Felix – Dirac gave a talk at the Kapitza Club about his just-published paper on quantum mechanics. Two days later, he sent off for publication the proof that his theory reproduced Balmer's formula, the first of three papers on the new theory that he wrote in the first four months of the year. In these first papers on quantum mechanics, Dirac was trying both to understand the theory and to apply it. Puzzled by the symbols in Heisenberg's theory, he spent months unsuccessfully trying to relate them to projective geometry; none of his ideas worked. He was using mathematics that was unknown or at least unfamiliar to most of his colleagues, yet he rarely gave details of the mathematical techniques he was using or the experimental observations he was trying to explain. He thus managed to perplex both physicists and mathematicians. Nearly fifty years later, Dirac admitted that his attitude to mathematics was cavalier:

I did not bother at all about finding a precise mathematical nature for [some of my symbols] or about any kind of precision in dealing with them. I think you can see here the effects of an engineering training. I just wanted to get results quickly, results which I felt one could have some confidence in, even though they did not follow from strict logic, and I was using the mathematics of engineers, rather than the rigorous mathematics which had been taught to me by Fraser.[4]

Those words would have puzzled Dirac's peers in the spring of 1925. Most of them would have been hard pressed to identify in his papers any remnants of an engineer's training, nor did his writings flaunt the quick-and-dirty approach to calculations favoured by engineers. Rather, Dirac's papers appeared to be impenetrable to all but the mathematically adept. One reason why Dirac's approach was so puzzling was that he was an unusual hybrid – part theoretical physicist, part pure mathematician, part engineer. He had the physicist's passion to know the underlying laws of nature, the mathematician's love of abstraction for its own sake and the engineer's insistence that theories give useful results.

Wearing the hat of the physicist, Dirac knew that, for all the mathematical elegance of quantum mechanics, it had yet to make a single prediction whose confirmation would demonstrate its superiority over Bohr's theory. Such a test of the new theory was not easy to find. The best that Dirac could do was to use the theory to describe the most-investigated example of subatomic collision – the scattering of a photon (a particle of light) by a single electron. This process always involves particles travelling at extremely high speeds, close to the speed of light, so any theory that seeks to describe it must be relativistic – consistent with Einstein's special theory of relativity. The problem was that Heisenberg and Dirac's theory of quantum mechanics was not relativistic, and it was unclear how to incorporate relativity into the theory. Dirac made a start on this by tweaking the theory to improve its consistency with relativity and then used it to make testable predictions, using the ideas he had developed at home in Bristol soon after he received Heisenberg's original paper. The theory was rough and ready, but it enabled Dirac to make the first prediction of quantum mechanics: using a graph, he compared observations of electron scattering with his 'new quantum theory' and showed that it was in better agreement than the classical theory.

Quantum mechanics was still only a rudimentary theory. Much remained to be clarified about the interpretation of its mathematical symbols: what did they really mean? And was it possible to say any more about the motion of subatomic particles? How could the theory be applied to atoms more complicated than hydrogen, containing more than one electron? In later life, Dirac liked to point out that quantum mechanics was the first physical theory to be discovered before anyone knew what it meant. He spent months on the problem of interpreting its symbols and came to see that the theory was mathematically less complicated than he had first thought. Born pointed out to Heisenberg that each array of numbers in his quantum theory was a matrix, which consists of numbers arranged in horizontal rows and vertical columns that behave according to simple rules spelt out in textbooks. Heisenberg had never heard of matrices when he discovered the theory, as Born often reminded his colleagues, adding that he was the one who had ensured that Heisenberg's egg was properly hatched and that its contents were nurtured into infancy.

It seemed to many physicists that Dirac was working in a private language, and this inaccessibility made his work unpopular. In Berlin, long the global capital of theoretical physics, the consensus was that the approach of the Göttingen group – Heisenberg, Born and Jordan – was the most effective. In the United States, then way behind Europe in developing quantum mechanics, the practically minded theoretician John Slater later recalled his frustration with Dirac's writings. In Slater's view, there are two types of theoretical physicist. The first consists of people like himself, 'the prosaic, pragmatic, matter-of-fact sort, who [. . .] tries to write or speak in the most comprehensible manner possible'. The second was 'the magical, or hand-waving type, who like a magician, waves his hands as if he were drawing a rabbit out of a hat, and who is not satisfied unless he can mystify his readers or hearers'. For Slater and many others, Dirac was a magician.[5]

*

Dirac's academic stock rose further in the spring of 1926, during his final term as a postgraduate. He was no longer just another of Cambridge's many brilliant but unfulfilled loners but was recognised as an extraordinary talent. Fowler arranged for him to give two series of lectures on quantum theory for his fellow students. Fowler was also in the audience, aware that his most brilliant protégé had overtaken him.

Although Rutherford affected to scorn highfalutin theory, he kept abreast of the latest news about quantum physics. At his request, Dirac gave a presentation at the Cavendish about the welter of quantum discoveries that had been made at Göttingen, but it was a poor, hastily prepared talk.[6] His audience almost certainly included Oppenheimer and also Kapitza and Blackett, who were – beneath a veneer of amity – increasingly at odds. The tensions were rooted in their relationships with Rutherford. Kapitza shamelessly flattered and courted him, who in return gave favours and even friendship, to the extent that Kapitza was sometimes described as the son Rutherford never had. None of this went down well with Blackett, who admired Rutherford's creative running of the laboratory but had no time for his authoritarianism. Blackett, too, was an object of envy. In the early autumn of 1925, he tutored Oppenheimer at the laboratory bench, teaching him the craft of experimental physics, for which Oppenheimer had little aptitude, as he well knew. With the peculiar logic of neurosis, Oppenheimer decided to get his own back by anonymously leaving on Blackett's desk an apple poisoned with chemicals from the laboratory.[7] Blackett survived but the authorities were outraged and Oppenheimer avoided expulsion from the university only after his parents persuaded the university not to press charges but to put him on probation, on the understanding that he would have regular sessions with a psychiatrist. A few months later, he switched to theoretical physics – a much more congenial field for him – and worked in the same circle as Dirac, who was busy hammering out his vision of quantum mechanics. Oppenheimer recalled that 'Dirac was not easily understood, not concerned with being understood. I thought he was absolutely grand.'[8]

Dirac probably did not notice the intrigues among his friends and acquaintances or their personal problems; even if he did, he would probably have ignored them. He worked all day long and took time off only for his Sunday walk and to play chess, a game he played well enough to beat most students in the college chess club, sometimes several at the same time. Nor did Dirac take much interest in politics. He was an onlooker during the General Strike that almost brought the UK to a halt for nine days in early May 1926 and led many to fear that a Bolshevik revolution was imminent. King George V urged moderation, while in the Government, Churchill demanded 'unconditional surrender' from the workers ('the enemy') who were supporting the demands of the Miners' Union. Some students thought the strike was a national crisis, but to others it was an opportunity to drive a tram or to play at being a docker or a policeman. Almost half the university's students took part in strike-breaking activities, so the authorities had no choice but to postpone the end-of-year examinations, prolonging the merriment.[9] Dirac heard from his mother that trams and buses in Bristol were still running, a relief to his father, so weakened by grief that he could not walk the mile between his home and the Merchant Venturers' School. Fate was about to bring Charles even more sorrow: he heard from Geneva in early March that his mother had died.[10]

The collapse of the General Strike was important in the development of political thought in Cambridge. The strength of opposition to the strike in the university demonstrated the unwillingness of its dons to disrupt the political status quo; even some of its socialist academics had been strike-breakers. The humiliation of May 1926 was one of the main motivations of a few Marxist scientists who were determined to establish radical politics in Cambridge and then to spread the word across the country. The most effective of the proselytisers was the young crystallographer Desmond Bernal, an energetic and charismatic polymath, who had joined the Communist Party after he graduated in 1923.[11] He had a vision of a just and well-informed collectivist society, with all policy decisions taken according to scientific principles and with the benefit of expert technological knowledge. Scientists were his ideal society's elite, to the extent that he suggested that they might be granted the freedom to form 'almost independent states and be enabled to undertake their largest experiments without consulting the outside world'.[12] The theoretical basis for Bernal's thinking was supplied by Marxism, which seemed to him and his friends to provide a framework for the solution of every social, political and economic problem.

Bernal and his colleagues at first made slow progress in converting colleagues to Marxist thinking, partly because of resistance by moderates such as Rutherford, who despised Bernal more than anyone else in Cambridge for his activism and, apparently, for his open sexual promiscuity.[13] The suspicion of card-carrying Communists was so intense that Bernal apparently decided in 1927, when he began a period of working full-time in the Cavendish, that it would be better to let his membership of the party drop. After that, it appears that none of his colleagues officially joined the party.[14]

Kapitza did not make the error of alienating senior colleagues: although he shared many of Bernal's political views, he was careful not to offend Rutherford by talking politics in the laboratory. However, Kapitza will have shared his vision of society with Dirac, who had arrived in Cambridge a political innocent and so heard for the first time the claim that Marxism offered an all-embracing scientific theory that could do for society what Newton had done for science. According to this vision, every economy could be the test bed for a theory that promised a brighter future, with intelligent planning taking the place of the sometimes cruel, invisible hand of market forces. Dirac may have noted the strong support Marxists gave to education and industrialisation and the contempt they poured on religion – themes that emerged soon afterwards in his perspective on aspects of life he was discovering outside physics.

During the General Strike, Dirac was absorbed in writing his Ph.D. thesis, a compact presentation of his vision of quantum mechanics. Confident though he was of his understanding of the theory, he knew as he wrote his thesis that it was not the whole story, for he had recently heard that an alternative version of quantum theory had appeared, one that looked completely different from Heisenberg's. The author of the new version was the Austrian theoretician Erwin Schrödinger, working in Zurich. He was thirty-eight years old, a generation older than Heisenberg and Dirac, with a formidable reputation in Europe as a brilliant polymath.

Schrödinger had discovered his quantum theory independently of Heisenberg and a few weeks later, by building on de Broglie's wave theory of matter, which Dirac had admired but had not taken seriously. In the Christmas vacation of 1925, during an illicit weekend with a girlfriend in the Swiss mountains, Schrödinger discovered an equation that described the behaviour of quanta of matter in terms of their associated waves, and then applied the theory in a series of dazzling papers. His achievement was to generalise de Broglie's idea: the young Frenchman's theory applied only to the special case of matter with no overall force acting on it, but Schrödinger's theory applied to all matter, in any circumstances.

The great virtue of Schrödinger's theory was that it was easy to use. For the many scientists intimidated by the abstract mathematics in Heisenberg's approach, Schrödinger offered the balm of familiarity: his theory was based on an equation that closely resembled those most physicists had mastered as undergraduates, when they were studying water and sound waves. Better still, in Schrödinger's theory, the atom could be, at least to some extent, visualised. Roughly speaking, the energy levels of an atom correspond to the waves that can be set up on a piece of rope, held fixed at one end and shaken up and down at the other. The shaker can set up a single half-wavelength (like a crest, moving up and down) on the rope, or, by shaking more vigorously, two half-wavelengths, or three half-wavelengths, or four, or five, and so on. Each of these wave patterns corresponds to a definite energy of the rope, just as each possible Schrödinger wave of an atom corresponds to an atomic energy level. The meaning of these Schrödinger waves was unclear: their discoverer suggested unconvincingly that they were a measure of the spread of the electron's charge around the nucleus. Whatever the true nature of these waves, they were more intuitively appealing than Heisenberg's matrices to those who lacked mathematical confidence. They, along with everyone else, were relieved when Schrödinger gave a preliminary proof (completed two years later by others) that his theory gave the same results as Heisenberg's. The frightened sceptics could then ignore those intimidating matrices.

At first, Dirac was annoyed by Schrödinger's theory, as he resented even the thought of suspending work on the new quantum mechanics and starting afresh. But in late May, as he was finishing the writing of his Ph.D. thesis, he received a persuasive letter from Heisenberg urging him to take Schrödinger's work seriously. This wise advice was ironic coming from Heisenberg, an opponent of the rival theory, who had written to Wolfgang Pauli in early June, 'The more I reflect on the physical portion of Schrödinger's theory the more disgusting I find it. What Schrödinger writes on the visualizability of his theory is probably not quite right. In other words, it's crap.' Schrödinger gave as good as he got, dismissing the mathematical arcana of Heisenberg's theory and the idea of quantum jumps. The two theorists clashed unpleasantly when they first met a month later at a packed seminar in Munich, the first skirmish in what was to be a long and acrimonious dispute.[15]

Dirac ignored Schrödinger's theory in his Ph.D. thesis, 'Quantum Mechanics', the first to be submitted anywhere on the subject. The thesis was a great success with his examiners, including Eddington, who took the unusual step on 19 June of sending him a short handwritten letter on behalf of the Degree Committee of the Mathematical Board, congratulating him on 'the exceptional distinction' of his work.[16] Dirac disliked celebrations and formality, so he was almost certainly not looking forward to the ceremony. He could have taken the degree without attending it but decided to be there in person for the sake of his proud parents, especially his father, who had given him the money that enabled him to begin his Cambridge studies.

Dirac's parents and his sister Betty set off at four in the morning, in good time to take the train to Cambridge via Paddington to see Paul be awarded his degree in the setting of the university's grand Senate House. Every detail of the proceedings harked back to the University's monastic origins. The ermine-collared Vice Chancellor presided and, like the other officials, spoke only in Latin, ensuring that Dirac understood scarcely a word. Wearing evening dress with a white bow tie, a small black cap and red silk gown, he knelt on a velvet cushion, placed his hands together and held them out to be grasped by the Vice Chancellor, who delivered a prayer-like oration. Dirac arose, a doctor.[17]

It was the wettest June in Cambridge for five years, but on that day the rain held off. The town was at its most relaxed, teeming with students and their families. Dirac had not learned the local practice of punting, so he and his family could only watch as others steered their flat-bottomed boats along the Cam, through the lawns and fields, past the gorgeous colleges and chapels.

The Dirac family arrived home at 4 a.m. on Sunday. It had been a happy trip, though its cost had upset Charles. Flo wrote to her son: 'Pa said it cost him £8, so that will be our summer holiday.'[18] It was to be the highlight of her summer, though she was worried that her son was looking drawn and emaciated: 'I wish you would have a nice rest & feed up & get strong. Do try!' As usual, he took no notice. Like his father, he had no need of holidays – the long vacations were not for relaxing but for hard work. The university was about to hibernate for the summer and would be virtually devoid of social distractions for the few scholars who remained. It was the perfect environment for Dirac to concentrate even more intensively on his work. Heisenberg and Schrödinger had knifed a sack of gemstones, and the race was on to pick out the diamonds.

Dirac moved out of his lodgings and into a college room, where he worked at his desk through a sweltering July, producing what would prove to be one of his most enduring insights into nature.[19] He realised that he had been wrong to be wary of Schrödinger's work. Dirac saw that he could have derived Schrödinger's equation using his theory if only he had not been quite so fixated on the links between classical and quantum mechanics. Now, having set aside his prejudice, he could proceed with new gusto. He explained how to generalise Schrödinger's first version of his equation, which applied only to cases that stayed the same as time progressed, to situations that *did* change with time, such as an atom in a fluctuating magnetic field. Quite independently, Schrödinger wrote down the same general equation, which is now named – not entirely fairly – only after him.

Within a few weeks of mastering Schrödinger's equation, Dirac used it to make one of his most famous contributions to science. It concerned the most basic particles that exist in nature, usually described as 'fundamental' because they are believed to have no constituents at all. Classic examples are photons and electrons. Today, two established experimental facts form the bedrock of studies about fundamental particles. First, for each type of fundamental particle, every single one of them in the universe is the same and identical to all other particles of the same type – every electron in every atom on Earth is indistinguishable from every electron in galaxies millions of light years away, just as all the trillions of photons given out each second from a light bulb are the same as the photons given out by the most distant star. For electrons and photons, if you have seen one, you have seen them all. Second, the types of fundamental particles fall into one of two classes, much as almost all human beings can be classified as males or females. The first class is exemplified by the photon, the second by the electron. In 1926, no one knew that there were two such classes.

The differences between the behaviours of electrons and photons exemplify the sharp contrast in behaviour between the two known classes of particle. For a collection of electrons, say in an atom, each available energy state can usually accommodate no more than *two* electrons. The situation is quite different for photons: each energy state can host *any number* of them. One way to visualise this difference is to imagine a pair of bookcases with horizontal shelves arranged vertically above one another in ascending order of energy – the higher the shelf, the higher the energy to which it corresponds. The shelves of the 'electron bookcase' represent the energy states available to electrons, while the shelves of the 'photon bookcase' correspond to the states available to photons. For the 'electron bookcase', each shelf can accommodate at most two books: once the shelf is occupied, it is full and no others can join it. The 'photon bookcase' is different because its shelves can each house any number of books. It is as if electrons are unsociable, whereas photons are gregarious.

Pauli first realised the aversion of electrons to their own company in 1925 when he suggested his exclusion principle. This explained the puzzle of why all the electrons in an atom do not all orbit the nucleus in the same, lowest-energy orbit: it is because the electrons simply are not allowed to fit into the same state – they are forced by the exclusion principle to occupy higher-energy states. This is why the different types of atom – manifest as different chemical elements – behave so differently. In common experience, neon is a gas and sodium is a metal, yet the atoms of neon gas are very similar to the sodium atoms: outside their nuclei, they differ only in that a sodium atom contains one more electron than a neon atom. That additional electron determines the differences between the two elements, and the Pauli exclusion principle explains why sodium's extra electron does not simply join the others and form an almost identical type of atom; rather, it occupies a higher-energy quantum state that is responsible for the differences between the behaviour of the two elements. For the same reason, if there were no exclusion principle, the world around us would have none of the huge variety of forms, textures and colours that we take for granted. Not only would our senses have nothing to perceive, they would not exist. Nor, indeed, would human beings or even life itself.

Dirac was aware of the exclusion principle's power. But he knew that there was much more to do before theorists could understand, at an atomic level, what was going on in the chemistry experiments that he had done at Bishop Road School. There, chemistry was about describing how the elements and other substances behaved: the prize was to move beyond these descriptions to explanations in terms of universal laws. Quantum mechanics promised to do just this, but in 1926 it was not even possible to apply it to atoms that contain more than just one electron, the so-called 'heavy atoms'.

In his college room, Dirac reflected on how Schrödinger waves might describe heavy atoms and the importance of the Pauli exclusion principle. At the back of Dirac's mind was Heisenberg's tenet that theories should be set up only in terms of quantities that experimenters can measure. He thought about the Schrödinger waves that describe two electrons in an atom and wondered whether each wave would be any different if the electrons swapped places. No experimenter could tell the difference, he concluded, because the light given out by the atom would be the same in each case. The way to describe the electrons was, he realised, in terms of waves with the property that they change sign (that is, are multiplied by minus one) when any two electrons are switched. In a few pages of algebra, he used this idea to work out how energy is shared out by groups of electrons as they fill the available energy states. The formulae Dirac derived that summer are now used every day by researchers who study metals and semi-conductors; the flows of heat and electricity in them are determined by their electrons, collectively dancing to the tunes of his formulae.

Yet the practical applications were of no interest to Dirac. He was concerned only with understanding how nature ticks at the most fundamental level and why there is such a sharp contrast between the waves that describe electrons and those that describe photons. He concluded that, while the wave describing a group of electrons changes sign if two electrons swap places, the corresponding wave describing a group of photons behaves in the opposite way – if two photons swap places, the wave remains the same.

This tied in neatly with the abortive work he had done on black-body radiation and led him to explain one of the most puzzling problems of quantum mechanics, a problem that was beyond the ken of Einstein. As Dirac had first heard in Tyndall's lectures in Bristol, quantum theory had begun in the closing weeks of 1900 when Max

Planck suggested that energy is delivered in quanta. The problem was that no one understood how the new theory of quantum mechanics explained Planck's formula. In the months of grief after Felix's death, Dirac had lost the scent of the solution because his theoretical tools were inadequate.[20] Now he had discovered the tool he needed to explain the black-body radiation spectrum: the waves that describe the photons remain unchanged when any two photons are switched. Two pages of calculations in Dirac's notebooks had brought to an end a research project that had been going on for twenty-five years. He must have known he had done something special, but he did not intend to share it with his parents. On 27 July, the message he wrote on his weekly postcard was 'There is not much to say now.'[21]

At the end of August, Dirac sent off an account of his new theory to the Royal Society. He had every reason to be pleased with himself, but disappointment was in store, as he had again been beaten into print. At the end of October, a month after his paper was published, he received a short, typewritten letter from a physicist in Rome who had published a quantum theory of groups of electrons eight months before. The letter was from Enrico Fermi, an Italian physicist a year older than Dirac. In a short note, written in Berlitz-enhanced English, Fermi drew attention to his paper, which he presumed that Dirac had not seen, and concluded without rancour: 'I beg to attract your attention to it.'[22] But Dirac *had* seen Fermi's paper several months before and thought it was unimportant; it had slipped his mind. Although Dirac's paper was very different in approach to Fermi's, their predictions for energies of groups of electrons were identical.

It later turned out that another physicist had also done work similar to Fermi's. In Göttingen, Pascual Jordan had independently derived the same results, had written them up in a manuscript and had given it to his adviser Max Born to read during a trip to the USA. Born put the paper at the bottom of his suitcase and forgot all about it until he returned to Germany several months later, but it was too late. Today, physicists associate the quantum description of groups of electrons only with Fermi and Dirac – in this project, Jordan was, unjustly, a loser.[23]

In September 1926, Dirac was preparing to leave Cambridge to spend a year in Europe funded by his scholarship from the 1851 Commission. His preference was to spend his first year as 'an 1851 man' with Heisenberg and his colleagues in Göttingen, but Fowler wanted him to go to Bohr's Institute for Theoretical Physics in Copenhagen. They agreed on a compromise: Dirac would spend half the time in each, beginning with six months in Denmark.

Dirac arrived in Copenhagen exhausted, having spent much of the sixteen-hour voyage across the North Sea vomiting.[24] The experience led him to a surprising resolution: he would keep sailing in stormy seas until he had cured himself of the weakness of seasickness. His colleague Nevill Mott was flabbergasted: 'he is quite indifferent to cold, discomfort, food etc. [. . .] Dirac is rather like one's idea of Gandhi.'[25]

Eight

MR PRALINE: [. . .] I wish to complain about this parrot what I pur-
chased not half an hour ago from this very boutique.
PET SHOP OWNER: Oh yes, the, uh, the Norwegian Blue . . . What's,
uh . . . What's wrong with it?
MR PRALINE: I'll tell you what's wrong with it, my lad. 'E's dead,
that's what's wrong with it!
   Monty Python's Flying Circus, script by JOHN CLEESE and
   GRAHAM CHAPMAN, 1970

Monty Python's famous sketch uncannily resembles a parable
Rutherford told Bohr soon after Dirac had arrived in Copenhagen.
'This Dirac,' Bohr grumbled, 'he seems to know a lot of physics, but
he never says anything.' This will not have been news to Rutherford,
who decided that the best way of answering Bohr's implied criticism
was to tell a story about a man who went to a pet store, bought a
parrot and tried to teach it to talk, but without success. The man
took the bird back to the store and asked for another, explaining to
the store manager that he wanted a parrot that talked. The manager
obliged, and the man took another parrot home, but this one also
said nothing. So, Rutherford continued, the man went back angrily
to the store manager: 'You promised me a parrot that talks, but this
one doesn't say anything.' The store manager paused for a moment,
then struck his head with his hand, and said, 'Oh, that's right! You
wanted a parrot that talks. Please forgive me. I gave you the parrot
that thinks.'[1]

Dirac did a lot of thinking in Copenhagen, mostly alone. No one
at Bohr's institute had ever seen anyone quite like him – even by the
standards of theoretical physicists he was profoundly eccentric, a
retiring figure, happiest when he was alone or listening in silence. His
predisposition to answer questions with either yes or no reminded
Bohr of Lewis Carroll's description, in *Alice through the Looking
Glass*, of the frustration involved in talking to cats: 'If they would
only purr for "yes" and mew for "no", or any rule of that sort, so
that one could keep up a conversation! But how can one deal with a

person if they always say the same thing?'[2] Once in a while, however, Dirac did extend his binary vocabulary of response. When Bohr or one of his friends fussed over him or pressed him to state his preference about something or other, he would bring the interrogation to an end with a curt 'I don't mind.'[3]

Perhaps surprisingly, Dirac thrived in the friendliness and informality of the institute, a world apart from the chilly formalities of Cambridge.[4] Bohr had taken great care to nurture this congeniality since the opening of the building in 1921. Located on the Blegdamsvej, a wide straight road on the north-western edge of the city, from the outside the institute looked anonymous, much like every other new building in the city. But inside, the institute's atmosphere was unique: for most of the day, it hummed with high-minded debate, most of it free of pomposity; individuality was prized, but collaboration was supported; the administration was efficient, free of asinine bureaucracy. Bohr encouraged his colleagues to relax together – to play silly games, to commandeer library tables for ping-pong tournaments, to spend the occasional evening at the cinema, followed by boozy discussions late into the night. Quantum physics was being forged by this generation of physicists, and they knew it. Every researcher was seeking to put their own stamp on the emerging quantum mechanics, nervous of producing trivialities, hopeful that they would come up with insights that would be of lasting value. Their research articles were news that aspired to be history.

Bohr was a national hero in Denmark, though he scarcely looked the part. An unassuming but commanding presence, he looked as if he had absconded from the captaincy of a herring trawler. His depth and versatility enormously impressed Dirac, proving to him it was possible to be a premier-division physicist while taking an active interest in the arts, the stock market, psychology and just about any other subject. Like his mentor Rutherford, Bohr had both an eerily sound intuition about the workings of nature and a real talent for getting the best out of his young colleagues. When a special visitor arrived, Bohr would take him or her on a walk among the beech trees of the Klampenborg Forest, just outside the city, to take the measure of his new colleague and give a sense of his non-mathematical approach to physics. Most of the young physicists came under the spell of Bohr, as he had come under Rutherford's.

Bohr and his queenly wife Margrethe oversaw life at the institute like the manager and manageress of a hostel, doing their best to make their guests feel at home. Bohr spent most of the day practising the art of talking and lighting his pipe at the same time, conversing with his colleagues alone or in groups, encouraging them and putting their ideas through the mill. Polite to a fault, his refrain when he cross-examined his young charges was 'Not to criticize, just to learn.'[5] Bohr was the Socrates of atomic physics and he made Copenhagen its Athens.

Dirac was billeted in a boarding house in the heart of the city. As he had done in Bristol and Cambridge, he lived life according to a strict routine: every day except Sunday, he took the thirty-minute walk to the institute, past the ducks and swans on the row of artificial lakes on the north-western rim of the city, returning to his lodgings for lunch.[6] On Sundays, he went on long strolls through the local woods or along the coast to the north of the city, usually alone but sometimes accompanied by some colleagues or just with Bohr.[7] Among the new acquaintances he made there, he got on well with Heisenberg – as likeable in person as he was as a correspondent – but apparently not with Pauli. Although prodigiously talented, Pauli was not the most endearing character in physics: he liked the sound of his own voice and routinely meted out casual verbal violence even to his friends, though he was widely admired for his candour, even by his victims. 'You are a complete fool,' Pauli would repeatedly tell his friend Heisenberg, who later said this joshing helped him to raise his game.[8] But Dirac had no taste for it, and Pauli repeatedly broke through the firewall of his self-confidence. However, Dirac showed no sign of discomfort: whether being praised or condemned, he looked straight ahead with his thousand-yard stare, his entire bearing powerfully radiating his unwillingness to speak or even to be approached.

Dirac's behaviour was apparently not a complete surprise to Bohr. A few years later, when describing Dirac's first visit to a journalist, Bohr echoed the gravedigger in *Hamlet*: 'in Copenhagen [we] expect anything of an Englishman'.[9]

The most pressing problem for quantum theorists remained: what did the symbols in their equations mean? During the summer, Max Born in Göttingen had interpreted Schrödinger's waves by abandoning the classical principle that the future state of any particle can always, in principle, be predicted. Born had pictured an electron being scattered by a target. He argued that it is impossible to predict precisely how much the electron will be deflected and that it is possible to know only the *probability* that the electron will be scattered around any given angle. This led him to suggest that when a particular wave describes an electron, the probability of detecting it in any tiny region follows from a simple calculation that involves, loosely speaking, multiplying the 'size' of the wave in that region by itself.[10] According to Born, the wave is a fictitious, mathematical quantity that enables the likelihood of future behaviour to be predicted. This was a dramatic break with the mechanistic certainties of Newton's picture of the universe, apparently putting an end to the centuries-old notion that the future is contained in the past. Others had the same idea, including Dirac, but it was Born who first published it, though at first even he does not seem to have fully recognised its importance: in the paper where he introduced the concept, he mentions it only in a footnote.

Born's quantum probabilities seem to have been news to no one at the institute, least of all Bohr, who remarked, 'We had never dreamt it could be otherwise,' though it is unclear why neither he nor any of his colleagues saw fit to publish the idea.[11] Whatever the origins of the probability-based interpretation of quantum mechanics, everyone in the physics community was talking about it in the autumn of 1926, and it was one of the themes of the first Bohr–Dirac 'dialogue'. Only weeks before Dirac's arrival, Schrödinger had been a visitor to the institute and made it clear that he found Born's interpretation of quantum waves and the concept of quantum jumps repugnant. On one occasion, after being grilled to a crisp by Bohr, Schrödinger retired sick to his bed, but there was to be no escape. Bohr appeared at his bedside and resumed the interrogation.[12]

Dirac would not have responded well to such intense questioning, but he made an effective sounding board for Bohr during their autumnal walks. Dirac hardly said a word while Bohr struggled to articulate one point after another, resolution always lying like a phantom, just beyond his grasp. It was on a Sunday hike in October that Bohr, perhaps speculating that Dirac might be interested in classic English literature, took him to the setting of *Hamlet*, the royal castle of Kronborg, overlooking the stretch of water between Denmark and Sweden. The Bard would have made comic hay from their verbal exchange, both from the clash of their conversational styles and their contrasting approaches to science and every other subject. Philosophy was an important, compulsory part of Bohr's education, and he took it seriously. Whereas Bohr sought understanding through words, Dirac thought they were treacherous and believed that true clarity could be achieved only in mathematical symbols. As Oppenheimer would later remark, Bohr 'regarded mathematics as Dirac regards words, namely as a way to make himself intelligible to other people, which he hardly needs'.[13]

There was never any hope that the two would collaborate, as became plain early in Dirac's stay when Bohr called him into his office to help him write a paper. This was Bohr's usual practice: he often dragooned one of his young colleagues into spending a few days as his scribe. The only reward was the honour of being asked and a daily lunch with the Bohrs in their apartment. But the process was not without its frustrations: no sooner would a sentence escape Bohr's lips than he would qualify, amend or delete it in favour of another form of words that might, or might not, be a closer approximation to his intended meaning. So, the tortuous process of dictation continued, never quite reaching a coherent conclusion. Dirac had better things to do than to spend hours disentangling Bohr's fractured locutions and rendering them into prose of exemplary clarity. 'At school', Dirac announced soon after the first session with Bohr began, 'I was always taught not to start a sentence until I knew how to finish it.' His employment as Bohr's amanuensis lasted about half an hour.[14]

In the evenings, most of the young physicists at the institute liked to relax in the cinema or in their lodgings with a plate of hot dogs and a few beers. But Dirac preferred to spend his nights taking long, solitary walks around the city. He would set out from his lodgings after dinner, take a tram to its terminus and walk the Copenhagen streets back to his digs, thinking about the problems of quantum physics.[15] He probably did not know that he was following in the footsteps of the nineteenth-century philosopher Søren Kierkegaard, pioneer of Christian existentialism and almost as famous among his fellow Danes for his eccentricities as his ideas.[16] Kierkegaard chewed over his ideas in his apartment, pacing back and forth for hours, and during the 'people bath' he took each day in the streets of his native city. For two decades from the mid-1830s, the people of Copenhagen saw the hunch-backed aristocrat walking around in his broad-rimmed hat, his umbrella folded under his arm. 'I have walked myself into my best thoughts,' he said, a remark precisely echoed by the elderly Dirac.[17] But they reacted differently to the people they passed in the street. Dirac said nothing to his fellow pedestrians, but Kierkegaard would startle some of them by interrogating them about some subject on his mind, following in the tradition of Socrates, whom he called 'the virtuoso of the casual encounter'.[18]

During the day, Dirac spent most of his time working in the library, occasionally pausing to read the latest publications in the adjoining 'journal room' and to attend a seminar. To Christian Møller, one of the young Danish physicists at the institute, Dirac appeared distracted and aloof:

Often he sat alone in the innermost room of the library in the most uncomfortable position and was so absorbed in his thoughts that we hardly dared to creep into the room, afraid as we were to disturb him. He could spend the whole day in the same position, writing an entire article, slowly and without ever crossing anything out.[19]

In the library, Dirac was cooking up what would turn out to be one of his most famous insights, the connection between the Heisenberg and Schrödinger versions of quantum theory. Everyone knew that the theories seemed to give the same results, yet they looked as different as Japanese and English. Dirac found the rules that allow the two languages to be translated into each other, laying bare the relationship between them and giving new clarity to the Schrödinger equation. It turned out that the Schrödinger waves were not quite as mysterious as they seemed but were simply the mathematical quantities involved in transforming a description of a quantum – an electron, or any other tiny particle – based on its energy values to one based on possible values of its position. Dirac's theory also accommodated Born's interpretation of Schrödinger's waves and showed how to calculate the probability of detecting a quantum. He began to realise that the knowledge an experimenter can have about the behaviour of a quantum is also limited. He wrote that 'one cannot answer any question on the quantum theory which refers to the numerical values for both [the initial position and momentum values of a quantum]', and he pointed out cryptically that one would expect to be able to answer questions in which only one of those initial values is known. He was within a split whisker of what would become the most famous principle in quantum mechanics, the uncertainty principle, soon to be snatched from under his nose by Heisenberg.

In the course of working out his theory, Dirac introduced a new mathematical construction that made no sense within conventional mathematics. The object, which he called the delta function, resembles the outer edge of the finest of needles, pointing vertically upwards from its base.[20] Away from that base, the numerical value of the delta function is zero, but its height is such that the area enclosed between the perimeter and the base is exactly one unit. Dirac knew but did not care that pure mathematicians would regard the function as preposterous as it did not behave according to the usual rules of mathematical logic. He conceded that the function was not 'proper' but added blithely that one can use it 'as though it were a proper function for practically all purposes in quantum mechanics without getting incorrect results'. It was not until the late 1940s that mathematicians accepted the function as a concept of unimpeachable respectability.

In an interview in 1963, he remarked that it was his study of engineering that led him to his new function:

I think it was probably that sort of training that first gave me the idea of the delta function because when you think of load in engineering structures, sometimes you have a distributed load and sometimes you have a concentrated load at the point. Well, it is essentially the same whether you have a concentrated load or a distributed one but you use somewhat different equations in the two cases. Essentially, it's only to unify these two things which sort of led to the delta function.[21]

But Dirac's recollections may have been wrong. It may well be that he first read about the delta function from Heaviside, who introduced the function with his customary belligerence in one of the books Dirac read as an engineering student in Bristol.[22] Today, the function is associated with Dirac's name, but he had not been the first to invent it – that appears to have been done in 1822 by Heaviside's favourite mathematician, the Frenchman Joseph Fourier, though several others later discovered it independently.[23]

Bohr was indifferent to mathematical rigour, so he would not have been perturbed by the delta function when he read about it in the draft Dirac submitted to him, following the understanding that Bohr had to approve each paper submitted from the institute. However,

Bohr and Dirac were soon in disagreement, like two poets in dispute over the syntax of a stanza. Bohr cared about every word and repeatedly requested detailed changes.[24] For Dirac, the words were there to give the clearest possible expression to his thoughts, and, once he had found the right words, he saw no need to change them. He would have agreed with T. S. Eliot: 'It means what it says and if I had wanted to say it any other way, I should have done so.'

Dirac was usually quick to attribute his success to luck, but not in this case – he referred to the paper as 'my darling'.[25] He later remarked that he was pleased to have solved the particular problem he set out to tackle, of laying bare the relationship between Heisenberg's theory and Schrödinger's. The main quality needed in its solution was technical skill and application; in his view, no special inspiration was involved. Another reason why Dirac was so fond of his 'darling' was probably that it was a success for his method of developing quantum mechanics by analogy with classical mechanics. During his reading about Hamilton's approach to classical mechanics, he had read how 'transformation theory' related different descriptions of the same phenomenon – by using this idea to find the connection between Heisenberg's theory and Schrödinger's, Dirac had shed light on both.

If he hoped that the paper would establish him as the leader in the field, he was soon to be disappointed. In the late autumn, before he had the proofs of his paper, he heard that Pascual Jordan had solved exactly the same problem. Although Dirac's approach and presentation were more elegant and easier to use, the two papers covered substantially the same ground and featured much the same conclusions. So although Dirac had made another distinguished contribution to quantum mechanics – his second within a year – he had yet to beat all his colleagues to a key innovation in the theory. He had, however, acquired some distinguished admirers, though most of them were struggling to understand his peculiar combination of logic and intuition. One of them was Albert Einstein, who told a friend: 'I have trouble with Dirac. This balancing on the dizzying path between genius and madness is awful.'[26]

One evening in Dirac's lodgings shortly before Christmas, the telephone rang. It was Professor Bohr, Dirac's landlady told him, as she passed the receiver to him. This was a new experience for him – he had never before used a telephone.[27] Knowing that Dirac was about to spend the holiday alone, Bohr was calling to ask if he would like to spend Christmas with him and his family. Dirac accepted, though he did not tell his parents. They had been shivering in an unseasonably cold autumn and recovering from the upheaval of having mains electricity installed. Dirac's mother persisted with her doomed campaign to persuade him to do less work and to eat more ('I hope you will take it easy & get nice and plump like Shakespeare's Hamlet') and, for the first time, confided in her son that she was unhappy and tired of the domestic routine. Desperate for a measure of independence, when Charles was out, she and the unemployed Betty sneaked out together to evening classes in French.[28]

The Dirac family was also preparing itself for its saddest Christmas: a year before, they had had three children at home for the holiday; now they would have only one. On 22 December, the ailing Charles wrote his son a letter, one of only two that Dirac kept from his father, possibly the only letters Dirac received from him in adult life.[29] No longer communicating with Dirac only in French, Charles wrote the four-page letter entirely in English and on black-bordered notepaper that signalled his continuing mourning for Felix.

My dear Paul
It will be a lonely time here without you – the first time since you came to us – not so very long ago it seems, but my thoughts are with you to wish you all the happiness a father can wish his only son.

If you can any time spare a few moments to give me some details of your life there and your work – nothing could please me more, except seeing you again. I should like to feel sure you take sufficient care of yourself – and do not let your studies make you forget your health.[30]

Charles goes on to say that he would like to buy his son a Christmas present, perhaps 'a set of chessmen', and he offers to do 'anything at all' he can to help him. He signs off 'Many kisses from your loving Father'. The note is a window on his grief, his loneliness, his desperation to be closer to his unresponsive 'only son'.

At midnight on Christmas Eve, Charles and Betty went to a service at a local church, where Felix's death had first been marked. Later, on Christmas Day, Dirac's mother wrote Dirac a fragmentary letter showing that she was as lonely as the man she was living with:

All we do, as you know, is work & then more work. [. . .] I am trying to get Pa to have [the front room] re-papered. He ought to after 13 years [. . .] He and Betty went up to Horfield Church at 12-midnight for a Service [. . .] This is the first Xmas Day you have been away from home. It is lonely without you.

She then asked him an unusual favour:

Would you like to send me a few pounds for a diamond ring? I want one so very much. I could wear it in the evenings & think what a darling you are. It is so monotonous doing housework all day long. I get so fed up with it.

Pa has pupils all the year round & gives me £8 a year for clothes and everything. It is worse than a servant.[31]

For the first time in her correspondence, she showed Dirac that he was not just her favourite son but her most intimate confidant and even a substitute for a gift-bearing lover. As her subsequent letters showed, she was in desperate straits, trapped in an unfulfilling marriage to a man who was highly regarded in the community but whom she regarded as an unsympathetic and insensitive brute. In the coming years, her life would unfold like an Ibsen tragedy.

Another of the out-of-the-blue ideas that Dirac apparently conceived in Copenhagen is now the basis of all modern descriptions of the fundamental constituents of the universe. Such descriptions are based on the nineteenth-century concept of a 'field', which had superseded Newton's vision that nature's basic particles move under the influence of forces exerted by other such particles, often over long distances. Physicists replaced the notion that the Sun and the Earth exert gravitational forces on each other by the more effective picture that the Sun, the Earth and all the other matter in the universe collectively give rise to a gravitational field which pervades the entire universe and exerts a force on each particle, wherever it is located. Likewise, an all-pervasive electromagnetic field exerts a force on every electrically charged particle. Maxwell's theory of electromagnetism and Einstein's theory of gravity are examples of classical 'field theory', each featuring a field that varies smoothly throughout space and time, not mentioning individual quanta. Such classical theories describe the universe in terms of a smooth, underlying fabric. Yet, according to quantum theory, the universe is fundamentally granular: it is ultimately made of tiny particles such as electrons and photons. Loosely speaking, the texture of the underlying fields should, according to classical ideas, be rather like a smooth liquid, whereas quantum theory suggests that it would be like a vast collection of separate grains of sand. To find a quantum version of Maxwell's classical electromagnetism was one of the theoreticians' most pressing problems, and Dirac's next innovation was to solve it.

Quite what put him on to the solution is something of a mystery. Although he was probably aware of the first steps taken a few months before by Jordan, Dirac later said that he first hit on the idea when he was playing with Schrödinger waves as if they were mathematical toys, wondering what would happen if they behaved not as ordinary numbers but as *non-commuting* quantities.[32] The answer began a new way of describing the quantum world.

Dirac found a way of mathematically describing the creation and destruction of photons, both commonplace processes. Particles of light are continually created in vast numbers all over the universe in stars and also here on Earth, when an electric light is switched on, a match is struck, a candle is lit. Likewise, photons are continually destroyed – annihilated – for example, when they disappear into human retinas and when leaves convert sunlight to life-giving energy. Neither of these processes of creation and annihilation can be understood using Maxwell's classical theory, which has no way of describing things that appear out of nowhere or disappear into oblivion. Nor did ordinary quantum mechanics have anything to say in detail about the processes of emission or absorption. Yet Dirac showed that this wizardry can be described in a new type of theory, a compact mathematical description of the creation and destruction of photons. He associated each creation with a mathematical object, a creation operator, which is closely related to but quite distinct from another object associated with annihilation, an annihilation operator.

In this picture, at the heart of modern quantum field theory, the electromagnetic field pervades the entire universe. The appearance of every photon is simply an excitation of this field at a particular place and time, described by the action of a creation operator. By a similar token, the disappearance of a photon is the de-excitation of the field, described by an annihilation operator.

Dirac had begun to set out a quantum version of Maxwell's unified field theory of electricity and magnetism. He had learned about that theory only three years before, in Cunningham's lectures in

Cambridge, and was now standing on Maxwell's shoulders. So far as Dirac was concerned, his theory put an end to the hand-wringing about the apparent conflict between two theories of light: a wave theory seemed to account for propagation, while a particle theory was needed to explain the interactions with matter. The new theory avoided the embarrassment of having to choose between the wave and particle descriptions and replaced the two sharply contrasting pictures with a single, unified theory. Evidently pleased with himself, Dirac wrote that the pictures were in 'complete harmony'. But he was not interested in sharing the good news with his parents, who read on their weekly postcard their son's familiar message: 'There is not much to say now.'[33]

In his paper, Dirac applied his theory and compared his results with the successful predictions Einstein had made a decade before, in 1916. Einstein had used old quantum ideas to calculate the rate at which atoms can emit and absorb light, producing formulae that appeared to describe these processes successfully. The question Dirac had to answer was: does the new theory compare favourably with Einstein's?

Einstein's theory had accounted for the interaction of light and matter in terms of three fundamental processes. Two of them were familiar enough: the emission and absorption of a photon by an atom. But Einstein also predicted a previously unknown way of 'persuading' an atom to jump from one energy level to a lower one, by stimulating it with another photon whose energy is exactly equal to the difference between the two energy levels. The result of this process of 'stimulated emission' is that two photons emerge from the atom: the original one and another one given out when the atom jumps to the lower energy level. This process takes place in the ubiquitous laser – there is at least one in every CD and DVD player and in every bar-code reader – and so is the most common technological application of Einstein's science. Dirac's theory produced exactly the same formulae as Einstein's and had the other advantages that it was more general and mathematically more coherent. As he probably realised, he had gone one better than Einstein.

At the end of January, as he was preparing to leave Copenhagen, Dirac posted his paper to the Royal Society. It turned out that he was the first to introduce the mathematics of creation and annihilation into quantum theory, though his results had been reached independently by John Slater, studying in Cambridge with Fowler. Slater was one of the many who admired Dirac's paper for its content but found its presentation perversely complicated: 'his paper was a typical example of what I very much distrusted, namely one in which a great deal of seemingly unnecessary mathematical formalism is introduced'.[34]

Dirac's time in Copenhagen was an unqualified success. The two theories that he had nurtured there had underlined his status as a leading player on the international stage of science. Although he was still the archetypal individualist, he had come to see the value of taking different approaches to his subject and of having his views cross-examined. Apart from Bohr, the interrogator who most fascinated him was Paul Ehrenfest, an intense and disturbed theoretician based at the University of Leiden in the Netherlands. Ehrenfest got on well with Dirac, who was almost half his age, the two no doubt especially comfortable in each other's company because – unusually among the Institute's members – they disliked both alcohol and smoking. Ehrenfest's aversion to smoking was in part due to his extremely sensitive sense of smell. One victim of this was the amiable Dutch graduate student Hendrik Casimir. Soon after he arrived in Leiden, Casimir had his hair cut before a meeting with Ehrenfest, who soon sniffed the perfume of the barber's dressing. Ehrenfest quickly became angry and shouted, 'I will not tolerate perfume here. Get out. Go home, get out. Get out. Get out.' A few days later, Casimir was dismissed.[35]

Ehrenfest was at his best during seminars. Unafraid of ridicule, he would politely but persistently interrupt speakers, seeking clarification of every unclear point. When he first met Dirac, Ehrenfest was uncomfortable with quantum mechanics and was worried that his close friend Einstein was unhappy about the central role played in the theory by probability. Einstein had been the first to identify that when an atom spontaneously jumps to a lower energy level, quantum theory cannot predict either the direction of the emergent photon or the precise time of its ejection. This was also true of ordinary quantum mechanics and of Dirac's new quantum field theory. Einstein was sure that a satisfactory theory had to do better than just predict probabilities: 'God is not playing dice,' he wrote to Max Born.[36] Dirac thought his hero worried too much about the philosophical issues of quantum mechanics. All that mattered to Dirac – true to his mathematical and engineering training – was that the theory was logical and accurately accounted for the results of experiments.

At the end of January 1927, Dirac was preparing to travel to Göttingen. Soon he would be leaving the company of Niels Bohr, whom Dirac would later describe as 'the Newton of the atom' and 'the deepest thinker that I ever met'.[37] But it was Bohr's warmth and humanity that most impressed Dirac. At Christmas – while Charles, Flo and Betty Dirac were going through the family rituals – Dirac had been welcomed into the Bohrs' loving fold and witnessed familial joy for the first time. Dirac had seen that it was possible to be both a great physicist and a dedicated family man and that perhaps – just perhaps – there might be more to life than science.

For Bohr, Dirac was 'probably the most remarkable scientific mind which has appeared for a very long time' and 'a complete logical genius'.[38] Also intrigued by Dirac's personality, Bohr never forgot one incident, during a visit to an art gallery in Copenhagen, of his young visitor's eccentricity. When they were looking at a French impressionist painting showing a boat sketched by just a few lines, Dirac observed, 'This boat looks as if it was not finished.' Of another picture, Dirac remarked, 'I like that because the degree of inaccuracy is the same all over.'[39] Such anecdotes became part of scientific lore, and physicists vied with one another to relate the most amusing instances of Dirac's verbal economy, his literal-mindedness, mathematical precision and otherworldliness. With no psychological framework available to help understand him, his personality became an object of collective amusement, through a myriad of 'Dirac stories'.

No one relished telling the stories more than Bohr, who entertained visitors with them over afternoon tea in his office. Four years before he died, he told a colleague that, of all the people who had visited his institute, Dirac was 'the strangest man'.[40]

Nine

[For young Germans after the great inflation they experienced in 1923] their aims were to live from day to day; and to enjoy to the utmost everything that was free: sun, water, friendship, their bodies.
STEPHEN SPENDER, *World Within World*, 1951

In Göttingen, Dirac made another of his unlikely friendships. This one was with Robert Oppenheimer, who had fled Cambridge and was flourishing in Max Born's Department of Theoretical Physics as a Ph.D. student of rare ability, self-confidence and superciliousness. Ever the intellectual peacock, Oppenheimer ensured that his colleagues knew he was thinking about more than physics: his eclectic reading list included F. Scott Fitzgerald's collection of short stories *Winter Dreams*, Chekhov's play *Ivanov* and the works of the German lyric poet Johann Hölderlin.[1] He was also composing verse, a hobby that puzzled Dirac. 'I don't see how you can work on physics and write poetry at the same time,' he remarked during one of their walks. 'In science, you want to say something nobody knew before, in words everyone can understand. In poetry, you are bound to say something that everybody knows already in words that nobody can understand.' For decades to come, Oppenheimer liked to recount this anecdote over cocktails, no doubt having polished Dirac's original phrasing to give it the bite of one of Wilde's paradoxes.[2]

Dirac kept normal working hours, while Oppenheimer was nocturnal, so the two young men could not have seen much of each other.[3] They boarded with the Cario family in a spacious granite villa on Giesmarlandstrasse, which led from the town centre out to the local countryside.[4] From the outside, the home appeared to be just another of the town's many lavish residences, but there was a bitterness and penury inside. During the unstable early years of the Weimar Republic, the Carios had been victims of the precipitate fall of the German currency: the number of deutschmarks that could be purchased with an American dollar rose from 64.8 in January 1920 to 4.2 trillion in November 1923.[5] Worse, the family's breadwinner, a doctor, had been disqualified for malpractice. Now that the

Republic had stabilised, the Carios made a living by turning their home into a guesthouse for the stream of foreign visitors, many of them American students visiting the Georgia Augusta University, one of the most prestigious academic addresses in Europe. With his fellow boarders, Dirac sat down every evening to a meal based on the local fare of potatoes, smoked meats, sausages, cabbages and apples.

It took Dirac and Oppenheimer only five minutes to stroll from their lodgings to Born's department in the Second Physics Institute, located in an ugly red-brick building with all the charm of a Prussian cavalry barracks. Born – a handsome, clean-shaven man, who looked younger than his forty-four years – was reserved but warmer than most of his professorial colleagues. He cultivated a competitive environment but was sensitive to the needs of the brightest students and tolerant of their peccadilloes. Dirac and Oppenheimer were among the many students Born invited to his villa on the Planckstrasse, a quiet road on the outskirts of the town. To be invited there was always a pleasure: dinner would be followed by good-humoured conversation and a concert in the huge front room, which contained two grand pianos.[6] Heisenberg, a close friend of the family, took every opportunity to display his pianistic skills in flamboyant renditions of Beethoven, Mozart and Haydn.[7]

Dirac lived just a few steps away from the historic centre of Göttingen, one of the best-preserved medieval towns in Lower Saxony: its half-timbered houses and shops, its churches and cobbled backstreets had remained virtually unchanged for centuries. Nor was it yet overrun by the motor car. Most people got around on foot or by bicycle, many of the cyclists sporting garishly coloured caps to show their affiliation to one of the clubs and societies.[8] Like Cambridge, Göttingen was a tranquil academic town, dominated by the needs and whims of its academics and students. Seniority and intellectual distinction were at a premium there. Its most revered citizens were the most venerable of its distinguished professors, including the gruff David Hilbert, sixty-three years old and the most celebrated mathematician alive.

Also like Cambridge, many of Göttingen's (mainly male) students were there not so much to be well educated as to spend a few hedonistic years in the fug and cacophony of the town's taverns and coffee bars.[9] No doubt having left Dirac to get his sleep, Oppenheimer and his friends spent many a night on the razzle; he happily picked up the tab after downing a few pints of *frisches Bier* in the Black Bear pub or dining on *Wienerschnitzel* at the four-hundred-year-old Junker Hall.[10] The atmosphere in the pub had hardly changed in generations: most evenings, the din of the students would often dissolve into bibulous choruses of favourite folk songs, while virile young men sloped off to put on their chain mail, don their swords and do some 'academic fencing'. When the combatants returned, their faces were 'decorated' with scars, each a bloody badge of honour.[11]

At weekends, Oppenheimer and other affluent students often took the two-and-a-half-hour train journey to Berlin, the city of Bertolt Brecht, Arnold Schönberg and Kurt Weill. But Dirac had no interest in broadening his horizons much beyond the towns and villages of Lower Saxony, where he went on long Sunday walks, if he was not snowbound. Within twenty minutes of leaving his lodgings, he was walking in the gently rolling countryside, following the fast-flowing rivers and pausing at the scattered monuments to Bismarck. By early spring, the walking conditions were perfect: almost all the winter snow had melted, and the linden trees, shrubs and flowers were scenting the air. He passed occasional groups of young men in the German Youth Movement but otherwise saw scarcely another person, which was just as he preferred – his empathies lay more with uncommunicative forms of nature than with human beings.

So Göttingen gave Dirac everything he wanted in a town – a great university with a world-leading physics department and comfortable lodgings close to walking country, where he could escape from other people. Göttingen was a German Cambridge, with hills.

In early February 1927, within days of Dirac's arrival in Göttingen, he had set Oppenheimer's imagination alight. Oppenheimer was completing his Ph.D., on the quantum mechanics of molecules, and looking to the future which appeared to lay in the direction that Dirac had opened up. Near the end of Oppenheimer's life, when he looked back on his career, he remarked that 'perhaps the most exciting time of my life was when Dirac arrived [in Göttingen] and gave me the proofs of his paper on the quantum theory of radiation'. While others found Dirac's field theory mystifying, to Oppenheimer it was 'extraordinarily beautiful'.[12]

Oppenheimer had been an outsider at Cambridge and Harvard and so he was pleased at last to feel part of the small community of

Göttingen physicists, gradually recovering from his clinical depression. Among his colleagues was Pascual Jordan, born a few weeks after Dirac and the youngest of the quantum innovators. Intense, haunted and private, his eyes stared out from behind elliptical glasses with lenses as thick as jam jars. Oppenheimer later remarked that Jordan's peculiarities may have led him to be underestimated: 'it was in part because he was really an unbelievably queer duck with tics and mannerisms and [. . .] apparent brutalities, which put people off very much.'[13] According to Oppenheimer, Jordan had a stutter so crippling that 'it was difficult to get through', though Oppenheimer may have to some extent admired it – he began to affect a stutter, muttering 'njum-njum-njum' before some of his finely crafted declamations.[14]

Although Jordan and his colleagues admired Oppenheimer's quick-fire intelligence – one of them likened him to 'an inhabitant of Olympus who had strayed among humans' – they found his arrogance irritating, to the point that it became unacceptable.[15] One morning, Born found on his desk a letter from several of his colleagues threatening to boycott seminars unless he stopped Oppenheimer from disrupting them with his continual interruptions. Always fearful of showdowns, Born chose to leave the letter – a large sheet of parchment lettered in ornamental script – on his desk for Oppenheimer to see. It did the trick. Relations between Born and Oppenheimer were superficially cordial, but Oppenheimer regarded Born as a 'terrible egotist' who continually complained that he had not been given enough credit for pioneering quantum mechanics.[16] Born had good reason to feel slighted. He had been one of the creators of quantum mechanics, having used his battery of mathematical skills to develop Heisenberg's initial idea. Most physicists gave the lion's share of the credit to Heisenberg, but Born believed that it was he who first fully appreciated the idea's potential and he who led its development in Göttingen.

By the time Dirac arrived there, Born was confident that he had found the right way to develop quantum mechanics, using Heisenberg's ideas, not Schrödinger's. Although Born knew of Dirac's reputation, he was not expecting his young visitor to be so adept and knowledgeable. The American physicist Raymond Birge, then visiting Göttingen, observed that 'Dirac is the real master of the situation [. . .] when he talks, Born just sits and listens to him open-mouthed.'[17]

Another colleague, the German theoretician Walter Elsasser, later wrote his impressions of Dirac: 'tall, gaunt, awkward and extremely taciturn. [. . .] of towering magnitude in one field, but with little interest and competence left for other human activities'. Elsasser remembered that although Dirac was always polite, his conversations were almost always stilted: 'one was never sure that he would say something intelligible.'[18] Another of Dirac's traits was his inability to comprehend anyone else's point of view if it didn't fit into his way of looking at things: colleagues would spend hours presenting their perspective on a physics problem, only for him to walk away after making a brief comment, apparently apathetic or bored. Oppenheimer was quite different: he would listen to a colleague's ramblings for a few minutes but would then interject with an eloquent summary of what he was probably trying to say.

Whereas Oppenheimer mixed freely with his colleagues, Dirac spent most of his time working in the library or in one of the empty classrooms. But he was not a complete loner: in Copenhagen, he had come to appreciate being with other physicists, provided they didn't put pressure on him to speak. Most mornings, he walked with fellow boarders at the Carios' to the Mathematics Institute, where he attended lectures that kept them abreast of the latest experimental findings. He also took the time to go to the often-combative afternoon seminars. When Ehrenfest was in town, he was their undisputed inquisitor-in-chief, deflating egos and revealing the crux of every new argument, having cut away the underbrush. In the previous June, he had brought along a Ceylonese parrot trained to say 'But, gentlemen, that's not physics' and recommended that it should chair all forthcoming seminars on quantum mechanics.[19]

Max Delbrück, one of the young Göttingen physicists, was probably not exaggerating when he later described the experience of walking into one of their seminars: 'you could well imagine that you were in a madhouse.'[20]

Word spread to Berlin that Dirac was a difficult man and that his work was impenetrable and overrated. The Hungarian theoretician Jenő (later Eugene) Wigner later said that, in the mid-1920s, his German colleagues were suspicious of 'the queer young Englishman who resolves [questions of physics] in his own language'.[21] Many Germans were put off by Dirac's manner. The English were known for their reserve – they acted as if everyone else was either an enemy or a bore, as John Stuart Mill had pointed out – but Dirac's frigidity was unlike anything they had ever seen.[22]

Born was one of the few Germans who warmed to Dirac, but even he had trouble understanding his new field theory and apparently thought it unimportant. His lack of foresight frustrated Jordan, who had begun to develop ideas on field theory very similar to Dirac's, only to be met with indifference.[23] It would have been fascinating to see what Dirac and Jordan could have achieved in quantum field theory, but Dirac had no interest in collaboration. He turned his attention to using field theory to understand what happens when light is scattered by an atom, normally visualised as being rather like a basketball bouncing off the hard rim of the basket. But, in the new field theory, things are not so straightforward. Dirac showed that, in the fleeting moment of a photon's scattering, it appears to pass through some strange, unobserved energy states. What makes these intermediate processes so odd is that they appear to flout the sacred law of conservation of energy. Although these subatomic 'virtual states' cannot be seen directly, experimenters were later able to detect their subtle influences on fundamental particles.[24]

Dirac's calculations also threw up a more troubling artefact. He found that his new theory kept generating bizarre predictions: for example, when he calculated the probability that a photon had been emitted after a given interval, the answer was not an ordinary number but was infinitely large. This made no sense. The probability that an atom would emit a photon must surely be a number between zero (no chance) and one (complete certainty), so it seemed obvious that the prediction of infinity was wrong. But Dirac chose to be pragmatic. 'This difficulty is not due to any fundamental mistake in the theory,' he wrote with more confidence than was warranted. The root of the problem, he speculated, was a simplistic assumption he had made in applying the theory; when he had identified his error and tweaked the theory, he implied, the problem would disappear. In the meantime, he dodged the difficulties using clever mathematical tricks, enabling him to use the theory to make sensible, finite predictions. But it would not be long before he saw that his optimism was misplaced: the lamb had caught its first sight of the wolf's tail.

*

Meanwhile, the debates about the interpretation of quantum theory had not abated, least of all in Copenhagen, where Heisenberg was struggling to understand the theoretical limits of what can be known about a quantum. He achieved this brilliantly with his uncertainty principle, which made him into the nearest the quantum fraternity had to a household name.

The principle emerged only after anguished and protracted gestation, which apparently began with a letter from Pauli during the previous October.[25] Heisenberg believed that the correct way to think about the quantum world was in terms of particles, and that the more popular wave-based ideas were merely useful supplementaries. Somehow, Heisenberg wanted to find a way of making definite statements about the measurements that could be made on quantum particles, especially about the limitations on what experimenters can know about them. Heisenberg had talked with Einstein about this, and, when Dirac was in Copenhagen developing transformation theory, he had also discussed it with him.[26]

The nub of what became known as Heisenberg's uncertainty principle is that the knowledge experimenters have of a quantum's position limits what they can know about its speed, at the same instant. The more they know about a quantum's position, the less they can know about its speed. So, for example, if experimenters know an electron's location with perfect precision, then it follows that they can know nothing whatsoever about its speed at the same moment; on the other hand, if they know the exact value of the electron's speed, they will be totally ignorant of its position. There is, Heisenberg argued, no way round this: regardless of the accuracy of the measuring apparatus or the extent of the experimenters' ingenuity, the principle puts fundamental limitations on knowledge. It turns out that even the most accurate knowledge imaginable of the location of an ordinary object puts only negligible constraints on knowledge of its speed (likewise with the location and speed reversed), so the principle is unimportant in everyday life. This is the root of the physicists' joke about the motorist who tries to con the traffic police by pleading not guilty of speeding on the grounds 'I knew exactly where I was, so I had no idea how fast I was travelling': the plea would be perfectly admissible if it were made by a sentient electron.

In his paper, Heisenberg explained his principle by picturing what happens when an experimenter uses a photon of light to probe the behaviour of an electron, demonstrating that the very act of probing disturbs the electron. An analysis of this thought experiment led Heisenberg to a mathematical expression that encapsulated the principle. He also derived the expression mathematically, using two of Dirac's innovations: transformation theory and the relationship between the non-commuting position and momentum.[27]

As spring set in, Dirac will probably have thought about the principle during his constitutional walks along the tree-lined path following the contours of what was once Göttingen's outer wall.[28] He was not especially impressed with Heisenberg's discovery, as he noted later: 'People often take [the uncertainty principle] to be the cornerstone of quantum mechanics. But it is not really so, because it is not a precise equation, but only a statement about indeterminacies.'[29] Dirac was similarly lukewarm a few months later when Bohr announced his principle of complementarity, apparently related to Heisenberg's principle. According to Bohr's idea, quantum physicists have to accept that a complete picture of subatomic events always involves descriptions that appear incompatible but that are actually complementary – both the wave and particle pictures are needed. In Bohr's view, this idea was part of an ancient philosophical tradition, in which truth cannot be pinned down using only one approach but needs complementary concepts: for example, a mixture of reason and feeling, analysis and intuition, innovation and tradition.

This principle was fundamental to Bohr's thinking, to the extent that he chose it in 1947 as the basis of the design of his coat of arms.[30] The design features the Chinese yin-yang symbol, which represents the two opposing but inseparable elements of nature, and the Latin motto below reads 'Opposites are complementary'. Many physicists thought that Bohr had uncovered a great truth, but Dirac was again unimpressed: the principle 'always seemed to me a bit vague', he later said. 'It wasn't something which you could formulate by an equation.'[31]

Dirac's opinion of Heisenberg's uncertainty principle was not shared by most scientists, including Eddington. In his acclaimed book *The Nature of the Physical World*, published in November 1928, he gave a sparkling account of 'the principle of indeterminacy', describing it as 'a fundamental general principle that seems to rank in importance with the principle of relativity'. Writing with his usual panache, Eddington introduced tens of thousands of lay readers to the new principle as one of the cornerstones of quantum mechanics.

Eddington writes that he is giving an outline of the theory only against his better judgement: 'It would probably be wiser to nail up over the door of the new quantum theory a notice "Structural alterations in progress – No admittance except on business", and particularly to warn the doorkeeper to keep out prying philosophers.'[32] Eddington's account of the theory was the clearest account of quantum mechanics for English-speaking lay readers and was the first widespread publicity for the new theory. If Bohr or another influential figure had taken a leaf out of Eddington's book and been savvy enough to provide a dramatic presentation of the uncertainty principle's discovery to well-briefed journalists, then quantum mechanics may well have become much better known, along with its creators.

With a hint of nostalgia, Eddington pointed out that modern physicists no longer thought about the universe as a giant mechanism, as Victorian physicists such as James Clerk Maxwell had done, but framed their accounts of the fundamental nature of things in the language of mathematics. The images of cogs and gearwheels were now passé, but Eddington believed there were dangers inherent in the new, mathematical way of thinking of fundamental physics:

Doubtless the mathematician is a loftier being than the engineer, but perhaps even he ought not to be entrusted with the Creation unreservedly. We are dealing in physics with a symbolic world, and we can scarcely avoid employing the mathematician who is a professional wielder of symbols; but he must rise to the full opportunities of the responsible task entrusted to him and not indulge too freely his own bias for symbols with arithmetical interpretations.[33]

Eddington had put his finger on the central conceptual challenge that made quantum mechanics so difficult for most professional physicists. The great majority of them still thought like engineers and were mathematically weak by the standards of Dirac and his peers. So, most physicists were still trying to visualise the atom as if it were a mechanical device.

The metaphor of nature as a colossal clockwork mechanism, popular since Newton's day, had long been apt for most purposes. But no longer. Quantum mechanics was based fundamentally on mathematical abstractions and could not be visualised using concrete images – that is why Dirac refused to discuss quantum mechanics in everyday terms, except in later life, when he began to use analogies between the behaviour of quanta and the way ordinary matter behaves. Yet Dirac often remarked that he did not think about nature in terms of algebra, but by using visual images. Since he was a boy, he had been encouraged to develop visual imagination in his art and technical-drawing classes, which were an ideal grounding for his studies of projective geometry. None of the other pioneers of quantum mechanics had been given an education in which geometric visualisation played such a prominent part. Five decades later, when he looked back on his early work in quantum mechanics, Dirac declared that he had used the ideas of projective geometry, unfamiliar to most of his physicist colleagues:

[Projective geometry] was most useful for research, but I did not mention it in my published work [. . .] because I felt that most physicists were not familiar with it. When I had obtained a particular result, I translated it into an analytic form and put down the argument in terms of equations.[34]

Dirac had a perfect opportunity to explain the influence of projective geometry on his early thinking about quantum mechanics at a talk he gave in the autumn of 1972 at Boston University.[35] Its philosophy department had invited him to give the talk to clarify this influence and had recruited the urbane Roger Penrose, an eminent mathematician and scientist who knew Dirac well, to chair the seminar. If anyone could prise the story out of Dirac, it was he. In the event, Dirac gave a short, clear presentation on basic projective geometry but stopped short of connecting it to quantum behaviour. After Dirac had batted away a few simple questions, the disappointed Penrose gently turned to him and asked him point-blank how this geometry had influenced his early quantum work. Dirac firmly shook his head and declined to speak. Realising that it was pointless to continue, Penrose filled in the time by extemporising a short talk on a different subject. For those who wanted to demystify Dirac's magic, his silence had never been so exasperating.

Ten

Hitler is our Führer, he doesn't take the golden fee
That rolls before his feet from the Jew's throne
The day of revenge is coming, one day we will be free [. . .]
    From an early Nazi marching song, c. 1927

As a Jew, Max Born had every reason to be alarmed and frightened by the rise of anti-Semitism in Göttingen. The atmosphere was 'bitter, sullen [. . .] discontent[ed] and angry and loaded with all those ingredients which were later to produce a major disaster', Oppenheimer remembered, a few years before he died.[1] The Nazis had set up one of their first branches in the town in May 1922. Three years later, the chemistry student Achim Gercke secretly began to compile a list of Jewish-born professors, to provide 'a weapon in hand that should enable the German Reich to exclude the last Hebrew and all mixed race from the German population in the future and expel them from the country'.[2]

Life among the Göttingen researchers did have its lighter side, however. Many of them gloated that their profession was for the young, and they mocked the sclerotic imaginations of their elderly professors, paid and revered much more for doing much less. As his later comments confirm, Dirac shared this dismissiveness, and, if an improbable Göttingen legend is to be believed, he wrote a quatrain about this for a student review:

> Age is of course a fever chill
> That every physicist must fear
> He's better dead than living still
> When he's past his thirtieth year[3]

Göttingen students had a penchant for silly songs and for choral renditions of American tunes, which were sung with special enthusiasm at Thanksgiving. The cosmologist Howard Robertson, who introduced Dirac to ways of describing the curvature of space-time across the universe, had brought to the taverns of Göttingen one of their most popular new songs, 'Oh My Darling Clementine'.[4] Dirac probably did not join in, but he took part in the infantile games that helped to sublimate the physicists' intense competitiveness. One of the games was 'bobbing for apples', when professors and students – often woozy, after a few glasses of beer – would try to sink their teeth into an apple floating on water or beer. Another activity involved running a race while trying to balance a large potato on a tiny spoon. After one of these races in Born's home, a student saw Dirac practising surreptitiously – a sight that would have stunned his colleagues in Cambridge, including the classicist John Boys Smith, who described Dirac as being 'childlike but never childish'.[5]

Dirac's stay in Göttingen ended in early June 1927. St John's wanted him back and had been wooing him to apply for a fellowship, an honour well worth pursuing. If successful, he would benefit from free board and lodging in college, as well as a modest income to supplement the continuing funds from his 1851 scholarship, which would run out in 1928.[6] A tenured academic post in the university's mathematics department would almost certainly follow, and he would be set up for the rest of his working life. In his letters, Dirac was even less forthcoming about his personal life than he had been when he wrote from Copenhagen. In a letter to the college official James Wordie, Dirac wrote just a single sentence about his activities in Göttingen: 'The surrounding country is very beautiful.'[7] Although he preferred Bohr's pullulating institute to Born's comparatively cool department, he told his mother that he preferred Göttingen, as it gave him the best opportunities for solitary walks.[8]

In his research, Dirac appeared to be showing signs of running out of steam. In early May 1927, he used quantum mechanics to predict what happens when light is scattered by an atom – a problem that led to no exciting conclusions. Oppenheimer later said that he was disappointed by Dirac's work in Göttingen and could not understand why he did not press on with the development of quantum field theory. Dirac wanted to take a long rest over the summer, he told Oppenheimer, and would then turn his attention to the spin of the electron, still not understood.

Dirac intended to begin his break from quantum theory when he returned to England, after he had visited Ehrenfest in Leiden, a small university town in the Netherlands. Dirac stayed in the room at the top of Ehrenfest's large Russian-style house, where he signed his name on the bedroom wall that already bore the signatures of Einstein,

Blackett, Kapitza and dozens of others. The house served as a local hostel for the cream of the world's physicists, who traded anecdotes of their lively conversations with Ehrenfest's wife – a Russian mathematician – and their three children, two daughters and a son who had Down's syndrome.

Oppenheimer was planning to join Dirac in Leiden and began to learn Dutch so that he could give a seminar in the language of his host. But first he had to defend his Ph.D. thesis in an oral examination held by James Franck, the distinguished experimenter, and Max Born.[9] Franck took only twenty minutes to question Oppenheimer, but that was enough. On leaving the exam room, Franck sighed, 'I'm glad that is over. He was on the point of questioning *me*.' Born was relieved that his brilliant but troublesome student was off his hands. At the end of a typewritten letter to Ehrenfest, Born wrote a postscript:

I should like you to know what I think of [Oppenheimer]. Your judgement will not be influenced by the fact I openly admit that I have never suffered as much with anybody as him. He is doubtless very gifted but without mental discipline. He's outwardly modest but inwardly very arrogant. [. . .] he has paralyzed all of us for three quarters of a year. I can breathe again since he's gone and start to find the courage to work.[10]

Dirac had not been part of this departmental paralysis, nor does he appear to have been aware of it. Oppenheimer was awed by him and showed him a diffidence he granted to few of his other colleagues. Their days in Göttingen were the beginning of a forty-year friendship.

Göttingen was too far away for Dirac's family to visit. 'Thank goodness, you are saying, I expect,' his mother wrote in a pained aside.[11] She made it clear to her son how much she envied him: 'You are a lucky fellow to be away from home. [Here,] it is all work, work.'[12] When her husband was out, she wore her new ring – seven diamonds set in platinum – which she had furtively bought with £10 of the money Dirac had sent her, considerably more than Charles allowed her to spend on herself in a year. That piece of jewellery was a private symbol of her most important relationship. She wrote to her son: 'Don't tell Pa [. . .] I expect he would tell me to put the money in the housekeeping, but it is giving me such a lot of pleasure to look at it and think what a darling you are.'[13] In the evenings, she would sit in the front room with photos of her son, re-reading his postcards, trying to imagine what he would be doing at every time of day.

The twelve-year age difference between Charles and Flo had never been more plain. She still had an upright posture, smooth skin and scarcely a grey hair; he was hunch-backed, white-haired and wizened. In public, she put on the traditional show as the loyal, uncomplaining wife; in private, she was resentful of being an unpaid servant, as she often wrote to her son. At the beginning of 1927, she was surprised when her husband went on a spending spree, probably funded by his mother's legacy. Dirac often condemned the tattiness of the family home, which had not been decorated for thirteen years, so it may well have been that Charles paid for the extensive wallpapering and the installation of a gas fire in every room, with the aim of making 6 Julius Road more attractive to his son. Charles did not entirely neglect his wife – he bought her one of the new vacuum cleaners to help with the housework: 'Pa likes to see them at work on our carpets giving free demonstrations.'[14]

Still in poor health, Charles consulted a herbalist who advised him to become vegetarian, presenting endless catering problems for his wife, who worried incessantly about his nutrition. She wrote to Dirac: 'Pa is getting ever so many pupils he has scarcely time for meals. I am sure he is working his brain too hard and now he is a vegetarian, there are so many little things to cook which are not substantial enough for him.'[15] Although she thought he was mean and ungrateful, she devoted herself to taking care of him, and her letters to Dirac betrayed no sign that the state of affairs was anything less than she should expect or deserve. But her patience was beginning to run out.

Charles Dirac's work ethic had been the making of one of his sons and possibly the death of the other, but it did not have much influence on his daughter. Betty had left school and was, according to her mother, 'too shy or perhaps too lazy [. . .] to want to do anything to earn her own living & she is not fond of housework either'.[16] Without a job, she lolled around the house mourning the death of her dog and went out with her mother to evening classes in elocution and French.[17] In early July, the family chased out the decorators and made sure everything in their house was spick and span, ready for the return of the itinerant son. The family had not spoken to him for nine months, but in that time had sent him weekly family bulletins, showering him with affection and pleas for news at his end. In return, he had sent his parents fewer than seven hundred words. He had not once asked after his family on his postcards, which each had the warmth of a stone.

When Dirac arrived at the door of 6 Julius Road at lunchtime on 13 July – a dull and overcast afternoon – it is easy to imagine the tearful flutterings of his mother and sister as they hugged his unresponsive frame and the stiff handshake with his father, who was probably no less pleased to see him, even if he was unable to show it. He was soon back in his routine, shutting out his family, working alone in his room. One of Charles's students, D. C. Willis, left an anecdote that offers an insight into the domestic environment at the Diracs' that summer. Willis was sent by Monsieur Dirac 'on his errands to his home during the dinner hour [. . .] as he was concerned about his son Paul who, rumour had it, was working in his bedroom, and would not come out, except to collect his food and use the toilet'.[18]

Dirac knew he had a filial duty to be with his parents but felt wretched whenever he was with them. 'When I go back to my home in Bristol I lose all initiative,' he sighed in a letter to a friend, a few years later.[19] He felt oppressed by both his parents – by his father's high-handedness and by his mother's suffocating affection. Although Dirac was twenty-five years old and internationally successful, he still felt himself to be writhing under his father's thumb. And he saw no imminent prospect of escape.[20]

In October 1927, Dirac returned to Cambridge to reacquaint himself with his friends in St John's and Trinity. He now had even fewer social distractions, as Kapitza had recently married. His new wife was the émigré Russian artist Anna Krylova, a dark-haired beauty whom Kapitza unaccountably called 'Rat', a nickname that nonplussed audiences in Cambridge theatres for years, whenever they heard him holler it across the stalls. She and Kapitza contributed to the design of the detached house that was being built for them on Huntingdon Road, near the city centre, complete with a huge back garden and a studio for her in the loft.[21] Later, this house would become almost Dirac's second home in Cambridge but, in the early autumn of 1927, he was working hard on his project, first mooted to Oppenheimer, aiming to combine quantum theory and Einstein's special theory of relativity in the simplest practical case: to describe the behaviour of a single, isolated electron. The quantum theories of Heisenberg and Schrödinger were deficient because they did not conform to the special theory of relativity: observers moving at different speeds relative to one another would disagree on the theories' equations. At stake here was the prestige of being the first to find the theory; would he be the sole winner of a scientific prize or would he, yet again, have to share it?

Dirac worked on the problem for the first six weeks of the term but without success. He took a break in late October to sit, for the first time, at the top table of international physicists at the Solvay Conference in Brussels.[22] The aim of these invitation-only conferences, funded by the Belgian industrialist Ernest Solvay, was to bring together about twenty of the world's finest physicists every few years to ponder the problems of quantum theory. The youngest star of the first conference in 1911 had been Albert Einstein, then emerging from obscurity and quick to point out the prejudices of older, more conservative minds. In 1927, Einstein was the uncrowned king of physics and entering middle age, still a popular and unassuming figure but showing signs of crustiness and disillusion. He was ploughing his own furrow, seeking a unified theory of gravity and electromagnetism without assuming that quantum mechanics was correct. Now it was Einstein who seemed inflexible and backward-looking.

The conference was to become a landmark in physics – the place where Einstein first publicly articulated his unease with quantum mechanics but failed to dent the confidence of Bohr and his younger colleagues. There is no sign of the lively conference atmosphere in the famous photograph taken outside the building where the sessions took place: the twenty-nine conference delegates all look expressionless, as though they are posing for a communal passport photograph. Einstein sits at the centre of the front row, with Dirac standing behind his right shoulder. Dirac was so proud of this photograph that, for once succumbing to vanity, he prompted the University of Bristol's physics department to have it framed and mounted on one of their walls.[23] This portrait, a dismal memento, was for decades the best visual evidence available of the meeting, but in 2005 more clues about the atmosphere of the meeting appeared, with the release of a home movie of the delegates during a break between the lectures.[24] What is most striking about this two-minute clip is the delegates' cheerfulness. Marie Curie, the only woman in the group, does a fetching pirouette; the beaming Paul Ehrenfest waggishly pokes out his tongue at the camera. Dirac, the youngest delegate, looks relaxed and happy as he talks with Max Born.

Heisenberg later remembered that the most intense discussions took place not during the conference sessions but over meals at the delegates' nearby Hotel Britannique, near the site of today's European Parliament.[25] At the epicentre of the debates about quantum theory were Bohr and Einstein's disagreements about Heisenberg's uncertainty principle, which Bohr defended successfully against Einstein's repeated onslaughts. Most of their colleagues were fascinated to hear the two men lock horns, but Dirac was an indifferent bystander:

I listened to their arguments, but I did not join in them, essentially because I was not very much interested [. . .] It seemed to me that the foundation of the work of a mathematical physicist is to get the correct equations, that the interpretation of those equations was only of secondary importance.[26]

Dirac and Einstein were poles apart, and neither was comfortable speaking the other's language. Dirac was twenty-three years younger, and his awe rendered him even more shy than usual. But probably the main reason why they did not engage was that their approaches to science contrasted so sharply, partly because they responded so differently to philosophical matters. They agreed that science was fundamentally about explaining more and more phenomena in terms of fewer and fewer theories, a view they had read in Mill's *A System of Logic*. Yet, whereas Einstein remained interested in philosophy, for Dirac it was a waste of time. What Dirac had retained from his reading of Mill, bolstered by his studies of engineering, was a utilitarian approach to science: the salient question to ask about a theory is not 'Does it appeal to my beliefs about how the world behaves?' but 'Does it work?'

At the conference, Dirac made his first recorded outburst on topics outside physics – religion and politics. Some four decades later, Heisenberg described the event, which took place one evening in the hotel's smoky lounge, where some of the younger physicists were lying around on the chairs and sofas. Dirac's youthful outspokenness needed to be indulged, the elderly Heisenberg said: 'Dirac was a very young man and in some way was interested in Communistic ideas, which of course was perfectly all right at that time.'[27] Most vivid in Heisenberg's memory was a rant from Dirac about religion, triggered by a comment about Einstein's habit of referring to God during discussions about fundamental physics. Like many of Heisenberg's accounts of incidents in the 1920s, this one is implausibly detailed – it consists of two speeches of several hundred words, quoted as if his memory were word perfect – but it is consistent with other accounts of Dirac's views. According to Heisenberg, Dirac thought religion was just 'a jumble of false assertions, with no basis in reality. The very idea of God is a product of the human imagination.' For Dirac, 'the postulate of an Almighty God' is unhelpful and unnecessary, taught only 'because some of us want to keep the lower classes quiet'. Heisenberg wrote that he objected to Dirac's judgement of religion because 'most things in this world can be abused – even the Communist ideology which you recently propounded'. Dirac was not to be deflected. He disliked 'religious myths on principle' and believed that the way to decide what was right was 'to deduce it by reason alone from the situation in which I find myself: I live in a society with others, to whom, on principle, I must grant the same rights I claim for myself. I must simply try to strike a fair balance.'[28] Mill would have approved.

During Dirac's assault on religion, Pauli had been uncharacteristically silent. When asked what he thought, he replied, 'Well our friend Dirac, too, has a religion, and its guiding principle is "There is no God and Dirac is his prophet".' It was an old joke, but everyone laughed, including Dirac.[29] The opinions he expressed here, with uncharacteristic forwardness, were entirely in keeping with Kapitza's views and would not have drawn comment from any of the intellectuals who were flirting with Bolshevism. Although Dirac never put any of his political views on paper, it was clear from his actions in the coming decade where his sympathies lay.

During the Solvay Conference, Dirac gave a talk on his new field theory of light. He annotated his draft script with rewordings and other changes in every paragraph – more than any other talk he gave in his entire life – indicating that he was on edge.[30] Afterwards, he heard that his idea had been taken up and extended in a way he could have easily foreseen. Pascual Jordan, working with Eugene Wigner, had produced a field theory of the electron to complement Dirac's theory of the photon. Although Jordan and Wigner's mathematics was similar to Dirac's, their theory did not appeal to Dirac, who could not see how their symbols corresponded to things going on in nature. Their work looked to him like an exercise in algebra, though later he realised he was wrong; his mistake stemmed from his approach to theoretical physics, which was 'essentially a geometrical one and not an algebraic one' – if he could not visualise a theory, he tended to ignore it.[31]

That was not the only surprise Dirac received in the lecture hall. Shortly before the beginning of a lecture, Bohr asked Dirac what he was working on. He replied that he was trying to find a relativistic quantum theory of the electron. Bohr was baffled: 'But Klein has already solved this problem,' he said, referring to the Swedish theoretician Oskar Klein.[32] The lecture began before Dirac could reply, so the question hung in the air, where it remained: Bohr and Dirac did not have the chance to talk further about it before the conference dispersed. Another three months would elapse before Bohr appreciated his error when he read Dirac's wondrous solution to the problem.

Eleven

[T]he true and the beautiful are akin. Truth is beheld by the intellect
which is appeased by the most satisfying relations of the intelligible:
beauty is beheld by the imagination which is appeased by the most
satisfying relations of the sensible.
  JAMES JOYCE, *A Portrait of the Artist as a Young Man*, 1915,
  Chapter 5

Dirac always felt out of place at fancy college dinners. Rich food,
vintage wines, antiquated formalities, florid speeches, the fetid
smoke of after-dinner cigars – all were anathema to him. So he was
probably not looking forward to the evening of Wednesday, 9
November 1927, when he was to be one of the toasts of a dinner to
celebrate the election of three new Fellows to St John's College. He
was now certifiably a 'first-rate man', with a permanent seat at the
college's high table and the freedom to gather after dinner with his
colleagues in their grand, candle-lit Combination Room, completed
in 1602. In Hall, beneath the portrait of Lady Margaret Beaufort,
Dirac celebrated his election to the fellowship in the traditional way,
by consuming an eight-course meal that included oysters, a consommé, cream of chicken soup, sole, veal escalope and spinach,
pheasant with five vegetables and side salad, and three desserts. For
him, the meal was not so much a celebration as a penance.[1]

After the dinner, Dirac walked to his rooms, close to the Bridge of
Sighs, a Gothic stone structure that crosses the river Cam in a brief
undulation, leaving just enough room underneath for the punters.
He probably went straight to bed, as his aim was always to be fresh
for the morning, when he did his best work. His study was devoid of
decoration, with only a folding desk of the sort used by school-
children, a simple chair, a coal fire and 'a very ancient settee', as one
visitor described it.[2] He worked at his little desk like a schoolboy in
an empty classroom, writing in pencil on scraps of paper, sometimes
pausing to erase an error or to consult one of his books.[3] Now that
he was a Fellow, he had a manservant (a 'gyp') on hand day and
night.

In these austere but comfortable surroundings, Dirac made his most famous contribution to science. St John's had created the best environment imaginable for him. He could work all day, taking breaks only to fulfil his modest lecturing duties, give the occasional seminar and visit the library.

He was now preoccupied with a single challenge: to find the relativistic equation that describes the electron.[4] He believed that the electron was 'a point particle' but, like theoreticians, could not understand why it had not one but two states of spin. Several other physicists had suggested candidate equations – all of them contrived and ungainly – and Dirac was not satisfied with any of them, including the one by Klein that Bohr believed had solved the problem. Dirac was sure Klein's theory was wrong, as it predicted, absurdly, that the chance of detecting an electron in a tiny region of space-time is sometimes *less* than zero.

Dirac knew that it was impossible to deduce the equation from first principles and that he would find it only through a happy guess. But what he could do was to narrow the options, by setting out the characteristics the equation *must* have and the characteristics it *ought* to have. Rather than tinker with existing equations, he took the top-down approach, trying to identify the most general principles of the theory he was seeking, before going on to express his ideas mathematically. The first requirement was that the equation conformed to Einstein's special theory of relativity, treating space and time on an equal footing. Second, the equation must be consistent with his beloved transformation theory. Finally, when the equation describes an electron moving slowly compared with the speed of light, its predictions must resemble extremely closely ones made by ordinary quantum mechanics, which had already proved its worth.

Those were useful constraints, but there was still too much room for manoeuvre. If he stuck to them, Dirac could still have written down any number of equations for the electron, so he needed to use his intuition to narrow the possibilities. Believing that the relativistic equation would be fundamentally simple, he thought it most likely that the equation would feature the electron's energy and momentum just as themselves, not in complicated expressions such as the square root of energy or momentum squared. Another clue came from the way he and Pauli had independently found to describe the spin of the electron, using matrices that each consisted of four numbers arranged in two rows and two columns. Might these matrices feature in the equation he was seeking?

Dirac tried out one equation after another, discarding each one as soon at it failed to conform to his theoretical principles or to experimental facts. It was not until late November or early December 1927 that he hit on a promising equation, consistent with both special relativity and quantum mechanics. The equation looked like nothing theorists had ever seen before, as it described the electron not using a Schrödinger wave but using a new kind of wave with *four* interconnected parts, all of them essential.

Although the equation had an appealing elegance, that would count for nothing if it did not relate to real electrons. What did the equation have to say, for example, about the spin of the electron and its magnetic field? If his equation contradicted the experimenters' observations, he would have had no choice but to abandon it and start all over again. But there was no need for that. In a few pages of calculations, Dirac showed that he had conjured something miraculous: his equation described a particle not only with the mass of an electron but with precisely the spin and magnetic field measured by experimenters. His equation really did describe the electron so familiar to experimenters. Even better, the very existence of the equation made it clear that it was no longer necessary to tack on the electron's spin and magnetism to the standard description of the particle given by quantum theory. The equation demonstrated that if experimenters had not previously discovered the spin and magnetism of the electron, then these properties could have been *predicted* using the special theory of relativity and quantum mechanics.

Although Dirac apparently showed his usual Trappist calm, he was jubilant. In a few squiggles of his pen, he had described the behaviour of every single electron that had ever existed in the universe. The equation was 'achingly beautiful', as theoretical physicist Frank Wilczek later described it: like Einstein's equations of general relativity, the Dirac equation was universal yet fundamentally simple; nothing in it could be changed without destroying its power.[5] Nearly seventy years later, stonemasons carved a succinct version of the Dirac equation on his commemorative stone in Westminster Abbey: $i\gamma.\partial\psi = m\psi$. When set out in full, in the form he originally used, the equation looked intimidating even to many theoreticians simply because it was so unusual, not that this would have disturbed Dirac: all that mattered to him was that it was based on sound principles and that it worked. It might even have crossed his mind that he had done something that John Stuart Mill had articulated as one of the aims of science – to unify disparate theories to explain the widest possible range of observations.

When Dirac was an old man, younger physicists often asked him how he felt when he discovered the equation.[6] From his replies, it seems that he alternated between ecstasy and fear: although elated to have solved his problem so neatly, he worried that he would be the latest victim of the 'great tragedy of science' described in 1870 by Thomas Huxley: 'the slaying of a beautiful theory by an ugly fact'.[7] Dirac later confessed that his dread of such an outcome was so intense that he was 'too scared' to use it to make detailed predictions of the energy levels of atomic hydrogen – a test that he knew it had to pass.[8] He did an approximate version of the calculation and showed that there was acceptable agreement but did not go on to risk failure by subjecting his theory to a more rigorous examination.

During November and December, he shared with no one the pleasure he took in his discovery or his occasional panic attacks. Not a single significant letter or record of a conversation with anyone exists from those months. He broke his silence only before he set off to Bristol for the Christmas vacation when he bumped into his friend Charles Darwin, a grandson of the great naturalist and one of Britain's leading theoretical physicists. On Boxing Day, in a long letter to Bohr, Darwin wrote: '[Dirac] has now got a completely new system of equations for the electron which does the spin right in all cases and seems to be "the thing".'[9] That was how Bohr learned that the remark he had made to Dirac at the Solvay Conference – that the problem of finding a relativistic equation for the electron had already been solved – was completely wrong.

Fowler sent Dirac's paper 'The Quantum Theory of the Electron' to the Royal Society on New Year's Day 1928, and a month later sent off a second paper that cleared up a few details. While the first paper was in press, Dirac wrote to Max Born in Göttingen, not mentioning his new equation except in a ten-line postscript, where he spelt out the reasoning that had led to it. Born showed these words to his colleagues, who regarded the equation as 'an absolute wonder'.[10]

Jordan and Wigner, who were working on the problem that Dirac had solved, were flabbergasted.[11] Jordan, seeing his rival walk off with the prize, sank into depression.

When the equation appeared in print at the beginning of February, it was a sensation. Though most physicists struggled to understand the equation in all its mathematical complexities, the consensus was that Dirac had done something remarkable, the theorist's equivalent of a hole in one.[12] For the first time in his career, he had shown that he was capable of tackling one of the toughest problems of the day and beating his competitors to the solution, hands down. The American theoretician John Van Vleck later likened Dirac's explanation of electron spin to 'a magician's extraction of rabbits from a silk hat'.[13] John Slater, soon to be a colleague of Van Vleck's at Harvard, was even more effusive: 'we can hardly conceive of anyone else having thought of [the equation]. It shows the peculiar power of the sort of intuitive genius which he has possessed more than perhaps any of the other scientists of the period.'[14]

Even Heisenberg, more confident than ever after his recent appointment to a full professorship in Leipzig, was taken aback by Dirac's coup. One physicist later recalled Heisenberg speaking of an English physicist – unquestionably Dirac – who was so clever that it was not worth competing with him. Heisenberg was, however, concerned that despite the equation's beguiling beauty, it might be wrong: he was one of many who underlined a problem that Dirac had pointed out in his first paper on the equation – it made a strange prediction about the values of energy that an electron can have.

The background to the problem with the equation was that, like time, energy is a relative quantity, not an absolute one. The energy of motion of a free electron – one that has no net force acting on it – can be defined as zero when the particle is stationary; when the particle gathers speed, its energy of motion is always positive. Dirac's problem was that his equation predicted that, in addition to perfectly sensible positive energy levels, a free electron has *negative* energy levels, too. This arose because his theory agreed with Einstein's special theory of relativity, which said that the most general equation for a particle's energy specifies the *square* of the energy, E^2. So if one knows that E^2 is, say, 25 (using some chosen unit of energy), then it follows that the energy E could be either +5 or −5 (each of them, when multiplied by itself, equals 25). So, Dirac's formula for the energy of a free electron predicted that there were *two* sets of energy values – one positive, the other negative. In classical physics, the negative-energy ones could be ruled out, simply because they are meaningless, but this cannot be done in quantum mechanics as it predicts that a positive-energy electron could always jump into one of them.

No one had observed such a jump, so the Dirac equation was in serious trouble. Despite this unsightly canker, however, the consensus was that his theory of the electron was a triumph. Yet Dirac seemed to take no pleasure from his success and showed none of the relief and elation that Einstein had demonstrated after he published his equation of general relativity. Dirac's younger colleague Nevill Mott later described the extent of Dirac's detachment from his fellow physicists in Cambridge. Mott was – like hundreds of other theorists – concentrating not on extending quantum mechanics but on applying it.

According to Mott, no one in the Cambridge mathematics department knew anything about Dirac's equation until they read his paper in the library. Dirac was, Mott said, passive and forbidding, the kind of expert no one quite dares to consult. Dirac did not seem to appreciate the narrowness of his understanding of companionship: he liked to be among fellow physicists, when they were friendly – as they were in Bohr's Institute – but felt no obligation to talk to them about his work or even to disclose his first name. Charles Darwin had known him for six years before writing him a postcard asking him about his signature: 'What does P. A. M. stand for?'[15]

Whereas at Copenhagen and Göttingen there were many premier-league quantum physicists, Fowler and Darwin were the only ones in Cambridge, so Dirac believed that it was his duty to deliver his seminars and lectures on the basics of quantum mechanics.[16] But that, in his view, was where his departmental teaching obligations ended. But, surprisingly for a young research scientist, he did agree to write a textbook on quantum mechanics, scheduled to be the first publication in the 'International Series of Monographs on Physics', edited by Kapitza and Fowler. The series was the brainchild of Jim Crowther, the science reporter of the *Manchester Guardian*, the unofficial writer-in-residence at the Cavendish Laboratory and the only journalist Dirac regarded as a friend. A passionate Marxist, Crowther had joined the Communist Party in 1923 and managed to be close to both Bernal and Rutherford – sworn enemies – making the most of

the talents and influence of each of them.[17] By subtly cultivating relationships with all the finest young scientists in the Cavendish, including Dirac, Crowther became an influential bit-part player in the
emerging group of radical scientists in Cambridge. One of his
strengths was his sensitivity: he will have realised quickly that, to
make friends with the great young theoretician, he had to overcome
Dirac's reluctance to have anything to do with importuning journalists. Dirac just wanted to be left in peace.

Dirac's family knew nothing of his equation. For Charles, always
keen to find out about Dirac's work, his son's unwillingness to share
his science was cruel. In April 1928, when he read an anonymous
article in *The Times* about quantum physics, Charles may have been
discouraged by the conclusion: 'Far past is the day when the scientist
could talk to the layman as man to man [. . .] the world loses much
when science has got into such deep waters that only a Channel
swimmer can follow it.'[18] When Charles pressed his son to explain
something of his new physics – as he surely did – Dirac almost certainly gave his usual response of shaking his head or remarking
unhelpfully that the new quantum theories 'are built up from physical concepts which cannot be explained in words at all'.[19] Although
Dirac used his visual imagination to think about quantum mechanics, he declined every request to describe images of the quantum
world. As he would later remark: 'To draw its picture is like a blind
man sensing a snowflake. One touch and it's gone.'[20]

To judge from the letters Dirac received from his mother, relations
between her and Charles had settled down now she was spending
more time out of the house. She went to talks on Tennyson's poetry,
saw shows at the Hippodrome theatre with Charles and Betty and
visited the cinema, including a trip to see one of the last great silent
films, *Ben Hur*. But the Dirac family's favourite novelty was the
motor car, the most exciting of the new mass-produced technological
innovations. One of Charles's private tutees owned a car and treated
the Dirac family to afternoon joyrides to the coast and to countryside
teashops, keeping to the speed limit of 20 mph. Images of trips like
these – carefree families, cutting loose from worldly concerns for a
day – symbolised the prosperity of Britain in the third quarter of the
1920s. For the majority, life had never been better.

But when Dirac was not at home, his mother's life was empty.

Always in search of a plausible excuse to visit him, she invited herself to Cambridge in mid-February to see the Lent boat races, sheepishly asking if he had the time to see her when she was in town ('I shall be dressed quite nicely & shall not be any trouble').[21] He often ignored such requests, but this time he agreed, and she arrived in a foggy Cambridge at lunchtime to spend a few hours talking with her son, who apparently gave no sign that he was living through one of the most exciting times of his life and that some of his peers were beginning to talk of him as the heir to Newton.

Dirac appeared also to resemble Newton in having no interest in forming romantic relationships with women. Many of Dirac's colleagues had the impression that he was frightened of women of his own age and they could scarcely imagine that he would ever marry. But he did have a close friendship with one woman, the fifty-six-year-old mother of his friend Henry Whitehead, a promising mathematician at Oxford University. Isabel Whitehead, a tall, solidly-built Scot, was the wife of the Right Reverend Henry Whitehead, nineteen years her senior and formerly the Bishop of Madras in India. The couple had spent almost twenty years living there, before retiring to the UK in 1923. Among her fellow expatriates, Mrs Whitehead was notorious: according to an authoritative account of the Christian community in India, she was imperious 'even by the domineering standards of the many British memsahibs'.[22]

The Whiteheads lived in a half-wood, half-brick cottage in Pincent's Hill, near Reading, about three hours' drive from Cambridge. Always accompanied by their dogs, they led a leisurely life, taking just an hour or two each day to run a small farm with pedigree Guernsey cattle and a few chickens. Both Isabel and Henry were Oxford-educated mathematicians, but it seems from Mrs Whitehead's letters that the two of them talked less about science with Dirac than about other matters, especially Henry's enthusiasm for cricket and their adventures in India, including the week they spent in their home entertaining Gandhi. In the coming years, Mrs Whitehead's correspondence with Dirac also makes it clear that she robustly challenged his atheism and that he trusted her with his most private thoughts about his family. Pincent's Hill became a favourite weekend retreat for him and Mrs Whitehead became his second mother, giving him not only support and affection but also something his own mother could not provide – intellectual stimulation.

During the early spring of 1928, Dirac was planning his next journey. His six-month itinerary would begin in April and take him back to Bohr's Copenhagen and Ehrenfest's Leiden, on to Heisenberg's Leipzig and Born's Göttingen, and finally his first visit to Stalin's Union of Soviet Socialist Republics. Dirac had heard much about this country; now he would be able to judge for himself.

Twelve

See how physical science, which is Reason's trade
And high profession, booketh ever and docketeth
All things in order and pattern.
    ROBERT BRIDGES, *Testament of Beauty*, 1929

Paul Ehrenfest could be a moody and demanding colleague, but he was a charming and generous host. In April 1928, when he realised that he would not be able to greet Dirac at Leiden railway station at the beginning of his visit, Ehrenfest arranged for a phalanx of his assistants to be waiting for him on the platform when his train steamed in shortly after 10 p.m. The problem was that none of them knew what Dirac looked like. Ehrenfest's solution was to ensure that, outside every train door facing the platform, there was a student waving a reprint of 'The Quantum Theory of the Electron'. The plan worked.[1]

One member of the welcoming party was Igor Tamm, a thirty-two-year-old Soviet theoretician, soon to become one of Dirac's closest companions. Tamm was famously restless: in group photographs, while others appeared in sharp definition, he would be a blur.[2] A Marxist even before he went to university, he joined the Social-Democratic Workers' Party in 1915 and, during the subsequent years in Moscow, Kiev, Odessa and Elizavetgrad, studied science while being a part-time activist for the Bolsheviks. He tired of their fanaticism and, when they declared all other political parties illegal in the summer of 1918, was concentrating on science. He became the first Soviet theoretician to use quantum mechanics.[3] In January 1927, he arrived in Leiden and, a year later, electrified by the Dirac equation, was looking forward to meeting its discoverer. Tamm wrote to his wife in Moscow that he wanted to see if there was any truth in rumours that 'it costs a tremendous effort to get a word from [Dirac], and that he talks only to children under ten'.[4]

The two men soon clicked. In Tamm, Dirac had found another intellegent and entertaining Russian extrovert; in Dirac, Tamm found a companion who was surprisingly agreeable, provided he was under no pressure to speak. The two men spent the spring afternoons strolling around the town's cobbled streets, watching the traffic on the interlocking network of canals and occasionally walking out to the nearby tulip fields.[5] Tamm taught Dirac to ride a bicycle, Dirac taught Tamm physics, and they talked about matters outside science, probably including politics and Tamm's favourite hobby of mountain climbing. Tamm was humbled by Dirac's erudition: 'I feel like a little child next to him,' he wrote to his wife.[6]

As was customary for visitors to Leiden, Dirac gave a series of lectures. He had much improved his technique as a public speaker: when he strode towards the blackboard, he seemed to change from being a pitiful wallflower to the Demosthenes of quantum mechanics. Standing quite still, he looked into the eyes of his audience and talked plainly and articulately, with the force of an advocate, not letting a pause or hesitation break his rhythm. He did not read from a prepared text but knew exactly what he wanted to say; once he had decided on the clearest way of expressing an idea, he would not deviate from it, from one lecture to another. When Ehrenfest asked for further explanation, Dirac would respond by repeating what he said, almost word for word.[7]

In mid-June 1928, Dirac moved on with Tamm to Leipzig to spend a week at a conference co-organised by Heisenberg, who was agonising about the Dirac equation. Darwin and others had demonstrated that it perfectly reproduced previously successful formulae for atomic hydrogen's energy levels, but this news cut no ice with Heisenberg. He was troubled by the equation's absurd prediction that a free electron can have negative energy – and it had become clear that no subtle tinkering with the equation could change it. For Dirac, this was simply the next problem to be addressed. For Heisenberg, it was evidence that the equation was sick. A month after Dirac departed from Leipzig, Heisenberg wrote to Bohr: 'I find the present situation quite absurd and on that account, almost out of despair, I have taken up another field, [trying to understand magnetism].'[8] A month later, Heisenberg was even more depressed when he wrote to Pauli: 'The saddest chapter of modern physics is and remains the Dirac theory.'[9] Dirac knew Heisenberg's criticisms were well founded and that the onus was on him to demonstrate that the theory was more than a beautiful mirage.

Among the scientists Dirac met for the first time in Leipzig was

Heisenberg's student Rudolf Peierls, just turned twenty-one. Wiry, bespectacled and with a pronounced overbite, Peierls oozed vitality and ambition. His professors asked him to take Dirac to the opera, a challenge that his guest's Cambridge colleagues regarded as all but impossible. They could scarcely imagine him sitting through any kind of drama: the artifice, the focus on speech or lyrics and the often contorted plotting would surely have no appeal to his literal mind. Decades later, Peierls could not remember the play or his guest's reaction to it but squirmed at the thought of Dirac's insistence on following the English custom of taking his hat with him to the performance, pointedly refusing to follow the German practice of leaving headwear in the theatre cloakroom. Peierls, whose formal Prussian education had given him a strong sense of politesse, found Dirac's behaviour mortifyingly crude.[10] Dirac, probably oblivious of his colleague's discomfiture, often behaved like this: he was a stickler for English conventions of courtesy and saw no reason to deviate from them in other countries. Flexibility was not his forte.

After the conference, Dirac travelled with Tamm to Göttingen. Its theoretical physics department was losing its edge as its leader, Max Born, struggled to maintain his momentum. Overworked, worried that younger and fresher minds were leaving him behind, depressed by marital problems and the Nazis' 'blood and soil' anti-Semitism, he slid into a nervous breakdown.[11] His colleague Jordan was openly a conservative nationalist but in private was writing reactionary articles in the journal *Deutsches Volkstrum* ('German Heritage'), under the cover of a pseudonym.[12]

Göttingen was, however, still on the itinerary of every young theoretician. During this visit, Dirac began his long friendship with two other visitors, who embodied his taste for the company of both introverts and extroverts and who were to lead him to his first close relationships with women of his own age. At the flamboyant extreme was George Gamow, a Russian theoretician two years Dirac's junior, destined to be the court jester of quantum physics. Variously nicknamed Johnny, Gee-Gee and (by Bohr) Joe, he was a six-foot three-inch, 220-pound giant and close to being Dirac's polar opposite: loquacious, a passionate smoker and drinker, relentlessly jocular.[13] Shortly before his visit to Göttingen, he had made his name by being one of the first to use quantum mechanics to explain the type of radioactive decay in which an alpha particle can be ejected from types of atomic nuclei (impossible, according to classical mechanics). Dirac, probably to Rutherford's frustration, had attended many Cavendish seminars about new findings in nuclear physics but showed no interest in trying to understand them.[14] As theoreticians, Gamow and Dirac were entirely different: Gee-Gee did not try to come up with fundamental new ideas but preferred to apply ones discovered by others. Yet the two men got along well and often dined together, Dirac listening expressionlessly as his new friend told of how he had learned Euclidean geometry under artillery bombardment and other such stories, most of them more impressive for their colour than their accuracy.[15]

At the other end of the personality spectrum was Eugene Wigner, who had recently arrived in Göttingen after spending years with Einstein in Berlin, having switched to physics after being trained as an engineer. The scion of a wealthy Jewish family, Wigner and his two sisters had been raised by a governess in a grand apartment in one of the most exclusive residential areas of Budapest, overlooking the Danube. He loved to reminisce about his boyhood home: the formal family dinners, the scurryings of the two uniformed servant girls, the scent of freshly cut roses.[16] Unlike Dirac, the young Wigner was politically alert and acutely aware of the instability of his country. Since the break-up of the Austro-Hungarian Empire in 1918, Hungary had been through a bloody Bolshevik revolution led by Béla Kun and the White Terror organised by nationalist and anti-Semitic forces. Wigner was fearful of the future of the country, then under Admiral Horthy's authoritarian regime.

Despite all the political upheavals, Wigner had an exceptionally fine school education in mathematics and science, even more thorough than Dirac's. Historians still debate why Budapest in the early twentieth century produced so many intellectual innovators, including John von Neumann, whom Dirac would later rate as the world's finest mathematician, and Wigner's friends Leó Szilárd and Edward Teller, both to do important research into the first nuclear weapons.[17] The success of this cohort of Hungarians is partly due to their education, shortly after the war, in Budapest's excellent high schools and partly to the vibrancy and ambition of the city's Western-focused culture.[18]

Wigner was one of the shyest and most uncommunicative of the quantum physicists but, compared with Dirac, he was gregariousness itself, so conversation during their evening meals together was probably strained. They had to find a common language – Dirac did not know Hungarian, hated to speak French and spoke fractured German with a bitumen-thick accent, while Wigner's English was weak, and he liked to converse in German or French. They probably settled on German. No record remains of the details of their early conversations, but it is likely that Wigner mentioned his politics and youthful experiences of anti-Semitism: since he was sixteen, he had followed his father in ideologically opposing Communism, and his views had hardened a year later during Kun's regime, in which his father was thrown out of his job as director of a tannery.[19] For a few months, the Wigners had fled to Austria but returned after the Communists were overthrown.

Dirac would have been content to listen to as much of Wigner's life story as he was willing to tell. But when Wigner turned his attention to physics, he quickly saw that Dirac had no interest in sharing his thoughts and ideas. The moment Wigner began to probe, Dirac withdrew into himself like a frightened hedgehog.[20] Igor Tamm knew how to avoid this kind of defensiveness: keep conversation to a functional minimum, avoid personal questions and never risk wasting breath on trivialities. Tamm and Dirac's relationship flourished partly because they had complementary talents: intellectual leadership was provided by Dirac, while the social impetus came from Tamm. It was he who introduced Dirac to what would be one of the greatest pleasures of his young life: mountain climbing. In one long trip east, the two journeyed out to the wooded Harz – ablaze with fireflies in the evenings – and they climbed the challenging peak of Mount Brocken (1,142 metres).[21] Dirac was smitten: apart from equations, nothing did more to stir his sense of beauty than mountains.[22]

At the end of July 1928, Dirac was preparing for his first visit to Russia, a two-month stay that combined the chores of lecturing with the pleasure of relaxing with Kapitza. Dirac's mother was fearful: 'If you go to Russia, do take care of yourself. We hear such dreadful accounts of the Bolshevists in the papers. There seems to be no law and order anywhere. I expect you know more about the facts than we do, though, as you are so much nearer.'[23] Since 1918, the British press had reported on the Soviet regime's growing repressiveness, which increased with the rise of Stalin to absolute power in 1926. The British Government did not officially recognise the Soviet Union, but profitable trade between the countries was easing relations between them, culminating in the Labour Prime Minister Ramsay MacDonald's restoration in 1929 of full diplomatic relations.[24]

After his arrival in Leningrad on 5 August, Dirac's hosts introduced him to caviar, one of the few luxury foods for which he had a taste. Dirac blossomed in Russia – the scenery, the architecture, the museums and the art galleries – as he wrote in a long and chatty letter to Tamm:

I spent the first two days in Leningrad with Born and his [Göttingen colleague] Pohl and we saw the sights and visited the Hermitage and the Museum of Russian Art and the Natural History Museum and also the Roentgen Institute [for physics research] [. . .] I found Leningrad a very beautiful place, and was more impressed by it than by any other town during the journey, particularly as I came up the river in the steamer and first saw the large number of churches, with their gilded domes, quite different from anything I had ever seen [. . .].[25]

Moscow still resembled the city of Anna Karenina, with its squat wooden houses, multicoloured cupolas, horse-drawn cabs driven around the sprawl of zigzag streets by peasants in blue robes, bearded traders sipping vodka and eating cucumbers in the Slovenski Bazaar.[26] Dirac was there to attend the no-expense-spared Congress of Russian Physicists, at his hosts' expense. Physicists in the Soviet Union had been quick to realise the importance of quantum mechanics and wanted to learn from the innovators in western Europe. Of the one hundred and twenty physicists who attended the Congress, about twenty were foreign. Dirac was the star of the occasion, but he arrived in Moscow too late to give his talk, scheduled for the opening session. When he should have been giving his presentation, he was walking around one of the royal palaces on the outskirts of the city; in the evening, he went to a performance of Japanese theatre. The next day, Dirac went with the conference delegates to the Kremlin before setting off alone to walk the streets until sundown.

The venue for the second part of the Congress was a steamer that sailed down the Volga to Stalingrad. During the week-long cruise, Dirac gave a talk on his theory of the electron and met the leaders of Soviet physics, including his admirer Lev Landau, a twenty-year-old graduate student, soon to be his country's greatest theoretician – the most accomplished but least mature. Mangy and undernourished, he was so tall that in most company people could see his long, thin face standing out, topped with dark wavy hair that was piled on the right of his head like a burnt crest of meringue. As a critic, he was so aggressive that he made Pauli look demure; as a colleague, so socially inept that he made Dirac look suave.

After the Congress, Dirac took a two-day train journey to the Caucasus. He stayed with Kapitza and joined a party of sightseers for a six-hour hike up a glacier near Vladikavkas. Dirac described his adventures in a letter to Tamm but did not mention that, during his time with Kapitza, he experienced an incident that was, in some way, his sexual awakening.[27] Forty-five years later, he remembered that he first saw a naked young woman in the Caucasus: '[she was] a child, an adolescent. I was taken to a girls' swimming pool, and they bathed without swimming suits. I thought they looked nice.' He was twenty-six years old.

Dirac was in no hurry to return to Bristol: the journey took him almost a month.[28] The disparity between the excitement of his work and the dreariness of his home life had never been so stark. He was lionised by many of his colleagues, he was financially independent, and he was benefiting from international travel at a time when it was a luxury. Charles, Flo and Betty, on the other hand, were locked in their routine and left their hometown only rarely. Betty was happy to do nothing at all when she was not looking after her new dog; Charles was overworked and run down; Flo was trying to make the most of every opportunity to leave the house. At her elocution classes, she wrote and practised giving speeches, including one opposing the notion that there might one day be a woman prime minister. She rehearsed her speech on the Bristol Downs, beginning with the flourish 'I rise to oppose the motion of a woman prime minister – to oppose most decidedly and definitely.' For one thing, Flo argued, women do not have sufficiently strong constitutions to take on such a responsibility: 'As regards physique – women today are wonderful: but none can say when a woman may faint! None when she may scream! Is it becoming for a Prime Minister to suddenly fall to the ground, or to burst into hysterics at a crucial moment?'[29]

Although Flo was not in the vanguard of feminism, Dirac knew that underneath his mother's apparent submissiveness lay stoicism and an independence of spirit. These qualities would, over the next three years, be tested to breaking point.

When Dirac returned to Cambridge in October 1928, he knew that the onus was on him to cure the sickness of his theory of the electron. Somehow, he needed to find a rational explanation for the negative-energy states which were undermining confidence in the Dirac equation; some of his colleagues were becoming worried that the equation might not be right after all.[30]

That autumn, he was, unusually, working on several projects at the same time: his hole theory, his textbook and a brief paper on one of his favourite subjects – the relationship between classical mechanics and quantum mechanics. The paper was based on the ultra-rigorous work of von Neumann, who had derived one result that caught Dirac's eye. Von Neumann had found a way of describing the overall behaviour of an enormously large number of non-interacting quantum particles, when nothing is known about their individual behaviour. It turned out, surprisingly, that the statistical description given by quantum mechanics is just as simple as the account given by classical mechanics; in both, the behaviour of the individual particles averages out to a smooth overall pattern, just as the behaviour of a swarming crowd can be described without referring to any of its individuals. In this bijou paper, Dirac developed von Neumann's ideas and laid bare the precise analogy between the classical and quantum understandings of vast numbers of particles. This was a divertimento composed during a holiday from fixing his troublesome symphony.

In those politically tranquil times, the favourite topic of conversation in Cambridge was poetry.[31] The eighty-five-year-old poet laureate Robert Bridges had written the most talked-about poem of the year, *A Testament to Beauty*, 5,600 lines about the nature of beauty. It is now read only rarely, but then it struck a chord with tens of thousands of lay readers and some literary critics, including one in the *Cambridge Review* who described it as 'a high philosophical explanation of Keats's "Beauty is truth, Truth beauty"'.[32] To some extent,

Bridges was reacting against modernist art – such as Arnold Schönberg's atonal music, Picasso's cubism, Eliot's fragmented poetry. Bridges sought beauty and found it not only in music, art and nature but also in science, food and even in football matches. Dirac knew, too, that beauty was about much more than art and nature. He had seen it in Einstein's equation for the general theory of relativity and he now had an equation of his own that was no less of a contribution to aesthetics. But aesthetic judgements like that count for nothing in science if a theory fails to agree with experiment. Unless someone could explain the meaning of the negative-energy solutions to the Dirac equation, it was doomed to be remembered only as just another scientific fad.

A few of Dirac's colleagues in Cambridge would not have been distraught if fortune had clipped his wings: his ascending reputation had led, inevitably, to envy. No longer were the two leading lights of the university's experimental and theoretical physics cited as Rutherford and Eddington, but as Rutherford and Dirac. Eddington's star was waning, and he knew it. Meanwhile, the old guard of Cambridge physics looked pitifully out of touch. The proud Irishman Sir Joseph Larmor, holder of the most prestigious chair in Cambridge, the Lucasian Professorship of Mathematics, once held by Newton, was living in the past, unable to understand relativity theory and disdainful of quantum mechanics. He and his friend J. J. Thomson wandered the streets of Cambridge, each of them wearing a bowler hat, a black three-piece suit and an immaculate white shirt, and each wagging a stick behind his back. When they peered into one of the shop windows on Trinity Street, the two superannuated professors looked like a pair of penguins.

The two men knew that their views counted for nothing among physicists who were once their admiring students and who were now running physics. No one symbolised the new generation's ascendancy more powerfully than Dirac, but he still did not have a permanent job. He had turned down Arthur Compton's offer of a post in Chicago and had later declined an offer of a professorship in applied mathematics at Manchester University, commenting that 'my knowledge of and interest in mathematics outside my own special branch are too small for me to be competent [in such a post].'[33] If the spurned mathematicians in Manchester found his modesty hilarious, Dirac would have been uncomprehending, as he was simply being candid. As Mott said:

'He is quite incapable of pretending to think anything that he did not really think.'[34]

If Dirac and Fowler were away, Cambridge University would struggle to teach quantum mechanics, as Harold Jeffreys virtually admitted when he wrote to Dirac in March 1929, pleading with him to set the questions on quantum mechanics for the summer examinations. Jeffreys and his fellow 'ignorant and philistine' faculty colleagues were in the embarrassing position of having to admit that 'the candidates know more than we [do]'.[35] Fowler led the campaign to ensure that Dirac remained in Cambridge, and he soon had some success: in June 1929, St John's College awarded Dirac a special lectureship, though it was funded for only three years.[36] Dirac's loyalty to Cambridge was to be tested, repeatedly.

As Dirac was getting nowhere with his top priority of sorting out the difficulties with his equation, he decided to devote himself to other things. In late 1929, he spent most of his time drafting his book and working on another research project, the theory of heavy atoms. This was by no means his favourite branch of physics, but it was closer to the work of the great majority of quantum theorists, who were applying the theory to complicated atoms and molecules. Dirac was, however, in no doubt that quantum mechanics would be successful:

The underlying physical laws necessary for the mathematical theory of a large part of physics and the whole of chemistry are thus completely known, and the difficulty is only that the exact application of these laws leads to equations much too complicated to be soluble.

Those words became one of the clarion calls of reductionists, who believe that complex things can be explained in terms of their components, right down to the level of atoms and their constituents. Extreme reductionism implies, for example, that quantum mechanics lies at the bottom of an inverted pyramid of questions that begins, for example, with 'Why does a dog bark?' A reductionist seeks to answer the question by understanding the chemical reactions going on inside the dog's brain, and those reactions are ultimately understood by the interactions of the chemicals' electrons, whose behaviour is ultimately described by quantum mechanics. Although popular with many scientists, the approach does not describe *how* to make the links between the layers of explanation.

In his paper, Dirac applied quantum mechanics to atoms that contain more than one electron, such as carbon atoms. Such atoms are much harder to describe than hydrogen atoms because, in every multi-electron atom, the complicated and unwieldy interactions between all the electrons have to be taken into account. Dirac found a way of describing these interactions approximately and investigated the consequences of the fact that it is impossible to detect experimentally if two of the electrons swapped places. As usual, Dirac left it to others to work out the theory's consequences: the American theoretician John Van Vleck, based in Minneapolis, quickly saw the potential of Dirac's ideas and spent years using them to explain the origin of magnetism, the various ways that atoms can bond to form molecules and the patterns of light emitted by multi-electron atoms. This was to be the main legacy of Dirac's excursion into atomic physics – his first paper on the subject, and his last.

At the end of term, he visited his family briefly and then, in what was becoming a ritual, set off on another long journey. At Southampton, on the freezing Wednesday morning of 13 March, he boarded the liner *Aquitania* with his travelling companion, Isabel Whitehead's son Henry. In the crowd at quayside was Florence Dirac, who by then had got the message: her only son wanted to spend as little time at home as he could. Just as she must have dreaded, he would be away for as long as his teaching obligations in Cambridge allowed, on his first visit to the United States of America. His reputation had preceded him.

Thirteen

[I]n England there is something very like a cult of eccentricity. [. . .]
With us [Americans], as more than one European has said, the trait is
more distinguishable nationally than individually.
GARDNER L. HARDING, *New York Times*, 17 March 1929

In every branch of science, theorists vie with experimenters to set the
agenda. Since Heisenberg's publication of his path-breaking paper
in the autumn of 1925, theoreticians had been pointing the way
ahead in physics. Yet the foundations of some of the new theoretical
ideas had not even been checked experimentally: according to
Schrödinger's quantum theory, for example, every material particle
has an associated wave, but no experimenter had been able to prove
the idea or to refute it. So there was an almost palpable sigh of relief
among quantum physicists back in early 1927 when news reached
Europe that the American experimenters Clinton Davisson and his
student Lester Germer had shown that the electron could indeed
behave like a wave. Dirac, often believed to regard experiments with
a high-minded insouciance, belied his reputation by arranging to visit
Davisson's laboratory on West Street in south Manhattan, a few
blocks from the meatpacking district, the first stop on his itinerary.[1]

This was Dirac's first sight of New York, then booming with
wealth and new technology. The Jazz Age was, according to the man
who named it, F. Scott Fitzgerald, past its 'heady middle age', though
Americans were still enjoying 'the most expensive orgy in history'.[2]
The hurried pace of American life was not at first to Dirac's taste: it
was somehow fitting that during the first night Dirac spent in his
hotel on Seventh Avenue, he was kept awake until the small hours by
revellers in an adjacent room.[3] As soon as he awoke the next day,
shortly before four o'clock in the afternoon, he realised he had
missed his appointment with Davisson. Rather than waste the late
afternoon, he spent it strolling around rush-hour midtown
Manhattan, teeming with four-square black automobiles navigating
around the skyscrapers, each of them a powerful symbol of
America's soaring prosperity.

In Davisson's laboratory the next day, Dirac saw the ingenious apparatus that first persuaded the electron to reveal its wave nature. Davisson and Germer had fired beams of electrons towards a nickel crystal and found that the number of electrons they detected at different angles had alternating peaks and troughs. These variations were impossible to understand if the electron is simply a particle: the only explanation was that the electrons behave as waves which are bent ('diffracted') by the crystal, like two waves combining on the surface of pond, forming peaks when the waves reinforce one another and troughs when they cancel each other out. Physicists had no choice but to conclude that the electron behaved sometimes like a particle and sometimes as a wave – a 'wavicle', as Eddington had dubbed it – precisely as quantum theory had supposed.

Dirac quickly headed off on his five-month journey across North America, travelling mainly on the railroad. He kept a record of his trip in terms of numbers, not words: his diary contains no descriptions of his experiences, just a cumulative record of the number of nights he had spent on a train and on board ship.[4]

After paying brief visits to Princeton and Chicago, Dirac travelled to Madison, capital of the Midwestern state Wisconsin. Like Göttingen, Madison was his sort of town, with a good university and surrounded by countryside offering plenty of opportunities for walks. He was the first foreign guest of John Van Vleck, newly appointed to the university faculty. Slightly older than Dirac, Van Vleck excelled at applying quantum physics and had no interest in the fundamentals of quantum mechanics. They spent hours together walking in the vast fields overlooking Lake Mendota, one of the four lakes around the town. For Dirac, Van Vleck was the perfect walking companion – fit, uninterested in small talk and content to say nothing for hours. Perhaps Van Vleck mentioned his passion for railroads and his feat of memorising the passenger railway timetable for the whole of Europe and the United States.[5] Like Dirac, Van Vleck was fascinated by technology, numbers and order.

Dirac's hosts were aware of his reputation for eccentricity, and they soon saw that it was well justified and that his sangfroid was extreme even by the standards of the English. He left them several Dirac stories, including a classic that appears to have been first spread around by a tickled Niels Bohr.[6] The story begins during one of Dirac's lectures, moments after he has finished talking, when the moderator asks if anyone has any questions. Someone in the audience says, 'I don't understand the equation on the top-right-hand corner of the blackboard.' Dirac says nothing. The audience shuffles nervously, but he remains silent, whiling away the time of day, looking unconcerned. The moderator, feeling obliged to break the silence, asks for a reply, whereupon Dirac says, 'That was not a question, it was a comment.'

Madison was also the venue of what would become the most widely quoted interview that Dirac ever gave, to the journalist Joseph Coughlin, known to everyone as Roundy owing to his substantial girth.[7] Well known in the town, he was one of Wisconsin's most popular columnists, delivering regular doses of homespun wisdom on sport and other topics in language that was often ungrammatical but always alive with quirky humour. Dirac kept a typed transcript of the four-page article, in which Roundy recounts verbatim his attempts to persuade his interviewee to utter more than one syllable at a time:[8]

ROUNDY: Professor, I notice you have quite a few letters in front of your last name. Do they stand for anything in particular?
DIRAC: No.
ROUNDY: You mean I can write my own ticket?
DIRAC: Yes.
ROUNDY: Will it be all right if I say that P. A. M. stands for Poincaré Aloysius Mussolini?
DIRAC: Yes.
ROUNDY: Fine! We are getting along great! Now doctor will you give me in a few words the low-down on all your investigations?
DIRAC: No.
ROUNDY: Good. Will it be all right if I put it this way: 'Professor Dirac solves all the problems of mathematical physics, but is unable to find a better way of figuring out Babe Ruth's batting average?'
DIRAC: Yes.

The interview continues for another page. According to the transcript, Roundy's interview was published in the 'P. A. M. issue' of the *Wisconsin Journal* on 31 April (*sic*). However, the records of the newspaper show that no such edition was published, so it appears that this much-anthologised interview is a spoof.[9] One possibility is that the typed document was a pastiche presented to Dirac by his Madison colleagues during his farewell dinner at the University

Club, where – as Van Vleck later wrote – they played an elaborate game to tease out of Dirac the names designated by his initials P. A. M.[10] Whatever the origins of the Roundy interview, it is an example of a probably apocryphal Dirac story that captures his behaviour so accurately that it somehow ought to be true.

Dirac left Madison with a cheque for $1,800, more than enough to cover his costs for the remainder of his trip.[11] In June, he combined business and pleasure, giving a series of lectures on quantum mechanics in Iowa and Michigan, also walking down and up the Grand Canyon and hiking in Yosemite National Park and the Canadian Rockies – his introductions to grand North American scenery, which he explored on foot during several trips in the coming decades.[12] He again demonstrated his interest in the latest experimental tools when, during a stay at the California Institute of Technology, he visited the Mount Wilson Observatory, near Pasadena, whose telescope was the largest in the world and by far the most productive source of new information about the universe.

A few months before, Heisenberg had proposed to Dirac that they should travel together to 'bring European life into the American hurry'.[13] When they met in early August at their hotel near the Old Faithful geyser, Heisenberg was surprised to find that Dirac had planned a route that would enable them to see the maximum number of geysers erupt.[14] Even his scenic walks were informed by mathematical analysis. Heisenberg had arranged for them to travel first class to Japan on the steamer *Shinyo Maru*, sharing a roomy cabin with a sea view.[15] Two leading theoreticians were about to spend weeks together, with every opportunity to talk and perhaps to crack the gnawing problem of how to interpret the negative-energy solutions to Dirac's equation. The clubbable Heisenberg would probably have been game for a collaboration, but not Dirac. Although he admired Heisenberg and regarded him as a friend, Dirac felt no obligation to share any of his thoughts about physics with him. His motto was: 'People should work on their own problems.'[16]

In the middle of August, after they had each given a series of lectures in Oppenheimer's department at the University of California at Berkeley, they set off from San Francisco on their two-week cruise to Japan.[17] On board, Heisenberg was a conventionally hedonistic tourist, honing his technique at ping-pong and dancing with the flapper girls.[18] Dirac looked on, probably bemused. It is easy to imagine Dirac at one of the evening balls, sitting at a table and gazing quizzically at Heisenberg as he jived on the dance floor. Heisenberg long remembered being asked by Dirac, 'Why do you dance?' After Heisenberg replied, reasonably enough, 'When there are nice girls it is a pleasure to dance,' Dirac looked thoughtful. After about five minutes of silence, he said, 'Heisenberg, how do you know *beforehand* that the girls are nice?'[19]

As their steamer approached Yokohama, a reporter sought an interview with the two famous theoreticians. Unfamiliar with Dirac's appearance but not with Heisenberg's, the reporter said to Heisenberg, 'I have searched all over the ship for Dirac, but I cannot find him.' Heisenberg knew how to handle this: he talked affably to the journalist, no doubt giving him the story he wanted and not mentioning that Dirac was standing next to him, looking in another direction.[20]

In Japan, the two physicists were greeted as heroes. Leading scientists in Japan knew that their science lagged well behind that of Europe and the USA, and physicists flocked from all over the country to see and hear two of the young founders of quantum mechanics. Dirac and Heisenberg were given round-the-clock obeisance and the full VIP treatment, their first taste of international celebrity. From the official photographs, it is clear that Heisenberg slipped easily into the role of the touring dignitary, looking poised and relaxed in the light summer suit he wore to stay cool in the searing heat. Looking less comfortable than his friend, Dirac made no such changes to his wardrobe: he wore the same three-piece suit and boots that he wore in the depths of the Cambridge winter.

The itinerary was the usual one for academics making a short trip to the country: a stay in Tokyo followed by a visit to the old imperial city of Kyoto, lecturing to packed, hushed audiences of respectful men wearing Western suits splashed with *jako* perfume, scenting the auditorium with the fragrance of geraniums.[21] The texts of the lectures were swiftly translated into Japanese and published as the Orient's first authoritative book on quantum mechanics, a bible for Japan's next generation of physicists, destined to make a huge impact. Dirac and Heisenberg, each of them only twenty-seven, were already training their successors.

At the end of their stay in Japan, Dirac and Heisenberg parted company. Dirac wanted to return by the fastest practicable route, by traversing Russia on the Trans-Siberian Railway. The construction of the 5,785-mile railway in Siberia – with brutal extremes of climate, little local labour available and dreadfully primitive supply routes – had been an engineering project that would have daunted even Brunel. It took twenty-five years to complete. Dirac boarded the train on 24 September at Vladivostock on the eastern coast and, nine days later, arrived in Moscow. He met up with Tamm, and they went on a long walk to see the sights of the city, including the sixteenth-century St Basil's cathedral, later converted into one of the country's many anti-religion museums.[22] Dirac then headed back to England after taking what seems to have been his first flight, from Leningrad to Berlin. This was probably not the most agreeable of experiences: for the next few decades, he preferred to admire aviation technology from a secure vantage point on the ground.

While he was away, his family were 'plodding along as usual', as his mother put it.[23] The highlight of the year had been the General Election in June. For Flo, new technology had taken much of the thrill out of politics: 'The Election is being conducted mainly by "Wireless",' she wrote to Dirac, 'so I don't get any fun out of meetings.'[24] She and Charles supported Lloyd George's Liberal Party, which was trounced in Bristol by the Labour Party, consistent with the national swing that put Ramsay MacDonald back into 10 Downing Street.

Dirac's father, in better health than he had been for some years, was drifting further away from his wife and ever closer to Betty. While Charles and his favourite child played with the family dog in the garden, Flo was left inside, dreaming of *her* favourite child thousands of miles away. She imagined him touring the Hollywood studios and riding a donkey down the Grand Canyon in a Panama hat, though she was disappointed to hear that he had done neither. Flo and Charles, having not seen their son for six months, were hoping to see him before the beginning of term and prepared the house for his visit. But in early October, Dirac perfunctorily informed them that he was back in Cambridge and mentioned no plans to visit Bristol.[25]

He and other theoreticians had made virtually no progress with the problem of negative-energy electrons. Although most physicists wanted to be rid of them, the Swedish physicist Ivar Waller had shown a few months before that they were indispensable to the theory. Waller had found a strange result when he analysed what happens when a photon is scattered by a stationary electron: Dirac's theory could reproduce the successful classical prediction at low energies only if the electron had access to negative-energy states. There could be only one conclusion for Dirac: his equation would survive only if someone could understand these negative-energy electrons.

As he settled down for the new term, Dirac was aware that the critical chorus had swelled from a whisper to a roar. In the opinion of its most dominant soloist, Pauli, the equation's sickness was incurable and its agreement with experiment was a fluke.[26] The onus was on the equation's discoverer, refreshed after almost six months' vacation, to rescue it. So he set about the problem again.

At the end of October, news broke from New York of the event that ended the calm of late-1920s politics and began the descent into global economic catastrophe. The Dow Jones index had reached its historic peak a month before. Then panic struck when the bubble burst. On Friday, 25 October, the newspapers in the St John's common room all featured reports that made clear the scale of the crisis: the *Manchester Guardian* wrote of 'Wild selling in record turnover of 13,000,000 shares'; *The Times* wrote, 'a Niagara of liquidation took place on the American stock market today'. Four days later, on 'Black Tuesday', Wall Street all but melted down, and, as F. Scott Fitzgerald later noted, the decade of unparalleled prosperity had 'leapt to its spectacular death [. . .] as if reluctant to die outmoded in its bed'.[27]

Britain braced itself for the aftershock. Dirac kept abreast of the news, but he was focusing mainly on solving the mystery of the negative-energy electrons. Why had no one observed jumps of the familiar, positive-energy electrons into negative-energy states? After a few weeks, he had found an answer. He imagined all the electrons in the universe gradually filling up the energy states: the states with negative energy will be populated first, because they have the lower energies. Only when they are full will electrons occupy positive energy states. Because the negative-energy states are full, there are no vacancies into which these positive-energy electrons can jump. It is ironic that the crucial idea that underpinned the theory was supplied by Dirac's harshest critic, Pauli: according to his exclusion principle, every negative-energy state can be occupied by only *one* electron.

This prevents each negative-energy state from being filled ad infinitum with electrons.

The bizarre upshot of the theory is that the entire universe is pervaded by an infinite number of negative-energy electrons – what might be thought of as a 'sea'. Dirac argued that this sea has a constant density everywhere, so that experimenters can observe only departures from this perfect uniformity. If this view is correct, experimenters are in rather the same position as a tribe that has spent its entire life hearing the unchanging background sound of a single musical note: this would not seem like torture because people are aware only of *changes* to their environment.

Only a disturbance in Dirac's sea – a bursting bubble, for example – would be observable. He envisaged just this when he foresaw that there would be some vacant states in the sea of negative-energy electrons, causing tiny departures from the otherwise perfect uniformity. Dirac called these unoccupied states 'holes'. They would be observed, he reasoned, only when they are filled by an ordinary electron, which would then emit radiation as it makes the transition. It should therefore be possible to detect a hole in the sea when an ordinary positive-energy electron jumps into it. But what characteristics do the holes have? They mark the absence of a negative-energy electron. Within the general scheme of the 'electron sea', the *absence* of negative energy amounts to the *presence* of positive energy (two negatives make a positive: when *debt decreases* by £5, *wealth increases* by the same amount). Furthermore, a negative-energy electron is negatively charged, so its absence is equivalent to the presence of a positive charge.

It follows that each hole has positive energy and positive charge – the properties of the proton, the only other subatomic particle known at that time. So Dirac made the simplest possible assumption by suggesting that a hole *is* a proton. What he could not explain was why the proton is almost two thousand times as heavy as the electron. That was a problem for the theory, he conceded, a 'serious deficiency'.

The provenance of the hole theory is not entirely clear. The mathematician Hermann Weyl and others suggested that protons were related in some way to the negative-energy electrons, but their thinking was too woolly for Dirac. He later remarked that 'it was not really so hard to get this idea [of the hole theory]' as he was simply drawing an analogy with the theory of how atoms emit X-rays (high-energy light).[28] This theory says that an electron close to the nucleus can be knocked out of the atom, leaving a gap into which another electron falls, accompanied by the emission of an X-ray. It is also possible that Dirac had acquired the germ of his idea when he was sailing down the river Volga fifteen months before. At the Russian Congress, he met the Soviet theoretician Yakov Frenkel: someone snapped a photograph of them lying on the deck of the steamer, in their dress suits. In 1926, Frenkel had produced a theory of crystals in which 'empty spaces' in the regular lattice structure of the crystal would behave like particles – again, precisely analogous to Dirac's hole theory. Frenkel may have mentioned this theory to Dirac only for him to forget it and retrieve it later from his subconscious. But Dirac had no such recollection.[29]

Whatever the origins of the theory, there is no doubting the boldness of Dirac's application of the idea. Nowhere in the paper does he pause to comment on the theory's credibility. The crucial point for him was that he now had the beginnings of a viable theory of matter, based on an appealing equation and solid principles. Who was going to accept that the universe was full of unseen negative-energy electrons, an infinite sea of negative electrical charge? Yet his short paper 'A Theory of Electrons and Protons' bears no sign that he was expecting his idea to be greeted with incredulity. He wrote the article in his uncluttered style but with fewer equations than usual, free of the windiness that would have been excusable in the first presentation of a theory that suggested a new way of looking at the material universe.

Although Dirac never admitted to being nervous about the reception of his hole theory, he often talked of anxiety as the handmaiden of scientific daring.[30] So it is likely that he feared his theory contained a humiliating fallacy, a concern stoked by a letter he received in late November from Bohr, who had heard about the hole idea on the grapevine. For Bohr, the existence of negative energy levels in Dirac's theory of the electron undermined confidence in the entire concept of energy, a problem that – Bohr observed – also occurred in explanations of why some types of atomic nucleus can sometimes spontaneously eject a high-energy electron, a process known as radioactive beta decay. It seemed that energy was not conserved in this process – there was less energy before the decay than there was afterwards – so energy appeared to emerge out of nowhere. This was serious: Bohr was questioning quantum mechanics and even the law of conservation of energy. Dirac thought his mentor was overreacting and, in a roundabout way, recommended him to calm down. Dirac had already told Bohr that he believed that the law of conservation of energy had to be preserved at all costs and that, to keep it, he would be prepared to abandon the idea that matter consists of separate atoms and electrons. And Dirac thought it premature to be pessimistic about quantum mechanics, which had only just passed its fourth birthday:

I am afraid I do not completely agree with your views. Although I believe that quantum mechanics has its limitations and will ultimately be replaced by something better (and this applies to all physical theories) I cannot see any reason for thinking that quantum mechanics has already reached the limit of its development. I think it will undergo a number of small changes, mainly with regard to its method of application, and by these means most of the difficulties now confronting the theory will be removed.[31]

Dirac concluded by reiterating – almost word for word – his reasons for believing in his hole theory. Although his defence could be regarded as stubborn, he does make it clear that he expected his theory to be superseded; the task in hand was to develop the theory as far as it could be taken. Bohr's criticisms do not seem to have shaken him in the least – he would need this thick skin during the coming barrage of scepticism and derision.

A week after he wrote to Bohr, Dirac gave his first public presentation of the hole theory to an audience in Paris, at the Henri Poincaré Institute. He will not have taken much pleasure from giving the lecture, as he reluctantly agreed to give it in French, bringing back abhorrent memories of meals with his father. When he returned to Bristol for Christmas, he had no choice but to speak French again. After his absence for nine months, his family was desperate to see him and to show him their latest plaything – the 'Gramaphone' (sic).[32] But Dirac was, as always, downhearted even at the thought of returning to his enervating Bristol routine, his mother endlessly fussing over him, his father still intimidating him simply by his presence. Although Dirac appears to have told none of his physicist friends, he believed that his home life had stultified him as a child and was still grinding him down. He appears to have first shared the full extent of his pain only a few years later with a friend who was not one of his academic associates. In a letter, he wrote, 'going to see my parents will change me very much, I am afraid, and makes me feel like a child again and unable to do anything for myself'.[33] For now, like all his other emotions, his suffering was hidden.

Fourteen

O hear the sad petition we electrons make to you
To free us from the dominion of the hated quantum view
For we are all abandoned to its dread uncertainty.
Except by you, our champion. O we pray you, set us free!
Once in a pleasant order our smooth-flowing time was spent
As the classical equations told us where to go, we went.
We vibrated in the atom, and a beam of light was freed;
And we hadn't any structure – only mass and charge and speed.
We know not if we're particles, or a jelly sort of phi,
Or waves, or if we're real at all, or where we are, or why,
To protons – holes in ether – according to Dirac.
ANON.[1]

Those anonymous lines are from an ode to the electron, pinned to a noticeboard in the Cavendish Laboratory around 1930. Only the most hard-headed theorist could fail to sympathise with the poet's nostalgia. A decade before, atomic physics had been a matter of common sense: electrons were just tiny particles, and they behaved predictably, according to straightforward laws of nature – the same ones that described everything else in the universe. How quaint those ideas now seemed: the classical laws that had held sway for a quarter of a millennium were now, in the atomic domain, obsolete, as Dirac liked to point out, the idea Jonathan Swift explored in *Gulliver's Travels* – that no one would notice if naturally occurring things expanded or contracted in the same proportion – was wrong.[2] The laws of the everyday world cannot be scaled down to the atomic domain: things are different there. Theorists could now reject every attempt to picture the electron as meaningless and therefore fraudulent. The particle did not even behave predictably: physicists were calculating odds like croupiers at nature's gambling table, using waves that no one believed were real. To cap it all, Dirac had the temerity to argue that common-or-garden electrons, with positive energy, are outnumbered by negative-energy ones that cannot even be observed.

It was probably a Cavendish experimenter, one of many who were suspicious of hole theory, who wrote the anonymous poem. Only a few theoreticians, including Tamm and Oppenheimer, took the theory seriously, and even they soon found it wanting. In February 1930, Oppenheimer showed that the average lifetime of an atom was about a billionth of a second according to Dirac's hole theory, because the atomic electron would quickly fall to its death in the negative-energy sea. Soon afterwards, Tamm and Dirac independently arrived at the same conclusion. Pauli suggested what became known as his Second Principle: whenever a physicist proposes a new theory, he should apply it to the atoms in his own body.[3] Dirac would be the first victim.

Pauli's jest appealed to Gamow, who was staying in Cambridge in the first academic term of 1930, mainly to work with Rutherford and his colleagues. Dirac was charmed by Gamow's non-stop good humour and sense of fun: no one did more to show Dirac what he had missed in his youth. Gamow taught Dirac how to ride a motorcycle (and filmed him doing it), gave him a taste for Conan Doyle's detective novels and apparently introduced him to the high jinks of Mickey Mouse, who first appeared on the screen two years before, in *Steamboat Willy*.[4] Dirac adored Mickey Mouse films, the animated successors of the cartoons he had seen as a boy in the penny weeklies. A few years later, he made a point of attending a day-long festival of the films in Boston, though it seems that he kept this innocent pleasure secret from his highbrow Cambridge colleagues.[5] He was self-aware enough to know that his standing in the St John's common room would not be increased if he were too enthusiastic in his praise of Peg-Leg Pete or Horace Horsecollar.

More respectable at High Table was Dirac's appetite for mathematical games and puzzles that served no purpose at all beyond entertainment. Once, he gave a devastating performance in a game that had been introduced at Göttingen in 1929. The challenge was to express any whole number using the number 2 precisely four times, and using only well-known mathematical symbols. The first few numbers are easy:

$$1 = (2 + 2)/(2 + 2),$$
$$2 = (2/2) + (2/2),$$
$$3 = (2 \times 2) - (2/2),$$
$$4 = 2 + 2 + 2 - 2.$$

Soon, the game becomes much more difficult, even for Göttingen's finest mathematical minds. They spent hundreds of hours playing the game with ever-higher numbers – until Dirac found a simple and general formula enabling *any* number to be expressed using four 2s, entirely within the rules.[6] He had rendered the game pointless.

On 20 February 1930, Dirac sent his parents the usual newsless weekly postcard, consisting of a ten-word summary of the Cambridge weather.[7] The day after his mother received it, she visited the library and was astonished to read in a newspaper that her son had been elected a Fellow of the Royal Society, one of the highest honours in British science. Excited and flushed with pride, she dashed out to the post office and sent him a congratulatory telegram, keeping in check her annoyance that he had not mentioned the news on the card.[8] Dirac was a 'naughty boy', she told him two days later in a letter, enquiring whether the society was organising a ceremony of induction. '<u>Do</u> <u>tell</u> <u>me</u>,' she wrote, stressing each word in frustration.[9]

Dirac could now put the initials FRS after his name, letters that render all other academic qualifications redundant. The Society, which then had 447 Fellows, usually gave the honour to scientists in their forties and fifties, after they had been nominated and passed over several times, so it was extraordinary for Dirac to be appointed the first time he had been put up for election, when he was only twenty-seven. As the news spread among the High Tables and common rooms of Cambridge, it would not have escaped the dons' notice that he had been elected a Fellow at a younger age than any of his senior colleagues.[10]

The announcement appears to have made Dirac's parents realise how rapidly the reputation of their son had risen. 'How hard you must have worked to get to the top of the tree like that,' his mother wrote. 'No wonder you didn't take any interest in the Boat Racing.'[11] The news was a welcome fillip for Flo, whose morale was low. Now that her husband was about to retire, her prospects were pitiable: only fifty-two years old, all she had to look forward to were years cooped up at home with a sick man whom she regarded as a browbeating ingrate and who, she knew, saw her as an inadequate nurse and servant. At school, Charles Dirac's colleagues queued up to offer their congratulations, and he received several letters to congratulate him on raising such a successful son. Paul's engineering teacher Andrew Robertson pointed out that he believed Dirac was the first

Bristol graduate to have been elected an FRS; Ronald Hassé, who first steered Dirac towards a career in theoretical physics, wrote to say how much he was looking forward to Dirac's first public speech in Bristol in September. The city was to host the annual meeting of the British Association for the Advancement of Science, where scientists and members of the public got together to hear a week of lectures on the latest science.[12] At the Cotham Road School – formerly the Merchant Venturers' School – they celebrated by taking a day off. Charles never quite knew when to expect the next plaudit: once, during a lesson, two complete strangers knocked on his classroom door, entered, complimented him on his son's great achievement, and left.[13]

Perhaps to celebrate his latest success, Dirac took his mother's advice and splashed out almost £200 on his first car, a Morris Oxford Tourer, capable of a then-impressive 50 mph.[14] There was no driving test: after completing the sale, the garage owner gave him a short demonstration drive around Cambridge and then handed him the keys. He was then free to take his chances on the roads. With the scrapping of the 20 mph speed limit that year, the highways became even more dangerous, not least because of Dirac's presence. A colleague laughed that 'Dirac's car has two gears, reverse and top.'[15] Only Mott left an account of being driven by Dirac, to London on an icy March day when 'Dirac ran – very gently – into the back of a lorry and smashed a headlamp.'[16] Like Kapitza, Dirac was a wild driver, and this appears to have been due both to his poor handling of the vehicle – his appreciation of machines always exceeded his competence at using them – and to the virtual absence of a highway code. Dirac was a stickler for obeying rules that he believed were rational and obviously for the common good, so, in the absence of regulations, he was free to drive as he wished.

Dirac was, at last, showing signs of mellowing. Leisure was not reserved only for Sundays: at lunchtimes, the bulk of his day's work done, he would often motor out of Cambridge to the Gog Magog Hills, park his car near a tall tree and climb it, still wearing his three-piece suit.[17] He wore it whatever the weather, whatever the occasion, and took it off only during his drives out to secluded sites by the river Cam and in the fens north-east of the city, where he bathed, as Lord Byron had done 125 years before. Later, when he returned to college or to his desk, he would do only the lightest of tasks. He was taking a leaf out of the book of G. H. Hardy, who believed that the longest a mathematician can profitably spend doing serious work is four hours.[18]

Of all the months in the Cambridge academic calendar, June was the most relaxed. The examinations over, it was time for the students to leave the university, but only after the catharsis of the summer ball. The intoxicating mix of music and dancing, free-flowing champagne, gorgeous frocks and sharply cut dinner suits could cheer up the most abject examinee. Dons could put on their summer suits and wind down to the 'long vac', when they had no administrative duties and were free to spend the long, languid afternoons doing nothing except sit in a deckchair and watch a game of cricket. Dirac was nonplussed by the appeal of an activity that involved twenty-two men spending hours – sometimes days – playing a game that often ended in a draw, which devoted spectators would often deem exciting. The game had no more ardent admirer than G. H. Hardy, for whom it was akin to pure mathematics: all the more beautiful for its lack of useful purpose. A few years later, he gave pride of place in his study to a photograph of the Australian batsman Donald Bradman, one of Hardy's three greatest heroes (the others were Einstein and Lenin).[19] Hardy was probably looking forward to Bradman's first Test appearances on English soil, but the prospect will have left Dirac unmoved; he was busy preparing to spend the summer climbing and hillwalking with friends. He needed a break and some fresh inspiration if he was to sort out the problems with his hole theory and so answer his critics, including the mocking Pauli and the privately scornful Bohr. Several of Dirac's colleagues would be lining up to attend his public lecture at the Bristol meeting at the end of the summer, he knew, to see if he had cracked the problem of negative-energy electrons.

Preparing for his second trip to the Soviet Union, Dirac read in the British press that Stalin was tightening his grip, forcing through his programme of collective farming, squeezing the peasants in order to pay for a crash programme of industrialisation and persecuting political opponents and religious minorities. Some newspapers were in no doubt of Stalin's malevolence – the *Daily Telegraph* wrote regularly of his 'Reign of Blood' and his 'war on religion' – but others, including the *Manchester Guardian*, gave him the benefit of the doubt.[20]

The *New Statesman* – the house journal of leftist intellectuals in Britain and favourite reading of Kapitza's in the Trinity common room – insisted that Stalin should be given a fair hearing. Dirac agreed: one of the few things that would draw him into conversation were comments that he perceived to be unfairly hostile to the Soviet Union. Rudolf Peierls later recalled: 'At a time when everything Russian was anathema, he questioned why each particular item was wrong, and this often caused raised eyebrows.'[21] Wanting to see life there for himself, he again ignored the fears of his mother: 'I do hope it is safe in Russia. One hears dreadful stories about it.'[22]

During his trip, Dirac felt the arm of the Soviet military on his shoulder: en route to Kharkov, when he attempted to cross the Soviet border at a place not mentioned in the visa that Tamm had obtained for him, border guards held him at the crossing point for three days before releasing him.[23] By early July, he had heard that Soviet law forbade foreigners who stayed in the country for more than a month to take out either Soviet money or foreign currency. So he left the USSR in late July, within a month of his arrival, having cancelled his plans to hike in the Caucasus. His vacation foreshortened, he soon returned to England, to what most scientists would regard as the media highlight of their life.

In September, Hardy was praising Bradman's devastating performances in the Ashes, and Bristol was preparing to host the British Association meeting. Almost three thousand delegates – including George Bernard Shaw – attended, each of them having paid a pound for the privilege.[24] Jim Crowther told readers of the *Manchester Guardian* that the public delegates were young and dressed informally, many of the women in sleeveless and flowered voile frocks, the men in alpaca jackets and grey flannels. The ticket price had not changed since the meetings began almost a century before, when the Association's leaders were choosing the most appropriate word to describe the participants. They considered 'savants', 'nature peepers' and 'nature pokers', but finally settled on 'scientists', coined in 1834 by William Whewell, one of John Stuart Mill's philosophical adversaries. Though many hated the new word – Michael Faraday disliked it almost as much as the triply sibilant 'physicist' – it had caught on by the time Dirac was in junior school.[25]

The organisers, probably fearing that Dirac would give a technical talk of limited public appeal, scheduled him to speak in a modest room in one of the university's new physics laboratories, funded by the tobacco manufacturer H. H. Wills. At 11 a.m. on Monday, 8 September, Dirac stood up without fanfare to address a crowded room on the subject of 'The Proton'.[26] Never confident when he spoke at public meetings, he may have been particularly apprehensive at this one: this was the first time he had agreed to address a lay audience and the first time he had spoken to many of the teachers who had seen him flower. If Charles was there, as is likely, he will have had a full heart as he had not heard his son speak in public before: Paul Dirac would now have no choice but to talk about his science to his father.

Dirac entered into the spirit of the British Association. Speaking with his usual directness, in lilting Bristol tones, he talked about his research in a way that might almost have passed as colloquial, though with none of Eddington's flair. To ensure that he was intelligible to people with no science training, he began with the statement that 'matter is made from atoms', and quickly went up the gears, ending with his idea that the proton is a hole in the negative-energy sea of electrons. This implied, he pointed out, that there is only one fundamental particle, the electron, adding that such an economy in nature was 'the dream of philosophers'. For many in his audience, this will have been an exciting revelation, but not for Gamow and Landau, who were at the back of the room, sitting on wooden benches. The two of them had roared down to Bristol on Gamow's motorbike, Landau perched behind him on the luggage carrier. They travelled to the meeting, partly as Bohr's unofficial emissaries, specifically to see if Dirac had anything new to say about his theory. During the talk, Gamow and Landau craned their necks to see the speaker, hanging on his every word, Landau, as usual, unable to resist making snide asides.[27] After twenty minutes of reiterating arguments he had already published, often using the same words as he had used in his papers, Dirac drew to a close, and they realised that he had said nothing new. Their trip to Bristol had been a wild goose chase.

Dirac's theory of negative-energy electrons nevertheless captured the imagination of journalists, and the British newspaper reports gave him more publicity than he had ever known. After his presentation, the representative from the American Science News Service wired Washington: 'This new theory may prove to be as important and interesting to the public as Einstein's theories have been.'[28] The *New York Times* picked up the story and reported that Dirac's 'acclaimed' theory 'upset all present conceptions of space and matter', adding that 'These physical scientists have a more exciting life than Columbus.'[29] But Dirac's peers were unimpressed. On the way back to Cambridge, Landau and Gamow stopped at a post office. Landau sent Bohr a telegram consisting of a single word: 'Crap'.[30]

The telegram reached Bohr soon after he received from Dirac a copy of his textbook, *The Principles of Quantum Mechanics*. Even if the author's name were not on the cover, his identity would have been obvious to Bohr from a quick flick through: the unadorned presentation, the logical construction of the subject from first principles and the complete absence of historical perspective, philosophical niceties and illustrative calculations. This was the vision of a mathematically minded physicist, not an engineer. Dirac's peers marvelled at its elegance and at the deceptively plain language, which somehow seemed to reveal new insights on each reading, like a great poem. Many of the students – especially the less able ones – were bemused, dissatisfied and sometimes even dispirited.[31] The book had been written with no regard for his readers' intellectual shortcomings, without the slightest sign of emotion, with not a single leavening metaphor or simile. For Dirac, the quantum world was not like anything else people experience, so it would have been misleading to include comparisons with everyday behaviour. He scarcely mentioned empirical observations except at the beginning, where he described an experiment that demonstrates the failure of classical theory to account for matter on the atomic scale and, hence, motivates the need for quantum mechanics. In its 357 pages, *The Principles of Quantum Mechanics* featured neither a single diagram, nor an index, nor a list of references, nor suggestions for further reading. This was, above all, a personal view of quantum mechanics, which is why Dirac – usually someone who abjured personal pronouns – always referred to it as 'my book'.

Physicists immediately hailed it a classic. *Nature* published a rhapsodic review by an anonymous reviewer who – to judge by the eloquence and sharp turn of phrase – may well have been Eddington.

The author made it clear that this was no ordinary account of quantum mechanics:

[Dirac] bids us throw aside preconceived ideas regarding the nature of phenomena and admit the existence of a substratum of which it is impossible to form a picture. We may describe this as the application of 'pure thought' to physics, and it is this which makes Dirac's method more profound than that of other writers.[32]

The book eclipsed all the other texts on quantum mechanics written at about the same time – one by Born, another by Jordan – and became the canonical text on the subject in the 1930s. Pauli warmly praised it as a triumph and, although he worried that its abstractions rendered the theory too distant from experiment, described the book as 'an indispensable standard work'.[33] Einstein was another admirer, writing that the book was 'the most logically perfect presentation of quantum theory'.[34] *The Principles of Quantum Mechanics* later became Einstein's constant companion: he often took it on vacation for leisure reading and, when he came across a difficult quantum problem, would mutter to himself, 'Where's my Dirac?'[35]

But some of Dirac's undergraduate students were not pleased to find that the book was largely a transcript of his lectures: why, these students wondered, was it worth bothering to go and listen to him? Yet others found the course uniquely compelling.[36] He would enter the lecture theatre punctually and in full academic garb, wearing the traditional uniform of gown and mortarboard. Otherwise, there was nothing else theatrical about him. He would clear his throat, wait for silence, then begin. For most of the lecture, he would stand still and erect, enunciating each word, addressing what one of his students described as his 'personal unseen world'.[37] At the blackboard, he was an artist, writing calmly and clearly, beginning at the top left-hand corner, then methodically working downwards, writing every letter and symbol so that someone at the back of the room could see it clearly. The audience was usually quiescent. If a student asked a question, he would dispatch it with the economy of a great batsman and then move on, as if nothing had disturbed his flow. After precisely fifty-five minutes he would draw his presentation to a close and then, unceremoniously, gather his papers together and walk out.

One of the new students who were impressed by Dirac's course in the autumn of 1930 was Subrahmanyan Chandrasekhar, later a leading astrophysicist but then a wide-eyed student just arrived from Bombay. For him, the course was 'just like a piece of music you want to hear over and over again'.[38] During his time in Cambridge, he attended the entire course four times.

Dirac probably knew he had disappointed his colleagues at the British Association meeting by failing to say anything new. He was about to go to his second Solvay Conference, aware that few of the physicists took seriously his unified theory of electrons and protons; his proposal that protons were holes in the negative-energy sea was beginning to look not just implausible but untenable. One of the blows he suffered came shortly after the Bristol meeting when Tamm wrote to tell him that Pauli had proved that the holes have the same mass as the electron. Experimenters had not detected such a particle, which is probably why Tamm added a sympathetic comment: 'I would be very much pleased to hear that Pauli is wrong.'[39]

This Solvay meeting was later remembered for being the one where leadership of the community of theoreticians passed from Einstein to Bohr. Einstein was looking out of touch, downcast after Bohr had bested him in one of their tussles about quantum mechanics and its meaning. For Einstein, the theory was fundamentally unsatisfactory as it did not even claim to describe physical reality, only the probabilities for the appearance of a particular physical reality on which an observing experimenter's attention is fixed. Such a theory may be good at explaining experimental results, but it is certainly not complete, Einstein argued.[40] Disillusioned, and uninterested in much of what his colleagues had to say, he consoled himself by playing after-dinner violin duets with the Queen of Belgium, one of his new friends.

Unlike the previous Solvay Conference in 1927, the atmosphere at this one was heavy with forebodings about the world outside physics, where the recession was ravaging most industrialised nations and providing fertile ground for political extremists. A month before the conference, Hitler's National Socialists had taken second place in Germany's election, followed by the Communists. Göttingen was now bedecked with Nazi flags, many of its shops displaying trinkets decorated with swastikas. Einstein was sick of the anti-Semitism in Berlin and despised Germany's emerging leader: 'If the stomach of Germany was not empty, Hitler would not be where he is.'[41]

As Dirac kept his politics almost exclusively to himself, most of his Cambridge colleagues mistakenly believed he had no interests at all, that he was as one-dimensional as the lines in his projective geometry. He was privately alarmed by the rise of Hitler and broadly supportive of Stalin's project in the USSR, especially its commitment to mass literacy and education. Aware of Dirac's interest, Tamm wrote to him about the radical experiment in 'brigade education', in which students studied intensively, alone and in groups, with no lectures, but with a professor on standby for consultation:

I never thought it possible for a large body of students to work as hard as our students do now. Our [brigades, each of five students, work and study together] 9 days out of 10 [. . .] from 9am to 9pm with a 2-hour interruption for a meal (research work included, which is of course conducted individually by each student). Yesterday, speaking with a brigade, I found them troubled by the fact that they have 'lost without cause' six out of 270 working hours of the last month![42]

Although Dirac was interested in the Soviet experiment, it was of only marginal interest to him compared with theoretical physics. By late autumn, he had every reason to be dissatisfied with his progress as his hole theory was in deep trouble. Oppenheimer and Weyl had independently come to the same conclusion as Pauli – that Dirac had no theoretical justification for believing that his holes were protons. The implication was that the theory was incorrect; something was amiss with the Dirac equation. But he was convinced that it was correct – what was needed was the correct interpretation of its mathematics. The American theoretician Edwin Kemble later put his finger on the kind of faith Dirac had in his equation: '[He] has always seemed to me a good deal of a mystic [. . .] he thinks every formula has a meaning if properly understood.'[43]

Towards the end of term, Dirac went through his annual chore of refusing most invitations to Christmas parties, though he did occasionally attend the annual dinner of the Cavendish Physical Society, a boisterous evening of eating, drinking and singing.[44] After Kapitza attended the dinner for the first time in December 1921, he wrote incredulously to his mother, observing how quickly even a moderate amount of alcohol freed the inhibitions of his English colleagues and made their faces 'lose their stiffness and become lively and animated'.[45] By the end of the meal, after the cheeseboard and port had been passed round, the air was thick with cigar smoke and everyone was shouting to be heard above the din. The ritual was not yet over: the next stage was a series of facetious toasts (one had been 'To the electron: may it never be of any use to anybody'[46]) alternating with off-key renditions of popular tunes such as 'I Love a Lassie', their lyrics rewritten as a jokey commentary on the past year at the laboratory.[47] At the climax, the portly Rutherford, Thomson and everyone else stood on their chairs, linked together with arms crossed and belted out 'Auld Lang Syne' and then, finally, the National Anthem 'God Save the King'. After the bacchanalia ended, usually well after midnight, it was up to those left standing to take their drunken colleagues to their homes.

In 1930, Dirac did not attend the dinner but will probably have heard later that Kapitza was the focus of attention that night. Rutherford, then President of the Royal Society, had secured a professorship for his favourite colleague and funding for the construction of a new building to accommodate him and his laboratories. At the end of the seven-course dinner, while the sixty guests were chewing their mince pies, Darwin reminded them of the experience of entering Kapitza's laboratory: 'you had to ring to be admitted by a "flunkey" and became confronted not with men working in their shirt sleeves, but with Prof Kapitza seated at a table, like the arch criminal in a detective story, only having to press a button to do a gigantic experiment'.[48]

The laughter at this image of Kapitza, apparently a forerunner of a James Bond villain, will have been hearty, and it is safe to guess that knowing glances will have passed among his colleagues, many of them envious of his relationship with their laboratory's director. Blackett was not there. Rutherford had no time for petty jealousy but was not above making a thinly disguised attack on his recently retired colleague Sir James Jeans, whose The Mysterious Universe had been a best-seller since it first appeared in the bookstores the month before. Rutherford was as down to earth and, at the same time, as snobbish as anyone in science. As the recorder of the dinner wrote: Sir Ernest Rutherford 'deplored the writing of popular books by men who had been serious scientists, to satisfy the craving for the mysterious exhibited by the public'.[49] This was a common opinion in Cambridge. A few months later, his idoliser C. P. Snow – a scientist about to become a writer – sneered at science popularisers for doing a job that was just too easy: 'there is no argument and no appeal, just worshipper and worshipped'. The result was, Snow declared, a 'great evil'.[50] Within three years, Snow published his semi-autobiographical novel *The Search*, the first fiction to bring to a wide audience the atmosphere of Rutherford's laboratory, and to feature Paul Dirac.[51]

A week after Christmas, Rutherford was ennobled at the end of his five-year stint as President of the Royal Society. But the pleasure the honour gave him was eclipsed by a family tragedy: his daughter and only child, Fowler's wife, died in childbirth two days before Christmas. Lord Rutherford, grieving as he approached his sixtieth birthday, must have thought his years of glory were over. He was not doing much research of his own, so his remaining hopes of being involved in more of the ground-breaking discoveries that he longed for were in the hands of his 'boys'.

Dirac showed none of the confidence that might be expected of a young man at the top of his game. Chandrasekhar wrote home to his father that he was disappointed that Dirac did not show a bit more swagger: '[Dirac is a] lean, meek shy young "Fellow" (FRS) who goes slyly along the streets. He walks quite close to the walls (like a thief!), and is not at all healthy. A contrast to Mr Fowler [. . .] Dirac is pale, thin, and looks terribly overworked.'[52]

Work was not Dirac's only concern. Having read his mother's letters, he may have sensed that his parents' relationship, tense and unstable, was fast approaching a flashpoint. Charles Dirac, dreading retirement, was pleading with the Bristol education authorities to be allowed to stay on in his job, but they were resisting. Betty, now with a car of her own, was doing little except chauffeur him three times a day to and from Cotham Road School. Dirac was watching his sister become another of his father's servants.

Meanwhile, Flo knew that, in only a few months, she would be spending most of her life at home alone with her husband: 'It simply won't bear thinking about.'[53]

Fifteen

Russian politics like opium seems infallibly to provoke the most fantastic dreams and imaginings on the part of the people who study them.

E. A. WALKER, British Embassy, Moscow, 1931

In Cambridge, during the spring of 1931, Dirac happened upon a rich new seam of ideas that would crystallise into one of his most famous contributions to science. In the thick of this project, he received a letter from his mother, beginning:

27 April 1931

My dear Paul

Pa and I had quite a row yesterday all about some wine upset on some cheap stamps. He got in the most awful rage for a few minutes & then said he had had enough of me & should *go* if I did anything more to upset him.

I apologised most humbly as usual but on thinking it over, I am pretty certain he meant it.

In three pages of brief, matter-of-fact sentences, she described to Dirac – apparently for the first time – the charade of her marriage. She told him of a young woman who had visited the family when he was a baby, stayed to supper and had been escorted home by Charles to Bedminster. Flo had written to her that she 'wouldn't have it any more and thought it was all finished'. But she was deluding herself, as she realised when she visited Charles's Esperanto exhibition at Bishop Road School and saw that the woman who was presenting it with him, wearing a huge pair of tortoise-shell glasses, was the young woman who had visited them decades before. 'Fancy if they have kept up the acquaintance for 29 years,' Flo wrote. By this account, his father had been cheating on the woman who had spent most of her life looking after him. Her conclusion was: 'She has nothing to do but humour him, I have to keep the house clean, dress him, bath him & worst of all find something to feed him on.'[1]

As usual, Dirac appears to have said nothing of this to anyone, even to his close friends. In the early months of 1931, a quiet time for his fellow theoreticians, he was working on the most promising new theory he had conceived for years.[2] The theory broke new ground in magnetism. For centuries, it had been a commonplace of science that magnetic poles come only in pairs, labelled north and south: if one pole is spotted, then the opposite one will be close by. Dirac had found that quantum theory is compatible with the existence of *single* magnetic poles. During a talk at the Kapitza Club, he dubbed them magnons, but the name never caught on in this context; the particles became known as magnetic monopoles.[3]

The idea arose accidentally, he later said, when he was playing with equations, seeking to understand not magnetism but electrical charge.[4] The American experimenter Robert Millikan had demonstrated that this charge exists only in discrete amounts, each of them exactly equal to a whole number multiplied by the size of the electron's charge, usually denoted by e. So the electrical charge of a piece of matter can be, for example, five times the charge of the electron ($5e$) or minus six times its charge ($-6e$), but *never* two and a half times its charge ($2.5e$). The question Dirac wanted to answer was: *why* does electric charge come only in discrete amounts?

At first, Dirac worked in traditional ways, with quantum mechanics and Maxwell's equations of electromagnetism. Then, like a jazz musician working with two intertwining melodies, he began the riff that led to the monopole. Dirac pictured the magnetic lines of force that end on a quantum particle, much like the ones that terminate on the pole of a bar magnet, usually displayed by patterns of iron filings, each of them obediently aligned to the magnetic force acting on it. He asked: if quantum mechanics and Maxwell's equations of electromagnetism are assumed to be true, what can be said about the magnetic field associated with a quantum particle? To answer the question, he used an innovative combination of geometric thinking – picturing the possible waves in space and time – with powerful algebraic reasoning. He found a way of building on the existing structure of quantum theory, without changing any of its essential foundations and preserving all the rules that governed the interpretation of the theory. If quantum mechanics can be likened to a house of playing cards – with a fragile balance between its interconnected parts – then Dirac can be said to have added a few more cards, preserving the structure's balance, while extending its range to include a new type of particle. The theory furnished a new connection between electricity and magnetism, an equation that relates the smallest-possible electrical charge with the weakest-possible magnetic charge.

The equation enabled him to draw some startling conclusions. First, the strength of the magnetic field of a monopole is quantised – it can have only certain allowed values, whole-number multiples of the minimum quantity, whose value he could easily calculate. It turned out that two monopoles of opposite sign are hard to separate: the force pulling them together is almost five thousand times the force that attracts an electron to a proton.[5] This, Dirac suggested, might be why magnetic poles of opposite sign have never been separated and therefore appear in pairs.

His second conclusion was still more striking: the observation of just one monopole anywhere in the universe would explain why *electrical* charge is quantised – the very thing Dirac had set out to understand. Having checked his final calculations and having found no errors, he came to a bold conclusion: if an experimenter happens on a single monopole anywhere in the universe, the new theory can explain why nature had chosen to apportion electric charge *only* in discrete amounts.

Dirac's theory did not guarantee the existence of monopoles but did show that quantum mechanics can describe such particles *if* they occur in nature. Centuries earlier, other scientists had speculated that monopoles might exist, but those ideas were just hunches, with no logical underpinning.[6] Dirac was the first to give clear reasons *why* such particles might be observed. He may well have thought that the idea was too beautiful to be wrong, but he followed the convention of presenting his conclusion as an understatement: 'one would be surprised if Nature made no use of it'. And he chose not to go the whole hog by trumpeting the magnetic monopole as a prediction of his theory. Like all physicists at that time, he accepted that experimenters had found the need for only two fundamental particles – the electron and the proton – and that it was not the job of theorists to complicate matters by proposing new ones. Ironically, the first physicist to buck the trend was an experimenter, Rutherford, when he proposed in 1920 that most atomic nuclei contain a hitherto undetected particle, roughly as heavy as the proton. He called the new particle 'the neutron'.

Yet, in his paper on the monopole, Dirac implied for the first time that he no longer believed there are only two fundamental particles. In the introduction, he declared that he had suggested that a proton is a hole in the negative-energy sea of electrons: Oppenheimer and Weyl had convinced him that the hole must have the same mass as the electron (he did not mention Pauli, who had also come to the same conclusion). So Dirac followed the logic of Sherlock Holmes: 'When you have eliminated all which is impossible, then whatever remains, however improbable, must be the truth.'[7] The conclusion was that each hole corresponded to a new, hitherto undetected type of particle with exactly the same mass as the electron:

A hole, if there were one, would be a new kind of particle, unknown to experimental physics, having the same mass and opposite charge to an electron. We may call such a particle an anti-electron. We should not expect to find any of them in nature, on account of their rapid rate of recombination with electrons, but if they could be produced experimentally in high vacuum they would be quite stable and amenable to observation.

Again, Dirac is surprisingly circumspect. Although he states the properties of his new particle and even names it, he seems less keen to stress the inevitability of its existence than the difficulty of detecting it. If Dirac had been confident, he would have included a plain-spoken sentence such as 'According to this version of hole theory, the anti-electron should be detectable,' but he held back. Paradoxically, he did underline a radically new interpretation of protons: they were nothing to do with electrons, he suggested, but have their own negative-energy states, 'an unoccupied one appearing as an anti-proton'. Within twenty lines of prose, he had foreseen the existence of the anti-electron and the anti-proton.

Though chary about predicting new particles, Dirac showed no timidity at all when he introduced what amounted to a new way of doing theoretical physics. In two paragraphs, consisting of 350 words and no equations, he argued that the best way to make progress was to seek ever-more-powerful mathematical foundations for fundamental theories, not to tinker with existing theories or look to experiment for inspiration. He envisaged the future of physical science as an unending series of revolutions, driven by mathematical imagination, not by opportunistic responses to the latest announcements from experimenters. This was tantamount to a new style of scientific investigation: seeking laws of ever-greater generality – as Descartes, John Stuart Mill and others had recommended – but relying on mathematical inspiration to find them, rather than taking their cues mainly from observations.

He began by pointing out that before Einstein used non-Euclidean geometry as the basis of the general theory of relativity and before Heisenberg used non-commutative algebra in quantum mechanics, these branches of mathematics were 'considered to be purely fictions of the mind and pastimes for logical thinkers'. The solution to the hardest problems in fundamental physics, Dirac inferred, will 'presumably require a more drastic revision of our fundamental concepts than any that have gone before'. He set out his manifesto with the blazing confidence of a young scientist at the height of his powers:

Quite likely these changes [to our fundamental concepts] will be so great that it will be beyond the power of human intelligence to get the necessary new ideas by direct attempts to formulate the experimental data in mathematical terms. The theoretical worker will therefore have to proceed in a more indirect way. The most powerful method of advance that can be suggested at present is to employ all the resources of pure mathematics in attempts to perfect and generalise the mathematical formalism that forms the existing basis of theoretical physics, and after each success in this direction, to try to interpret the new mathematical features in terms of physical entities . . .

His message was clear: theorists should concentrate much more on the mathematical foundations of their subject and much less on the latest bulletins from the laboratories – to abandon centuries of tradition. No wonder Dirac became known as 'the theorist's theorist'.[8]

Early in May 1931, when Dirac was writing his paper, Tamm arrived in Cambridge to spend a few months in St John's College, having left his wife and children in Moscow.[9] He had no trouble securing permission to work in the UK, as Dirac was officially a favoured scientist in the Soviet Union, having been elected a corresponding member of the USSR Academy of Sciences three months before.

For once, Dirac was willing to share his ideas and briefed Tamm on his magnetic monopole theory, suggesting that he use the new theory to calculate the energy values and quantum waves that describe an electron in the vicinity of a monopole. Apart from when he was asleep, Tamm worked non-stop for three and a half days and finished just in time for Dirac to include his results – less exciting than Dirac had hoped – in the paper. In college, Tamm fraternised easily with the dons, including a few who had become friends with Dirac, having broken through his crust of reserve. Among them were the mathematician Max Newman and the Cavendish experimentalist John Cockcroft, both five years older than Dirac.[10] The Yorkshire-born Cockcroft was a trained engineer and a natural manager, intensely focused to the point of near silence and with a flair for helping Kapitza and his other colleagues to solve technical problems. He was 'a sort of scientific dogsbody of genius', Crowther said.[11]

Only four days after Tamm arrived, Dirac organised a breakfast in his room to talk about Russia with Tamm and the classicist Martin Charlesworth. Dirac's gyp will have delivered the food, probably plates of bacon, eggs and fried bread, served with a pot of tea, toast and marmalade. The three men talked for four and a half hours.[12] Dirac wanted to learn about the Soviet economy, but he was uneasy when there was any sign that Tamm might present his Marxist views in public, as he showed when Tamm told him that he had been invited to speak on 'Higher Education in the Soviet Union' in London. Dirac remarked pointedly to him that he hoped the talk would be on education, not politics.[13]

From the tone of the letters he wrote to his wife in Moscow, Tamm was surprised that so many Cambridge dons were interested in the Soviet experiment. When he had lived in Britain eighteen years before, the university was known for its conservatism, but around the time he arrived there this time, the Marxist Bernal and his colleagues had established a nucleus of left-wing thought and activity among the academics.[14] As Dirac will have heard, it was standard Marxist practice to praise the successes of the Soviet Union and not to dwell on its failures, but to draw attention to the millions of victims of unemployment and imperialist wars and the economic waste that could allegedly be prevented by a properly planned cooperation.[15] The comments Tamm makes in his letters give the impression that Dirac was then no more than an interested observer of the Marxist proselytisers; his passion was physics, though he was now more relaxed about taking time off to pursue other interests. After lunch, Dirac would often drive Tamm out into the countryside, sometimes pausing by a roadside tree so that Tamm could teach Dirac the elements of rock climbing and help him to overcome his fear of heights; in return, Dirac taught Tamm to drive and even helped him pass the recently introduced driving test.

In late June, near the end of Tamm's visit, he and Dirac headed north to the more challenging terrain of Scotland, where they spent a week in the mountains of the Isle of Skye with the industrial chemist James Bell. An expert climber, he had been a friend of Tamm's since their student days in Edinburgh and was a close follower and sceptical supporter of the Soviet experiment, steering a moderate course between Soviet propaganda and the anti-Soviet articles in the British press.[16] Skye provided just the kind of scenery and company Dirac loved, and his vacation gave him an excuse to delay his return to Bristol.

That year, the summer days of Cambridge did not have their usual languor. They were rudely interrupted by a political frisson whose unlikely source was the Science Museum in London, the location of the second International Congress on the History of Science and Technology.[17] For a few days in early July 1931, a red flag flew over South Kensington. Such gatherings usually attracted no attention, but this one was special: it was attended by a high-powered Soviet delegation that included Nikolai Bukharin – formerly one of Lenin's closest associates, now a colleague of Stalin's – and by several leaders of the Soviet scientific community, notably Boris Hessen. A few weeks before, Stalin had announced the end to almost eighteen months of political warfare between the Soviet state and its intelligentsia, so this conference offered an opportunity to present the Soviet outlook on science and technology in a favourable light. Bukharin had been the darling of the Bolshevik Party but had been pilloried in 1929 when he opposed forced collectivisation of farming and the crash industrialisation of the economy. A year later, he was sacked as the editor of *Pravda*, but remained loyal to Stalin and gave a full-throated presentation of the Marxist view of science to his audience in the museum. Bukharin stressed the historical context of science and the influence of social and economic conditions on scientific development, dismissing the traditional emphasis on the achievements of outstanding individuals, such as Newton and Darwin. The Soviets knew the right way forward, Bukharin concluded – by developing science as part of a unified plan for the whole of society:

The building of science in the U.S.S.R. is proceeding as the conscious construction of the scientific 'superstructures': the plan of scientific works is determined in the first instance by the technical and economic plan, the perspectives of technical and economic development. But this means that thereby we are arriving *not only at a synthesis of science, but at a social synthesis of science and practice.*[18]

At the end of Bukharin's lecture, there was silence, followed by coughs and shufflings. But the talk was a success: it was reported in several British newspapers and magazines and made an indelible impression on many of the delegates. Desmond Bernal called the gathering 'the most important meeting of ideas [. . .] since the [Bolshevik] Revolution'.[19] Dirac did not participate in the meeting but will have heard about it from Tamm, who accompanied the Soviet party to visit Marx's grave in Highgate Cemetery, and from Kapitza, who organised a lunch in their honour at Trinity College.[20]

That MI5 was carefully monitoring Bukharin's activities during his visit to Britain would not have surprised Kapitza, but he would surely have been taken aback if he had known that, since January, Special Branch had been opening, checking and sometimes copying mail sent to him from Moscow and Berlin. Armed with folders bulging with vaguely incriminating reports – all of them scientifically inaccurate, sometimes to the point of illiteracy – MI5 were concerned that he had access to sensitive military information and suspected 'that he may be sending [it] abroad'.[21] The search revealed nothing and the government warrant to intercept his mail was suspended on 3 June. But MI5 kept its tabs on him.

Dirac was shortly to travel to the United States for another hiking vacation and a sabbatical term in Princeton, but he was duty-bound to visit Bristol first. He disliked confrontations, so he must have been steeling himself in late July as he prepared to spend a week in 6 Julius Road.[22] Everyone was even more unhappy than they had been when he had last seen them, as Dirac knew from his mother's letters. Betty, unable to afford to run her car, sold it for a knockdown price. Charles, bitter that he was being forced to retire, consoled himself by spending the evenings with his friends Mr and Mrs Fisher at their bungalow in Portishead. Flo, suspicious that Mrs Fisher was one of his mistresses, was hoping he would leave to set up home with her or his girlfriend in the Esperanto group: 'I can't help it anyhow, he is tired of me and likes someone younger.'[23]

Dirac thought his family home was a disgrace – it was in a state of seedy disrepair, as his father refused to have maintenance work done and his mother disliked housework more every year.[24] According to Flo, the atmosphere inside was toxic, thick with resentment. She despised Charles, and it would not be surprising if he were upset that she had exploited their marital problems by thickening the wedge between him and his son. It would have been out of character for Dirac to do anything other than to keep his head down and to depart after putting in a token appearance. He did just that, driving back to Cambridge after a few days to give a talk. But he could not escape quite so easily: on the day before the seminar, another harrowing letter from his mother arrived:

19 July 1931

My dear Paul,
I don't know if this will surprise you but your father & I are going to part (as his own father & mother did.)

It is his own idea; he says he has hated me for 30 years. I know I could never please him but didn't know it was quite so bad as that.

He will give me £1 a week or more (it will have to be more) & I am to clear out.

I don't mind, if I have never pleased him. I sent one of his lady friends away when you were just born because she came in every night & he took her home to Bedminster & returned nearly 12 P.M. She has kept in with him ever since & he says he wishes he had married her. She is a nurse now & I suppose will come & look after him.

Otherwise, he sits in the waiting room at Zetland Road with Mrs Fisher from Portishead & she comes up here pretty often, or he is always out. Betty says she will stay with him as they are both after his money.

I am going to see a lawyer Fred [my brother] knows, to-morrow morning & will get it settled before he leaves school on Friday or he may clear out.

Do you know of a tiny cottage or bungalow near the sea up your way? It would be a complete change & I love the sea. I expect Louie or Nell would come along occasionally & I should not meet anyone I know.

If you could find me a tiny place anywhere I should be so grateful. I wouldn't interfere with you in the least but you could come & see me in your car whenever you had time.

We are not having any row about it – it is not dignified so you need not stay away if you care to come along earlier. I'll post this while they are at Church.

With love from Mother[25]

Dirac could now understand a scene that had haunted him since he was a child: his parents bawling at each other in the kitchen while he, Felix and Betty were locked outside in the garden. The phrase 'he has hated me for 30 years' probably struck home in Dirac's mind, constantly in search of numbers to process: as he was only twenty-nine years old, she had, in effect, told him that he had not been conceived in a loving relationship, let alone raised in one.

Flo did not wait for her son's advice. She went straight to her lawyer, who advised that Charles could not legally throw her out unless she was with another man, otherwise he would lose his pension. As soon as she was alone, she wrote to Dirac: '[Charles and I] don't speak, but never did much, but I guess it better to stick to Betty. Two of us ought to manage him.'[26]

Ten days after he received his mother's most recent letter, on 31 July 1931, Dirac sailed from Liverpool to North America, then in the tightening grip of economic depression. He took his mother with him for the first part of the journey, apparently to give her a short break from the acrimony in 6 Julius Road (she appears to have returned home immediately).[27] After another long hiking vacation with Van Vleck, in the Glacier National Park, Dirac arrived in Princeton – a little over an hour's drive from both New York and Philadelphia – then stirring after the long torpor of the summer vacation.[28] The mathematician Malcolm Robertson, who arrived there at the same time, later remembered being overwhelmed when he drove through the town for the first time at dusk:

This was my first glimpse of the charming college town that was to play such a large part in my life, and a joyful and exhilarating experience it was indeed. I have never forgotten that first encounter, and my feeling of excitement and awe at the lovely stately homes among the old trees, the magnificent university campus with both new and old stone buildings, acres of well-kept lawns, and even a lake and a peaceful golf course.[29]

Soon after Dirac arrived there at the end of August, he was given a handsomely appointed office in Fine Hall, home of the university's mathematics department, the newest building on the campus. It was largely the initiative of the tweed-suited Princeton mathematician Oswald Veblen, who oversaw every detail of the building's opulent design, right down to the locations of the electrical sockets.[30] Almost a third of its budget for internal decorations had been allocated to rugs woven from seamless Scottish chenille. Throughout the new building, there was other evidence of his Anglophile tastes, with a firm nod to the ambience in Göttingen: the hall's faux Oxbridge architecture and furnishings, its freshly varnished oak-panelled walls, even the ritual of taking afternoon tea. In the common room used for special occasions, Veblen had arranged for Einstein's aphorism *Raffiniert ist der Herr Gott, aber boshaft ist Er nicht* (God is cunning, but He is not malicious) to be engraved in German on the rim of the huge stone fireplace.[31]

On the morning of Wednesday 1 October, Dirac walked to Fine Hall from his lodgings near the town centre through the blaze of red and orange foliage, dried-out leaves crackling underfoot. A few hours later, for the first time in his career, he was to co-present a seminar, and with the least likely of his colleagues, Wolfgang Pauli. For Princeton University's physicists, walking to the hall through the connecting corridor, and other faculty members, crossing the campus in the biting chill of the late afternoon, this was an exciting start to the new academic term, an opportunity to see two of the subject's luminaries talking about some of their freshest ideas. The occasion was, Pauli wrote to Rudolf Peierls, 'a first national attraction'.[32]

Each speaker was going to present what amounted to a prediction of a new particle: Dirac presented the monopole, Pauli another hypothetical particle, later called the neutrino. The event marked the dawn of a new culture in physics, in which theory could pre-empt experiment. The figures and demeanours of the two speakers contrasted comically. Dirac was thin as a reed, distant and serene, with the smooth and unblemished skin of a young man but, incongruously, with a pronounced stoop. The overweight Pauli was two years Dirac's senior but his waistline made him look older. When sitting, he looked like a judge deep in reflection, his arms folded over his belly, his bulbous torso rocking rhythmically back and forth. At the seminar, he probably looked troubled and in some pain, having broken his left shoulder when he fell downstairs a few months before, the worse for drink.[33]

Many in the audience will have read about Dirac's prediction, but Pauli's had not appeared in an academic journal, though attentive readers of the *New York Times* read about it in an article published a few months before.[34] Pauli had first proposed the existence of his new particle in a private letter to a meeting of experts on radioactivity.[35] There, he tentatively suggested that the existence of the particle could explain the problem that Bohr had identified with energy conservation when a radioactive nucleus ejects an electron. The essence of the problem was that electrons from these nuclei did not all have the same energy; rather, the electrons had a continuous range of energies. Pauli put forward a 'desperate' explanation for this spectrum of energies: the electron in each radioactive decay was ejected with another particle – hitherto undetected – so that the two particles shared their total energy in proportions that varied from one decay to the next. According to Pauli's theory, the new particle should have no electrical charge, the same spin as the electron and only a tiny mass. Few of Pauli's peers liked the idea: for Wigner it was 'crazy', for Bohr it was implausible and Dirac thought it was simply wrong.[36] Pauli later described the neutrino as 'that foolish child of the crisis of my life', referring to his troubled psychological state. His problems had begun earlier in the year, following a series of tragedies – the suicide of his mother three years before, the remarriage of his father to a woman Pauli loathed, and the ending of his brief first marriage, when his wife had the impertinence to leave him for a scientific mediocrity ('such an average chemist').[37]

The next day, Pauli left Princeton to return to Europe, but Dirac stayed to give a six-lecture course on quantum mechanics, ending with a presentation of his hole theory. In the closing few minutes, he affirmed more clearly than ever in public that anti-electrons should be detectable because:

[they] are not to be considered as a mathematical fiction; it should be possible to detect them by experimental means.[38]

Dirac repeated his suggestion that the idea could be tested experimentally by arranging for pairs of ultra-energy photons to collide: if the theory were correct, in some of these collisions the photons would disappear and an electron would appear with an anti-electron. But he was pessimistic. So far as he could see, it would not be feasible for experimenters to test the idea in the next few years.

He did not realise that the solution to his problem lay in the columns of the *New York Times*. Dirac read it regularly and must have seen the articles on the investigations of cosmic rays being carried out by Millikan, who had given them their catchy name in 1925. The rays had been discovered in 1912 but were still a mystery: all that was known for sure was that they had extremely high energy, typically thousands of times higher than particles ejected from atomic nuclei on Earth.[39] Millikan developed a religion-based theory of the cosmic rays and, by 1928, regarded it as 'fairly definite' that they were the 'signals broadcast throughout the heavens [. . .] the birth cries of infant atoms', clear evidence for divine benison.[40]

Dirac must have known that high-energy cosmic rays could produce anti-electrons if the rays collided with other particles on Earth. Yet it seems that he was never much interested in these particles, perhaps because he was influenced by modish opinion in the Cavendish Laboratory in the mid-1920s, when no one there studied the rays. Rutherford's deputy James Chadwick had sighed when he came across another of Millikan's research articles on cosmic rays: 'Another cackle. Will there ever be an egg?'[41] But that was six years before, and by the autumn of 1931 the attitude to the rays at the Cavendish was changing. The first of its scientists to latch on to their importance was Blackett, who was at a crossroads in his career, casting around for a new research topic.[42] This subject must have had a special appeal to the independent-minded Blackett as it would distance him from Rutherford, whose ego was becoming overweening.

Blackett was in the audience at a special Cavendish seminar on Monday, 23 November, when Millikan presented the latest photographs of cosmic rays to be taken at the California Institute of Technology (Caltech). The photographer was Carl Anderson, until recently Millikan's Ph.D. student, only twenty-six years old and already touted as one of the brightest experimenters in the United States. Three weeks earlier, he had pointed out to his boss that the new photographs showed 'Very frequent occurrence of simultaneous ejection of electron and positive particle'.[43] Anderson was trying to take images of the charged particles produced by cosmic rays using a cloud chamber, which enables the tracks of electrically charged particles to be photographed as they travel through a cloud of water vapour. Anderson had built his own cloud chamber and, at Millikan's suggestion, arranged for the entire chamber to be bathed in a strong and uniform magnetic field, which would deflect the paths of the charged particles as they hurtled through it. Each track contained crucial information: from the density of droplets along each track, Anderson could determine the particle's electric charge, and he could calculate the particle's momentum from the deflection caused by the magnetic field.[44]

It required great skill for Anderson to take any photographs at all. Most of his images were blank, but by early November he had obtained some 'dramatic and completely unexpected' images, which he sent to Millikan in Europe.[45] The photographs made no sense in terms of the theory they were using. In a puzzled letter to Millikan, Anderson remarked that many of the photographs featured the track of a negatively charged electron with a positively charged particle, two particles appearing at the same time, presumably when a cosmic ray strikes an atomic nucleus in the chamber.

When Millikan presented Anderson's inexplicable subatomic images in his seminar at the Cavendish, Blackett was fascinated. Here was a cloud-chamber expert with a talent that everyone knew was great but unfulfilled. Here was a new field in a mess. And here was the perfect opportunity for him to make his name.

Millikan's audience in the Cavendish seminar did not include Dirac, who was still in Princeton. Many of his colleagues, including Martin Charlesworth in St John's, feared they were about to lose him to one of the higher-paying American universities. Charlesworth wrote to Dirac saying how much he missed his 'kindly irony', imploring him 'Don't let them persuade you to stay in the USA. Here is your home.'[46] Charlesworth was right to be concerned, for Veblen was energetically wooing Dirac. Even before the carpenters and decorators had put the finishing touches to Fine Hall, Veblen had begun to work with the educator Abraham Flexner, who was trying to set up an institute for advanced study, where world-class thinkers could study in peace, free of all distractions. Einstein was at the top of their wish list, but they were competing with others, including the wily Millikan, at Caltech.[47]

Charlesworth may have worried, too, that Dirac might not be looking forward to returning home. From newspaper and radio reports, Dirac knew that his homeland was plunging into difficult times. On 21 September, the Government removed the pound from

197

the gold standard and allowed the currency to settle down to whatever price the money-market dealers were prepared to pay for it. It was a national humiliation. The economy plunged deeper into crisis: unemployment continued to escalate, and soon the pound had been devalued by 30 per cent, making Dirac's $5,000 fee for his single-term stay look even more generous. The inevitable General Election returned a stabilising coalition government, but the economic privations continued: that year, one in every two British industrial workers had been unemployed for over four months.

Yet the depression was still more serious in the United States, even in affluent Princeton. At the university, many students struggled to pay their fees. Around the town, young vagrants were walking the streets, some of the two million roaming the country in search of work. About thirty million Americans, a quarter of the population, had no income at all. Many people who had money were so frightened of losing it that they hoarded their dollars under mattresses or buried it in the garden. Even President Hoover – long in denial about the extent of the depression – realised that ordinary people were losing faith in the American way of life.[48]

As Dirac will have been aware, unemployment was said to be zero in the USSR. The admirers of Stalin's Five Year Plan in the press included the *New York Times*'s Moscow correspondent Walter Duranty, who called the plan a 'stroke of genius' and won the Pulitzer Prize the next year for his reports.[49] Yet Dirac's friends in the Soviet Union suffered terribly when Stalin's attitude towards science changed abruptly, from a subject worthy of study for its own sake to a weapon for fighting capitalism. Tamm and Kapitza supported the new Soviet line, at least in public, but Dirac heard the other side of the story from Gamow, who had been exasperated by the change in the Government's attitude when he returned to Russia in the spring of 1931. The Communist Academy had declared Heisenberg's version of quantum mechanics anti-materialistic, incompatible with the state's increasingly rigid version of Marxist philosophy. During a public lecture at the university on the uncertainty principle, Gamow experienced the full force of state censorship when a commissar, responsible for supervising moral standards, interrupted him and told the audience to leave. A week later, Gamow was forbidden to speak again about the principle in public.[50]

Since the mid-1920s, Gamow and Landau had been two leaders of the informal group of young Soviet theorists nicknamed the 'Jazz Band'.[51] In its seminars, the group discussed new physics, the Bolshoi Ballet, Kipling's poetry, Freudian psychology and any other subject that took their fancy. The Jazz Band was mastering the new quantum physics much more quickly than their professors – 'the bisons' – whom they teased unmercifully, while taking care to remain within the bounds of decorum. The Band overstepped the mark in 1931, however, when they ridiculed a new encyclopedia article on relativity theory, edited to conform to the Party's views on the subject. The butt of the Jazz Band's barbs was the Director of the Physics Institute in Moscow, Boris Hessen, a thoughtful Marxist who had fended off several of the Government's attempts to make orthodox theories of physics conform to 'dialectical materialist' principles, the philosophical basis of Stalinist Marxism, which accords much higher priority to concrete matters than to abstractions. Hessen had only a meagre knowledge of quantum mechanics and general relativity, so he was ill equipped to defend them against ideological interference from Stalin's officials.[52] This ignorance led him to write a ludicrous article in the *Greater Soviet Encyclopedia* about the ether, declaring it to be 'an objective reality together with other material bodies', contrary to Einstein's teaching. Gamow, Landau and three colleagues sent a mocking note to Comrade Hessen and were put on trial as saboteurs of Soviet science. Landau was temporarily banned from teaching at the Moscow Polytechnic, and the miscreants were banned from living in the five largest cities of the USSR, though the ban was not enforced. According to Gamow, the offending physicists had been found guilty by a jury of machine-shop workers.

Even Dirac fell foul of the censors when the Russian translation of his book was being edited, when his publishers objected that his quantum mechanics was in conflict with dialectical materialism. The book eventually appeared in bookstores after an uneasy deal between the publisher and the editor, Dmitry 'Dimus' Ivanenko, a Jazz Band leader and another of Dirac's effervescent Russian friends. In the awkward opening to the book, it is easy to see reflections of the delicacy of the deal: Ivanenko's preface is conventionally laudatory, but it is preceded by an apologetic note from the 'Publishing House', arguing feebly that although the material in the book is ideologically unsound, Soviet scientists need to use its methods to advance dialectical materialism.[53] A 'counterflow' of ideologically correct science will then follow, the publishers hoped.[54] In a simpering conclusion, Ivanenko thanked Dirac, 'a sincere friend of Soviet science'.

Censors were also scrutinising science in Germany, where the Depression was wreaking economic mayhem. Scruffy buskers, match-sellers and bootlace salesmen walked the streets in the hope of being paid a few *pfennig* to buy a loaf; tens of thousands of the unemployed queued outside Nazi offices, waiting for the storm troopers to reward them with a mug of hot soup. The once-peaceful Göttingen, where Born was Dean of his faculty, was now seething with political tensions: in the physics library he saw Communist leaflets, while outside the Nazis greeted each other ostentatiously with a click of their heels and a 'Heil Hitler' salute.[55] The Nazis, the majority party in the local government and student congress, were insisting that Einstein's 'Jewish physics' was wrong and pernicious. Born was beginning to think that he had no alternative but to emigrate.

To most people who came across Dirac, he seemed to be no more engaged with world affairs than an automaton. With no need to share his thoughts with others, unless they were close friends, he gave the impression that he was indifferent to the fate of others. He appeared to have none of the usual need to be warmed by the good opinion of other human beings.

At work in his office in the new Fine Hall, he was putting into practice the philosophy that he had preached earlier in the year, learning advanced topics in pure mathematics in the hope that they would find application in theoretical physics.[56] He had also returned to field theory, a subject he had co-founded four years before. The theory seemed fated to generate predictions that were not ordinary numbers but infinitely large. While Dirac was preoccupied with his ideas, Heisenberg and Pauli had been developing a full-blown theory of how electrons and photons interact with one another, a quantum theory that accounted for the spontaneous creation and destruction of particles, consistent with the special theory of relativity. Heisenberg and Pauli's theory was also consistent with both quantum theory and experiment, but it was ugly and unwieldy. Oppenheimer later described it as 'a monstrous boo-boo'.[57]

Unconvinced that this was the right way to describe nature at a fundamental level, Dirac sought a superior description, one that was logically sound and not plagued with infinities. The more Dirac looked into the Heisenberg–Pauli theory, the more he disliked it. In his view, it was not even consistent with the special theory of relativity because it describes processes throughout space using time measured by a single observer, whereas Einstein had taught that no single time could suffice for all observers, as they make different measurements of time. Dirac spent hours in Fine Hall examining the Heisenberg–Pauli theory and coming to terms with the problem of curing the sickness of field theory. The challenge would obsess him for the rest of his life.

By the end of the autumn, as Dirac's sabbatical was ending, it was clear that the industrialised world was sliding into its worst-ever economic crisis, and there was a disturbing new militarism in Germany, Japan, Italy and throughout much of east-central Europe. In Britain, everyone was talking about the possibility of another war. The spirit of the age was no longer caught in the freewheeling, life-affirming bravura of *Rhapsody in Blue* but in the headlong, ominous prelude to *Die Walküre*.

In Bristol, it had been a sombre autumn at 6 Julius Road. In her letters, Dirac's mother told him that she and his father had recovered from their climactic row and were back to their routine: she waited on him almost full-time, feeding him his vegetarian meals, washing his clothes and spending hours helping him dress. Each Sunday, she would give him – in silence – the 'ninety-degree' bath that he insisted was good for his rheumatism. After one of them, he had a heart attack. The family doctor told her soon afterwards that her husband 'is a man accustomed to his own way & will not take advice [. . .] He may live 20 years or he may go suddenly.'[58]

By September, the family were feeling the pinch of the economic crisis: Charles cut his tuition fees and insisted that they could no longer afford to run the car. When Betty told the family's bank manager this, he laughed, Flo told her son. She believed Charles had plenty of money stashed away, although he was spending virtually nothing. Earlier, when Flo tried to claim the small amount of money Felix had left six years before, the authorities sent her a form for her husband to sign as the law specified that the funds must be paid to him. She told Dirac: 'I tore up the form.'[59]

Dirac did not return in time for Christmas. Three days before the holiday, his mother wrote to him: 'I am always so grateful that you broke away from our narrow little life.'[60]

Dirac was about to have one of his most exhilarating years. The word on the physicists' street was that Chadwick was on to something important at the Cavendish Laboratory.[61] Chadwick – a lean, severe figure – was usually busy overseeing his colleagues' work, dispensing the paltry annual budget for equipment. But he had temporarily put administration to one side. Soon after the Christmas vacation, Chadwick had read an article that he suspected might lead to the neutron, a particle whose existence Rutherford had predicted.[62] In the article, two French experimenters – Frédéric Joliot and Madame Curie's daughter Irène – reported from their Paris laboratory that they had fired helium nuclei at a target made of the chemical element beryllium and found that particles with no electrical charge were ejected. They argued that these particles were photons, but Chadwick believed they were wrong and that the particles were Rutherford's elusive neutrons. Rutherford agreed. Having just turned forty, Chadwick may have sensed that this could be the last chance for him to make his name, to emerge from the shadow of his imperious leader. He hungrily grabbed the opportunity, working alone night and day, borrowing apparatus and radioactive samples from colleagues all over the laboratory, making new equipment, filling his notebook with data and calculations. Oblivious of the freezing Cambridge midwinter, he was in a world of his own, as his colleagues saw. After three exhausting weeks, he had nailed the neutron. He proved to his satisfaction, and Rutherford's, that his results made sense only if a particle with no charge and about the same mass as a proton is ejected in the nuclear collisions he observed. But when he wrote a report on his work for the journal *Nature*, he gave it the cautious title 'Possible Existence of the Neutron'.

On 17 February, Chadwick sent off his paper to *Nature*, which rushed it into print. Six days later, after a good dinner in Trinity College with Kapitza, he presented his results to his colleagues at the Kapitza Club. Relaxed and emboldened by a few glasses of wine, Chadwick confidently described his experiments, giving appropriate credit to his colleagues, and finally set out the powerful arguments for the existence of the neutron. It was a coup for Chadwick and for the Cavendish Laboratory, which had at last come up with the kind of ground-breaking result that Rutherford longed for – one that put nature into fresh focus, clarifying the very nature of matter. The audience gave him the unusual accolade of a spontaneous ovation. After the meeting, he asked 'to be chloroformed and put to bed for a fortnight'.[63]

The discovery gave fresh impetus to the notion that new types of subatomic particle might be predicted before they were detected. The ability to foresee the different types of grain in nature's fabric was a challenge to even the greatest scientists: Einstein had, in effect, predicted the existence of the photon but occasionally lost confidence in his idea before he was proved right; Rutherford – the experimenter's experimenter – had actually been more consistent, never wavering in his belief in the reality of neutrons. Perhaps Dirac's anti-electron and Pauli's neutrino were worth taking seriously, after all?

Sixteen

> I hope it will not shock experimental physicists too much if I say
> that we do not accept their observations unless they are confirmed by
> theory.
>
> SIR ARTHUR EDDINGTON, 11 September 1933[1]

The character of Paul Dirac first appeared on stage in a special version
of *Faust*, the *Hamlet* of German literature. Goethe's drama is the literary antithesis of Agatha Christie's penny-plain narratives that Dirac
wolfed down in the evenings. He had no taste for epic plays, but he
will have been absorbed in this *Faust*, a forty-minute musical parody
of the twenty-one-hour play, written as a physicists' entertainment.[2]

The authors, the cast and the audience were the physicists at Bohr's
spring meeting in April 1932, and Dirac was there. In the oasis of the
institute, physics had not looked more exciting for years, in hideous
contrast to the world outside. Chadwick's discovery had revitalised
interest in the atomic nucleus, whose detailed structure was a mystery to theoreticians. They had a wealth of other problems to solve,
too, including the status of quantum field theory and of the predicted
anti-electron, monopole and neutrino – each controversial, none yet
detected. As Bohr liked to point out, science often flourishes quickest
when it faces problems and contradictions; the Princeton physicist
John Wheeler once went so far as to spell out the central idea of the
institute as 'No progress without paradox'.[3]

The version of *Faust* performed at the Institute was in the tradition
of office Christmas parties, with their licensed burlesque and private
jokes that stay close to the boundaries of good taste but carefully
avoid crossing them. The journalist Jim Crowther was among the
audience of twenty-odd conference delegates who entered into the
spirit of the occasion, happily indulging the manifold crimes against
artistic taste.[4] Bohr, represented in the play by the Lord Almighty, sat
in the middle of the front row of the audience, convulsed with laughter as one of his colleagues mimicked his tortured oratory.

In Goethe's original play, the sharp-tongued Mephistopheles
seduces Faust, discontented with his limited wisdom, into a bargain

that grants him universal insight and the love of the beguiling virgin Gretchen. The main theme of the Copenhagen version is the story of the neutrino and of Pauli's attempts to persuade Ehrenfest of its existence. Pauli (not at the meeting) was represented by Mephistopheles, Ehrenfest by Faust, and the neutrino by Gretchen, whose songs Heisenberg accompanied at the piano. The original version of the play opens with speeches from three archangels, and the Copenhagen version began in the same way, except that the trio was represented by the English astrophysicists Eddington, Jeans and Milne, who stood on the almost room-wide desk of the main lecture theatre, declaiming in rhyming doggerel about the latest theories of the universe.

Ehrenfest's leg was pulled unmercifully. He was played as a character who lay on the couch with his trousers in disarray, meditating on the vanity of science and life. This probably struck some participants, including Dirac, as being too close to home: Ehrenfest was morose, deeply uneasy about the state of physics and losing his spark. At the meeting, when Darwin approached him with a question, he rebuffed him, saying only, 'I'm bored with physics.'[5]

In the second half of the playlet, Dirac comes under the spotlight. His monopole is a singing character, treated with respectful curiosity, in contrast to his hole theory, portrayed as bizarre and not wholly serious. In a few revealing lines, the character of Dirac describes the state of his subject:

> A strange bird croaks. It croaks of what? Bad luck!
> Our theories, gentlemen, have run amuck.
> To 1926 we must return;
> Our work since then is only fit to burn.

These few words accurately capture Dirac's despondency about the state of quantum field theory. He had tried to produce an improved version of Heisenberg and Pauli's relativistic version of quantum field theory but had found out during the meeting that his theory was no improvement at all: both field theories were shot through with infinities. The root of the problem appeared to lie in 'singularities', particular points in the theory where the mathematics become ill defined or even incomprehensible. It was a deft decision of the authors of the Copenhagen *Faust*, headed by Max Delbrück, to arrange for Dirac to exit the stage chased by the actor playing a bit part, Singularity.

The jibes about hole theory were not confined to the entertainment; throughout the meeting, Dirac had to put up with Bohr's hostile questioning and the taunts of other colleagues. Dirac appeared to take it all on the chin; according to one colleague, during the meetings that week he did not utter a word.[6] In the final session of the meeting, Bohr lost patience and put him on the spot: 'Tell us, Dirac, do you really believe in that stuff?' The room went silent, and Dirac stood briefly to intone his twelve-word reply: 'I don't think anybody has put forward any conclusive argument against it.' Although outwardly loyal to his interpretation of hole theory and to his proposal of the anti-electron, the absence of the particle was sapping his morale. Soon, even he stopped believing in his hole theory, he later told Heisenberg.[7]

Just less than three weeks after the Copenhagen meeting, news broke from the Cavendish of another experimental sensation: the atom had been split. It was the work of John Cockcroft and the dishevelled Irishman Ernest Walton, an expert in engineering hardware. Together, the two men had built the largest machine ever constructed in the Cavendish, capable of accelerating protons through 125,000 volts and smashing them into a metal target.[8] Quantum mechanics predicted that the accelerated protons should have enough energy to break up the nuclei at the heart of the lithium atoms, but it was a challenge to prove it. Cockcroft and Walton increased the intensity of their beam until it was high enough to stand a chance of splitting some of the atoms in their lithium target. After eight months of work, when the beam was delivering a hundred trillion protons per second, telltale flashes on the detector in Cockcroft and Walton's darkened laboratory told them that they had split lithium nuclei into two nuclei of a different element, helium. Here, on the nuclear scale, Cockcroft and Walton realised the dream of alchemists by transforming one type of element into another. For the second time in three months, Rutherford was overseeing the announcement of a great experiment. He was not best pleased when Crowther's news-management skills faltered and the story leaked to the press and broke in the popular Sunday newspaper *Reynolds's Illustrated News*, which trumpeted the latest Cavendish finding as 'Science's Greatest Discovery'.[9] Other newspapers soon followed, including a nervous *Daily Mirror*: 'Let it be split, so long as it does not explode.'[10]

When the discovery was announced, Einstein happened to be in Cambridge to give a lecture. On 4 May, at the height of public interest in the experiment, an intrigued Einstein paid a private visit to the Cavendish Laboratory for a demonstration.[11] He must have been gratified to see that Cockcroft and Walton's results were consistent with his most famous equation: the total energy of the particles involved in the nuclear reaction is conserved only if energy and mass are related by $E = mc^2$. Cockcroft and Walton had been the first to verify the equation.

Eddington – ready, as ever, with a down-to-earth analogy – linked Cockcroft and Walton's fragmentation of the nucleus to what appeared to be the fissuring of society. He observed that splitting the once-indivisible atom had become the ordinary occupation of the physicist since 1932 and that the social unsettlement of the age seemed to have extended to atoms.[12] By 1932, Cambridge University's political centre of gravity had moved sharply to the left. Only six years before, the great majority of students worked to break the General Strike; by May 1932, the Cambridge Union – bellwether of student opinion – supported the motion that they saw more hope in Moscow than in Detroit.[13] The students were fearful of another war, angry that the spirit of the Locarno Treaty was being mocked by events. Another war was beginning to look all but inevitable.

The Cavendish triumphs demonstrated the quality of Rutherford's leadership of experimental physicists in Cambridge. By comparison, the university's theoreticians were embarrassingly unproductive – their titular head was the Lucasian Professor Sir Joseph Larmor, then seventy-five and about to retire, not before time. To no one's surprise, the authorities announced in July that his successor was Dirac, who was not quite thirty and just a few months older than Newton's age in 1669 when he took the Chair. As soon as the authorities announced his appointment, he left Cambridge for a while to escape the clamour of congratulations.[14]

Dirac knew that the Chair was more than an accolade: it was a vote of confidence but also a challenge. He was expected to continue to be a leader, to set the pace in his field, to leave a legacy that scientists would talk about for centuries. By no means all the holders of the Lucasian Chair had justified their promise: William Whiston, John Colson and Isaac Milner are in no one's list of great mathematicians or scientists. Dirac still had more to prove. He was confident in the durability of his early work on quantum mechanics, though he had good reason to fear that his later ideas – field theory, hole theory, the monopole – might one day be regarded as honourable failures. Worse, he worried that he was becoming too old to come up with original theoretical ideas: earlier in the year, soon after Heisenberg's thirtieth birthday, Dirac told him: 'You are now past 30 and you are no longer a physicist.'[15]

Rutherford wrote to congratulate Dirac, hoping that he 'will still continue to be a frequent visitor to the Cavendish', probably an allusion to Larmor, who rarely set foot in the Laboratory. One of Dirac's colleagues summed up the mood when he told the new professor: 'I don't think any recent election to a professorship can have been more popular.'[16] Only Larmor was sniffy about his successor's appointment, later cattishly remarking that Dirac was 'an ornament of the German school [. . .] though a minor one.'[17]

Dirac did not look the part of the distinguished Cambridge professor. Shy as a mouse, he had so little gravitas outside the lecture theatre that in the streets of Cambridge he passed for a tyro graduate student. He was nervous in the company of women of his own age, so many of his colleagues assumed he was gay, that he would die a bachelor and had no interest in having children. Yet Kapitza knew better. He came to know Dirac well during their relaxed conversations in the Kapitzas' house, a noisy den that always seemed to be teetering on the edge of familial anarchy. Dirac was at ease there, talking with Kapitza and Rat over a Russian-style meal, playing chess and larking about with their two rumbustious sons. The contrast between the dysfunctional household of 6 Julius Road and the happiness he saw in the Kapitzas' home could scarcely have been plainer. Perhaps Dirac was already longing for the vibrant family life that Kapitza and Bohr had shown him, an environment in which sourness and unkindness were rare, not the norm.

By the standards of British academics, Dirac was wealthy. When he took up the Lucasian Chair, his annual salary rose sharply, from £150 to £1,200, supplemented by his annual college 'dividend' of £300. The modern value of his salary at the end of 1932 is £256,000. He had seen the last of penury, though for him frugality was too ingrained to be anything other than a way of life.[18] So far as he was concerned, a suit and a tie were all he needed, and he wore them indoors and outdoors, rain or shine, until most men would regard them as being fit only for the bin. His mother, perpetually chivvying him to smarten up, thought it was high time she bought some new clothes for herself and asked him to pay for them: 'If you have a really substantial salary in the autumn you may be able to treat your mother to a winter coat.'[19]

Charles and Flo were the toast of the city for producing its most famous scientist, but the old quarrels continued. Worried that Charles was planning to convert their daughter into a nun, Dirac's mother suggested that he pay for Betty to take a degree in French at the university. There was not much chance that Charles would pay for it as he believed that higher education should be a male preserve. Betty sensed this, as she told her brother in a letter: 'I haven't actually asked Pa for financial assistance, but he takes no interest in it and doesn't seem willing to help in any way.'[20] But Betty was not resentful: she accepted it as part of her father's character and, besides, most other men felt the same way.

In Betty's letters to Paul around this time, she seems conventionally affectionate to him, but nothing of substance is known about their relationship. It seems safe to conclude that he thought well of her, however, because in July 1932 he generously offered to pay for his sister's fees and expenses for the next four years.[21] Although she struggled before successfully crossing the first hurdle of gaining a mandatory pass in Latin, she was a contented student. In a touching letter to her brother she assured him, 'I will do my best to give you value for your money, and I am honestly working, for the first time in my life, I believe.'[22] Her educational liberation seems to have disheartened Charles, now a stooped and tottering invalid. He was slowly losing his grip on his family, Flo reported to her son: during a routine domestic stand-off about the use of their car, he huffily agreed to give in to her and Betty, but only after an hour's sullen reflection. It was a momentous moment, the first time in thirty-two years of marriage that she could remember him backing down.[23] He may well have wondered how his life had come to such a pass. Perhaps he would have sympathised with Fatty Bowling, the narrator of *Coming Up for Air*, George Orwell's satire on 1930s suburbia. Like Charles, Bowling was a hostage to his ungrateful family, tied by convention and financial convenience to a slattern he despised. Unlike Bowling, however, Charles took pleasure from his friends and his work: language students still traipsed up to 6 Julius Road for his tutorials, and he was still active in the local Esperanto Society.

By early August, Charles was planning to visit his family in Geneva. As usual, he did not tell his wife about his travel plans but disclosed them to his son, in a letter written almost entirely in French (only the final line was in English). He trod carefully:

7 August 1932

My dear Paul

I suppose that you are very busy so I will only take a few minutes of your time to tell you how happy and proud I am of your great success. All the newspapers have given us the details. Several friends and acquaintances have asked me to congratulate you on their behalf.

Will this new position change your plans to go to Russia? I would like to know the date when you have decided because as soon as I am strong enough to undertake the journey I should go to Switzerland to sort out some family matters and I do not want to be away from Bristol when you are here.

Obviously if you could come with me that would please me more.

My fond good wishes and may God prosper you.

Father[24]

But Charles was to be disappointed. His son was planning another vacation in the Soviet Union, this time with Kapitza in Gaspra, a mountainous coastal resort in the Crimea. In Stalin's time, it was a place for the scientific elite to take breaks, away from the forced migrations of peasant farmers, the food shortages and rationings and all the other disasters of the Five Year Plan and collectivisation.

Dirac had begun his trip at a conference in Leningrad, where he spoke about his field theory of electrons and photons. After Boris Podolsky – an American of Russian-Jewish blood – and Vladimir Fock told him that they were studying the same problem, Dirac agreed to work with them. During his stay in Kharkhov, Dirac collaborated with his Russian colleagues, and, after a long exchange of technical correspondence, they produced a surprisingly simple proof that Dirac's field theory is equivalent to Heisenberg and Pauli's and more transparently consistent with the special theory of relativity.[25] This project was another sign that Dirac was no longer quite so insular: early in the year, he had written a modest paper on atomic physics with one of Rutherford's students and now here he was, working on quantum fields in equal harness with Soviet theoreticians. But Dirac remained wary of collaboration: visiting theoreticians who were not previously acquainted with him found him distant, utterly uninterested in sharing his ideas.[26] When Dirac was visited by one of them, Leopold Infeld, the young Pole found him friendly and smiling but unwilling to respond to any statement that was not a direct question. After twice receiving a reply of just 'No', Infeld managed to phrase a technical query that drew from Dirac an answer consisting of five words. They took Infeld two days to digest.[27]

When Dirac was relaxing on the Crimean coast, he was unaware that the story of the anti-electron was approaching its conclusion more speedily than he had dared to believe possible. Many of the characters in this strange denouement, including Dirac, behaved in ways that are now barely comprehensible, even bearing in mind that hardly any physicists in 1932 took Dirac's hole theory seriously and few were even vaguely aware of his prediction of the anti-electron.

The end of the story began shortly before Dirac's vacation, at the end of July 1932 in Pasadena, not far from the Hollywood Bowl, where the Los Angeles Olympic Games were just beginning. It would be a welcome opportunity for the people of the city and millions of radio listeners to have some respite from the economic gloom and political manoeuvrings in advance of the coming presidential election.[28] At Caltech, many of the scientists were on vacation. But in a comfortably warm room on the third floor of the aeronautics laboratory, Carl Anderson was hard at work on the effects of cosmic rays within his cloud chamber. By the end of the first day of August, a Monday, all he had to show for his latest experiments were blank photographs, but, on the following day, he struck lucky.[29]

Anderson managed to take a photograph of a single track, just five centimetres long. It looked rather like a hair. The density of bubbles around the track seemed to indicate that it had been left by an electron, but the curvature of the path suggested otherwise – it had been left by a *positively* charged particle, so it could not possibly have been an electron. Still not quite believing his eyes, Anderson spent an hour or two checking that the poles of his magnet were correct and that they had not been switched by jokesters.[30] Convinced he was not the victim of a prank, he was elated, though his euphoria was cooled by an icy trickle of panic: was this really a discovery or some stupid mistake?[31] To clinch the existence of the positive electron Anderson needed more evidence, but by the end of the month he had found only two more examples of his unusual tracks, neither as cut and dried as the first. Millikan was not persuaded.

After the Olympic pageant had folded and the Caltech staff had returned after the summer break, Anderson wrote a short description of his experiment for the journal *Science*. Like Chadwick's presentation of his apparent discovery of the neutron, Anderson's account was cautious: he examined every conceivable reason why the track might not be a new particle. Even more circumspect than Chadwick had been, Anderson couched his claim to a discovery in a paper that he entitled 'The Apparent Existence of Easily Deflectable Positives', hardly an eye-catching phrase. Readers who reached the end of the article were rewarded with a sentence that qualifies as a masterpiece of scientific conservatism: 'It seems necessary to call upon a positively charged particle having a mass comparable with that of an electron.' According to one report, Anderson was so worried by his failure to find more good examples of the track that he thought of writing to *Science* to withdraw his paper. But it was too late: the article was at the printers.[32]

Here, under Anderson's nose, was clear evidence for Dirac's anti-electron – a particle with the same mass as the electron but with the opposite charge. Anderson had earlier spent several evenings a week struggling through Oppenheimer's evening lectures on Dirac's hole theory, so it is practically certain that he knew about the part played by the anti-electron within it.[33] Yet he did not make the connection, probably because he was directing his attention almost exclusively to the cosmic-ray theory of his boss.[34]

Anderson sent off his paper on 1 September, and it appeared in the libraries of American physics departments about eight days later, to be greeted with indifference and disbelief. His finding was 'nonsense', one of his Caltech friends told him. Millikan still believed that something was wrong with Anderson's experiment and so did almost nothing to promote it. Anderson, worried that he had not found another track like the one he detected in early August, spoke publicly about the need to be cautious.[35] Oppenheimer was almost certainly among the thousands of physicists who read the article, and he wrote soon after to his brother that he 'was worrying about [. . .] Anderson's positive electrons'.[36] But Oppenheimer failed to put two and two together. Perhaps he was blinkered by a narrow interpretation of Dirac's sea of negative-energy electrons: Dirac had always believed that this sea would contain some holes, whereas Oppenheimer assumed that the electron sea was always completely full, so that the concept of the hole was redundant. It beggars belief that Oppenheimer never pointed out the connection between Dirac's theory and Anderson's experiment to Dirac, to Anderson or to anyone else. Yet that appears to be what happened.

One of Anderson's colleagues did, however, take his result seriously. Rudolph Langer – a Harvard-trained mathematician, talented but not noteworthy – had read Dirac's work on the anti-electron and talked with Anderson and Millikan about the new cosmic-ray photographs. The day after *Science* published Anderson's paper, Langer sent a short paper to the journal, making connections between the new observations and Dirac's theories. Showing none of Anderson's restraint, Langer concluded that Anderson had observed Dirac's anti-electron. He did not stop there; he went on to build an imaginative new picture of matter, suggesting that the photon is a combination of an ordinary electron and a negative-energy electron, that the monopole is built from a positive and negative monopole and that the proton 'of course' comprises a neutron and a positive electron. The paper looks impressively imaginative today, but it made no impact in 1932, probably because Langer was not sufficiently respected to command attention and because it was simply not done to speculate with such abandon. His insight left no trace in Anderson's memory and was soon forgotten.

By early autumn, Anderson's 'easily deflected positive' appears to have been a minor query in the minds of most Caltech physicists, a rogue result to be refuted or possibly a puzzle to be solved. In Cambridge, no one seems to have been aware of Anderson's experiment or of Langer's article. The journal *Science* arrived in the Cambridge libraries by early November, but neither Dirac nor any of his colleagues appear to have read it. But, by then, Blackett was hot on Anderson's trail.

Rutherford had agreed that Blackett could begin a new programme of research into cosmic rays. But Blackett's patience with his boss's despotic style had worn thin, as a graduate student saw when Blackett returned from Rutherford's office white-faced with rage and said, 'If physics laboratories have to be run dictatorially [. . .] I would rather be my own dictator.'[37] Blackett carved out a niche in the Cavendish, working with an Italian visitor, Giuseppe Occhialini, a light-hearted Bohemian commonly known by his nickname 'Beppo'.[38] Ten years younger than Blackett, Occhialini was an expert experimenter who tended to rely on his intuition, rarely pausing to write down an equation, preferring to spell out the steps in his reasoning with an impressive range of accompanying gesticulations. When Occhialini arrived in Cambridge the year before, in July 1931, he had already been involved in experiments to detect cosmic rays and brought to the Cavendish years of experience working with Geiger counters, only recently introduced to Cambridge. These counters were delicate and unreliable, Blackett later remembered: 'In order to make it work you had to spit on the wire on some Friday evening in Lent.'[39] For Occhialini, Blackett was a jack of all trades in the laboratory:

I remember his hands, skilfully designing the cloud chamber, drawing each piece in the smallest detail, without an error, lovingly shaping some delicate parts on his schoolboy's lathe. They were the sensitive yet powerful hands of an artisan, of an artist, and what he built had beauty. Some of my efforts produced what he called 'very ugly bits'.[40]

Occhialini often visited Blackett at home in the evening. The two of them would relax in the front room and review their day's work over glasses of lemonade and a plate of biscuits, while Blackett fondled the ears of his sheepdog. During their conversations at home and in the Cavendish, they came up with a clever way of getting cosmic rays to take photographs of themselves: the trick was to place one Geiger counter above their cloud chamber and another counter below it, so that the chamber was triggered when a burst of cosmic rays entered both the upper and lower counters. By the autumn of 1932, Blackett and Occhialini had used this technique to take the art of photographing cosmic rays from a time-wasting matter of pot luck to a new era of automation. Soon, word circulated round the Cavendish corridors that something special was emerging from the Anglo-Italian duo. Even the reserved Blackett, the quintessence of the upper-crust Englishman, was excited.

Soon Blackett and Occhialini were ready to treat their colleagues to the clearest batch of cosmic-ray photographs ever taken. At their seminar, Dirac was in the audience. This was surely his moment: he could quite reasonably have suggested that Blackett and Occhialini had discovered the anti-electron and, therefore, vindicated his hole theory. But he stayed silent. The mention of the possible presence of positive electrons drew Kapitza to turn to the new Lucasian Professor, sitting in the front row, exclaiming, 'Now, Dirac, put that into your theory! Positive electrons, eh! Positive electrons!' Kapitza had spent hours talking with Dirac but had evidently not even heard of the anti-electron. Dirac replied, 'Oh, but positive electrons have been in the theory for a very long time.'[41] Here, unless electrons really were shooting upwards from the Cavendish basement, the anti-electron seemed to be showing its face. Yet Dirac's colleagues so mistrusted his theory that none of them was prepared to believe that it could predict new particles. Nor, it seems, did Dirac try hard to persuade them, perhaps because he believed that there was still a chance that every positive electron in his colleagues' photographs was in some way a mirage. This was reticence taken to the point of perversity.

At that time, Dirac was not concentrating on his hole theory but on one of his favourite subjects: how quantum mechanics can be developed by analogy with classical mechanics. In the autumn of 1932, he found another way of doing this, by generalising the property of classical physics that enables the path of any object to be calculated, regardless of the nature of the forces acting on it. Newton's laws could also do this job, and gave the same answer, but this technique – named after the French-Italian mathematician Joseph Louis Lagrange – was more convenient in practice. Dirac had first heard about this method when he was a graduate student, from lectures given by Fowler: it had taken some six years for the penny to drop.[42]

Although the technique is usually easy to use, it sounds complicated. At its heart are two quantities. The first, known as the Lagrangian, is the difference between an object's energy of motion and the energy it has by virtue of its location. The second, the so-called 'action' associated with the object's path, is calculated by adding the values of the Lagrangian from the beginning of the path to its end. In classical physics, the path taken by any object between two points in any specified time interval turns out, regardless of the forces acting on it, to be the one corresponding to the smallest value of the 'action' – in other words, nature takes the path of least action. The method enables physicists to calculate the path taken by any object – a football kicked across the park, a moon in orbit around Saturn, a dust particle ascending a chimney – and, in every case, the result is exactly the same as the one predicted by Newton's laws.

Dirac thought that the concept of 'action' might be just as important in the quantum world of electrons and atomic nuclei as it is in the large-scale domain. When he generalised the idea to quantum mechanics, he found that a quantum particle has not just one path available to it but an infinite number, and they are – loosely speaking – centred around the path predicted by classical mechanics. He also found a way of taking into account all the paths available to the particle to calculate the probability that the quantum particle moves from one place to another. This approach should be useful in relativistic theories of quantum mechanics, he noticed, because it treats space and time on an equal footing, just as relativity demands. He sketched out applications of the idea in field theory but, as usual, gave no specific examples; his concern was principles, not calculations.

Normally, he would submit a paper like this to a British journal, such as the *Proceedings of the Royal Society*, but this time he chose to demonstrate his support for Soviet physics by sending the paper to a new Soviet journal about to publish his collaborative paper on his field theory. Dirac was quietly pleased with his 'little paper' and wrote in early November to one of his colleagues in Russia: 'It appears that all the important things in the classical [. . .] treatment can be taken over, perhaps in a rather disguised form, into the quantum theory.'[43]

Even if Crowther had wanted to publicise this idea, he would have found it hard to get his article published in the *Manchester Guardian*: it was too technical, too abstract. The 'little paper' appears to have been too abstruse even for most physicists and so remained on library shelves for years, a rarely read curiosity. It was not until almost a decade later that a few young theoreticians in the next generation cottoned on to the significance of the paper and realised that it contained one of Dirac's most enduring insights into nature.

In the closing months of 1932, the news from Germany was that Hitler stood a fair chance of being elected chancellor in the impending elections: if Dirac's later comments on the Führer are anything to go by, he will have been uneasy at the prospect. Einstein, sick of the political climate and the violent anti-Semitism, fled to the USA and agreed to join Abraham Flexner's Institute for Advanced Study in Princeton, while Born hung on in Göttingen, where the Nazis were the largest single party: half its voters now supported them.[44] In the USSR, Stalin was showing ever-greater intolerance of academic freedom. In the USA, Franklin D. Roosevelt had been elected by a landslide, but the country remained in desperate economic straits. In the UK, unemployment rose to unprecedented levels, and there were mass demonstrations about unemployment benefits all over the country.

In the normally calm centre of Bristol, near the Merchant Venturers' College, hundreds of protestors were baton-charged by the police.[45] A mile away, the Dirac household was again a battlefield. With Betty spending most of her time at university, her parents were left to explore every crevasse of their fractured marriage. Flo told Dirac that his father, becoming more aggressive, was still trying to throw her out of the house. Charles was incensed when he heard that she had given a pupil wrong information about his tuition fees and threw a glass of hot cocoa at her, she reported to Dirac. Yet, to most of the people he knew, Charles looked like a model of the contented retiree. At the Cotham School prize-giving, the Headmaster praised him for his son's success, and they talked over tea and cakes about Dirac's recent trip to Russia. Flo wrote to her son, 'Really, he is quite a gossip outside his own home, where he only condescends to scold.'[46]

The Dirac family was together for what promised to be a torrid Christmas. But Charles and Flo ceased hostilities, and the family had what Flo described as 'quite the best Xmas we have had for years'.[47] Part of the reason for this may have been that Dirac was in a good mood, as news he had wanted to hear for eighteen months had just arrived.

Seventeen

Einstein says that he considers Dirac the best possible choice for another chair in the Institute [for Advanced Study]. He would like to see us try for D[irac] even if the chance of getting him is very small. He rates him ahead of everyone else in their field. He places Pauli of Zurich second, apparently.

Letter from OSWALD VEBLEN to ABRAHAM FLEXNER, 17 March 1933[1]

It seems that it was not until mid-December 1932 that Dirac was confident that the anti-electron exists. Later, memories were too hazy for the date to be made precise: Dirac recalled that he 'probably' heard the news from Blackett, who never said publicly when he was sure of the new particle's existence. It may be that he discovered it independently of Anderson, though Blackett was always careful to give credit to his American rival for being the first to put his observation into print. Blackett and Occhialini probably learned of Anderson's photographs in the autumn through the grapevine, but they read his article on 'easily deflectable positives' only in January, three months after its publication, when they were taking cosmic-ray photographs by the dozen every day.[2] In this bitterly cold Cambridge winter, Blackett and Occhialini had to trudge each morning to the entrance of the Cavendish through snow, slush and ice; inside, the laboratory was buzzing with the thrill of the new cosmic-ray photographs. It seemed that another success was in the offing, but there was a problem: no one was sure precisely what the images were showing.

The photographs featured a 'shower' of cosmic rays, with tracks that curved both to the left and to the right, emanating from a single location. In several of the snaps it was plain that Blackett and Occhialini had observed positively and negatively charged particles of about the same mass as they zipped through the cloud chamber: these appeared to be electrons and anti-electrons. Blackett asked Dirac to help interpret the data, and soon he was in the laboratory, doing detailed calculations using his hole theory. The most likely explanation was, they concluded, that incoming cosmic rays were breaking up nuclei and that in the vicinity of some of these break-ups, pairs of positive and negative electrons were being created. It was a classic application of Einstein's equation $E = mc^2$: the energy of the collision was converted into the masses of the particles. Dirac's calculations persuaded the hyper-cautious Blackett that the photographs were strong evidence for anti-electrons that behaved just as the Dirac equation predicted.

When Blackett and Occhialini were preparing to make their results public, Dirac was also reading about events in Berlin. In the November election, the Nazis had lost over two million votes and had seen their representation in the Reichstag fall, but on 30 January, after weeks of chicanery by Hitler and his supporters, he was appointed Chancellor. The following night, Göttingen was ablaze with torchlight as a procession of uniformed Nazis wended its way through the streets of the old town, singing patriotic songs at the tops of their voices, waving their swastikas and making anti-Semitic jokes. Hitler dashed naive hopes that he would moderate his policies on coming to power, swiftly implementing a dictatorship. On 6 May, the Nazis announced a purge of non-Aryan academics from universities, and, four days later, book-burning ceremonies were held all over Germany, including Göttingen and Berlin. Even before Hitler rose to power, Einstein had left Germany, and he quickly announced that he would not return.

Hundreds of other Jewish scientists were desperate to emigrate. Dozens were rescued by Frederick Lindemann, Rutherford's counterpart at Oxford University, a prickly and sarcastic snob who had toured universities in Germany in his chauffeur-driven Rolls Royce offering threatened academics a safe haven in his laboratory. Cambridge University did not openly recruit potential refugees but waited for them to apply: from scientists, it received thirty such applications every day.[3] One of them was Max Born, who was given a short-term academic appointment and – partly as a result of Dirac's support – an honorary position at St John's. In November, his colleague Pascual Jordan became one of three million storm troopers and proudly wore his brown uniform, his jackboots and his swastika armband.[4]

Although Heisenberg never joined the Party, he remained in Germany and was pleased that Hitler had come to power, if an anecdote related by Bohr's Belgian student Léon Rosenfeld is correct. Soon after Hitler became Chancellor, Bohr commented to Rosenfeld

that the events in Germany might bring peace and tranquillity, insisting that the situation 'with those Communists' was 'untenable'.
When pressed by Rosenfeld, Bohr remarked: 'I have just seen
Heisenberg and you should have seen how happy [he] was. Now we
have at least order, an end is put to the unrest, and we have a strong
hand governing Germany which will be to the good of Europe.'[5]

Although Dirac was privately appalled by Hitler's appointment,
his outward response was so discreet as to pass unnoticed except by
a few colleagues, including Heisenberg: Dirac vowed never again to
talk in German.[6] He had learned two foreign languages but now
wanted to speak neither of them.

International politics were not Dirac's only distraction. He was also
turning his attention to moral philosophy, probably as a result of
talking with the formidable Isabel Whitehead. 'Don't despise
philosophers too much,' she had counselled him after one of his visits, 'a great deal that they say may be useless, but they are after something which matters.'[7] Mrs Whitehead had been on the receiving end
of one of Dirac's tirades against the only academic discipline he
openly disdained. One of his *bêtes noires* was the internationally
admired Trinity College philosopher Ludwig Wittgenstein, regarded
by many as one of the cleverest academics in Cambridge. Several
decades later, Dirac remarked that he was an 'Awful fellow. Never
stopped talking.'[8]

Dirac's disenchantment with philosophers had degenerated into
hostility when he read the ignorant comments several of them made
on quantum mechanics; in a book review, he had already noted that
it had taken the Heisenberg uncertainty principle to awaken the dozy
philosophers to the revolutionary implications of quantum mechanics.[9] The philosophers who least offended Dirac and other theoretical physicists were the logical positivists, who held that a statement
had meaning only if it could be verified by observation.[10] There are
traces of this philosophy in three pages of notes Dirac wrote out by
hand in mid-January 1933, the raw and unpretentious jottings of a
young man who wants to take stock and clarify his thinking about
religion, belief and faith.[11] He had recently told Isabel Whitehead, 'I
am mainly guided in my philosophical belief by Niels Bohr', but
these notes indicate that mainstream philosophers influenced Dirac
more than he knew.[12]

Dirac begins by considering belief. Some of the things a person believes in, he remarks, are not based on evidence but simply because they promote happiness, peace of mind or moral welfare. Such things constitute a person's faith or religion. In the only example he gives to illustrate this, he considers suicide, pointing out that most people believe that it 'is not a good thing, although there is no logical reason against it'. He was still haunted by Felix's demise and by the feeble purchase of logic on grief.

When Dirac focuses on the transience of life, he is driven to an important moral conclusion: 'A termination of one's life is necessary in the scheme of things to provide a logical reason for unselfishness. [. . .] The fact that there is an end to one's life compels one to take an interest in things that will continue to live after one is dead.'

This, he says, is quite different from the unselfishness preached by orthodox religion, which he characterises as sacrificing one's interests in this life for one's interests in the next. Although he regards such a sacrifice as wrong-headed, he concedes – with uncharacteristic condescension – the argument made by many an imperial missionary that 'Orthodox religion would be very suitable for a primitive community whose members are not sufficiently developed normally to be taught true unselfishness.'

Although Dirac rejects religious faith, he accepts that another faith is needed to replace it, something to make human life, effort and perseverance worthwhile. This leads him to his credo, one that would later influence his thinking on cosmology:

In my case this article of faith is that the human race will continue to live for ever and will develop and progress without limit. This is an assumption that I must make for my peace of mind. Living is worthwhile if one can contribute in some small way to this endless chain of progress.

At the end of his notes, Dirac turns to belief in God. This notion is so vague and ill defined, he says, that it is hard to discuss with any rigour. He first gave his views on the subject in his diatribe at the 1927 Solvay Conference, and is no less scathing here: 'The object of this belief is to cheer one up and give one courage to face the future after a misfortune or catastrophe. It does this by leading one to think that the catastrophe is necessary for the ultimate good of the people.'

Perhaps Dirac had at least partly in mind his father's rediscovery of his childhood Catholicism after the death of Felix. Dirac himself had no such solace and had to try to cope with the tragedy entirely without a spiritual crutch. Unable to fathom what he takes to be the religious justification for how a benevolent deity could condone natural disasters – they are part of God's plan, ultimately to the good of humanity – Dirac concludes by dismissing the idea that religion has any place in modern life: 'Any further assumption implied by belief in a God which one may have in one's faith is inadmissible from the point of view of modern science, and should not be needed in a well-organized society.'

The entire document reveals that Dirac's thinking about morality and religion is suffused with two principal concerns: how these types of knowledge square with scientific observations and how they can be used as a guide to living. This is consistent with the approach of John Stuart Mill, who would have applauded Dirac's suggestion that a personally rewarding faith was sometimes needed to replace the untenable belief in eternal life and for everyone to feel that they are contributing in some way to human progress. Some of Dirac's turns of phrase – his reference to 'a well-organized society' in particular – might be a result of the influence of Mill's French colleague and friend Auguste Comte, the founder of positivism.[13] More likely, Dirac was taking the Marxist line that religion is 'the opium of the people'.

On Thursday 16 February, dozens of scientists made their way through the London fog in the fast-fading light of the late afternoon. They were heading for the grand Piccadilly home of the Royal Society, in the East Wing of Burlington House, on the site of today's Royal Academy of Arts. This was the headquarters of British science, a stone's throw from many of the city's finest shops and restaurants, a few minutes' walk from the West End theatres.[14] The audience, including Cockcroft and Walton, probably hoped that the first of the five talks that they would hear would be more exciting than its title: 'Some Results of the Photography of the Tracks of Penetrating Radiation'. Unusually for formal presentations like this, the audience included a posse of journalists – no doubt tipped off by Crowther – most of them probably wondering whether they were wasting their time. If there really was a good story here, why announce it so close to their deadline? It is likely that the newshounds hoped, too, that the handsome speaker at the front of the room was more excited than he looked. Shortly after four-thirty, Blackett rose.

His talk was sensational.[15] He described his experiment and showed vivid photographs of the showers of charged particles that continually rain down on the planet and yet, until these experiments, had never been recorded on film. Blackett had almost no sense of theatre, but when he projected the photographs of cosmic-ray showers – revealing the hitherto unnoticed showers of particles bombarding the planet from outer space – mouths fell open in disbelief. Although cautious in his interpretation of his pairs of positive and negative particles, Blackett said that they fitted 'extraordinarily well' with the Dirac hole theory. Here, in front of the audience's eyes, was plain evidence for particles emerging out of nothing and for the opposite process, in which electrons and anti-electrons annihilate one another as soon as they meet. Blackett described this as their 'death compact'.

After the talk, when the applause had faded, Blackett agreed to give interviews to journalists. Always the perfect gentleman, he stressed that the discoverer of the positive electron was Carl Anderson and that the best theoretical interpretation of the photographs had been given by Dirac. Where, then, was Dirac? He was giving a seminar in another part of Burlington House, unavailable for comment.[16]

The newspaper reports reflected the excitement of the briefing. Of all the London newspapers, the *Daily Herald* featured the story most prominently: the headline 'Science Shaken by Young Man's Researches' and 'Greatest Atom Discovery of the Century' was followed by a breathless account of the experiment. It made no mention of Dirac's theory. The anonymous writer excised Occhialini from the story, as did Crowther in the same morning's *Manchester Guardian*, where he interpreted the discovery using Dirac's theory and used Millikan's colourful term 'cosmic rays'. The *New York Times* also featured the story on the Friday morning and included a wary quote from Rutherford: 'there seems to be strong evidence of the existence of a light positive particle corresponding to the electron. But the whole phenomenon is exceedingly complex and a great deal of work will have to be done on it.' The reporter did well to extract this quote, as Rutherford did not attend the meeting, having made clear that he mistrusted Blackett and Occhialini's use of Dirac's ideas, which Rutherford believed were nonsense.

Not since Eddington's solar-eclipse announcement thirteen years before had a talk at the Society made such a splash in the international press. Eddington's shrewd handling of the press had made Einstein an international star, but Blackett's presentation was never going to do the same for Dirac. He had no wish at all to be a celebrity; the very thought of it would have revolted him. And, after Rutherford's guarded comments, few journalists will have been motivated to draw Dirac out of his carapace.

After the press reported Blackett's announcement, Anderson was on edge. Most physicists had not read or even heard of his paper on the 'easily deflectable positives', and he had not yet published his photographs in a professional journal. He had not even given the new particle a name. For several months, he and his Caltech colleagues had considered contracting the term 'positive electron' to 'positron' and, at the same time, suggested that the ordinary, negatively charged electron might be renamed the negatron. Other names were forthcoming, too: the astrophysicist Herbert Dingle in London recalled that Electra in Greek mythology had a brother Orestes and so suggested that the positive electron should be called the oreston. It was Anderson, hurriedly completing a long paper on his discovery, who chose the name that stuck: the positron.[17]

The debate about the positron rumbled on for months. Bohr thought the particle might not be real but caused 'by air current drift' in the cloud chamber. Only after Heisenberg and colleagues went on a skiing vacation in Bavaria with Bohr and took one of Anderson's cloud-chamber photographs did Bohr begin to believe that the positron existed. In California, Anderson wavered and Millikan refused to believe that electrons and positrons were produced in pairs, because the observations did not agree with his theory of cosmic rays. Even in Cambridge, the question was controversial for several months. Rutherford, uncomfortable with the idea that abstract theory could predict a new particle, liked his physics done bottom-up: 'I would have liked it better if the theory had arrived after the experimental facts had been established.'[18]

Although few theoreticians accepted Dirac's hole theory, many interpreted the positron's detection as another personal triumph, some once again wearily despairing that it was impossible to compete with him.[19] Tamm, writing to Dirac from Moscow, was unstinting in his praise and even implied that Dirac had given up hope that his prediction would be verified: 'your prediction of the existence of the [positron] [. . .] seemed so extravagant and totally new that you yourself dared not cling to it and preferred to abandon the theory.'[20] Dirac, privately pleased that his controversial theory had been vindicated by experiments, showed no emotion. He remarked thirty years later, with a detachment that went beyond the Olympian, that he derived his greatest satisfaction not from the discovery of the positrons but from getting the original equations right.[21] In case Dirac should be in the least pleased with himself, Pauli was as ready as ever to bring him down to earth: 'I do not believe in your perception of "holes" even if the anti-electron is proved.'[22]

It was only by the end of 1933 that the majority of quantum physicists accepted that the positron existed, that electron–positron pairs could be created out of the vacuum that the positron had figured in Dirac's hole theory before its detection. Only Millikan, almost alone in standing by his 'birth cry' theory of cosmic rays, held out against the pair-creation idea.[23] But by early 1934, the evidence for the new particle was incontrovertible: the number of positrons detected annually had risen, owing mainly to Blackett and Occhialini's technique, from about four in the previous year to a new annual total of thirty thousand.[24] More importantly, experimenters at the Cavendish and at other laboratories had demonstrated that positrons could be produced at will using radioactive sources on the laboratory bench rather than only as a consequence of showers of cosmic rays bombarding the Earth.[25] Again, Dirac monitored the experimenters' results to see if they agreed with his theory's predictions.

In hindsight, it was clear that if physicists had taken the Dirac hole theory seriously, the positron would have been detected several months earlier. Anderson later remarked that any experimenter who took the theory at face value and who was working in a well-equipped laboratory 'could have discovered the positron in a single afternoon' using radioactive sources.[26] Blackett agreed.[27] As Dirac appeared to realise later, he must shoulder most of the responsibility for this, as he never advocated strongly that experimenters should hunt for the anti-electron or suggested how they might detect it using apparatus readily available to them. Thirty-three years later, when asked why he did not speak out plainly and predict the anti-electron, Dirac replied: 'Pure cowardice.'[28]

Although Dirac believed he had predicted the positron, and talked about it publicly from 1933 onwards, some commentators have objected that 'prediction' is too strong a word.[29] Even Blackett wrote in 1969 that 'Dirac nearly but not quite predicted the positron,' words that will probably have stung Dirac if he read them.[30] The consensus among today's scientists, however, is that Dirac's role in foreseeing the existence of the positron is one of the greatest achievements in science. In 2002, shortly after the centenary of Dirac's birth, the theoretical physicist Kurt Gottfried went further: 'Physics has produced other far-fetched predictions that have subsequently been confirmed by experiment. But Dirac's prediction of anti-matter stands alone in being motivated solely by faith in pure theory, without any hint from data, and yet revealing a deep and universal property of nature.'[31]

During the past seven years, theoreticians had driven most of the progress in physics, but there were now clear signs – particularly from the Cavendish and Caltech discoveries – that experimenters were in the driving seat. Disillusioned with quantum field theory, and having worked for two years without coming up with what he regarded as a strong new idea, Dirac joined Kapitza in his laboratory. It was another unlikely pairing: the most reserved, cerebral theoretician working with the most outgoing, practically minded experimenter. Yet they were like brothers at play.

They were among the first users of the state-of-the-art facilities in the Mond Laboratory, which Rutherford had arranged to be built for Kapitza in the courtyard of the Cavendish, with funds from the Royal Society. Its opening in early February 1933 was a grand occasion, dozens of trilby-hatted journalists scribbling on their notepads as the procession passed, adding flashes of colour to the grey midwinter afternoon. Dirac was there, in his scarlet gown, watching the proceedings led by Stanley Baldwin, the university's Vice Chancellor and Deputy to the Prime Minister Ramsay MacDonald. During one of the ceremonies, Kapitza pointed to the body of a crocodile carved into the brickwork of the laboratory's main entrance by the modernist sculptor and typographer Eric Gill. Inside the laboratory foyer there was another Gill commission, a bas-relief of Rutherford, a carving that exaggerated the size of Rutherford's nose, making him look like a brother of Einstein. Some artistically conservative authorities in Cambridge were so upset by Gill's depiction that they spent three months trying to have it removed; their anger was diffused only after Bohr declared the carving to be 'most excellent, being at the same time thoughtful and powerful'.[32] During the furore, Rutherford remained indifferent, claiming that he did 'not understand anything about art'.[33]

Dirac and Kapitza conceived a new and potentially revealing experiment to probe the nature of light and electrons. As Dirac had seen for himself in Davisson's Manhattan laboratory, when a crystal is struck by a beam of electrons, their paths are bent, demonstrating that electrons can behave as waves. Thus, electrons and light resemble one another in that both behave sometimes as waves, sometimes as particles. Dirac and Kapitza hit on the idea of replacing the crystal with light. Their idea was to reflect light back and forth between two mirrors so that only a whole number of half-wavelengths of light can exist between the mirrors, analogous to the number of half-wavelengths on a rope that is held down at one end and swung at the other. Just as the crystal consists of a regular three-dimensional arrangement of atoms, the reflected light has a regular pattern of allowed wavelengths, so both should be able to bend the path of a beam of electrons. Such an experiment should be a unique probe of the wave-like and particle-like behaviour of both electrons and light. Dirac's calculations showed that it should be possible to detect the electron beam's bending but only if the reflected light is extremely bright, brighter than the best-available lamps. So the state of lighting technology had thwarted the first plans of Dirac and Kapitza to do experiments together. It would not be long, however, before they were back in the laboratory.

In spring 1933, the *Cambridge Review*, sober chronicler of the university's affairs, published an anonymous article pointing out that 'the young are now more concerned [with politics] than they have been for a long time past'.[34] The hedonism of the late 1920s had all but disappeared, giving way to alarm about the national economic malaise and the threat of war. Hitler, Mussolini and Stalin were shaking the English out of their indifference to political extremes. Winston Churchill, in the political wilderness, repeatedly warned of the need to rearm, but he was ignored.

At the Cambridge Union in late February, despite a barnstorming performance from the Fascist Sir Oswald Moseley, the motion 'This House Prefers Fascism to Socialism' was heavily defeated, another sign that the students favoured Stalin over Hitler.[35] The dons were also turning left, many of them dissatisfied with the unscientific approach taken by politicians to social issues and revolted by the harsh treatment meted out to the unemployed. A few political leaders emerged among the academics, egged on by Jim Crowther, who cleverly promoted his Marxist views without ruffling the feathers of the many scientists who were wary of political commitment. The ones who emerged as the socialist leaders were all workaholic males, able to combine high-flying academic careers with an energetic commitment to politics and, in some cases, effective popularisation. Quietest among them was Blackett, not a Communist but a firm supporter of the Labour Party. He was horrified to see that 'the whole structure of liberalism and free trade is collapsing all over the world', and was struck by 'the paradoxical situation in which so many starve in the midst of so much plenty'. Scientists and engineers had, in Blackett's view, 'produced the technical revolution which has led to this situation', and so '*must* therefore be directly concerned with the great political struggles of the day'.[36]

Most influential of all was Bernal, 'the Saint Paul of the science and society movement of the thirties', as one of his colleagues later described him.[37] He later remembered how he was inspired by the Soviet experiment:

[T]here was no mistaking the sense of purpose and achievement in the Soviet Union in those days of trial. It was grim but great. Our hardships in England were less; theirs were deliberate and undergone in an assurance of building a better future. Their hardships were compensated by a reasonable hope.[38]

Although Dirac talked politics with Kapitza and Blackett, he seems to have been one of the fellow travellers with the socialist and Communist scientists, never in the vanguard. The political activists were becoming impatient with Dirac's indifference to sharing new knowledge with people outside science: in a short article 'Quantum Mechanics and Bolshevism' in the *Cambridge Review*, the anonymous author reported on Soviet displeasure with the 'completely non-political character of his work, and its detached tone, divorced from problems and questions of the present day'.[39] In the summer, Bernal included Dirac in his list of intellectual 'culprits' – including Joyce, Picasso and Eliot – who were 'tending to a private dream world', indifferent to the popular accessibility of their work.[40] Dirac would have pleaded guilty as charged as he regarded it as his job to seek better theories of fundamental particles, not to inform the public about the search. Although he did not attend the annual meeting of the British Association for the Advancement of Science in September 1933, he agreed with its conclusion: scientists have a duty to contribute to public debate and should promote the importance of science and technology in getting the country back on its feet.[41] The community was leaning on Dirac and other scientists of his soloist ilk to speak out.

Dirac appears not to have bothered to tell his parents about his success with the positron. Their first excitement that year was a spring visit to Paris, where Betty was studying for her degree. She did not write to her mother but sent regular letters to her father, who was so thrilled when he heard that she might be heading for Geneva that he decided to drop everything and join her. Soon after 5 a.m., on the day after Betty's letter arrived, Charles and Flo headed down to the railway station via the tram, Flo carrying her husband's laden suitcase.[42] She returned home to receive a letter from Dirac inviting her to spend a day with him in Cambridge, and he later paid for her to take a ten-day cruise round the Mediterranean. 'Won't it be funny', she wrote to him from her cabin like a truant schoolgirl, 'if I get home and Pa doesn't know anything about it?'[43] So it turned out: Charles and Betty arrived back at 6 Julius Road in the middle of September, having cabled her in advance, the first communication Flo had received from her husband in eight weeks. This act of abandonment seems to have annoyed Dirac. For at least eight years, he had addressed his postcards home to both parents but, from then on, he addressed them only to his mother.[44]

Dirac had spent the summer in Cambridge, trying to understand the infinities that plagued his field theory of photons and electrons and reflecting on the work he had done during the previous year. He had proved the equivalence of his theory to Heisenberg and Pauli's, had discovered the action principle in quantum mechanics, had seen his prediction of the positron verified and had begun a promising laboratory project with Kapitza. This was one of the most distinguished years of work by any scientist in modern times, but Dirac was disappointed. He wrote to Tamm, who had complained that he was going through lean times: 'I am like you in feeling dissatisfied with my research work during the past year, but unlike you in having no external reasons to blame it on.'[45] He needed a vacation.

After hiking and climbing in Norway, Dirac was to attend a conference at Bohr's institute before moving on to Leningrad for the first Soviet Conference on Nuclear Physics, where he was sure to be feted as a star. But it turned out that he would be in no mood to savour the acclaim.

The atmosphere at Bohr's annual meeting in 1933 was tense and uneasy. It hardly felt right to enjoy a spirited debate about the positron or a cathartic game of ping-pong while Jewish colleagues in Germany were being hounded out of the country. But, with most physicists now convinced of the existence of the positron, Dirac could feel that his confidence in hole theory had been rewarded. Pauli, not wanting to be there to see it, skipped the meeting and went on vacation to the south of France.[46]

Bohr organised the usual week-long programme, combining talks at the institute and gatherings at his new home, a mid-nineteenth-century mansion in the south-west of Copenhagen, in the grounds of the local Carlsberg brewery.[47] Set in hundreds of acres of immaculate gardens, this was a grace-and-favour residence, a gift of the Government, who offered it, whenever it became vacant, to the person considered the most distinguished living Dane.

The physicists at the meeting were in buoyant mood, though Ehrenfest was in poor spirits. Pudgy-faced and overweight, he was losing his grip on physics; for him, the succession of research reports were now a dispiriting agglomeration of detail. Convinced that his own work was worthless, he was looking for a new, less prominent academic position where he could motor in the slow lane.[48] But he had not given up completely: during the discussions, he was still the unselfconscious inquisitor, pressing every speaker towards complete clarity, helping to draw attention away from irrelevancies and towards the saliencies of the new ideas. He was especially close to Dirac at this meeting, and they spent hours talking, keeping a few breaths away from the smokers' fug.[49]

After the closing speeches in Bohr's home, the physicists put their luggage in the entrance hall and said their goodbyes.[50] It was the usual bitter-sweet parting, but one delegate seemed especially out of sorts: Ehrenfest, about to catch a waiting taxi, looked flustered and unhappy. When Dirac thanked him for his contributions to the meeting, he was speechless and, apparently to avoid responding, hurried over to Bohr to say farewell. When he returned, Ehrenfest was bowing and sobbing: 'What you have said, coming from a young man like you, means very much to me because, maybe, a man such as I feels he has no force to live.' Ehrenfest should not be allowed to travel home alone, Dirac thought, but he changed his mind. Abandoning his usual assumption that people mean exactly what they say, he concluded that Ehrenfest meant to say not 'maybe' but 'sometimes' – he *sometimes* felt that life is not worth living. Trying to say the right thing, Dirac stressed that his compliment was sincere. Still weeping, Ehrenfest held on to Dirac's arm, struggling for words. But none came. He climbed into the taxi, which speedily made its way round the small grassy roundabout in front of the mansion, through the gardens, under the arch of the Carlsberg building and on towards the railway station.

A few days later, Dirac was sailing to Helsinki, playing deck games and relaxing in the sun, en route to the Soviet Union. Since Hitler came to power, the attitude of the USSR towards scientists from other countries had changed: Stalin no longer encouraged his own scientists to mix with foreign colleagues, and such liaisons became a crime, except for Dirac and a small number of other friends of the Soviet Union. Dirac was keen to make light of this when he wrote an ambassadorial letter to Bohr a month before, assuring him of a 'warm welcome from Russian physicists' and noting that the economy there was not depressed: 'the economic situation there is completely different from everywhere else'.[51] Like many other gullible guests, Dirac had virtually no idea of the extent of the starvation and economic tribulations in the Soviet Union since the beginning of the Five Year Plan and the adoption of the collectivisation programme: people went round with string bags in their pockets on the off chance that they should come across a queue.[52] In 1933, the privations were at their worst: the Soviet diet included little milk and fruit, and only a fifth of the meat and fish consumed thirty years before. Almost the only people to eat well were state officials and visiting dignitaries, such as Dirac, who was almost certainly unaware of the cost of the collectivisation programme: about 14.5 million lives during the previous four years, a higher death toll than the Great War.[53] But Dirac knew that times were hard and that even basic items of clothing were not in the shops: when Tamm said that he would not be able to buy a heavy coat he needed for the coming months of freezing cold, Dirac gave his own coat to him and spent the next winter in England without one.[54]

This conference was shaping up to be a highlight in Dirac's career, until he heard some appalling news from Amsterdam. Lunchtime in the city's Vondelpark on the last Monday in September had been like any other on an early autumn weekday: the mothers teaching their little children to feed the ducks, the cyclists whooshing past the strolling pedestrians, a few picnickers in the last of the bright afternoon light. But suddenly the calm was shattered by gunshots. A few onlookers gathered round a horrifyingly violent scene: a young boy with Down's syndrome, fatally wounded but still breathing, lying next to a man in his fifties, dead, part of his head blown away. The man was Paul Ehrenfest. Moments before, he had shot his son Wassik but had not quite summoned the will to kill him. Two hours later, the boy died.[55]

In countless confused seminars on the new quantum ideas, he had done more than anyone else to pick out the diamonds from the mud. He had now been drowned by the wave he had helped to create. Dirac, needing to clarify his own thoughts and feelings, wrote Bohr a four-page letter, describing his last moments with Ehrenfest.[56] Of all Dirac's surviving letters, this is among the longest and most emotionally direct. With the fluency of a novelist, he recalls every detail of his last meeting with Ehrenfest, more sensitive to emotional nuance than most of his colleagues would have believed. He lamented to Mrs Bohr that he should have taken Ehrenfest's last words to him more literally – a shortcoming of which no one thought Dirac capable – and that he should have advised her husband to keep Ehrenfest in Copenhagen. Dirac concluded that he 'could not help blaming himself for what happened'. Mrs Bohr replied with consoling words, thanking him for doing 'so much to make Ehrenfest's last days here as happy as his sad mood allowed'. She added, 'he loved you very much.'[57]

Ehrenfest had written a suicide note a month before the Copenhagen meeting – to Bohr, Einstein and a few other close colleagues, though not to Dirac. After declaring that his life had become 'unbearable', he concluded:

In recent years it has become ever more difficult for me to follow develop-ments [in physics] with understanding. After trying, ever more enervated and torn, I have finally given up in DESPERATION [. . .] This made me completely 'weary of life' [. . .] I did feel 'condemned to live on' mainly because of economic cares for the children [. . .] Therefore I concentrated more and more on ever more precise details of suicide [. . .] I have no other 'practical' possibility than suicide, and that after having killed Wassik. Forgive me.[58]

Ehrenfest never sent this terrible note. It was tragic that he did not live to take his place a few weeks later at the Solvay Conference, the climax of almost a decade of research into matter at its most elemen-tary level. Originally scheduled to be about the applications of quantum mechanics to chemistry, the organisers had decided in July 1932 – in the wake of the Cavendish discoveries that year – to switch the theme to the atomic nucleus. It was probably expected that Rutherford would be the cock of the walk at the meeting, but by autumn 1933 nuclear physics had moved on and was aflame with new discoveries, new ideas, new techniques. Rutherford, never one to avoid the limelight, may well have felt eclipsed as he saw the focus of attention turn to others: to America's most flamboyant young experimenter, Ernest Lawrence, and his invention of a high-energy particle accelerator so compact that it fitted on a desktop; to Enrico Fermi and his discovery that slow neutrons could induce some nuclei to undergo radioactive decay artificially; to Heisenberg and his new picture of the typical atomic nucleus as a combination of protons and neutrons, but no electrons.

Dirac's intuition was not as sure-footed in this subatomic realm: he disagreed with Heisenberg's view of the nucleus – soon to be in text-books – just as he did not believe in the existence of Pauli's neutrino. Dirac was most at home when he was teasing out the implications of quantum mechanics, and he was able to do so at the conference, but only after the organisers had been pressed to give him a slot by Pauli.[59]

This was to be another of Dirac's seminal talks. Having pointed out that the discovery of the positron had renewed interest in the existence of a sea of negative-energy electrons, he argued that the presence of these background particles forces physicists to rethink the concepts of the vacuum and of electrical charge. As Oppenheimer and one of his students had independently suggested, the vacuum was not completely empty but was seething with activity, vast numbers of particle–antiparticle pairs continually bubbling up out of nothing and then annihilating each other, in fractions of a billionth of a second. These processes of creation and destruction are so brief that there is no hope of detecting them directly, but their existence should cause measurable changes in the energies of atomic electrons. Likewise, Dirac suggested that the charge of an ordinary positive-energy electron should be affected by the presence of the negative-energy sea: the electrical charge of an ordinary electron should be slightly less than the value it would have if the background were absent.

But the theory was still replete with infinities. Dirac suggested ways of coping with this, using special mathematical techniques to make testable predictions. The audience could see that this was the work of a master, if one who was too clever by half. Pauli despaired of the theory ('so artificial'), while for Heisenberg it was 'erudite trash'.[60]

Dirac probably agreed with Pauli and Heisenberg more than he let on, for he knew as well as anyone that his techniques involved the sort of procedures results-hungry engineers would be happy to use but that would make any self-respecting mathematician blanch. Convinced that any fundamental theory worth its salt must make perfect mathematical sense, he was becoming seriously disenchanted with quantum field theory. This Solvay talk would be the last time he used the theory to probe the inner workings of the atom: he would go on to make other fundamental contributions to science, but this presentation marks the end of his golden creative streak, which he had sustained for eight years.

Midway through the autumn term in Cambridge, on Thursday, 9 November, Dirac received the telephone call that most first-rate physicists hope for, if only in secret. A voice from Stockholm told him that he was to share the 1933 Nobel Prize for physics with Schrödinger for 'the discovery of new and productive forms of atomic theory'; the deferred 1932 prize went to Heisenberg. Dirac was surprised by his own award but not by the other two, certainly not by the one given to Heisenberg – the principal discoverer of quantum mechanics, in Dirac's opinion.[61] Nervous of the inevitable press attention, Dirac considered refusing the prize, but he soon took

Rutherford's advice: 'A refusal will get you more publicity.'[62] The Dirac family first heard the news on the day of the announcement, soon after ten at night, when a note was slipped through their letter-box by Charles's friend Mrs Fisher.

The Nobel Prize for physics had been instituted in 1901, when it was awarded to the German experimenter Wilhelm Röntgen for his discovery of X-rays. The institution of the prize for physics – and also for chemistry, literature and physiology – was the idea of the Swedish inventor, Alfred Nobel, whose legacy funded the prize in perpetuity. Since the first year, the status of the prizes had grown, and, by 1933, the annual announcements of the winners were featured in newspapers all over the world. As some of the reports noted, Dirac was a special winner: at thirty-one, he was the youngest theoretician ever to win the prize for physics.[63]

Most English national newspapers mentioned Dirac's prize on the day after it was announced.[64] The *Daily Mail* squeezed in a short report about the award to the 'silent celebrity' next to a long article on 'Hitler's homage to fallen Nazis'. Readers of *The Times* also read of Dirac's award alongside a report from Germany, where Hitler's deputy, Rudolf Hess, had issued regulations to ensure that electioneering is 'conducted in a dignified manner'. None of the hurriedly prepared articles mentioned the discovery of the positron or captured Dirac's personality; it was left to the *Sunday Dispatch* later in the month to publish an overheated but insightful description of Britain's newest Nobel laureate. The anonymous author noted that 'more than publicity, [Dirac] fears women. He has no interest in them, and even after being introduced to them, cannot remember whether they are pretty or plain.' Dirac was 'as shy as a gazelle and modest as a Victorian maid'.[65]

The first congratulatory note to arrive in Dirac's pigeonhole was a telegram from Bohr. Dirac replied with forgivable sentimentality:

I feel that all my deepest ideas have been very greatly and favourably influenced by the talks I have had with you, more than with anyone else. Even if this influence does not show itself very clearly in my writings, it governs the plan of all my attempts at research.[66]

In the Cavendish, the announcement of the prizes was welcomed by everyone except Max Born, bitter that he had been passed over in favour of Dirac.[67] Others in Cambridge were preoccupied with the most dramatic event to take place in the town for years: on Armistice Day, three days after Dirac heard from Stockholm, the Socialist Society organised a march of hundreds of students through the centre of Cambridge, seeking 'to provoke clashes, to make a stir [. . .] to put politics on the map and into university conversation; to bounce, startle, or shock people into being interested'.[68] In a normal Armistice Day march, a carnival of several hundred undergraduates walked through the city centre, selling blood-red paper poppies to passers-by in order to raise money for survivors of recent wars and to commemorate the lives of soldiers who had fallen in battle. The tragic aspect of the proceedings was often lost in hilarity, making the occasion ripe for subversion. On that grey Sunday afternoon, the pavements of Cambridge were lined with crowds, jeering as they were passed by marchers, some of them holding the banner pole of the Socialist Society, others bearing a wreath inscribed 'To the victims of the Great War, from those who are determined to prevent similar crimes of imperialism'. The second phrase should be removed, the police escorts insisted, as it might provoke a breach of the peace. By the time the marchers reached the entrance to Peterhouse College, an eruption was inevitable. Onlookers threw flour and white feathers over the students and pelted them with rotten eggs, tomatoes and fish; the marchers retaliated by using a car as a battering ram to push back their tormentors.

The university authorities panicked. Away from the public posturing, students and dons debated round college firesides whether the marchers had desecrated the day of remembrance or had restored seriousness to what had become a maudlin carnival. The event had marked the beginning of a militant student socialist movement in Cambridge.

In his rooms in St John's, the Lucasian Professor probably watched the events carefully and pondered how he could make his feelings heard.

Eighteen

Few misfortunes can befall a boy which bring worse consequences
than to have a really affectionate mother.
W. SOMERSET MAUGHAM, *A Writer's Notebook*, 1896

It has often been said that Dirac hated his father so much that he
denied him an invitation to attend the Nobel ceremony.[1] Plausible
though the story sounds, it is probably untrue. The Nobel
Foundation invited the laureates each to bring only one guest, but
they could bring others if the prize-winner paid for their travel and
accommodation.[2] Heisenberg took his mother, and Schrödinger
brought his wife, so it did not look at all odd that Dirac was accompanied only by his mother. She gave her husband a dose of his own
medicine by not telling him about her trip until a few days before she
set off, determined to make the most of her time away. She knew
that, in only eleven days, she would back at the kitchen sink, the
Cinderella of 6 Julius Road.[3]

Early on the Friday evening of 8 December 1933, Dirac and his
mother were in the Swedish port of Malmö, waiting for the night
train that would take them to Stockholm in time for breakfast. A few
reporters spent several hours hunting for them all over Malmö and
eventually tracked them down to a station café, which became the
unlikely scene of a press conference. The journalists' persistence was
rewarded with a newsworthy interview with two prize eccentrics, 'a
very shy and timid boy' and 'a lively and talkative lady'.[4]

'Did the Nobel Prize come as a surprise?' asked one journalist. 'Oh
no, not particularly,' Dirac's mother butted in, adding, 'I have been
waiting for him to receive the Prize as hard as he has been working.'
She was so curious about Sweden that one reporter found himself
answering her questions rather than asking his own – here was a
woman who revelled in the attentions of the press. Dirac did not stay
silent but was unusually forthcoming when the journalist from
Svenska Dagbladet asked him how quantum mechanics applies to
everyday life and was rewarded with a stream of insights into his
unapologetic philistinism:

DIRAC: My work has no practical significance.

JOURNALIST: But might it have?

DIRAC: That I do not know. I don't think so. In any case, I have been working on my theory for eight years and now I have started developing a theory that deals with the positive electrons. I am not interested in literature, I do not go to the theatre, and I do not listen to music. I am occupied only with atomic theories.

JOURNALIST: The scientific world that you have built during the past eight years, does it influence the way you look at everyday occurrences?

DIRAC: I am not that mad. Or rather, if it did [have such an influence] then I would go mad. When I rest – that is when I am at sleep of course also when I am taking a walk or when I am travelling – then I make a complete break with my work and my experiments. That is necessary so that there is no explosion here. (*Dirac points to his head*).

The story of the interview was on the news-stands in Stockholm station when the Diracs arrived shortly before eight o'clock in the morning. A quarter of an hour later, Heisenberg, Schrödinger and their guests stepped off the train and were met by a posse of dignitaries, all of them concerned that Dirac and his mother were nowhere to be seen. But when the photographers asked for the laureates and guests to pose, Dirac and his mother stepped forward into the flashes of the awaiting cameras. The welcoming committee was apparently too stunned to ask where they had been and only later heard what had happened: after Dirac's absent-minded mother had failed to wake up when the train reached the station, she had been ejected by a guard, who had thrown her clothes, hairbrush and comb out of the carriage window.[5] After the kerfuffle, the Diracs had made their way to the warm waiting room and had sat apart from the party of officials. When the group left the room, the Diracs followed them like a pair of ducks, without saying a word.

Heisenberg and Schrödinger obliged the press with interviews, but Dirac wanted to escape to the hotel as quickly as politeness allowed.[6] He and his mother were accompanied on the short chauffeur-driven journey to their hotel by the Nobel Foundation's attaché Count Tolstoy, a grandson of the novelist and a polished diplomat. His first challenge was to sort out the Diracs' accommodation in the 500-room Grand Hotel, overlooking the harbour. The staff must have thought they had done Dirac a favour by putting him and his mother in the bridal suite, but Flo was having none of that and demanded a room of her own. After making plain his displeasure, Dirac – about to pocket his prize money, approximately £200,000 in today's money – took the cost on the chin.

While Heisenberg and Schrödinger were relaxing in their baths, Dirac escaped the gaggle of journalists by leaving the hotel surreptitiously, taking his mother with him. They were then free to walk anonymously around the chilly city, in its best suit for the Nobel celebrations, a pre-Christmas festival unique to Stockholm. It looked like fairyland when darkness fell, the firs and Christmas trees lit up with coloured electric lights, the murmurings of the crowd accompanied by the tinkling of lounge pianists and the occasional cry of a seagull overhead.

Flo was not going to be deprived of press attention for much longer. While Dirac was resting, she held court with four journalists, inviting them separately to her suite to talk about her son and to show them the frocks, furs and jewellery he had bought her. The reporters already knew she was a colourful character, but they were not prepared for her torrent of maternal ardour, delivered in words that resembled 'shattering beads of quicksilver', as the *Svenska Dagbladet* put it. In the interviews, her eyes darted around as she delivered a disjointed, stream-of-consciousness lecture, as if she had been given two minutes to convince them that her son was Superman. One of her targets was the Nobel authorities, who had shamefully credited her son only as 'Dr Dirac' when he is 'the top professor in the world!'

Asked about life at home, Mrs Dirac laid into his father, 'the domestic tyrant', a man who hated wasting time and whose motto was 'work, work, work'. Not mentioning Felix, she described how Charles leant heavily, and unnecessarily, on the young Paul to study, not allowing him to play with other boys: 'If the boy had shown any other tendencies they would have been stifled. But that stifling was not necessary. The boy was not interested in anything else.'

As a result, Dirac had never known what it was to be a child. None of the journalists appears to have asked her if she took any responsibility for this; it was all the fault of her husband, she thought. When a reporter enquired whether Dirac's father was happy about his son's success, Flo replied disingenuously: 'I would not say so. The father has been surpassed and he doesn't like it.' What of her son's interest in the opposite sex? 'He is not interested in young women [. . .] despite the fact that the most beautiful women of England are in

Cambridge.' The only women he cares for are his mother, his sister and 'perhaps ladies with white hair' (she may have been referring to Isabel Whitehead).[7] Since Flo had vetoed the visit of Felix's girlfriend a decade earlier, possibly before, Dirac had known that his mother feared that young women would be attracted to him, and her attitude had not changed.

On the following day, the Stockholm news-vendors sold newspapers with headlines that included 'Thirty-One-Year-Old Professor Dirac Never Looks at Girls'.

Early on Sunday evening, hundreds of coiffed men and women packed the galleries at the Stockholm Concert Hall to witness the King's presentation of the prizes. At 5 p.m. sharp, a blazing chorus of trumpets silenced the crowd before the opening of the two huge doors into the room where the prizes would be awarded. Each of the laureates, escorted by one of the Swedish hosts, marched to their separate armchairs by the platform, covered in red velvet and decorated with banks of pink cyclamen, maidenhair ferns and palms. The national flags of the new laureates hung overhead alongside Sweden's. The prize-winners were in the customary starched white shirt and bow tie, and all of them wore dinner suits, except Dirac, who won the sartorial booby prize by wearing a pitifully old-fashioned dress suit. He bowed low to the King before accepting his medal and certificate and then bowed several times to the crowd amid tumultuous applause. Compared with Heisenberg, Dirac looked pallid and sickly: he looked 'far too thin and stooping', one reporter worried, adding that 'All the motherly ladies warmly hoped that he should feed up and get the time to exercise and enjoy himself a bit.'[8]

After the ceremony, the laureates were driven back to the Grand Hotel to attend the Nordic midwinter feast of the Nobel Banquet, in the winter garden of the Royal Salon. Even by the standards of Cambridge this was a spectacular setting for a dinner: the tables, lit with hundreds of bright-red candles in silver holders, were arranged in a horseshoe shape around the water fountain in the centre of the room. There were three hundred guests, every woman in her most scintillating gown, every man in a dinner jacket, except Dirac.[9] At the top table, men were seated alternately with women.[10] On a balcony above, liveried musicians played, in competition with canaries chirruping in their cages near the glass roof.

After the speeches, a silent toast to the memory of Alfred Nobel and the singing of the Swedish national anthem, a fleet of waiters began to deliver the first course from a menu that featured game consommé, sole fillets with clams and shrimps and fried chicken with vegetable-stuffed artichokes. The climax was the chef's *pièce de résistance* dessert: ice-cream bombes that shone in the dark after they had been doused in alcohol and set alight.[11] Afterwards, each laureate was expected to make a short speech, customarily a few pieties of gratitude and reflection, laced with self-deprecating wit. After the first speech – given by Ivan Bunin, winner of the prize for literature – Dirac rose from his seat and walked to the rostrum, where, as usual, he shed his shyness. After paying his compliments to the hosts, he declared that he was not going to speak about physics but, instead, wanted to outline how a theoretical physicist would approach the problems of modern economics. This was just the kind of applied thinking that Bernal and his colleagues had been urging Dirac to do, but they might have expected him to choose a different venue for his first public comment on social and economic affairs. Nervous glances were exchanged round the great hall as he leaned over the rostrum and presented an argument that all the economic troubles of the industrialised world stemmed from a fundamental error:

[W]e have an economic system which tries to maintain an equality of value between two things, which it would be better to recognise from the beginning as of unequal value. These two things are the receipt of a certain single payment (say 100 crowns) and the receipt of a regular income (say 3 crowns a year) through all eternity. The course of events is continually showing that the second of these is more highly valued than the first. The shortage of buyers, which the world is suffering from, is readily understood, not as due to people not wishing to obtain possession of goods, but as people being unwilling to part with something which might earn a regular income in exchange for those goods. May I ask you to trace out for yourselves how all the obscurities become clear, if one assumes from the beginning that a regular income is worth incomparably more, in fact infinitely more, in the mathematical sense, than any single payment?

Without bothering to suggest how his explanation could be tested, he concluded with a Rutherfordian swipe at science popularisers, informing the diners that once they had done their homework, they will have 'a better insight into the way in which a physical theory is fitted in with the facts than you could get from studying popular books on physics'.[12] After thanking the audience for its patience, he returned to his seat. A spatter of clapping gradually gathered into firm applause, many of the diners laughing nervously and apparently wondering what to make of Dirac's speech. Heisenberg and Schrödinger did not follow suit by talking about economics and politics; speaking in German, they gave speeches that followed the convention of steering clear of anything that might be politically controversial.

Dirac's reasoning puzzled Schrödinger and his wife, and Anny described it as a 'tirade of communist propaganda'.[13] But if the written record of Dirac's speech is accurate, she was being unfair: Dirac was addressing a topic of theoretical economics that transcended politics. He was also wrong: his theory is approximately correct only when interest rates are always low, but he had not taken into account that it makes good sense to take the lump sum if interest rates are high and remain so.[14] If Dirac had bothered to consult a professional economist, such as his Cambridge colleague John Maynard Keynes, he would have been spared posterity's judgement that in his first foray outside his own field he had talked nonsense. And he had done so in the glare of the Nobel spotlight.

Dirac's fallacy seems to have gone unnoticed or, at least, unremarked in the after-dinner levity. Flo watched Heisenberg and Schrödinger closely as they laughed and joked with the other guests, while Dirac strained to make conversation and occasionally disappeared from gatherings, as if vanishing into thin air. Flo kept a sharp eye on Schrödinger, not caring much for his braggadocio: by far the oldest of the trio of physics prize-winners, he kept trying to assert himself as their leader, though Heisenberg and Dirac declined to follow him. She also noticed that Schrödinger and his wife 'terribly resent' that he had to share his prize with her son. More to her liking was the genial Heisenberg and his mother, dressed like a Dresden shepherdess. Flo admired Heisenberg for having 'no swank at all', although she thought him a 'terrible flirt', like her son, and she complained that both of them cruised the circles of adoring ladies before they ran 'back to [their] poor, tired mother[s] whenever they have had enough'.[15] She had not previously seen Dirac in the company of admiring young women, and she did not like it: whether or not she noticed, he was drifting away from her.

The lavish hospitality continued for four days, unabated. Dirac's only task was to give his Nobel lecture on the Tuesday afternoon, traditionally an opportunity for the laureates to present their work to other academics. Dirac spent most of his twenty-minute presentation on 'The Theory of Electrons and Positrons', describing how quantum mechanics and relativity made possible 'the prediction of the positron'. This was the first time he had referred to his speculation about the positron as a prediction, and he went on to repeat another of his speculations, with more confidence than usual: 'It is probable that negative protons can exist.' Finally, after pointing out the apparent symmetry between positive and negative charge, he hinted that the universe might consist of equal amounts of matter and anti-matter:

[W]e must regard it as an accident that the Earth (and presumably the whole solar system), contains a preponderance of negative electrons and positive protons. It is quite possible that for some of the stars it is the other way about, these stars being built up mainly of positrons and negative protons. In fact, there may be half the stars of each kind.[16]

He had glimpsed a universe made from equal amounts of matter and anti-matter in which, for some unknown reason, human experience is confined almost entirely to matter. But was this a speculation or a prediction? The audience had good reason to be unsure.

Dirac appears to have been unaware that he was not the first to imagine a universe made of both matter and anti-matter. In the high summer of 1898, soon after J. J. Thomson had discovered the electron, the Manchester University physicist Arthur Schuster had hatched a similar idea. In a light-hearted article in a summer edition of *Nature*, he conceived a universe made of equal amounts of 'matter and anti-matter', based on the bizarre idea that atoms are sources of invisible fluid matter that flow into sinks of anti-atoms.[17] But Schuster's whimsy lacked substantial underpinnings from reason or observation and so remained a 'holiday dream', as he termed it. Within a decade, it was forgotten.

After the Nobel festivities, most of the prize-winners usually return home. But Dirac, Heisenberg and their mothers moved on to yet more celebrations, in Copenhagen. Bohr, probably wanting a piece of the action, threw a grand party in their honour on the Saturday evening at his mansion. Schrödinger, not a member of Bohr's inner circle, declined his invitation and returned to Oxford, where he was living, having fled Germany a few months before. His colleagues in England looked askance at his personal life – he lived with his wife and his mistress – and he, in return, despised the colleges as 'academies of homosexuality'.[18]

Dirac's mother had heard many stories about the agreeable life at the court of Bohr, and she was not disappointed. Bohr's was a 'commanding' presence, Flo observed, and she was charmed by his wife Margrethe, whose donnish air was lightened by her daring dress, a green morning frock trimmed with leopard skin and yellow beads.[19] The Bohr residence was looking resplendent: the sprays of winter flowers and ferns, the statues, the cubist painting hanging above the grand piano, the huge windows overlooking acres of garden and woodland. For Flo, this opulence had done nothing to spoil the family, least of all the Bohrs' five playful but well-behaved boys.

Bohr was out during the guests' first evening at the house and returned to find that Dirac had been the first to retire to bed. Unwilling to lose precious time, Bohr bounded up to Dirac's room and brought him downstairs for a discussion that lasted into the small hours. She could now see why Dirac held Bohr in such affection: here was an older man, authoritative but not authoritarian, forceful but not intimidating, able to bring out the best in everyone. It may well have crossed Flo's mind that Bohr would have been the perfect father for her son.

The Bohrs' party would not have disgraced one of the Nobel Foundation's receptions. In the mansion's main hall, three hundred guests sat at tables under the huge glass roof, drinking the endless supplies of champagne, beer and wine and eating the food from the generous buffet. When everyone had eaten, Bohr stood in the centre of the hall and gave a speech in English, subtly ensuring that no one overlooked his contribution to the achievements of his 'young pupils'. Heisenberg replied, in German, but Dirac said nothing; throughout the speeches, he stood behind a pillar. After the toasts, Bohr steered the party into the drawing room for a cabaret from a pink-frocked American singer accompanied by the Danish virtuoso Gertrude Stockman and, inevitably, by Heisenberg at the piano.

Dirac will probably have found the celebrations a chore and will have been relieved to spend the next day only with people he knew, a relaxed family Sunday. The many in Cambridge who saw Dirac as a shadow of a man, with no sense of fun, would have been surprised to see him at ease in the Bohrs' nest, playfully squirting water from an indoor fountain over his mother and Margrethe, both of them laughing and protesting as they tried hopelessly to shield themselves from the dousing. Dirac's Cambridge acquaintances would not have expected, either, that he would happily spend a day larking around with Bohr, his boys and Heisenberg, playing badminton and sleighing on the hills near Copenhagen. In the evening, Dirac reverted to his usual stand-offishness: he sloped off to bed early, not bothering to wish anyone goodnight. But Bohr wanted Dirac to talk shop and so yanked him back downstairs.

On her return to Bristol late on Monday, Flo was met at the railway station by Betty, who was up until the small hours listening to her mother's account of her 'great and wonderful adventure'. Charles was nowhere to be seen.

For the rest of his life, Dirac was curious about how he came to win the prize with Heisenberg and Schrödinger. The Nobel Foundation, always the essence of discretion, releases the papers concerning each year's prize only after keeping them under lock and key for fifty years. Dirac never did find out about the political machinations that led to the first prizes for quantum mechanics; he eventually learned only that the English crystallographer William Bragg had nominated him and that Einstein had not.[20] Only after Dirac died did it come to light that he had been fortunate to win the prize so young.

In the first three decades of the prize, the committee that decided the Nobel Prize for physics was biased against theoretical contributions, probably because of Alfred Nobel's wish that his prizes should reward practical inventions and discoveries. The committee, not always well informed about theoretical physics, issued a statement in 1929 that the theories of Heisenberg and Schrödinger 'have not yet given rise to any discovery of a more fundamental nature'.[21] Behind the scenes in Stockholm, a long and involved battle was being fought about when to award a prize for the new theory and who should receive it. The Foundation was still arguing about this in 1932, when nominations for Heisenberg and Schrödinger were accumulating by the month. By early 1933, the pressure to award a prize for the theory was overwhelming, but there were still disagreements about how to share it. Dirac's name had barely registered with the committee.[22]

By the time the committee met in September 1933, after the discovery of the positron had become widely accepted, his name was
much more prominent. The Swedish physicist Carl Oseen, the most
influential member of the committee, had heard from his student Ivar
Waller of the quality of Dirac's work. More important, the positron's
discovery was viewed as 'an actual fact', an observation that illustrated the utility of Dirac's theory. At the end of the meeting, the consensus was that Heisenberg, Schrödinger and Dirac were head and
shoulders above the other candidates, including Pauli and Born, and
that Heisenberg deserved special recognition for being the first to
publish the new theory.

Today, the committee's judgements appear capricious. It would,
perhaps, have been fairer to award Heisenberg and Schrödinger
individual prizes in 1932 and 1933, leaving Dirac to win his own
prize a year later, an outcome that Dirac himself would almost certainly have regarded as just. None of this really matters; today, no
one doubts that the three physicists honoured in Stockholm in
December 1933 deserved their Nobel status. Dirac, Heisenberg and
Schrödinger are now among the select group of winners that give all
Nobel Prizes their special lustre.

Nineteen

To fast, to study, and to see no woman –
Flat treason 'gainst the kingly state of youth.
WILLIAM SHAKESPEARE, *Love's Labour's Lost*,
Act IV, Scene III

At the age of thirty-two, Dirac appeared to have everything he could wish for. He was in excellent health, was recognised as one of the best theoretical physicists in the world, had plenty of money and could not have been in a more agreeable job. Apart from worries about his home life, his only problem was that all his friends were men. Most people seemed to take it for granted that Dirac would spend the rest of his life being cosseted in the all-male bastion of St John's College and would die a bachelor. Over the next three years, he would surprise them all.

As several theoretical physicists guessed, their subject was coming to the end of a golden age. The toolkit of quantum mechanics was now available to solve almost all the practical problems encountered by scientists studying atoms and nuclei. In that domain, the theory worked wonderfully well. But for Dirac and others at the forefront of research, the subject was far from finished: most pressing was the need to find a field theory of electrons, positrons and photons – a theory known as quantum electrodynamics – that is free of infinities.

Based in California, Oppenheimer was an international leader in the field, which he studied when he was not immersed in the *Bhagavad Gita* and a dozen other books. Early in 1934, Oppenheimer and one of his students had dealt a heavy blow to Dirac's hole theory when they proved that quantum field theory accommodates the existence of anti-electrons without assuming the existence of a negative-energy sea. Oppenheimer sent Dirac a copy of his paper, but heard nothing in reply. In Europe, Pauli and his young student Vicki Weisskopf proved that particles with no spin also have anti-particles, flatly contradicting Dirac's theory, which implied that spinless particles should not have anti-particles because they do not obey the Pauli exclusion principle. Pauli was proud of what he called his 'anti-Dirac paper' and pleased that he was 'able again to stick one on my old enemy – Dirac's theory of the spinning electron'.[1] Pauli and Weisskopf rendered the concept of the negative-energy sea redundant, and it gradually fell into disuse, as physicists became inured to the idea that each positron was just as real as the electron – there was no need to treat the positron as the absence of anything. But Dirac did not accept this – there are no spinless fundamental particles, he noted unconvincingly, so Pauli and Weisskopf's arguments were academic. For this reason, he continued to use the hole theory, which yielded precisely the same results as theories that dispensed with the sea. His authority ensured that many other physicists followed him, and the hole theory continued to be used, if only as a heuristic device.[2]

Whichever version of quantum electrodynamics physicists used, it was plain that the theory was in trouble. However hard Dirac and his fellow physicists tried, they could not find a way of removing the infinities in the theory, to make rigorous calculations possible. Theoretical physics was 'in a hell of a way', Oppenheimer groaned, though he remained optimistic that either Pauli or Dirac would find a way of rescuing the theory by the following summer. If not, they would have to agree with many others that the theory was beyond salvation.[3]

Visitors to Cambridge, including Heisenberg and Wigner, found that Dirac was not working on quantum field theory but doing experiments with Kapitza in his new laboratory. Dirac was trying to solve a practical problem for some Cavendish colleagues, who needed pure samples of chemical elements. Each atom of every element contains the same number of electrons and protons, but the nuclei do not all have the same number of neutrons: the different varieties of nuclei, each with a characteristic number of neutrons, are known as the element's isotopes. There are, for example, three isotopes of hydrogen: most hydrogen nuclei contain no neutrons at all, but there exist others with one and two neutrons. Rutherford's colleagues needed pure samples of some isotopes for their experiments, but this was difficult, as atoms of naturally occurring samples of elements are a mixture of isotopes, extremely difficult to separate because they behave almost identically in chemical reactions. Dirac thought of a neat way of separating a mixture of two isotopes in a gas, using apparatus with no moving parts. His idea was to force a high-pressure jet of gas to follow a spiral path: the heavier, more sluggish molecules should tend to aggregate on the outside of the rotating mass of gas, while the lighter ones should hog the inside track. Dirac designed his apparatus for this 'jet stream method of isotope separation', then rolled up his sleeves and built it, having borrowed one of the compressors in Kapitza's store. Once again, he was trying his hand at being an engineer.

He was surprised by the results. The apparatus did not separate the isotopes efficiently but produced what he later described as 'something like a conjuring trick'.[4] When he pumped gas at six times ordinary atmospheric pressure into a small copper pipe, he found that, after the gas had undergone its spiral motion, it separated into two streams with very different temperatures – one stream was hotter than the other by about one hundred degrees Celsius. During a visit to Cambridge in May 1934, Wigner saw the apparatus and asked Dirac questions about it, but Dirac's replies were terse and unhelpful, causing the mannerly Wigner to take umbrage. Wigner understood that Dirac did not want to speak about the apparatus until he knew what he was talking about and that Dirac was unaware of the convention of parrying ignorance with a polite remark. Dirac thought the temperature difference was caused by the differences in the resistance to flow of the two gases, though it is more likely that the rotational motion tends to separate the faster gas molecules from the slower ones. Dirac spent months collaborating with Kapitza under the approving eye of Rutherford, who thought it augured well for theoretical physics that the Lucasian Professor was soiling his hands in the laboratory.[5]

During his discussions with Dirac, Kapitza will have talked a good deal about his friends at Trinity College High Table and the interdisciplinary wanderings of their conversations. What Kapitza did not know was that, from March 1934, one of his acquaintances, whom he and Anna often welcomed to their home, was an MI5 informant. Codenamed 'VSO', the colleague was convinced that 'it would be impossible for a Soviet citizen to go backwards and forwards to Russia unless his value to the Soviet authorities in this country were greater than his value in Russia'. The reports submitted by VSO, flecked with jealous asides about Kapitza's scientific reputation, contained no proof that he was a spy but enough circumstantial evidence to worry the security services. Why was Kapitza so sheepish about admitting, even to a friend, that he held a Soviet passport? The rest home for scientists in the Crimea was open only to Communist Party members, so why was Kapitza allowed to stay there if, as he claimed, he was not a member?[6] Most suspicious were the clandestine meetings Kapitza had near Cambridge with the new Soviet Ambassador in London, Ivan Maysky.[7] So far as MI5 were concerned, Kapitza was now one of their top suspects.

Yet Dirac seems to have aroused no suspicion at all, probably because – to most people – he seemed to be a perfect embodiment of the apolitical, head-in-the-clouds don. If VSO had been as diligent as he was suspicious, he might have wondered why Dirac was able to join Kapitza in the exclusive rest home in the Crimea. But Dirac appears to have entirely escaped the attention of MI5; if they kept a file on him, there remains no public record of it.

The brutality of Hitler's regime was now clear from press reports, though it seems that Heisenberg made light of them when he visited Cambridge in the spring of 1934 for what turned out to be a fruitless attempt to engage Dirac on the future of quantum electrodynamics. Heisenberg stayed in Born's home and tried to persuade him to return to his homeland.[8] During an afternoon walk in the garden with his host, he mentioned that the Nazi Government had agreed that Born could return to Germany to continue his research but not to teach. His family would not be allowed to go with him. Born, indignant that a close family friend could even contemplate conveying such a message, was furious and broke off the conversation. Only much later could Born bear to listen to Heisenberg describing the privations of trying to be a decent citizen amid the Nazi barbarities.

Conditions were no better in the USSR for scientists unwilling to toe the Stalinist line. George Gamow, worried that his support of orthodox quantum mechanics would result in his deportation to a Siberian concentration camp, used his invitation to the 1933 Solvay Conference as a way to escape. He persuaded the Soviet Prime Minister Vyacheslav Molotov to grant him and his wife Rho exit visas and then fled, leaving the Soviet authorities livid. The Gamows arrived in Cambridge in early 1934 and were soon a popular couple, delighting all comers with their friendly vivacity. Rho was a strikingly attractive brunette, with a Garboesque presence that could light up a roomful of the dourest dons. Stylishly dressed with smart accessories, all colour-coordinated with her lipstick, she sometimes looked as if she had walked off a photo shoot for *Vogue*.[9] She smoked one cigarette after another, but this did not put Dirac off; he adored her. The feeling was mutual, and they soon found ways of doing things together that entailed being alone with each other: she would teach him Russian in exchange for his teaching her to drive. Dirac made steady progress with learning his fourth language, as Rho recorded in the coming months by plotting a graph showing a gradual fall in his 'error index', an undefined concept, Dirac could not help noting.[10] After spending just a few weeks in Cambridge, the Gamows departed for Copenhagen, leaving Dirac bereft.

According to private comments Dirac made a few years later, he was not in love with Rho.[11] Nonetheless, their affectionate notes bounced back and forth across the North Sea for months, in a rally of infatuation. 'Please read my letters alone,' she pleaded. She returned the letters he had written in Russian, each one marked with a grade and with his errors neatly corrected in red ink. Hoping that he would approve of her cutting down on her smoking, she asked how many times each day he would like her to think of him; he worried that her memories of him were even slightly harmful to her. They were like cooing teenagers, each desperate not to offend the other and constantly seeking forgiveness. When Rho apologised if she had appeared to be insolent, Dirac reassured her that he was not in the least upset and that, in any case, he 'was not expecting Russian women to be as boring as English ones'.[12] Impatient to see each other again, it would not be long before their wish was fulfilled.

In the meantime, Dirac continued to learn Russian with a woman teacher who gave him hour-long lessons on Saturday mornings in Cambridge. Her name was Lydia Jackson, a Russian émigré poet known as Elisaveta Fen before her ill-fated marriage to Meredith Jackson, a Fellow at St John's. Romantic and strong-willed, she felt out of place in Cambridge – no place for assertive women, she thought – and made a living by teaching the language of her homeland. At a gathering of one of London's literary circles, she introduced George Orwell – probably one of her lovers – to the woman who became his first wife.[13] Jackson liked to talk about the Soviet Union with Dirac, and, by her tantalisingly vague account, it seems that she was more sceptical than he was about Stalin's regime.[14] He rarely spoke about science but did once exchange a few words with her about mathematics: she thought it was a human invention, while Dirac maintained it had 'always existed' and had been 'discovered' by humans. 'Doesn't that mean that it was created by God?' she asked. He smiled and conceded, 'Perhaps animals knew a little mathematics.'[15]

Her familiarity with Dirac is clear from her letters to him. In one, she commends him for being down to earth, not one of his most lauded qualities: 'I know that you are not as absent-minded as all professors and mathematicians are supposed to be: there must be quite a large chunk of an engineer still in you.' After referring teasingly to a spot of nude bathing she had done in a pond on Hampstead Heath, she gives him some stout advice for the sabbatical he was about to take in Princeton:

By the way, will you try and *not* forget all your Russian in the barbarous United States. Please try and read a little from time to time. [. . .] And do remember what I told you about not marrying an American: it would be a fatal mistake! An English girl, of firm but tactful disposition will be most suitable for you. As for a Russian – they are a handful under any circumstances [. . .].[16]

Determined that no one else would read Dirac's letters to her, she routinely burned them. Their opinions about the Soviet Union, as well as the evidence of whether the relationship became physically intimate, were probably destroyed in those flames.[17]

Dirac arrived in Princeton at the end of September, after another hiking vacation with John Van Vleck, this time in the mountains of Colorado.[18] Once again, Dirac provided his friend with more stories of his strangeness, including one in Durango where he was wandering around the town at night, probably wearing what might be kindly described as functional clothing, and was mistaken for a tramp. This would not be the last time Americans would mistake the Lucasian Professor for a vagrant.

In Princeton, Dirac was working at the Institute for Advanced Study, then a suite of offices in Fine Hall. He and his colleagues in Fine Hall liked to eat at one of the modest restaurants in Nassau Street, the rod-straight road that separates the university buildings on one side from the shops on the other. A faculty favourite was the Baltimore Dairy Lunch, known locally as the Balt, which served wholesome food at low prices, though only to white customers.

One of Dirac's preferred dining companions was his new colleague Eugene Wigner, the courtly Hungarian who was on a mission to bring modern quantum mechanics into Princeton. Inexplicably parsimonious, he declared proudly to visitors to his two-bedroom apartment that its furnishings had set him back less than $25, as if it were not obvious.[19] On the day after Dirac arrived in Princeton, neither Wigner nor any other Fine Hall colleague was free for lunch, so Dirac set off alone on the five-minute walk into the town centre. When he entered the restaurant, probably the Balt, he saw Wigner sitting with a woman.[20] Well-groomed and slightly younger than Wigner, and with an infectious cackle of a laugh, she looked rather like him, her face similarly long and angular. She spoke faltering English with the same thick accent, though with none of his reserve, and smoked her cigarettes using a long black holder.

The woman was Wigner's sister Margit, known as Manci to her friends and family. She was struck by the sight of the slender, vulnerable-looking young man who walked into the restaurant, later remembering that he looked lost, sad and disconcerted. 'Who is that?' she asked her brother. Wigner told her that he was one of the town's most distinguished visitors, one of the previous year's Nobel laureates. When he added that Dirac did not like to eat alone, she asked, 'So why don't you ask him to join us?' Thus began a lunch that changed Dirac's life. His personality could scarcely have contrasted more sharply with hers: to the same extent that he was reticent, measured, objective and cold, she was talkative, impulsive, subjective and passionate – she was the kind of extrovert Dirac liked. They occasionally had dinner together but were not officially dating, perhaps partly because he was distracted by Rho Gamow, who was staying in Princeton, having been left in the care of Dirac by her evidently trusting husband.[21] But these social matters were a sideline: he spent most of his time hard at work in his office in Fine Hall and in the rooms he rented in a grand house on one of the leafy avenues close to Nassau Street.[22] So far as his colleagues could see, for all the interest he showed in women, he could have been a eunuch.

In Fine Hall, Dirac was accommodated on the same corridor as Einstein, their offices separated only by Wigner's. Einstein was the town's most famous celebrity, after Veblen the first faculty member of the institute. He and his wife had arrived in October 1933 and lived in an apartment before settling in a modest detached house in Mercer Street, about five minutes' walk from the centre of the town, which he described as a 'quaint ceremonious village of puny demigods on stilts'.[23] Although grateful to be in a safe haven and 'almost ashamed to be living in such peace while all the rest struggle and suffer', he could see his new home town was not free of racism and may have discussed this in his meetings with Paul Robeson, the town's most famous son.[24]

Then fifty-four, Einstein looked older: he shambled around the town in his plain raincoat and woolly hat, avoiding eye contact with fellow pedestrians, especially ones who recognised him.[25] On the day he arrived in Fine Hall, newspaper photographers and a crowd of hundreds gathered to catch a glimpse of him through an open library window. The authorities had to smuggle him in and out of the hall through a back entrance.[26]

Veblen and his colleagues were licking their lips at the thought of Einstein and Dirac working together, but it soon became clear that this was only a dream. The two men respected each other, but there was no special warmth between them, no spark to ignite collaboration. They were studying the same subject, but their approaches were quite different: Dirac was developing quantum theory and was deaf to its alleged philosophical weaknesses; Einstein admired the success of the theory but mistrusted it (during the spring of 1935 he completed his collaboration with his younger research associates Boris Podolsky and Nathan Rosen on a paper that cast serious doubts on the conventional interpretation of the theory).[27] Whereas Einstein was a conservative scientist, Dirac was always ready to discard well-established theories, even ones he had helped to create. Language was another barrier: with only weak English, Einstein preferred to talk in his native tongue, which Dirac spoke only with difficulty (in the company of refugees from Hitler's regime, Dirac relaxed his rule of not speaking German). And Dirac tended to avoid smokers, although Einstein temporarily removed that barrier in late November when he gave up his pipe for a few weeks, to demonstrate his willpower to his wife, who disapproved of the habit. 'You see,' he complained to a neighbour, 'I am no longer a slave to my pipe, I am a slave to dat vooman!'[28]

Dirac spent much of this sabbatical writing the second edition of *The Principles of Quantum Mechanics*, making it less mathematical and less intimidating. The completed version preserved the structure of the original and was more accessible than the first edition, though for all but the most gifted students it was aspirational reading. Most students who wanted to use quantum mechanics to do actual calculations used more practically minded texts, secure in the knowledge that the underlying beauty of the subject was nowhere clearer than in this book, sometimes described as 'the bible of modern physics'.[29]

Still believing that mathematics offered the royal road to the truth about the fundamental workings of nature, Dirac spent much of his time in Princeton learning more mathematics. This led him to find a new way of writing his equation for the electron, by describing its behaviour in a space-time whose geometry is not the standard Euclidean type (in which the sum of the angles of a triangle is one hundred and eighty degrees) but is of a more exotic variety developed by the Dutch mathematician Wilhelm de Sitter. Perhaps this would enable the quantum theory of the electron to be harmonised with the general theory of relativity? The result was a sumptuous piece of mathematics, though one that failed to yield new insights into nature. Dirac had yet to show that his idea – that fundamental physics could be gleaned from promising mathematics – was fertile. No other leading theoreticians had taken much notice of it: they remained pragmatic, taking cues from experiment and trying to learn from the weaknesses and loose ends of the best-available theories.

One of the most intriguing topics for theorists was radioactive beta decay, in which an unstable nucleus spontaneously ejects a high-energy electron. Early in 1934, Fermi underlined his talent as a theoretician once again, this time by setting out the first quantum field theory of beta decay and giving a clearer understanding of the role of the neutrino. He gave a clear mathematical description of how an atomic nucleus undergoes beta decay, one of its neutrons transmuting into a proton, which remains in the nucleus, while two other particles – an electron and a massless neutrino – are simultaneously created and ejected. This decay was caused by the weak force, a previously unidentified type of force that acts only over extremely short distances, unlike the familiar forces of gravity and electromagnetism. Although Dirac admired Fermi's theory, he did not follow him into the nucleus and its complexities. Dirac was adamant that the best way of making progress was to focus on nature's simplest particles, taking inspiration from the most beautiful mathematics. Time would decide whether such purism was wise.

*

Dirac's colleagues in Fine Hall saw that his fanatical dedication to work was on the wane. He spent most afternoons playing games in the two common rooms, each of them furnished in the style of the best-appointed Oxford University common rooms – plush curtains framing every window, deep-pile carpets on the floor, capacious leather armchairs and imitation-antique tables.[30] During the ritual of afternoon tea, he fruitlessly searched for a way that a king could pass eight opposing pawns and got thrashed by his colleagues in their favourite game, Wei Chi (also known as Go), which he had introduced into Fine Hall a few years before.[31] He was relaxed enough to channel some of his intellectual energy away from the toughest problems in science to games that had no point beyond personal pleasure. The impasse in quantum electrodynamics appears to have sapped his morale: he may have feared that he had fallen victim to the alleged 'Nobel disease', said to prevent prize-winners from repeating the quality of their best work after their return from Stockholm.

Over ice-cream sodas and lobster dinners, Dirac's friendship with Manci deepened.[32] She was a lively, big-hearted conversationalist, and, although she often struggled to find the right words in English, she had the rare ability to make him thaw. Between the long – but gradually shortening – silences, he told her of the pain of his youth, of his brother's suicide, of the father whom he believed had tyrannised him into his defensive silence. Manci also had plenty of private unhappiness to share, telling him that she was an unwanted child, less attractive than her sister, intellectually worthless compared with her brother. Mainly to get out of her parents' house, she married when she was only nineteen. Her Hungarian husband, Richard Balázs, turned out to be a playboy and philanderer, and the marriage was an eight-year calamity mitigated only by the birth of her son Gabriel and daughter Judy. She took the bold step of instigating divorce proceedings and had finally become single again two years before she set sail for Princeton.[33] There had been other men after Balázs, but none of them were around for long, and she was lonely and unfulfilled.[34] She was staying with Eugene for a change of scenery, having promised her children – in Budapest with their governess – that she would be home for Christmas. At thirty years old, she had never felt so free in her life.

Although a self-declared 'scientific zero', Manci took a lively interest in international ethics, morals and politics, often impressing experts with her knowledge but at the same time affronting them with her shameless lack of objectivity. Once she had made up her mind, facts alone were rarely enough to budge it; she seemed to think not just with her brain but with her heart. Religion caused her special anguish. Until 1915, when she was eleven, her family had subscribed half-heartedly to the Jewish faith, visiting the synagogue twice a year, but then had become Lutherans.[35] By the time she met Dirac, she was no longer devoutly religious but appears to have somehow yearned to believe in some kind of deity and did not like to hear religion slighted. She would probably not have welcomed Dirac's view that his religion was simply that 'the world has to improve'.[36]

Manci was a keen follower of the arts, and she chivvied Dirac into taking more interest in music, literary novels and ballet. In the evenings, like many people during the Depression, they joined the long cinema queues ready to pay their quarters for a few hours' harmless escapism. They may well have seen some films featuring one of Hollywood's new stars, Cary Grant, rapidly establishing himself as a versatile actor with a gift for playing both comedy and – having thoroughly suppressed his Bristol vowels – the charming, all-American gentleman.

About ten days before Christmas 1934, during a journey on the New York subway, Dirac read an unexpected and chilling piece of news.[37] He was in the city to buy an overcoat, to replace the one he had given Tamm fifteen months before. Dreading the Christmas throng of Manhattan and its noisy, bullying traffic, he did not hesitate when Manci offered to go along to keep him company. They agreed to meet in Fine Hall, before driving to Princeton Junction, where they would catch the train to Penn Station. After arriving first at the hall, she took a moment to look in his mailbox and found an airmail letter, which she hurriedly put in her handbag and forgot in the excitement of what was her first trip to the shopping capital of America. When she was sitting next to Dirac in a subway car, clattering and squealing its way towards the Midtown stores, she opened her bag to look for a handkerchief and saw the envelope, which she handed to Dirac. It was from Anna Kapitza in Cambridge, he saw, but it was not just another family chronicle. Manci watched Dirac as he read the typewritten letter, a little over a page long. He turned to her with alarming news – the Soviet Government had detained Peter Kapitza in Moscow.

Anna was desperate. She wrote that her husband's detention was 'a terrible blow to him, almost the severest he ever had in his life', and she pleaded with Dirac for help:

I am writing to you as a friend of K and of Russia and you will understand the impossible situation [. . .] People will talk and the last thing I want is the press to get hold of it. [. . .] I wonder if you could write a letter to the Russian Ambassador in Washington, I feel that is the only way to do anything [. . .].[38]

Earlier, Kapitza had boasted that he was the only Soviet citizen who had unrestricted passage across his country's borders.[39] He had scoffed at his colleagues' warnings that he was courting disaster by returning home each summer for his vacation. Irritated by the defection of Gamow and other Soviet scientists, Stalin's authorities were determined to secure the country's best brains to help build its future. During a trip to the USSR in late September with his wife and children, officials in Leningrad told Kapitza that he must stay in the Soviet Union for the foreseeable future, though his family was free to return to Cambridge. Furious, Kapitza tried to talk his way out of it, pleading unsuccessfully that he could not break faith with his colleagues in England, and was dispatched to Moscow, where he lived in a sparsely furnished room at the Hotel Metropole, with little to do except read, write desperate letters to Anna and go for walks – always under the surveillance of the security police.[40] Rutherford and the Foreign Office had kept the matter secret, in the hope that his detention could be resolved diplomatically.[41] No one, certainly none of the officials in the security services, had expected this: not for want of trying, MI5 had not found any hard evidence that he was a spy.

Dirac was still digesting the news when he was trying on overcoats in Lord and Taylor, one of the exclusive stores on Fifth Avenue. Manci had an uphill struggle to persuade him, devoid of dress sense, to take the purchase of the coat seriously. No doubt seeing an opportunity to refurbish his entire wardrobe, the salesman asked Manci discreetly whether Sir would also like a new suit, but Manci smiled and shook her head: to press him to buy more than he needed would be futile. The coat he bought there turned out to be a good investment – it lasted him to his death, a memento of the day he heard about Kapitza's plight and was moved to take political action for the first time in his life. Though he knew that he had none of the interpersonal skills and tact needed to be an effective diplomat, he became the de facto coordinator of the American-based campaign for Kapitza's release.

In Princeton the next day, Dirac urgently sought advice from the well-connected Abraham Flexner and from Einstein, who promptly agreed to help. Dirac was confident enough to write to Anna Kapitza in Cambridge to assure her that matters would 'all come right in the end'.[42] After the Christmas vacation, he would begin his campaign for Kapitza's release, but first he wanted to take a vacation in Florida. He was planning to go on his own, but Manci had other ideas: seeing an opportunity to spend some time alone with her new friend, she postponed her return to Hungary until after Christmas, breaking the promise she had made to her children.

Dirac and Manci motored down in early January from freezing Princeton to the warmth of St Augustine, a resort on the north-east coast of Florida. No one – except, possibly, Wigner – knew that they were together. The vacation appears to have been platonic. Their letters before and after the trip show that they were not yet close and still viewed each other differently – he regarded her only as an agreeable companion, but she saw him as a potential husband. They spent their week dodging the rainstorms and taking trips to the local tourist destinations, including a farm where Dirac spent a few dollars buying a baby alligator that he mailed anonymously to the Gamows in Washington, DC.[43] As Rho opened the package in their hotel room, the alligator jumped out and bit her hand – one of her husband's less amusing practical jokes, she thought. Gamow protested that he had nothing to do with the prank; he thought it was a crocodile, a symbol of his favourite experimenter, sent by someone with more playfulness than common sense. A month later, Dirac owned up, and the poor alligator languished, and a few months later died, in the Gamows' bath.

By the spring of 1935, the campaign for Kapitza's release was not going well. In Cambridge, Anna could see the vultures circling: several of her husband's colleagues in the town privately wanted to see Kapitza get his comeuppance after the years he had spent shamelessly fawning on the Crocodile. There were whispers that Kapitza was merely an engineer, that his experiments were leading nowhere and that he had received financial rewards in return for spying for the

USSR. Anna's reports drew from Dirac some uncharacteristically direct advice: 'You should not pay attention to stupid stories that no one believes in.'[44]

Kapitza's Marxist friends sat on their hands, while Rutherford led a discreet campaign for his release. Seeking advice from colleagues all over Europe and working closely with Soviet officials and with the British Foreign Office, Rutherford sought a face-saving solution. He sought to give Kapitza the option of working wherever he liked, though he confided in a letter to Bohr that he was certain Kapitza wanted to return to Cambridge, adding that he found the Soviet authorities particularly mendacious.[45] The first Cambridge scientist to visit Kapitza was Bernal, accompanied by his lover Margaret Gardiner, and they spent long afternoons trying to cheer him up over pancakes with caviar and soured cream, washed down with wine.[46] 'I feel like a woman who has been raped when she would have given herself for love,' Kapitza sulked. He used the phrase repeatedly.[47]

Gardiner had mixed feelings about Moscow, disturbed by the giant posters of Stalin all over the city and the quarter-mile queues that formed outside the shops the moment new supplies arrived. The Moscow hotels were just as bad as their reputation had led her to believe: rooms heated to a tropical swelter, shabbily dressed waiters pretending to be in a hurry, many of them cadging illegal gratuities. The Muscovites walked around their grey, freezing city 'wrapped shapelessly in their padded jackets and heavy fur coats', most of them treading the icy streets with their de-rigueur galoshes. Gardiner believed that the country's hopes lay in mass education, always an attractive vision for the English left. Decades later, she recalled seeing a platoon of young soldiers marching towards the Military Academy with exercise books under their arms. Her tour guide explained: 'They are having their illiteracy liquidated.'[48]

After Manci's departure in mid-January 1935, Dirac's routine in Princeton was unchanged. Each morning, he trudged through the snow from his rented home near Nassau Street to his room in Fine Hall, worked alone all morning, and had lunch at Newlin's restaurant with Wigner and with one of Princeton's most unusual visitors, the Belgian theoretician Abbé Georges Lemaître. He was an amateur scholar of the playwright Molière, an accomplished interpreter of Chopin and the only member of the physics department to wear a dog collar. Dirac had first seen him, but had apparently not met him, in October 1923, when he began his studies and when Lemaître was one of Eddington's postgraduate students. Four years later, Lemaître had introduced into science the idea that the universe had begun when a tiny egg, a 'primeval atom', suddenly exploded into the matter of the universe.[49] Quite independently, the Russian mathematician Alexander Friedmann had applied Einstein's general theory of relativity to the universe as a whole and demonstrated that some mathematical solutions of the equations correspond to an expanding universe, though his work was published only in Russian and at first went unnoticed.

The Friedmann–Lemaître picture of the universe's birth seemed to be at odds with the account of creation in Genesis, but this did not bother Lemaître, who believed that the Bible teaches not science but the way to salvation. The science–religion controversy 'is really a joke on the scientists', he said: 'They are a literal-minded lot.'[50] Dirac found Lemaître 'quite a pleasant man to speak with – not strictly religious as one might expect from an Abbé'.[51] It was probably during these conversations in Princeton's diners that Lemaître reawakened Dirac's interest in cosmology, the study of the entire universe and its workings, soon to become one of his main interests.[52] For now, he focused on mathematics and quantum physics, which he studied during the day, and he took it easy in the evening. After dinner, he would read one of the books Manci had recommended to him (including *Winnie the Pooh*) or go out, perhaps to a movie with the von Neumanns.[53] Probably as a result of Manci's encouragement, he had become much more interested in music: a highlight of the term for him was a university concert, where he heard a searching performance of Beethoven's last piano sonata by the Austrian virtuoso Artur Schnabel, another Jewish refugee from Hitler's Germany.[54]

Manci was with her children in Budapest. About once a week, in her spidery hand, she sent Dirac four lively pages of news and gossip, urging him to keep in close contact. Unaccustomed to receiving warm and attentive letters, Dirac struggled to respond: 'I am afraid I cannot write such nice letters to you – perhaps because my feelings are so weak and my life is mainly concerned with facts and not feelings.'[55]

Manci, 'very much upset' by this statement, knew that she would have to take the initiative if she were to stir in him the first quantum of romance.[56] Always wearing her heart on her sleeve, she wrote to

Dirac about her family and bombarded him with questions about his life in Princeton in all its minutiae. His reply was chilling: 'You ought to think less about me and take more interest in your own life and the people around you. I am very different from you. I find I can very quickly get used to living alone and seeing very few people.'[57]

He sent her lists of corrections to her English and answered her queries as tersely as a speak-your-weight machine. When she sent him photographs of herself, he was grateful but critical: 'I do not like this picture of you very much. The eyes look very sad and do not go well with the smiling mouth.'[58] After she complained that he did not answer all her questions, he re-read her letters, numbered them and sent her tabulated responses to every question he had ignored, including:

Letter number	Question	Answer
5	What makes me (Manci) so sad?	You have not enough interests.
5	Whom else could I love?	You should not expect me to answer this question. You would say I was cruel if I tried.
5	You know that I would like to see you very much?	Yes, but I cannot help it.
6	Do you know how I feel like?	Not very well. You change so quickly.
6	Were there any feelings for me?	Yes, some.[59]

When Manci received the list, she thought Dirac was jeering at her but eventually decided that it was 'quite funny'. Beginning to realise that Dirac did not understand rhetorical questions, she seethed: '*Most of them were not meant to be answered*.'[60] It is easy to imagine her tearing out her hair in frustration. But his answers gave her an opportunity to engage with his feelings, and she did not hold back: for his statement that she changed so quickly, she told him he should get 'a second Nobel Prize, in cruelty'. Manci was tough, but she made sure that Dirac was aware of her vulnerable and sensitive side: 'I am only a stupid little girl.'[61] With each letter, she flirted more audaciously, but Dirac made no comment until he realised that he was being targeted. He snapped: 'You should know that I am not in love with you. It would be wrong for me to pretend that I am. As I have never been in love I cannot understand fine feelings.'[62]

But Manci was not to be deflected. Although Dirac parried her repeated requests to join him during his forthcoming trip to Russia, she was determined to see him before the summer was over.

The news of Kapitza's detention first appeared in the British *News Chronicle* on 24 April 1935, after a leak. Soon, Kapitza's case was well known in the British media, and the newspapers featured long reports on the experiments he had been doing in Cambridge.[63] In interviews with journalists, Anna Kapitza was distraught. 'The whole affair has caused great mental pain to both my husband and myself,' she complained, adding that she was concerned about the effect of the upheaval on her highly strung husband: 'in his present state of mind he is not in a position to do any serious work'.[64] Yet she was underplaying his distress: 'Sometimes I rage and want to tear out my hair and scream,' he had written to her.[65] Life in the Moscow science community was dismal for him as most of his former friends there were shunning him until they knew officially, from Stalin's office, whether Kapitza was one of the 'enemies of the people'. His country's reward for his scientific success and for not making a fuss was, he wrote to Anna, to treat him 'like dog's excrement, which they try to mould in their own way'.[66] He knew his letters would be intercepted and read by the police, so he lambasted the agents of his captivity, not the Soviet system that employed them:

'Not only am I sincerely loyal, but I have deep faith in the success of the [plans for] new construction [in the Soviet Union] [. . .] But even in spite of my cursing, I do believe that the country will come out of all these difficulties victorious. I believe it will prove that the socialist economy is not only the most rational one, but will create a State answering to the world's spiritual and ethical demands. But, for me as a scientist, it is difficult to find a place during the birth pangs.[67]

But the Soviet Government had plans to keep Kapitza busy and to give him all the material goods he could wish for. It decided to set up a new Institute for Physical Problems, to make him the founding director, to give him a salary most academics would envy and then to throw in some generous perks, including an apartment in Moscow, a summer house in the Crimea for his family and a brand new Buick.[68] From the vantage point of the sofa in his hotel room, however, the future looked so bleak to Kapitza that he considered suicide. His depression was relieved only by trips to the theatre and the opera and by colour reproductions of his favourite modern art pinned to the blank walls. But Cézanne, Gogol and Shostakovich offered only meagre consolation: he longed to return to his experiments in the Mond Laboratory, to be with his family and friends in Trinity College.

On the day news of Kapitza's detention broke in the UK, Dirac was relaxing with the Gamows in Washington, DC.[69] On a fine warm day, the three of them took a forty-minute trip in an airship over the city and looked down on the cherry blossoms in the fullness of their second bloom and on Capitol Hill, where FDR was pushing through his controversial New Deal. Dirac was about to tread the streets of the capital as an unlikely lobbyist, having accepted Anna's suggestion that he should approach the first Soviet Ambassador to the USA, Stalin's friend Aleksandr Troyanovsky.

Dirac was officially in Washington to attend three consecutive conferences, where he spent most of his time publicising Kapitza's difficulties and collecting signatures to petition for his release. Every delegate approached by Dirac agreed to sign, including Léo Szilárd, who hatched a ludicrous plan to smuggle Kapitza out of Russia by submarine.[70]

Before Dirac could present the petition, some groundwork had to be done. He arranged for a letter to be written to the Ambassador from Karl Compton, brother of the famous experimenter and President of the Massachusetts Institute of Technology. Compton declared that Kapitza's absence from Cambridge 'is universally considered by scientists to be a major catastrophe' but suggested that his return 'would be universally acclaimed in the scientific world'.[71] The letter did its job: Troyanovsky quickly agreed to receive both Dirac and Millikan. Dirac later explained to Anna Kapitza why he wanted to be accompanied by Millikan: '[he] is known to be rather opposed to the Soviets but that would be counterbalanced by my being known to be rather in favour.'[72]

Thus, on the last Friday afternoon of April 1935, Dirac – for a decade regarded as an asocial misfit, out of touch with world affairs

– found himself walking to the Soviet Embassy with America's preeminent scientist-diplomat. The embassy, just north of the White House, was looking magnificent: Moscow museums had supplied antique furniture, paintings and rugs as contributions to its renovation.[73] After waiting in the reception room, dominated by a statue of Lenin, Dirac and Millikan shook hands with the lantern-jawed Troyanovsky, whose charm and accommodating manner had made him popular on the city's social circuit. The half-hour meeting was cordial and relaxed. Over a cup of tea, the Ambassador admitted that he had heard of Kapitza's case only when he read Compton's letter and described the Soviets' hurt when some of its most eminent citizens had failed to return home after travelling abroad. Millikan told him that Kapitza's health was deteriorating and suggested that the Soviet Union should bear in mind public opinion in other countries as well as its own. The continued detention of Kapitza would seriously damage relations between Soviet and American scientists, Millikan concluded. As the meeting drew to a close, Dirac spoke up and pleaded for Kapitza's release, in words he recalled the next day in a letter to Anna Kapitza: 'I have known Kapitza very well for a long time and I know him to be thoroughly reliable and honest [. . .] If he were let out under a promise to return he could be depended on to keep that promise.'[74] The Ambassador ended with an assurance that he would raise their concerns with the Soviet Government, so, Dirac told Anna, he left the meeting feeling hopeful.

Yet there was more to do. After the meeting, Millikan wrote to the Ambassador to reiterate the points he and Dirac had made, ratcheting up the diplomatic pressure. Dirac collected the last of the petition's sixty signatures, which included those of almost all the leading physicists in the USA, including Einstein. Flexner had agreed to send another petition, addressed to the American Ambassador in Moscow, who would be asked to present it to the Government. Dirac concluded his letter to Anna: 'I feel sure the Soviet Government will do something about it when they see how widespread is the feeling against them. If they don't, you may rely on me to do all I can when I am in Russia to get Kapitza out in any way.'[75]

A few days later, at the beginning of June, Dirac left Princeton. Compared with his most successful stays in Copenhagen and Göttingen, this sabbatical had been largely a scientific washout, but for good reasons. He had invested some time in his relationship with

Manci, but it was small compared with his commitment to secure Kapitza's release. Even at the cost of stalling his work, Dirac was not going to abandon his surrogate brother.

Twenty

STALIN: You, Mr Wells, evidently start out with the assumption that all men are good. I, however, do not forget that there are many wicked men.

'A Conversation between Stalin and [H. G.] Wells', *New Statesman*, 27 October 1934

Moscow was beckoning again. For the following four months, Dirac's diary was empty, and he was determined to spend most of that time with Kapitza. Dirac knew that the secret police read his letters to Anna Kapitza and that he would probably be followed when he was in Moscow. He told her, 'If anyone follows me around in Moscow he will get some long walks.'[1]

Dirac and Tamm had intended to spend the summer hiking and climbing together in the Caucasus, and Dirac hoped to see one of the allegedly productive factories and the new Dneproges hydroelectric power station, one of the proudest achievements of Soviet engineering. But when Anna Kapitza asked Dirac to cancel the trip in order to support her husband, Dirac shelved his plans and declared himself to be at the service of her and her husband: 'I am ready for anything.'[2] He travelled to Moscow via Berkeley, where Oppenheimer found that Dirac was as tight-lipped as ever about physics. Two of Oppenheimer's students were elated when he told them that their British guest was prepared to hear their ideas about quantum field theory, which built on his work. During their fifteen-minute presentation, Dirac said nothing. Afterwards, the students braced themselves for his perceptive comments, but there was an agonisingly long silence, eventually broken by Dirac when he asked them, 'Where is the post office?' The students offered to take him there and suggested that he could tell them what he thought of their presentation. Dirac told them, 'I can't do two things at once.'[3]

On the afternoon of 3 June 1935, Dirac waved goodbye to Oppenheimer and boarded the Japanese MS *Asuma Bura*.[4] He settled into his private cabin and prepared to sail through the mist to San Francisco – catching sight of the half-constructed Golden Gate Bridge – and then on to Japan, China and the USSR. Manci, meanwhile, was lounging around in Budapest, awaiting the arrival of her first car, a six-cylinder Mercedes Benz bought for her by her father.[5] She had persuaded Dirac to visit her in Budapest at the end of his trip. Her complaints that he didn't respond to her questions drew another tabulated response:

Have you played ping-pong with pretty girls?	With one pretty girl. Most of the passengers were Japanese, and Japanese girls do not play ping-pong.
Have you flirted?	No. She was too young (15 years old). But you ought not to mind if I did. Should I not make the most of what you taught me?
Why were you so derisive?	I am sorry, but I cannot help it at times.[6]

   Six weeks after he had set sail from the USA, Dirac arrived in Moscow railway station. Even he, with his Gandhian indifference to his surroundings, must have been struck by the contrast between the fresh, early summer air of Princeton and the stench of rotten eggs that hung over the Soviet capital. It was no longer the city that he had seen four years before but a reeking, overcrowded metropolis. The playwright Eugene Lyons described the 'viscous ooze of [Moscow's] dung-coloured people, not ugly but incredibly soiled, patched, drab; the odour and colour of ingrained poverty, fetid bundles, stale clothes'.[7] Dirac stayed there only briefly: he had arranged to spend most of his time in the more agreeable ambience of the Kapitzas' *dacha* (summer home) in the village of Bolshevo, thirty-five miles south of the city. Kapitza was looking forward to seeing his English friend, though the tone of his comments to his wife indicates that he did not fully reciprocate the intensity of Dirac's affection. But a day after Dirac's arrival, Kapitza appeared to have changed his mind. He wrote to her:

[We] came here with Tamm and have been walking, boating and talking ever since. I haven't had such a pleasant time with anyone up to now. Dirac treats me so simply and so well that I can feel what a good and loyal friend he is. We talk about all sorts of things and this has been very refreshing. [. . .] Dirac's arrival has revived my memories of the respect and reputation I enjoyed in Cambridge [. . .][8]

The two friends relaxed together for almost three weeks. Kapitza's abject morale had not improved when he heard that the Soviet authorities had, for unknown reasons, sent 'Dimus' Ivanenko into exile.[9] It was a familiar story, though no one dared to question Stalin's policy in public. Kapitza was considering giving up physics and changing the subject of his research to physiology so that he could work with Russia's most senior scientist, the elderly but still active Ivan Pavlov. Within the modest compass of Dirac's verbal skills, he tried to lift Kapitza's spirits, and in return Kapitza – evidently knowing nothing of Dirac's friendship with Manci – tried to fix him up with a young girl they met, a good-looking, English-speaking language student. Dirac did not respond.

During his stay, he met the Trinity College physiologist Edgar Adrian and other British colleagues asked by Rutherford to assess Kapitza's situation and his psychological state. The Soviet Government supported this visit, presumably to demonstrate their flexibility. But, by the time Adrian and his colleagues met with Kapitza, the die was cast: Kapitza had been forbidden to return to Cambridge, and it remained only to secure the best terms for him to work in his new institute. When Dirac left Moscow at the beginning of September, he knew that he had lost his first diplomatic battle; he would have to become accustomed to living in Cambridge without the man he thought of as his closest friend.

The final stage of his trip was an antidote to his disappointment: he was to visit Manci in Budapest. She was living with her children in an apartment in what had been Archduke Frederick's house, a short stroll from her parents' sumptuous residence opposite Count Batthyány Park. This was a world of plenty – fine food, exquisitely cut clothes, attentive servants and private concerts in the living room. Dirac's modest origins in Bishopston were part of another world. Manci took her material comforts for granted, but she was unhappy and longed to get away from her parents, who must have been taken aback by the arrival on their doorstep of an unkempt Englishman who knew hardly a word of Hungarian. They knew next to nothing about him and surely cannot have expected that their feisty, outspoken daughter would choose such a diffident man. But they liked him and could see that Manci and Dirac clicked during their nine days together, driving around the city in her new car, sightseeing and soaking in the famous indoor public baths.[10] When he returned to Cambridge, he wrote to Manci: 'I felt very sad when leaving you and still feel that I miss you very much. I do not understand why this should be, as I do not usually miss people when I leave them. I expect you spoil me too much when I am with you.'[11]

Manci was making progress. But three weeks later, her heart sank when she read the final entry in Dirac's latest table of unanswered questions: to her query 'Do you miss me a little?', he responded, 'Sometimes.'[12]

When Dirac returned to England in the early autumn of 1935, the country was still disfigured by unemployment and worried by Hitler's aggressive rearmament, Mussolini's sabre-rattling in East Africa and Japan's occupation of Manchuria. 'I would like to kill the politicians of middle Europe,' Manci fumed.[13] Dirac was soon back in his Cambridge routine, but the thrill had gone. Although he had not given up on quantum electrodynamics, he seemed to be getting nowhere. Dirac thought a revolution was needed and probably wondered whether he, now thirty-three, might be too old to be one of its leaders.

Rutherford had negotiated a deal that involved moving almost every item of Kapitza's equipment to the Institute for Physical Problems, enabling him to resume all his experiments there. Anna had made Dirac a guardian of the Kapitzas' sons, and he took his duties seriously, taking the boys out at the weekends for rides in his crumbling car and organising his first fireworks display for them on 5 November.[14] These were good times for Dirac, but he was preparing for yet more loneliness: the Blacketts had left for London, Chadwick for Liverpool, Walton for Dublin, and now the Kapitzas were about to depart for good. Dirac was not the self-sufficient eremite that many people believed him to be: he needed new companionship, and he knew it. Manci was eager to oblige, but he was wary of her forwardness, as he showed when she telephoned him one night late in November as he was preparing to go to bed.[15] She thought he would be delighted to receive an unexpected call from her but he was angry and shaken. The college telephone system was arranged so that the porters heard their stilted conversation, as he explained to her in a brusque note. Surely it was sufficient to communicate only by letter, he wrote, with all the warmth of a tax inspector. She swiftly replied, making clear what she thought of his secretiveness: 'ridiculous'.[16]

Incidents like that rattled him: could he live with someone who had so little sympathy with his need for privacy? He will have had no wish to be party to a disastrous marriage, like his parents', which he had seen in all its unpleasantness two months before, during another rain-soaked visit to Bristol.[17] Charles and Flo were living out their marriage contract in an unwinnable endgame of squabbles and recriminations. Divorce was out of the question for the born-again Catholic Charles, but when he read his copy of George Bernard Shaw's *Getting Married*, he may well have sympathised with the author's recommendation: 'Make divorce as easy, as cheap, and as private as marriage.'[18] Flo would probably have welcomed a divorce, but the shame would have been too much for her. So they both remained unhappily shackled to each other, with nothing to look forward to except more arguments. Flo told her son that her pleasures were limited to taking long walks on the Downs, sitting alone in the parks and attending meetings of the new Bristol Shiplovers' Society. 'I have made an awful mess of my life somehow,' she wrote, adding that she blamed herself: 'What we sow, we reap.'[19]

Dirac's mother appears to have had no more than a passing interest in his work, but his father struggled hard to understand it. Charles looked through the journals in the library, searching for readable accounts of quantum theory, hoping to absorb some of their content by writing out paragraphs of difficult technical prose, verbatim. He kept a record of his findings in a small, red notebook, on whose front cover he had written a two-inch-high letter P.[20] The desultory references and notes inside are heart-rending records of a keen but confused amateur, unable to make any headway in a subject he longed to understand. Charles had written, in his rheumatic hand, some of the most complimentary comments about his son, highlighting some of the most generous ones: '<u>Dirac stands out amongst his contemporaries in this field for his originality</u>.' Apart from a summary of one of Crowther's articles on 'New Particles', Charles had not tracked down any of the lively and accessible accounts of quantum mechanics by Eddington or any of the other accomplished popularisers. It seems that his son was giving him no help at all.

With Bristol's long tradition of adult education, it was easy for the city's citizens to find out about new science. Arnold Tyndall, who gave Dirac his first introduction to quantum theory, was a popular performer at the night classes on science organised by the university. During one of his courses, a male student caught the eye of the genial Tyndall. Much older than the other students, he always sat at the front, taking careful notes. At the end of the final lecture, he shuffled up to Tyndall to thank him. 'I am glad to have heard all this. My son does physics but he never tells me anything about it.' The student was Charles Dirac.[21]

In the early summer of 1935, Betty had finished her French course and had come bottom of her class, taking a third-class honours degree, as Felix had done.[22] She wanted to be a secretary and to get out of Bristol as quickly as she could. Charles was now open about his relationship with Mrs Fisher, Flo told Dirac: 'I wish he would go and live with her, folks are always seeing them about together and tell me [. . .] He has always had someone ever since I've been married: Betty says it is French.'[23]

Dirac's mother, preparing to go on another Mediterranean cruise alone, sensed that her daughter was growing away from her. In a few weeks, she would temporarily move to London, not leaving her mother a forwarding address. But first Betty went on an August vacation with her father, keeping their destination secret. They were travelling with a group of Catholic priests on a pilgrimage to Lourdes, in the French Pyrenees, where Charles may, to try to rid himself of his ailments, have bathed in its reputedly miraculous waters. He knew that his daughter would pray for him but that his wife and son were, at best, indifferent to his fate.

Dirac would probably have been happiest if, like Einstein, he had never supervised a graduate student. It was not until the 1935–6 academic year that Dirac first officially became a research supervisor, taking on two students Born left behind when he moved to Edinburgh to take up a professorship.[24] Dirac had almost none of the skills that he had seen in Fowler: the ability to set problems pitched at the right level for his students, to motivate them in lean times and to support them in the early stages of their career. Dirac believed his only obligation was to point his students towards an interesting theoretical concept and then to look over any work they produced in consequence; it was up to the student to take almost all the initiative. Only the cleverest and most independent-minded students could flourish under such a regime, as the Cambridge authorities knew. Dirac knew it, too, and showed no interest in recruiting apprentices. But several of the finest young minds sought his guidance, including the Indian mathematician Harish-Chandra and the Pakistani theoretician Abdus Salam, both part of a pattern – the great majority of Dirac's successful students were foreign.

Dirac encouraged his students to keep abreast of the latest publications in theoretical physics and also to keep an eye on the experimenters' latest findings. But his faith in the veracity of new experimental results was badly shaken by an incident that began in the autumn of 1935. Dirac heard that the Chicago experimenter Robert Shankland had found evidence that sometimes energy is not conserved, contrary to one of the fundamentals of science: when photons are scattered by other particles, he found the particles' total energy before the collision is not the same as it is afterwards. Setting aside his preference to be led by mathematics rather than experiment, Dirac smelt an impending revolution and in December wrote to Tamm, spelling out the consequences of Shankland's findings.[25] First, the neutrino would no longer be needed, as Pauli had based his entire argument for its existence on the energy-conservation law. Second, and more important, as Shankland's experiment involved light, his results might be a hint that energy is not conserved whenever particles collide at speeds close to the speed of light. If so, Dirac pointed out, it would be reasonable to retain the basic theory of quantum mechanics, which applies to comparatively slow-moving particles, though the relativistic extensions of the theory, such as quantum electrodynamics, would have to be abandoned. A few days later at the Kapitza Club – still meeting despite its founder's absence – Dirac gave a talk on the implications of Shankland's results. To most physicists, the experiments looked unreliable, and it seemed wise to wait for the results to be checked independently.[26] But Dirac could not wait: in January 1936, he set out the implications of Shankland's results in a short, equationless article in the journal *Nature*, addressing his comments to the entire scientific community. If Shankland was right, Dirac said, quantum electrodynamics would have to be abandoned, adding, 'most physicists will be very glad to see the end of it'.[27] Coming from one of the discoverers of relativistic quantum mechanics and field theory, these were striking words. Heisenberg privately dismissed Dirac's thoughts as 'nonsense'.[28] Einstein did not conceal his glee: 'I am very happy that one of the real experts now argues for the abandonment of the awful "quantum electrodynamics".'[29] Schrödinger, disillusioned with the conventional interpretation of quantum theory, was pleased that Dirac had apparently joined the malcontents.[30] Bohr, who in 1924 had been among the first to suggest that energy might not be conserved in every atomic process, was publicly less critical, though he took Shankland's results with a pinch of salt.[31]

Experimenters, including Blackett in London, downed tools, changed their plans and began programmes of experiments to investigate Shankland's claims. A few months later, however, it became clear that he had been wrong and that energy was indeed conserved. The false alarm made a deep impression on Dirac. A year later, he wrote ruefully to Blackett: 'After Shankland, I feel very sceptical of all unexpected experimental results. I think one should wait a year or so to see that further experiments do not contradict the previous results, before getting worried about them.'[32] Dirac's inclination to believe exciting new observations had been irreversibly undermined.

After another secret Christmas vacation with Manci and her children in Austria and Hungary, marriage was now on the cards.[33] But Dirac could not bring himself to commit. No one knew of his inner turmoil; all they saw was the familiar meditative Dirac, the prince of asceticism, going wordlessly about his business. But in private he was not quite as cold and detached as he seemed to be. On his mantelpiece, he kept a photograph of Manci in a swimsuit, but no one saw it: when there was a knock on the door of his college rooms, he took the photograph down and hid it in a drawer. When many of his associates thought he was working, he was sloping off to see Mickey Mouse films, taking the Kapitza boys out for runs in his new car and reading T. E. Lawrence's *Seven Pillars of Wisdom*. In a bid to make Dirac more self-aware, Manci recommended that he read Aldous Huxley's *Point Counterpoint* as she thought Dirac resembled the novel's character Philip Quarles: brilliant, solitary, emotionally 'a foreigner' and entrenched 'in that calm, remote, frigid silence'.[34] Not seeing the likeness, he wrote to Manci: 'I doubt whether I am really like Philip Quarles, because his parents are not really like mine,' underlining – perhaps unconsciously – the importance of his mother and father to his sense of identity.

Dirac wrote his letters to Manci before he went to bed, 'the best time for thinking about you'. He never mentioned his work, nor did she enquire about it, and he rarely referred to his colleagues, but he did make an exception in February, shortly before he was due to meet Bohr and his wife in London.[35] It was not long before Manci tired of the praise Dirac heaped on his elderly friend in one letter after another; 'Bohr, Bohr, Bohr,' she yawned. Dirac was surprisingly sensitive to these complaints and showed that he appreciated that her hair-trigger jealousy needed to be handled with care by toning down the complimentary references to colleagues he admired.[36] His tact was tested again shortly before the Easter vacation, when Manci was hoping to see him. He explained to her that he felt duty-bound to visit his parents, as he had not seen them for several months; the problem was that after his visit to Bristol he would be in no fit psychological state to meet her:

It really will change me very much when I go home; it will make me afraid to do anything for my own pleasure. I shall probably be afraid to think of you [. . .] I find it satisfies me to be able to think of you whenever I wish. Why cannot you be satisfied in the same way? You should cultivate your imagination [. . .] It would be no use for me to see you for one or two days because, as you know, I am never kind to you the first day or two when I meet you.[37]

Dirac pleaded with her to understand the paralysis that overcame him whenever he set foot in 6 Julius Road: 'If you cannot understand this, you will never understand me.'[38] But Manci showed no sympathy; he was selfish, she told him. She had no interest in cultivating her imagination – she was not asking the Earth; all she wanted was to see her man in the flesh:

You do not consider anything but from your point of view. We are very different in [that] you never think to help people or to make them happy in spite [of the fact] that you are in the lucky situation where it would be easy to do so . . . I like you less.[39]

She got her way. Shortly before Easter, Dirac returned to Bristol for a few days and, after taking a few days to recover, organised a vacation with Manci in Budapest. 'I cannot imagine being happier than I was with you,' she wrote to him. Finding it hard to express his joy, he assured her that the vacation left him 'not wanting feminine society at all'.[40]

After Easter, Dirac's colleagues in St John's were surprised to see him so sunburnt, and when they asked him where he had been, he replied, 'Yugoslavia.'[41] The first casualty of Dirac's secret love was his commitment to literal truth.[42]

During the first week of June 1936, Dirac was gathering together his rucksack, sleeping bag, ice axe, rope and crampons, preparing for his next climbing vacation in the USSR with Tamm.[43] Besides visiting Kapitza, he wanted to be in the Caucasus on 19 June to see a solar eclipse, the first he will have seen. Before leaving, he wrote to Manci, asking her not to write to him because if Tamm and Kapitza 'notice [that] you and I write very much to each other, then very quickly the news would spread to physicists all over the world and they would all gossip about us'.[44]

Kapitza was in better spirits, reading his subscription copies of the *New Statesman* and supervising the building of his new institute. Many of its rooms were replicas of ones in the Mond Laboratory, though Kapitza ensured that his new director's office was even grander, with an even larger footprint. After he demanded that every item of his laboratory equipment should be transferred, Rutherford complained that it seemed Kapitza would not be happy until the paint of the Mond Laboratory had been scraped off the walls.[45] The Soviet Union was still the talk of the Cambridge common rooms, and the *Cambridge Review* abounded with articles about it, including a sceptical review of Crowther's *Soviet Science*, a whitewash that declared the Stalinist state's interference in science to be minimal. The Trinity College scholar Anthony Blunt, later a distinguished art historian, wrote an article on how a gentleman traveller might make the most of Russian hospitality – the champagne and the caviar, if not the bed bugs.[46] Unknown to his colleagues, Blunt had recently become a Soviet spy.

Shortly before Dirac set off for Russia, he heard from his mother that his father was severely ill with pleurisy: every breath was painful and liable to be accompanied by a stabbing pain in his midriff. Flo wrote that the family doctor had ordered her husband to stay in bed for ten days but assured her that 'I'm not to worry as Pa is the kind of man to make the worse of anything just to keep me busy.'[47] From the tone of his mother's letter, Dirac sensed that his father was not seriously ill, and he knew his parents were supported by Betty, about to move permanently to London to become a secretary.[48] So he decided to set off on vacation and arrived in Moscow on Saturday, but within hours received a telegram from his mother, telling him that his father was dying.[49] He decided to head home, perhaps hoping to make one last effort to make his peace with his father, to achieve a reconciliation that had not been possible with Felix. Having left his hiking gear with Tamm, he caught the 7 a.m. flight from Moscow: he had twenty-two hours to find the right parting words.

Charles grumbled that he did not want to be confined to bed at home because his wife was not taking proper care of him. So his doctor arranged for a professional nurse to take up residence in 6 Julius Road at night and to supervise Charles's care during the day. But that was not enough: after a few days, he demanded to be moved to a nursing home on the perimeter of the nearby St Andrew's Park, where he chose a comfortable room whose bay window looked out on the beds of early summer flowers.[50] The staff soon realised that they had an awkward customer on their hands: the matron told Flo that Charles 'was an awful fidget so restless and fussy', and the nurses were instructed to leave him alone and to look into his room every half an hour. Struggling against pleurisy and the onset of pneumonia, he suddenly decided that he wanted to go home, but his doctor forbade it. Flo stopped visiting him, leaving him alone with his stabbing chest pains, his quarrels with the nurses and his reflections on the past sixty-nine years. One of his bitterest regrets must surely have been his estrangement from his son, 'Einstein the second', as the *Daily Mirror* had described him three months before. This adulatory article, which Charles is almost certain to have read, concluded by telling its readers that their great-grandchildren might one day talk about him, having forgotten Noël Coward, Henry Ford and Charlie Chaplin. One sentence in the piece will have taken Charles by surprise: the anonymous author wrote that Paul Dirac is only happy when he is in the lecture room, at the wheel of his sports car and 'in his home in Bristol, where he can talk with his father'.[51]

At the end, the only member of Charles Dirac's family to be standing by him was his daughter, and she was about to break his heart by moving to London. On the day she was due to start work, Monday 15 June, he died. The end came a few hours before his son arrived in Bristol: any hope of a deathbed reconciliation had been extinguished.

Two days later, on a warm and cloudy summer afternoon, Dirac was among the mourners at the funeral. It was a civic occasion, held in St Bonaventure's, the handsome Catholic church at the end of Egerton Road, near the family home. A few hours before, at eight in the morning, the choir had sung a requiem mass over Charles's open coffin near the altar. The funeral was scheduled to begin at 3 p.m. Shortly before, dozens of mourners made their way through the Bishopston streets – representatives from the Esperanto Society, the Merchant Venturers' Technical College, the French Circle and Cotham Road School, including several school-children. Also there was elderly Arthur Pickering, the man who had introduced Dirac to Riemannian geometry, still telling stories of how he had struggled to find challenges for the most precocious student he ever had.

The eulogy, the weeping, the sacred music, the lowering of Charles's coffin into the grave – together, they may have stirred Dirac to reflect on the good things his father had done for him. Charles had ensured that his younger son had an excellent education and had encouraged him to study mathematics. And it was Charles who had given him the money he desperately needed in order to begin his studies in Cambridge.

Straight after the funeral, Dirac gave vent to his feelings in a sin-gle-page letter to Manci. In the most expansive handwriting he ever used in his life, he wrote that he would return to Moscow after he had spent a week with his mother: 'I think that in Russia I can best get used to my new situation.' He wanted to see Manci again, he told her, but gave her firm instructions not to contact him: 'I would rather you did not wire me while I am in Bristol because my mother would probably open it.' Dirac concluded with some simple words of relief: 'I feel much more free now; I feel I am my own master.'[52]

Charles Dirac had left no will – he probably did not want to leave much to his wife and possibly could not face the thought that his true wishes would be known to all the people who revered him as a fam-ily man. Flo had long suspected that he had been squirreling his money away, but even she was stunned by the amount he had hoarded: the net value of his estate was worth £7,590 9s 6d, about

rac family, 3 September 1907

Left to right: Felix, Betty and Paul Dirac c.1909.
A French grammar book rests on Paul's lap.

ul Dirac, 17 August 1907

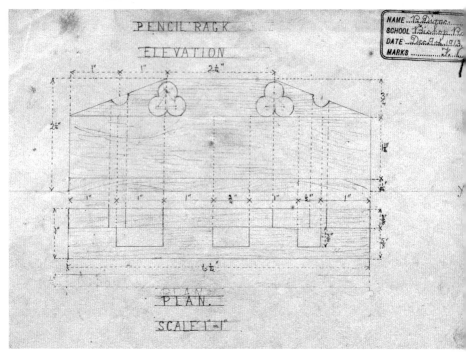

Technical drawing by Paul Dirac at Bishop Road School, Bristol, 9 December 1913

Bristol University Engineering Society's visit to Messrs Douglas' Works, Kingswood, 11 March 1919. Dirac is in the front row, fourth from the right.

Charles Dirac, *c.*1933

Felix Dirac, 1921

Julius Road, Bristol, where Dirac lived
with his family from April 1913 until he left
for Cambridge in 1923. He regularly returned
home and began his work on quantum
mechanics in his bedroom here.

Max Born (seated, central) with several
younger colleagues at his home in Göttingen,
spring 1926. Dirac is, as usual, diverted.
Oppenheimer is in the back row, fourth from
the left.

Some members of the Kapitza Club, after a meeting *c*.1925, in the room of Peter Kapitza, Trinity College, Cambridge. Kapitza is directly beneath the drawing of a crocodile on the ease

Patrick Blackett and Paul Ehrenfest,
c.1925

Isabel Whitehead with her husband Henry, an
their son Henry, 1922.

Dirac (standing close to the doorway) at a meeting in Kazan, Russia, 12 October 1928

Left to right: Heisenberg's mother, Schrödinger's wife, Flo Dirac, Dirac, Heisenberg and Schrödinger. They have just arrived at Stockholm railway station, 9 December 1933, for the Nobel celebrations.

Dear Manci,

Thanks for your 8th letter, which I received yesterday. It was a nice cheerful letter. You say I do not answer all your questions. I have read again your more recent letters and I give here the answers to the questions that I have not answered before.

letter number	question	answer
5	Have you seen Marietta's baby?	no.
5	What makes me (Manci) so sad?	You have not enough interests
5	Whom else could I love?	You should not expect me to answer this question. You would say I was cruel tried.
5	Isn't Gabor a dear little fellow?	Yes certainly, I expect so.
5	You know that I would like to see you very much?	Yes, but I cannot help it.
6	Are you "Dear Dirac"?	Sometimes.
6	Do you know how I feel like?	Not very well. You change so quickly.
6	Were there any feelings for me?	Yes, some.
6	Why should I miss your letters?	Because I have been writing so (too?).

Extract from a letter from Dirac to his friend Manci Balazs, 9 May 1935

Dirac and Manci on their honeymoon, Brighton, January 1937

The Dirac family in the garden of their Cambridge home, c.1946. Left to right: Dirac, Monica, Manci, Gabriel, Mary and Judy.

irac and Manci (on the far left) with a party during a crossing of the Atlantic on the
5 *America*, 2 April 1963

irac and Richard Feynman at a conference
n relativity, Warsaw, July 1962

Dirac at the Institute for Advanced Study,
Princeton, *c*.1958

The Diracs' home in Tallahasse, 223 Chapel Drive

Kapitza and Dirac at the Hotel Bad
Schachen, Lindau, summer 1982

One of the last photographs taken of Dirac,
Tallahassee, *c.*1983

fifteen times his final annual salary. Half of the legacy was shared by Paul and Betty, and the rest went to Flo, who quickly headed off on a restorative holiday in the Channel Islands, where she wrote to her son: 'I've won my liberty and shall keep it.'[53] Betty, apparently finding her mother's relief unseemly, departed for London and never lived in Bristol again but occasionally corresponded with her mother. Betty was piqued when she read that Flo had destroyed most of her father's papers in a bonfire in the back garden; the remainder of the papers she gave to Paul. From them we know that, somehow, several of his parents' love letters survived.

When Flo returned to Bristol, she arranged for Charles's gravestone in Canford Cemetery to be engraved with the words Paul had written for her:

In loving memory of
Our dear son
Reginald Charles Felix Dirac, B.Sc.
★ Easter Sunday 1900
† March 5th 1925
And of my dear husband
Charles Adrien Ladislas Dirac, B.ès.L
Father of the above
★ July 31st 1866
† June 15th 1936

Dirac was obviously determined that the tone of family memories of his father should owe more to propriety than honesty. His mother wrote to him: 'One doesn't mind after a few months.'[54]

When Dirac resumed his visit to Russia, he celebrated by attempting to climb Mount Elbrus, 5,640 metres above sea level, the highest peak in the Caucasus, a near wilderness.[55] With Tamm and a small party of his Russian colleagues, Dirac hiked through the forest to reach a base camp and then scaled the eastern side of the mountain, fearful of injury, sweat dribbling down his back and sunburned face during the day, shivering in the tent at night. Mount Elbrus yielded its rewards only grudgingly, as hundreds of defeated mountaineers had found, some as they fell to their deaths. After several days, Dirac and his fellow climbers saw Russia's most majestic glacial scenery, sights all the sweeter for the pain that must be suffered to win them. He only just made it; after reaching the top, he was spent and had to rest for a day before he could begin the journey back to base.[56] Never again would he attempt such an ambitious climb.

After recuperating, Dirac joined Kapitza, who was back to his buoyant best. The building of the institute was progressing well, and the first consignments of his equipment were about to arrive from the Cavendish. The authorities were taking care of him: although most Soviets suffered food shortages, Rutherford heard from Kapitza that he was eating oysters, caviar and smoked sturgeon of a quality that would make even the Trinity College 'gourmands at the high table dribble'.[57] In under three years, the Soviet authorities had won him round.

In the next stage of Dirac's hedonistic trip, he visited the two people he most wanted to see: Manci and Bohr. Having contemplated his bereavement for a few weeks, when Dirac saw Manci in Budapest, he confided his worries that he and his father were so similar: both devoted to work, both extremely methodical, both lacking in empathy. Apparently for the first time, he described how his father had treated his family so unspeakably. After he left Budapest, she urged him to put his resentments behind him: 'One has to try to understand and forgive.'[58] He will have been mulling over Manci's advice towards the end of September when staying with Bohr and his wife in their country retreat. The Bohrs were also recovering from grief, less complicated and probably much more painful than Dirac's: their eldest son Christian had died two years before, at the age of seventeen, in a freak yachting accident. Bohr had been on the deck with him and had been helpless as he watched him drown.[59]

At Bohr's suggestion, Dirac stayed in Denmark longer than he originally intended, to attend a special conference at the institute about a branch of science that Dirac knew almost nothing about: genetics. He learned, he wrote in a letter to Manci, that this 'is the most fundamental part of biology' and that there are 'laws governing the way in which one inherits characters from one's parents'. There was no escape from his father's genetic legacy – it was in Dirac's blood.[60]

When Dirac returned to Cambridge, his adventurous spirit was intact, and he changed his research topic from quantum physics to cosmology, refocusing his imagination from scales of a billionth of a centimetre to thousands of light years. Einstein's general theory of relativity provided the sturdy theoretical foundations of modern cosmology, but the subject was handicapped by a dearth of reliable data. As a result, theoretical cosmologists had more room for manoeuvre than was good for them and had to rely heavily on intuition.

Without question, the most successful observational astronomer was the former lawyer Edwin Hubble, an Anglophile American in his mid-forties, given to declaiming on conference platforms in a strangely affected English accent, akin to Oppenheimer's. Hubble had created a public sensation in 1929 when he suggested that galaxies (aggregates of stars and other matter) do not stay still with respect to one another but are always rushing apart. In what became known as Hubble's law, he used the data in his charts and tables to propose that the further a galaxy is away from the Earth, the faster the galaxy is moving away from it. This picture of galaxies dashing away from each other was consistent with Lemaître's 'primeval atom' theory of the origin of the universe, a precursor of the modern theory of the Big Bang.

Dirac's perspective on the subject emerged after a few months' gestation, when he was also contemplating one of the most important decisions of his life: should he marry Manci? Here was a warm, caring and cultured woman, the kind of extrovert he liked, one of the few with the patience to draw out his humanity. On the other hand, she was impulsive, hot-headed and overbearing. Could he be happy with a woman who had something of the controlling personality of his father? He knew it would be pointless to ask his mother, who wanted no competition for his loyalty. It would not be wise to seek the counsel of Wigner, as his loyalties would be divided; besides, he had problems of his own. Having felt undervalued at Princeton, Wigner had moved to the University of Madison, Wisconsin, and was contemplating marriage to his colleague Amelia Frank, one of the few female quantum physicists. When Wigner asked Manci to visit him and to size up his girlfriend, she jumped at the opportunity to sail from Southampton on the *Queen Mary*, the world's most luxurious liner, whose maiden voyage had taken place five months before.[61] When Manci asked Dirac if she could visit him in Cambridge before she sailed, he fobbed her off but quickly relented.[62] Still unsure whether he should commit to the relationship, he drove Manci over to see Isabel Whitehead for what Manci knew was an informal grilling. When he returned to Cambridge, he felt confident enough to forward some of Mrs Whitehead's views to Manci, excising points that might upset her:

Mrs Whitehead said she liked you. You are very unusual and have the simplicity of a child. I think this is what she meant by your being charming. [. . .] she said that I ought to make up my mind quickly, also that you and I would find it very difficult to get on together because we are so different.[63]

Yet Mrs Whitehead had second thoughts. Worried that Dirac was contemplating marriage without the spiritual commitment she believed was essential, she wrote Dirac a long and anguished letter, thundering like Lady Bracknell:

Would it be useful to go and talk to Prof. Eddington about spiritual things? I feel sad that you should have this limitation that you do not seem [?] to believe in God; and I am always afraid that I have failed to help you, how and when you need help.[64]

Mrs Whitehead pleaded with him not to make his decision when he was 'in a mood', a phrase he had used when they last met. This stung him into a rare candour about his state of mind. On 6 December, when Manci was preparing to sail from New York, he replied to Mrs Whitehead that he did not believe his decision depended on whether or not he believed in God. She had misunderstood his reference to his state of mind when he took his decision:

[By 'in a mood'] I meant only that I would need to be in a courageous mood to take an irrevocable step, after I had made up my mind what I ought to do. I think I err on the side of trying to be guided too much by reason and too little by feeling, and this makes me feel helpless when it comes to problems that cannot be solved by the clear-cut reasoning that one has in science [. . .] I have felt very favourably inclined to [Manci] for several months, with occasional relapses, which get less and less as time goes on.[65]

But Mrs Whitehead was not to be deflected; she wrote straight back to Dirac, insisting that 'married love comes to its highest perfection between people who know and love God'.[66] But these words were wasted on Dirac, for whom the concept of God had no precise meaning.

By the time he was among the dockside crowds at Southampton, waiting for Manci to arrive, he had made up his mind. During the drive to London in his sporty drop-head Triumph coupé, he steered his car to the kerbside and asked Manci, 'Will you marry me?'[67] She accepted immediately. When he told his mother the news, she was predictably shocked but summoned the grace to wish him and Manci well, offering to travel to London on the day before Christmas Eve to meet her future daughter-in-law. Dirac accepted, perhaps inadvertently giving his mother one last chance to persuade him to stay single.

Manci was staying in the smart Imperial Hotel in Bloomsbury, overlooking Russell Square. During their few hours together, Flo and Manci found a few moments to talk privately, leaving Manci puzzled.[68] As soon as Flo arrived home, she wrote to Dirac with a detailed account of the conversation:

FLO: You will be having twin-beds soon.
MANCI: Oh no, I must have a room to myself. I cannot allow Dirac to come in my bedroom.
FLO: What are you marrying him for?
MANCI: I like him very much and want a home.

Flo was astute enough to avoid outright condemnation. 'Manci was very nice indeed,' she wrote, before the inevitable qualification: 'I suppose you know she is only contracting a "marriage of convenience".'[69] His mother knew how to unsettle him. She had just seven days to make him reconsider the balance he had struck between reason and feeling.

Twenty-one

Pythagoras says that number is the origin of all things; certainly, the
law of number is the key that unlocks the secrets of the universe.
PAUL CARUS, *Reflections on Magic Squares*, 1906

On the morning of Saturday 2 January 1937, Dirac and Manci married in Holborn Registry Office in central London. He had wed his
anti-particle, a woman almost opposite to him in character and tem-
perament, as his father had done thirty-eight years before. That had
proved disastrous, resulting in something akin to mutual annihilation, so Dirac may have feared – at least at the back of his mind –
that history would repeat itself.

It was an overcast day, the crowds in London going about their
business after the Christmas holiday, girding themselves for the
harshness of winter. The wedding was a simple civil ceremony, with
only a few guests, including Dirac's mother and sister, the Blacketts,
Isabel Whitehead and her husband.[1] After lunching with them in a
restaurant near by, the couple returned to their hotel and drove to
Brighton. Dirac could not have picked a more conventional place for
his honeymoon: for decades, it had been the most popular seaside
venue in Britain for romantic trysts. It was a peculiarly raffish town,
famous for its two Victorian piers jutting imperiously out to sea, for
the pale green domes of its faux-oriental pavilion, its future-telling
robot and a host of other tacky attractions.

It appears that no photographs were taken of the wedding, but
Dirac took reels of them during the vacation, the best of them showing the newlyweds on a pebbled beach, smiling broadly, looking coy
and love-struck. Dirac looks comfortable lying on the beach in his ill-
fitting three-piece suit, pencils still protruding from the pocket of his
jacket. In some of the snaps, it is possible to see a string-operated
device that he devised to enable him and Manci to photograph themselves with no one else present.

After the honeymoon, while Manci was in Budapest with Betty,
Dirac looked around Cambridge for a permanent home and dis-
charged his duties as Lucasian Professor. Three weeks after Manci's

departure, rain lashing against the windows of his rooms in St John's, he was overcome with loneliness, sheltering from the wind and drizzle of the Cambridge winter. He wrote to his wife 'the first love letter I have ever written [. . .] Rather late to begin is it not?' In the two passionate letters he wrote in as many days, he revealed an almost Byronic expressiveness:

I realize more and more as time goes on that you are the only girl for me. Before we were married, I was afraid that getting married would cause a reaction, but now I feel that I will go on loving you more and more as I get to know you better and see what a dear, sweet girl you are. Do you think you will go on loving me more and more, or is it now as much as it can be?[2]

He had, at last, fallen in love. In the evenings, he read Bernard Shaw's *Getting Married* – retrieved from his father's library – and some books recommended by Manci, including John Galsworthy's sprawling *Forsyte Saga*.[3] But Dirac was spending most of his time in a Manci-obsessed reverie, counting the days to when she was due to return, dreaming of embracing her in bed under a new moon.[4] It was now Manci's turn to be sensitive about what others might think: brushing aside her worries that the censors in Hungary might be intercepting their mail, Dirac was uninhibited: 'You have a very beautiful figure, my darling, so round and charming – and to think that it all belongs to me. Is my love too physical, do you think?'[5] Struggling to find words equal to his passion, he continued:

Manci, my darling, you are very dear to me. You have made a wonderful alteration to my life. You have made me human. I shall be able to live happily with you, even if I have no more success in my work. [. . .] I feel that life for me is worth living if I just make you happy and do nothing else.[6]

Manci appears to have been no less intoxicated: 'If by any reason a war or anything would prevent me to see you again, I could never love anybody else.'[7] She and Betty were getting on well in Budapest, at the Moulin Rouge, skating on the rinks and doing the Charleston on the dance floor after a few glasses of champagne.[8] 'I am very very happy and being thoroughly spoiled,' Betty wrote to Dirac.[9] But she was depressed and mourning her father: 'he was the finest man I ever knew', she wailed.[10] In Betty's view, her parents had each been the victim of an unfortunate marriage, and she gave Manci a reason why her parents disliked each other, though this was too personal for Manci to spell it out explicitly in a letter to her husband.[11]

Manci decided to take Betty in hand and to find her a husband: '[Despite] her little faults, a bit of untidiness and unpunctuality, I shall try to [. . .] improve her and she will be a very good wife.'[12] Within days, Manci had decided that her Hungarian friend Joe Teszler was just the man for her sister-in-law: kind, gentle and – an essential requirement for Betty – a Roman Catholic. This was one of Manci's most effective pieces of social engineering: after a brief courtship, Betty married Joe – six years her senior – in London on 1 April 1937. In Bristol, Flo was now quite alone.

'Some say that I got married rather suddenly,' Dirac wrote to his wife.[13] One of the dons who were surprised by Dirac's marriage was Rutherford, who wrote to Kapitza: 'Our latest news is that Dirac has succumbed to the charms of a Hungarian widow with two children,' adding cryptically, 'I think it will require the ability of an experienced widow to look after him.'[14] A few days later, Dirac wrote to tell Kapitza the news: 'Have you heard that I was married during the vacation [. . .]?'[15] Kapitza was probably surprised as he thought he knew Dirac well but had not even known he was seeing a woman. Anna Kapitza quickly wrote to Manci, though she too had not met her:

Dear Mrs Dirac (it sounds very official but he did not even write us your name!)
I hope you will be very happy with that strange man, but he is a wonderful creature and we all love him very much. Do come to see us in the summer.
  Yours, Anna K[16]

After a second honeymoon in Brighton – only a month after the first – Dirac returned to Cambridge with Manci, who had left her children in Budapest. By late April 1937, they were still looking for a permanent home and living in a rented house in Huntingdon Road, a short stroll from the Kapitzas' former home. It is not recorded how Dirac referred to her when he introduced her to his university colleagues, but it is quite possible that he described her not as 'my wife' but by his favourite appellation as 'Wigner's sister' (this was a surprising choice of words for Dirac, usually fastidious in his choice of words to the point of pedantry: Manci was Wigner's *younger* sister).[17] She quickly established herself as one of the most colourful women in the university, holding dons spellbound as she passed on outrageous gossip about life in Princeton. Dirac looked on, adoringly.

For all her assertiveness, Manci was happy to be part of what she liked to call 'a very old-fashioned Victorian marriage'.[18] She regarded it as her duty to ensure that her husband's meals were ready on time, to put her husband's used clothes in the laundry basket every night, before laying out freshly ironed clothes for the next day.[19] She allowed Dirac to set out a few ground rules of the relationship, including an understanding that French must never be spoken conversationally in their home – he wanted to put to rest all memories of his father's linguistic regime. Perhaps surprisingly, she accepted that nothing in their domestic routine should ever interfere with Dirac's work. This apparently caused no friction when they were alone but it did, on at least one occasion early in their relationship, lead to an embarrassing tiff: Dirac had agreed to go with her to visit friends for afternoon tea but refused to leave his study because he had not finished thinking. Manci went alone, made excuses for her husband and had to put on a brave face when her host took offence.[20]

The wary British welcome given to Manci was made no more congenial by the inclement weather. The first few months of 1937 were one of the wettest periods Cambridge had seen for years. She felt unwelcome in the university, which seemed to be a place for men; spouses were meant to be agreeable ornaments – decorative but not obtrusive. Colleges did not allow wives to attend dinners, except on special occasions, so she had to sit alone with her novels and magazines while Dirac fulfilled his duty of dining in college at least once a week. Some of his colleagues thought that his marriage had lightened his character, though he was still as uncommunicative as ever, as the archaeologist Glyn Daniel found when he sat next to him at dinner in St John's:

The soup came and went in silence; halfway through the Sole Véronique I decided the effort must be made – the silence must be broken. But how? Not the weather. Not politics. Not the simple approach, 'My name is Daniel. I study megalithic monuments. Have you any views on Stonehenge?' I turned to Dirac, who was examining the grapes on his sole. 'Have you been to the theatre or the cinema this week?' I asked, innocently. He paused, turned to me with what I supposed was meant to be a kindly smile and said 'Why do you wish to know?' The rest of the meal was eaten in silence.[21]

By early September, the Diracs had moved into their grand new home, 7 Cavendish Avenue, a detached red-brick house south of the town, built sixty years before. It was in a quiet district – he had checked carefully that they would not be disturbed by the ringing of church bells – was a twenty-minute cycle ride from St John's College and had 'a beautiful garden' of almost two thirds of an acre.[22] In May, Dirac had written out a cheque for £1,902 10s. 0d., which paid for the property in a single transaction; unlike most newly married couples, they were unencumbered by a mortgage. The interior decor of the house reflected Hungarian tastes in the late 1920s. Manci imported much of the furniture from her Budapest apartment – heavy, dark wood sideboards and cabinets, capacious living-room chairs, gaudy side tables – though Dirac vetoed her most ornate items. Patterned, deep-pile carpeting and conventional landscape paintings helped to set the sober decorative tone.

Manci's children joined them in Cambridge and began to study at local schools, where they – with their uncertain, thickly accented English – had to work hard to integrate with other pupils. Although Dirac never legally adopted Judy and Gabriel, he raised them as if they were his own children and never referred to them as his stepchildren. But he also wanted biological children of his own.[23]

A few days after Dirac returned from his honeymoon, he completed his first contribution to cosmology. Had physicists known that he was working on this subject, they would probably have predicted a surprising new insight into the structure of the universe, or perhaps a fresh perspective on Einstein's theory of gravity. But he did neither. In a 650-word letter to *Nature* that included almost no mathematics, he set out a simple idea about the numbers that describe the universe on the largest scale. As soon as Bohr finished reading the letter for the first time, he walked into Gamow's room in the Copenhagen Institute and said, 'Look what happens to people when they get married.'[24]

Dirac's cosmological idea was not completely original, as it bore signs of having been strongly influenced by Eddington. Now perceived by many of his peers as a cocksure eccentric, Eddington had largely abandoned research in conventional cosmology and was spending his time trying to derive some of the most important numbers in science – such as the number of electrons in the universe – not by systematic reasoning but by pure thought. Most theoreticians, including Einstein, thought this was hokum: theoretical physics was about finding general principles, not about explaining numbers that arise in the search. In Rutherford's scabrous words, Eddington was 'like a religious mystic and [. . .] not all there.'[25]

In his *Nature* article, Dirac pointed out that the universe is characterised by several numbers that seem to be connected in a simple way. He focused on three numbers, each of them estimates:

1. The number of protons in the observable universe. Experimentally, this number is roughly 10^{78} (that is, 10 multiplied by itself 77 times).
2. The strength of the electrical force between an electron and a proton divided by the strength of the gravitational force between them. This turns out to be about 10^{39}.
3. The distance across the observable universe divided by the distance across an electron (according to a simple classical picture of the electron). Its value is approximately 10^{39}.

The first striking point about these numbers is that they are so much larger than any number that occurs anywhere else in science: 10^{39}, for example, exceeds the number of atoms in a human body by a factor of a hundred billion. The second point is that the largest estimated number, 10^{78}, is the square of the smaller one. This, Dirac believed, may not be a coincidence and suggested that these numbers might be related by extremely simple equations such as

$$\begin{array}{ccc} \text{distance across the observable} & & \text{linking number} \\ \text{universe divided by the distance} & = & \times \text{(strength of the electrical force} \\ \text{across an electron} & & \text{between an electron and a proton} \\ & & \text{divided by the strength of the} \\ & & \text{gravitational force between them)} \end{array}$$

and

$$\begin{array}{ccc} \text{number of protons in the} & & \text{another linking number} \\ \text{observable universe} & = & \times \text{(distance across the observable} \\ & & \text{universe divided by the distance} \\ & & \text{across an electron)}^2 \end{array}$$

Having noted that in both of these cases the linking number is about one, Dirac proposed a generalisation: this is always the case – *any* two of the huge numbers occurring in nature are connected by very simple relationships and linking numbers close to one. This is Dirac's large numbers hypothesis, a consequence of his faith that the laws underlying the workings of the universe are simple.

The suggestion has an intriguing consequence: because the size of the observable universe continuously increases as it expands, it follows that the ratio of this size to the radius of an electron cannot have always had its present value, 10^{39}, but has been increasing throughout time. If Dirac was correct to surmise that this number is connected to the ratio of the electrical force and the gravitational force between an electron and a proton, it followed that the relative strengths of these forces must have been changing as time progressed, as Milne had suggested a few years before. Dirac argued that one consequence of this is that the strength of the gravitational force withers proportionately as the universe ages: when the age doubles, the strength of gravity halves.

Dirac's decision to introduce his idea in such a short paper suggests that he believed he had hit on an important new principle and did not want to be beaten into print. If he was expecting the reception that greeted most of his papers, he will have been disappointed: this one was given a frosty reception. Yet none of the sceptics went public with their criticisms, with one prominent exception, the eccentric philosopher-astrophysicist Herbert Dingle. For him, the job of the theorist was to find laws based on experimental measurements, just as Dirac had done in quantum mechanics. Dingle spoke for many a more timid colleague when he wrote an article in *Nature* that condemned 'the pseudo-science of invertebrate cosmythology', and regretted that Dirac was the latest 'victim of the great Universe mania'.[26] Stung into a quick reply, Dirac repeated his earlier reasoning almost word for word, after prefacing his remarks with an uncontroversial comment about the nature of science: 'The successful development of science requires a proper balance to be maintained between the method of building up from observations and the method of deducing by pure reasoning from speculative assumptions.'[27]

In the same issue of *Nature*, Dingle resumed his offensive, stressing that he was not attacking Dirac personally: 'I cited Prof. Dirac's letter not as a source of infection but as an example of the bacteria that can flourish in a poisoned atmosphere; in a pure environment it would not have come to birth, and we should still have the old, incomparable Dirac.'[28]

Dirac was not deterred. However, after he had written at length about the implications of his hypothesis in a long paper – completed shortly after Christmas 1937 – he returned to quantum mechanics and did not revisit the hypothesis for another thirty-five years.

Although his idea influenced astronomers in the late 1930s, many of Dirac's peers regarded it as an aberration, joining Bohr in believing that Dirac had made a wrong move towards Eddington and Milne's quasi-mystical cosmology. But his status did not suffer significantly. In October, the Institute for Advanced Study in Princeton, still seeking to recruit the world's best theoretical physicists, put Dirac at the top of the list of the scientists they wanted to recruit, just above Pauli.[29]

Back in Bristol, Charles Dirac had left a surprise for his family: solicitors found, after months of delving through his accounts, that he had been a serial tax evader.[30] The authorities required Flo to pay six years of Charles's tax debt, the maximum they were allowed to reclaim, after making her swear affidavits that she knew nothing of his deception. 'No one knows how Pa managed to elude income tax on so many items,' she wrote to Dirac, who heard that his father had claimed £50 a year tax relief for educating Betty at university, while his son paid the bills.[31] But the nastiest revelation for Dirac was still to come, when he learned that the funds that enabled him to begin his studies at Cambridge had been provided not by his father but by the local education authority. Charles had pretended that he had stumped up the money. This petty and unpleasant deception was, for Dirac, the final straw. It negated everything that his father had done to nurture his career and revealed Charles in his true colours. This was why Paul Dirac told his closest friends, including Kurt Hofer, that he owed his father 'absolutely nothing'.[32] It was an understandable, if harsh, judgement.

After her marriage, Betty left England to live with her husband Joe, who owned and ran a flourishing camera shop in Amsterdam. Within a year they had a son, but their happiness was soon blighted by the news from Berlin, where Hitler was seeking 'living space' outside Germany and was thirsty for Jewish blood. It would not be long before the Teszlers would feel the full force of Hitler's ambitions.

At the High Table in St John's, everyone was talking about the German Chancellor and the pell-mell rush towards another global conflict. The only European country then openly at war was Spain, where Hitler supported Franco's fascist army; the British Government refused to take sides, outraging socialist opinion, particularly in

Cambridge, from where many idealists journeyed to support Franco's opponents. Dirac's eyes were, as usual, focused on the Soviet Union. That the country was suffering from an unconscionably bloody purge was clear to newspaper readers in Britain, but it appears that Dirac – like many others on the left – thought the reports were exaggerated. In Moscow, Kapitza was not aware of the extent of Stalin's murderous rampage; even so, he knew that several of his colleagues were being harassed and that he risked deportation to a labour camp if he complained, though the censors did not allow him to mention this in his letters.[33]

In the early summer of 1937, when the Diracs were in Budapest to see her family, Manci wrote to Oswald Veblen and his wife. 'Paul would like very much to go to Russia, but everybody advises him not to.'[34] Dirac insisted on making the visit and wanted to take his family, but Hungarian regulations allowed only Manci to accompany him. Kapitza confirmed the arrangements in a telegram intercepted by MI5, still checking mail he was sending to Cambridge.[35]

At the end of July, during an oppressively hot summer, the Diracs arrived at the Kapitzas' summer home days before Stalin authorised the torture of suspected enemies of the people. Only a short drive away, his henchmen were gouging out the eyes of their victims, kicking their testicles and forcing them to eat excrement. On the roads around Bolshevo, some of the trucks marked 'Meat' and 'Vegetables' hid prisoners on their way to be shot and buried in the forests to the north of the city which Dirac admired through his binoculars.[36] For many years, Soviet people would refer darkly to 'the year 1937', the height of the Great Purge, Stalin's chaotic and brutal campaign of mass intimidation, imprisonment and murder.[37] By the end of the year, the purge had claimed about four million lives. As Kapitza knew, one of the victims was Boris Hessen, a member of the delegation that had visited London and Trinity College six years before. Five of his fellow visitors would also soon be executed. Now confined to the Soviet Union at Stalin's behest, Kapitza had received all his equipment from the Cavendish Laboratory and had resumed his research.

The Diracs spent three idyllic weeks in Bolshevo with the Kapitzas in their modest summer house in the heart of a pine forest, with wild strawberries ripe for gathering and a fast-flowing river close by. They spent one languorous day after another lounging around on the covered veranda, telling off-colour jokes, the Diracs bringing the latest news on the Crocodile and his departing 'boys', the Kapitzas gossiping about life under Stalin. The two men took advantage of the cool mornings to do some manual labour – chopping down trees and clearing shrubs close to the house – and messing around with the boys. Manci, always as *soignée* as a duchess, wanted nothing to do with physical exercise and avoiding cooking anything more complicated than a boiled egg. Dismayed by the *dacha*'s lack of creature comforts, including toilet paper, she could scarcely believe that, for the first time in her life, she had to sleep outside, in a tent. But she was too polite to gripe: she shone in conversation and won over Kapitza, who saw that she had opened Dirac up. He wrote to Rutherford: 'It is great fun to see Dirac married, it makes him much more human.'[38]

Kapitza will almost certainly have enthused about the new institute being built for him. He was dealing adroitly with the authorities, bombarding them with complaints but always avoiding confrontation and keeping on the right side of the power brokers. In return, he was given unusual leeway to employ the staff he wanted and to allocate funds as he saw fit, with a minimum of bureaucracy.[39] In the following year, he was even able to hire Lev Landau as the institute's resident theoretician after he had been arrested in Moscow, having fled the Kharkov police, in fear of his life.[40] Kapitza had resumed the experiments he had begun in the Mond Laboratory and had successfully liquefied helium the previous February. Exciting new results were afoot.

Kapitza persuaded Dirac to demonstrate his support of the Russian experiment by sending his next paper to the *Bulletin of the Soviet Academy of Sciences*, in commemoration of the twentieth anniversary of the Bolshevik revolution. In the article, he investigated the symmetries underlying classical and quantum descriptions of matter, following the lead given by his brother-in-law Wigner. It was another elegant piece of work, though it produced no useful results and appeared to be more evidence that Dirac was losing his touch.

The Diracs and Kapitzas knew they were in uncertain times but could scarcely have guessed that they would not sit around the same dinner table again for another twenty-nine years.

At noon on 25 October 1937, Dirac stood among two thousand mourners in Westminster Abbey, probably wondering whether to join in the prayers and hymns or stay silent. He was at the memorial service for Rutherford. Nine days before, two weeks after the beginning of the autumn term, he had died after complications arising from surgery on his umbilical hernia: Cambridge was rife with rumours of a botched operation. Within days, government officials agreed that he was eligible to be commemorated in the 'science corner' of Westminster Abbey, alongside Newton, Darwin and Faraday. The funeral service was a national event, attended by a representative of the King, members of the cabinet, the former prime minister Ramsay MacDonald, eighty Cambridge scientists, and several foreign guests.[41] Bohr stayed with the Diracs and joined the Rutherford family party for the event, which ended when an official placed a small urn of the great experimenter's ashes a few inches from Newton's grave.

Two days after the service, Dirac wrote a consoling note to Kapitza, also grieving from the recent death of his mother. In his reply, Kapitza did not mention that the Crocodile's death occurred just as he was making his most exciting discovery – at sufficiently low temperatures, liquid helium could flow entirely without resistance to its motion. Such 'superfluid' helium could climb spontaneously up the walls of its container and behave in other strange ways that were beyond classical mechanics but which later were explained by applying quantum mechanics to the constituents of the fluid. *Nature* published Kapitza's results in a December issue, alongside a paper by two Mond experimenters who also announced the discovery of superfluidity: although Kapitza had spent two years without laboratory equipment, he had already caught up with the leaders in his field. It was no longer so easy for his detractors to sneer that he was really just a self-promoting lightweight.

Worried that the future of the Cavendish was in danger, Kapitza wrote to Dirac to enjoin him to take an active interest in securing the laboratory's future: 'I think that you who are now the leading personality in physics in Cambridge, you must take some serious interest in upkeeping the great traditions of the Cavendish Laboratory, so important for all the world.'[42]

But such a role was beyond Dirac – and, besides, he had no interest in it. The directorship of the Cavendish passed to the crystallographer Sir Lawrence Bragg, who steered the laboratory away from studies of the innermost structure of matter, partly because it could no longer keep up with the competition from the United States. With Rutherford's passing, the Cavendish had seen the last of its glory days as a place where experimenters probed atoms with the finest possible probes, though Bragg steered the laboratory's agenda into productive territory, culminating in Watson and Crick's discovery of the double-helix structure of DNA in 1953.

By the end of 1937, Dirac was bereft of the company of experimenters with similar interests in physics, and some of his most valued colleagues among the Cambridge theoreticians were in decline. Following a debilitating stroke, Fowler's health was failing, and, by early 1939, he had 'faded out', as he told Eddington.[43] In the sometimes gory seminars in the mathematics department, Eddington was timorous and unable to defend himself against pillory by his younger colleagues. Dirac looked on, unmoved and dissatisfied with his own research. Quantum field theory was virtually at a standstill, and even the best minds were finding it hard to make progress. Dirac often reflected on the contrast with only a decade before, when quantum mechanics had just been discovered: 'It was very easy in those days for any second-rate physicist to do first-rate work; it is very difficult now for a first-rate physicist to do second-rate work.'[44] These words resonated with the theoretician Fred Hoyle, an independent-minded Yorkshire man who had attended Dirac's undergraduate lectures and who had struggled in the late 1930s to find a subject ripe for development. Hoyle's bottom-up approach to physics was the antithesis of Dirac's style, but they got on well: the trick was, Hoyle said, to ask Dirac fewer questions than he asked you.[45] Hoyle was amused by Dirac's conversational eccentricities, though even he was stunned when he called Dirac to ask him a straightforward administrative question, only for Dirac to reply, 'I will put the telephone down for a minute and think, and then speak again.'[46] A few months later, Hoyle was told that he needed to find a supervisor, and Dirac took him on, partly because he was amused by the prospect of a relationship between a supervisor who did not want a student and a student who did not want a supervisor.[47]

Compared with many of the new ideas in quantum physics, the energy of an electron sounds a simple concept, but it was anything but simple to understand. This was because the energy that an electron has purely by virtue of its existence – its self-energy – turns out to be infinite. According to classical physics, the source of this embarrassment is the electric field of the electron (in some ways analogous to the gravitational field of a planet): the smaller the size of the particle, the stronger its field near by and the higher its energy. So if the electron were an infinitely small point, as it is usually assumed to be, its self-energy must be infinite. This makes no sense: how can a completely natural quantity have such an immeasurably huge value?

The theory of quantum electrodynamics, based on hole theory, had the same weakness: the self-energy of the electron was infinitely large. The most likely reason for this failure, Dirac believed, was that there was a fault in the classical theory on which his quantum theory was based: Maxwell's classical theory of electromagnetism. Dirac hoped that if he could remove the errors in the classical theory, he would be able to deduce a quantum theory of the electron that did not suffer from the disease of infinite self-energy. This was an unpopular view: most of his colleagues thought the classical theory was fine and that the challenge was to solve the problems with quantum theory. But Dirac, as usual, was unperturbed by popular opinion and spent several months in late 1937 and early 1938 working out a new classical theory and finding equations to describe an electron with a tiny but non-zero size. It was an immaculate theory but failed at its first hurdle: when Dirac tried to use it to find an infinities-free quantum version of the theory, he failed.[48]

He may have wondered whether he had lost his edge. Besides his work, he was now a family man with other priorities: a wife and two bickering children, the employment of a cook and several domestic helpers, and his dependent mother, now sixty, living a hundred and twenty-five miles away and with no telephone. Flo was, however, in good spirits: she was pottering around in her house, writing verse in bed, occasionally packing her suitcase and taking a Mediterranean vacation funded by her now healthy bank account.[49]

Manci still found it hard to settle and never felt completely comfortable in 7 Cavendish Avenue, a damp house that somehow always seemed cold, even in high summer. Disappointed that Dirac had turned down Princeton University's offer of a well-paid professorship, she thought Cambridge had nothing to commend it except its academic status and was beginning to dread the prospect of spending her life there.[50] She resented the snobbery of the Cambridge academics who patronised her from the moment they heard she did not have a degree. The Kapitzas were her sort of people – respectful, without side, full of life – but they were fifteen hundred miles away and in touch only irregularly. Always a thoughtful and generous friend, Manci inundated them with supplies to help them overcome shortages; Anna tactfully requested her to send only English books, coffee beans and good-quality pipe tobacco for her husband. She also encouraged Manci to be more positive about Cambridge: 'do you still feel lonely without your gay Budapest? If so, you are naughty and must not feel like this any more, because it worries people who like you and live with you (I mean Paul of course!)'[51]

Incessantly gloomy news bulletins on BBC radio about Hitler's increasingly transparent intentions did nothing to improve Manci's mood. In the spring of 1938, he had annexed Austria, where soldiers were welcomed with flowers and swastikas as they goose-stepped into towns. In late May, Dirac read an item in *Nature* that will probably have disturbed him: his friend Schrödinger was in Austria and appeared to be on Hitler's side. The article reported that Schrödinger had written to a local newspaper in March 1938, 'readily and joyfully' affirming his loyalty to the new regime, having 'misjudged up to the last the real will and true destiny of my land'.[52]

Dirac wanted to take his summer vacation in the Soviet Union, but this time the embassy in London refused his application and all others, in response to the British Government's denial of visas to Soviet citizens. So Dirac made more modest plans: in August 1938, he travelled to the Lake District in the north-west of England and went walking and climbing with his friend James Bell and with Wigner, still recovering from the tragically early death of his wife almost a year before, barely eight months after their marriage.[53] From their correspondence, it seems that Bell agreed with Wigner that the recent trials in the Soviet Union were frame-ups, though Bell thought they were no worse than ones organised by the English in their colony of India.[54] Meanwhile, Manci took her children and Dirac's mother to Budapest, where anti-Semitism was making her parents' life intolerable: they were beginning to see that they had no future in Hungary.

Soon, the Diracs' home became a popular hostel for physicists and their families fleeing Nazism. Among the first to arrive were the Schrödingers, who later settled in Dublin, after Schrödinger accepted a post at the newly created Institute for Advanced Studies.[55] During the stay, Schrödinger will have explained to the Diracs why he had earlier declared his support for the Nazis – he had been forced to make public his approval of the Nazi regime, he said, and had done this as ambiguously as he could.[56] Dirac appears to have accepted this explanation and not to have questioned that his friend's integrity had wavered for a minute.

The house guest whose courtesy Manci most admired was Wolfgang Pauli, en route to the Institute for Advanced Study in Princeton, where he spent most of the war. Dirac told Kapitza: '[Pauli] has got much milder after his second marriage.'[57]

Dirac agreed with the political left that the British Government had been weak and negligent in failing to tackle Hitler after his armies had invaded the Rhineland in March 1936. The left also, however, opposed rearmament and defence expenditure, a policy it would later regret. When Neville Chamberlain became British Prime Minister in 1937, he tried to mollify Hitler and waved away the warnings of his despised colleague Winston Churchill from the back-benches that the ambitions of the Führer would have to be opposed by force. The mood in Cambridge alternated from hope that a war could be avoided to fear that a conflict was inevitable.[58] Chamberlain brought about the most famous of these swings on 30 September 1938 when he returned from talks in Munich with Hitler, Mussolini and the French Prime Minister Édouard Daladier to declare 'peace for our time', having agreed that Hitler's troops would be free to enter Czechoslovakia. Crowds cheered Chamberlain's return until they were hoarse; the entire country was euphoric even after it became clear that Czechoslovakia had been betrayed. But Churchill thought the agreement was a travesty: '[The] German dictator, instead of snatching his victuals from the table, had been content to have them served course by course.'[59]

As he spoke those words, two German chemists, Otto Hahn and Fritz Strassman, were making a discovery that would change the course of history. The experiment they had done superficially looked recondite: when neutrons were fired at compounds of uranium, the new chemical elements that were formed were much lighter than had previously been thought. Within a few weeks, by the beginning of January 1939, it was clear that Hahn and Strassman had observed individual uranium nuclei breaking apart into two other nuclei, each with roughly half the mass of the original nucleus, as if a stone had split into two parts of about the same size. Analogous to cell division in biology, the process came to be called 'nuclear fission'. The key point was that the amount of energy released in the fission of a nucleus exceeds the energy produced when atoms change partners during the burning of gas, coal and other fossil fuels by a factor of about a million – this is energy release on a huge scale.

Eddington had long foreseen the possibility of harnessing nuclear energy and in 1930 looked forward to the time when there would be no need to fuel a power station with 'load after load of fuel' but that 'instead of pampering the appetite of our engine with delicacies like coal or oil we shall induce it to work on a plain diet of subatomic energy'.[60] Just over three years later, at the 1933 annual meeting of the British Association, Rutherford had ridiculed his colleague's vision as 'moonshine'. On the following day, after Leó Szilárd read about the prediction in *The Times*, it occurred to him as he traversed a pedestrian crossing in Bloomsbury that it might be possible to capture nuclear energy more easily than Rutherford had imagined: 'If we could find an element which is split by neutrons and which would emit *two* neutrons when it absorbs *one* neutron, such an element, if assembled in sufficiently large mass, could sustain a nuclear chain reaction.'[61]

When Szilárd heard about the discovery of fission, he realised that the chemical element he had in mind could be uranium. If more than one neutron was emitted when the uranium nucleus fissioned, those neutrons could go on to fission other uranium nuclei, which would emit more neutrons, and so on. Szilárd later recalled that 'All the things which H. G. Wells predicted appeared suddenly real to me.'[62]

The discovery of nuclear fission on the eve of a catastrophic conflict is one of history's most tragic coincidences. What made the prospect of nuclear weapons worrying for Dirac and other scientists who understood the implications of the discovery was that it had been made in Berlin, Hitler's capital.

Physicists and chemists were about to be drawn from the tranquillity of their offices and laboratories into a world of warfare, secrecy and power politics. The stakes could not have been higher, nor could the new work have been more troubling to their consciences. Scientists who regarded it as their duty to be open about their findings found themselves worrying that their results were too sensitive to be made public.[63] Szilárd believed that if uranium was in principle capable of sustaining a nuclear chain reaction, then the results should be kept secret from Hitler's scientists, including Heisenberg and Jordan.

The sometimes bad-tempered exchanges about whether to keep the fission properties of uranium secret involved most of the leading nuclear scientists, including Bohr, Blackett, Fermi, Joliot-Curie, Szilárd, Teller and Wigner. By early summer 1939, the campaign to keep the new science secret had failed. It was now public knowledge that uranium should be able to sustain a nuclear chain reaction: nuclear weapons were a practical possibility.

Dirac was only peripherally concerned with these discussions, having been asked by Wigner to support Blackett in the campaign to keep sensitive results confidential.[64] In Cambridge, the euphoria of Chamberlain's Munich agreement had faded into despair by the spring of 1939, when Hitler contemptuously absorbed previously unoccupied parts of Czechoslovakia into Nazi protectorates and client states. War now looked inevitable. During those grim early weeks of 1939, Dirac prepared his first lecture as a self-styled philosopher of science who professed no interest in philosophy. Although the two living scientists he most admired – Einstein and Bohr – were both accomplished at talking about science to wide audiences, Dirac had shown no interest in following their lead until the Royal Society of Edinburgh awarded him their Scott Prize and invited him to give the Scott lecture on their favoured theme of the philosophy of science to an audience that included many who knew little or no science.[65] Late on a Monday afternoon early in February 1939, he spoke for an hour on the relationship between the mathematician, who 'plays a game in which he invents the rules', and the physicist, 'who plays a game in which the rules are provided by Nature'.

Dirac's themes were the unity and beauty of nature. He identified three revolutions in modern physics – relativity, quantum mechanics and cosmology – and hinted that he expected them one day to be understood within a unified framework. Although he did not mention John Stuart Mill, Dirac was seeking to answer the same question posed in *A System of Logic*: 'What are the fewest general propositions from which all the uniformities existing in nature could be deduced?'[66] Whereas Mill never used the beauty of a theory as a criterion of its success, an appreciation for the value of aesthetics had been part of Dirac's education. He now gave vent to his feelings by proposing the principle of mathematical beauty, which says that researchers who seek the truly fundamental laws of nature in mathematical form should strive mainly for mathematical beauty. Ignoring centuries of philosophical analysis about the nature of aesthetics, he declared that mathematical beauty was a private matter for mathematicians: it is '[a quality that] cannot be defined, any more than beauty in art can be defined, but which people who study mathematics usually have no difficulty in appreciating'.[67]

The success of relativity and quantum mechanics illustrates the value of the principle of mathematical beauty, Dirac said. In each case, the mathematics involved in the theory is more beautiful than the mathematics of the theory it superseded. He even speculated that mathematics and physics will eventually become one, 'every branch of pure mathematics having its physical application, its importance in physics being proportional to interest in mathematics'. So he urged theoreticians to take beauty as their principal guide, even though this way of coming up with new theories 'has not yet been applied successfully'.

The physicists in the Edinburgh audience heard Dirac's enthusiasm for the discovery that the universe is expanding, which he said 'will probably turn out to be philosophically even more revolutionary than relativity or the quantum theory'. Focusing on how the universe developed from its birth, he suggested that classical mechanics will never be able to explain the present state of the universe because the conditions at the very beginning of the universe would be too simple to seed the complexity we now observe. Quantum mechanics might provide the answer, he believed: unpredictable quantum jumps early in the universe should be the origin of the complexity and 'now form the uncalculable part of natural phenomena'. Cosmologists rediscovered this idea forty years later, when it became one of the foundations of the quantum origins of the universe. While the world was heading into the gutter of war, Dirac was looking up at the stars.

In Cambridge, the students could not bring themselves to face the consequences of the expected war. In April, the students' sixpenny magazine *Granta* looked forward to another summer of croquet on the lawns, cucumber sandwiches, paprika salad and crème brûlées washed down with chilled Bollinger. For students wanting to wind down after the examinations, there were performances of Mozart's *Idomeneo* and more opportunities to see Disney's *Snow White and the Seven Dwarfs*.[68] The captain of the university cricket team knew that the party was soon to be over, though he said that he hoped to God that Hitler would not start a war before the end of the cricket season. But he was disappointed: after Hitler's invasion of Poland, Chamberlain declared war on 3 September, before the final overs had been bowled.

Ten days before, Dirac – on holiday with his family on the French Riviera – read that Stalin had signed a non-aggression pact with Hitler, a moment that George Orwell called 'the midnight of the century'. Stalin's opportunism was incomprehensible to Dirac. He still tended to expect politicians to practise with the consistency of mathematicians, and it is probably no coincidence that Dirac's disillusion with politics and politicians began that summer. From then on, he turned away from public affairs and concentrated on his family, which was about to expand – Manci was pregnant.

Twenty-two

As I write, highly civilized human beings are flying overhead, trying to kill me. They do not feel any enmity against me as an individual, nor I against them. They are only 'doing their duty' [. . .].
GEORGE ORWELL, *The Lion and the Unicorn*, 1941

Advances in aviation technology had made the aerial bombing of Britain inevitable, though some people in Cambridge could not believe the Germans would ever bomb a town of such beauty.[1] Nuclear weapons were being discussed, too, in newspapers and popular magazines, but most of the public and national leaders seem not to have noticed. Dirac, aware of the potential of nuclear fission, had an inkling of what might be in store: like many scientists, he would soon have to decide whether to drop his research and participate in the largest military programme the world had ever seen.

Soon the conflict would disperse Dirac's extended family across two continents. He waited every day for news of Betty in the Netherlands. Manci was worried about her Jewish relatives, especially her parents and sister, who had left Budapest and settled in New York State, assisted by Wigner and his new wife Mary. Although she strongly supported the war, Manci knew the pain of being suspected as an alien and smarted at the subtle signs of disapproval from strangers when she revealed her thick accent, which many took to be German. In her adopted country, she felt like a 'bloody foreigner'.[2]

When the Diracs ventured into the centre of Cambridge on the freezing nights of January 1940, they saw that much of the town looked just as it did in Newton's day. Under the moonlight, the architecture of the city – the College buildings, King's Parade, Senate House – had never looked more sublime.[3] The mood of the town was, however, becoming more apprehensive: thousands were bracing themselves for an attack, ready to flee to the new bomb shelters. Dirac and his family stayed indoors, carefully observing the 'blackout', preventing every shard of light from escaping into the night by covering their windows with black paper. By six o'clock each night, the town was as quiet as a village on a Sunday morning; by ten, it was almost deserted.[4] The church bells had been silenced, the street-lamps switched off.

At the beginning of the war, the population of the town had swelled by almost a tenth, to about eighty thousand. At the beginning of September 1939, trainloads of children had arrived from London and other towns that were expected to be the targets of enemy bombers. The evacuees, many with their home addresses written on luggage labels tied around their necks, were billeted with local families, many of which received them less warmly than sentiment now recalls.[5] The Diracs did not take in any of these children, though in the coming months they saw them virtually overrun the town.[6]

Everyone, including the dons, carried around a foul-smelling rubber gas mask. For the time being at least, academics in their gowns had lost their special status and were no more important than the thousands of volunteers and part-time workers who were preparing for war. The texture of day-to-day conversations changed: people talked more loudly, endlessly repeating catchphrases such as 'I'm doing my bit' and 'Don't you know there's a war on?' All over the town, posters warned that 'Careless talk costs lives', words that looked comically alarmist, as there were no signs of an imminent conflict: by March 1940, nothing much had happened since the collapse of Poland, and the restless public called it the Phoney War or, sometimes, the Bore War. Most of the evacuated children drifted home.

The university ticked over, though there were fewer dons as many of them had left to take up posts in government, the armed forces and war research establishments.[7] There were fewer students, too, but a skeleton programme of teaching continued, and Dirac gave his lectures on quantum mechanics as usual. A regular visitor to the college, he saw how much its atmosphere had changed: it now accommodated not only its staff and students but also uniformed members of the Army, Navy and Royal Air Force, who worked in the new buildings completed shortly after the outbreak of war. The college was one of the national centres of the Air Force, and hundreds of its cadets were trained there, mixing uneasily with the undergraduates, who had different catering facilities. The menus for college members were now much more modest: at High Table, about all the Fellows could expect was a ladleful of mutton stew and vegetables grown on college land. Gardeners had dug up the lawns to grow onions and potatoes.

At home, the Diracs lived like most others in Britain. They queued for their ration books and food coupons and took pots and pans to local collecting points to be melted down and turned into weapons.[8] Dirac had chopped down a tree in the garden for firewood, cultivated potatoes and carrots in a nearby allotment, and grew giant mush-rooms in his cellar. But Manci, well into her pregnancy, wanted sup-port. She would not dispense with her servants, and she fretted at the thought of losing even one of them. Dirac's mother in Bristol was counting the days to the birth of her second grandchild, hoping that the child would be a boy and that his parents would name him Paul.[9] But she was to be disappointed: the child was a girl, Mary, born on 9 February 1940, at London's Great Ormond Street Hospital.[10] As Manci wrote in her notebook, Mary was a 'daddy's girl', as she would remain. Dirac was a doting father, in his reserved way, dan-dling her on his knees, trying to entice her to play with a new doll sent by her godmother, Schrödinger's wife Anny.

Desperate to see her first granddaughter, Flo made a flying visit to see the baby and her mother. Flo's manner with the baby did not impress Manci, who complained to Dirac the next day:

It is awful of me to write about her, you never criticize my parents. But I never felt as much that she has neither heart nor feelings . . . She has no notion of how to handle a tiny thing as a baby but she picked her up. It was quite terrible to me.[11]

Dirac may have sensed that this would not be the last clash between the two women closest to him, each jealous of the other's place in his affections. But their disputes appear not to have spoiled his first few months of paternity. He now had the domesticity he craved, but it was soon disrupted by an urgent request to do something he had hoped to avoid: to join the scientists' war effort.

Rudolf Peierls was now in Birmingham, a professor of physics by day and volunteer fireman by night, equipped with a uniform, a helmet and an axe. Peierls had settled in England after fleeing Nazi Germany in 1933 with his Russian wife Genia, a former member of the Jazz Band of Soviet physicists. Like most scientists who had lived under Hitler, Peierls wanted him crushed, but the British authorities were slow to accept his offers of assistance: in early February 1940, Peierls and his wife were officially classified as 'enemy aliens'.[12] The couple's naturalisation papers came through later that month so he was eligible to work on secret projects, though the authorities still looked at him with suspicion and denied his request to work on the new radar technology.

In early February 1940, when Dirac was cradling his newborn daughter in his arms, Peierls was thinking about nuclear weapons. Like most scientists who were following the debate, he believed that such a weapon would not be possible after all. Niels Bohr and John Wheeler had apparently provided the clinching argument by proving that the fission of uranium by slow neutrons was due entirely to the rare isotope of uranium ^{235}U, containing a total of 235 nuclear particles, not to the much more common uranium isotope ^{238}U, which contains 238 particles. A little less than one part in a hundred of a typical sample of natural uranium is ^{235}U, and the rest is almost entirely ^{238}U. It followed that if a nuclear bomb were made using naturally occurring uranium, very few nuclei would undergo fission, so any chain reaction that started would soon fizzle out. But a loophole was spotted by one of Peierls' Birmingham colleagues, Otto Frisch, the scientist who had given fission its name and been the first to explain it, in collaboration with his aunt, Lise Meitner. Frisch was one of an almost unbroken string of bachelors who lodged with Rudolf and Genia Peierls and became part of the household, helping with the washing up and keeping their children amused during the blackouts.

The crucial question Frisch asked was: 'Suppose someone gave you a quantity of the pure 235 isotope of uranium – what would happen?' When Frisch and Peierls did the calculations, they found the amount of ^{235}U needed was about a pound, about the volume of a golf ball. Although it would be difficult and expensive to produce much of this rare isotope, the resources required, compared with the costs of running the war, would be chickenfeed. Frisch later recalled that when he and Peierls tumbled that the purification process could, in principle, be completed in weeks, 'we stared at each other and realized that an atomic bomb might after all be possible'.[13] Even more terrifying was the thought that the Germans might already have done their calculation and Hitler might be the first to have the bomb.

Frisch and Peierls secretly typed up two memos on the properties of a 'Super-Bomb' and the implications of building one, setting out their conclusions in a total of six foolscap pages, which they sent to the British Government, keeping just one carbon copy.[14] The authorities were grateful but asked them to understand that, as Peierls later recalled, 'henceforth the work would be continued by others; as actual or former "enemy aliens", we would not be told any more about it'.[15] If the Government wanted scientists to build a nuclear weapon, they would need to find a way to distil pure ^{235}U from mined uranium ore, which contains the mixture of ^{238}U and ^{235}U. Several groups were set up in the UK to investigate ways of separating the uranium isotopes, including ones at the universities of Liverpool and Oxford. Scientists in these groups knew that Dirac had invented one method of doing it: the centrifugal jet stream method of isotope separation, which he had investigated in the spring of 1934 but abandoned after the Soviets had detained his collaborator, Kapitza. By the late autumn of 1940, Dirac had heard that his long-discarded experiment might, after all, have important applications in developing material to make a nuclear bomb.[16] Soon he would be under pressure to resume his studies of the technique.

In the United States, Leó Szilárd – a close friend of Manci's brother Eugene Wigner – was trying frantically to persuade the Government to develop a nuclear bomb before the Germans. He was working at Columbia University in New York with his fellow refugee Enrico Fermi, the experimentalist best qualified to build a nuclear weapon if it were feasible. Progress was slow and funds were short, partly because few government officials took Szilárd's hectoring seriously. In the summer of 1939, Wigner, Szilárd and Teller persuaded Einstein to write to President Roosevelt, drawing his attention to the possibility of nuclear weapons and the danger that the Germans might produce one first.[17] After a long delay, Roosevelt invited Einstein to join a committee of government advisers but he brusquely declined and sat out the war at the Institute for Advanced Study in Princeton, where word spread that the Nazis were indeed working on a bomb. In the spring of 1940, Dirac's friends Oswald Veblen and John von Neumann wrote to the director Frank Aydelotte, urgently seeking his assistance to fund investigations into the chain reaction. In their letter, they mentioned a recent conversation with the Dutch physical chemist Peter Debye, who had led one of Berlin's largest research institutes until the German authorities sent him abroad in order to free his laboratories for secret war work.

[H]e made no secret of the fact that this work is essentially a study of the fission of uranium. This is an explosive nuclear process which is theoretically capable of generating 10,000 to 20,000 times more energy than the same weight of any known fuel or explosive [. . .] It is clear that the Nazi authorities hope to produce either a terrible explosive or a very compact and efficient source of power. We gather from Debye's remarks that they have brought together in this Institute the best German nuclear and theoretical physicists, including Heisenberg, for this research – this in spite of the fact that nuclear and theoretical physics in general and Heisenberg in particular were under a cloud, nuclear physics being considered to be 'Jewish physics' and Heisenberg a 'White Jew'.

There is a difference of opinion among theoretical physicists about the probability of reaching practical results at an early date. This, however, is a well-known stage in the pre-history of every great invention. The tremendous importance of the utilisation of atomic energy, even if only partially successful, suggests that the matter should not be left in the hands of the European gangsters, especially at the present juncture of world history.[18]

Aydelotte responded by helping Szilárd with his search for funding. The prime responsibility of Aydelotte and Veblen, however, was the Institute for Advanced Study, and they dreamt of setting up a wartime haven for the most eminent quantum physicists, including Bohr, Pauli, Schrödinger, Dirac and even Heisenberg.[19] But when the war intensified, it became unthinkable for most of them to concentrate on anything other than the war. The pursuit of the fundamental laws of physics was set aside.

In April 1940, the Nazis overwhelmed Norway and Denmark and launched a blitzkrieg on Belgium, Luxembourg and the Netherlands a few weeks later: the Phoney War was over. Dirac's sister Betty and her family were now living in an occupied country. Joe, like all the other Jews, lost much of his freedom: he was subjected to a curfew, forbidden to ride in trams or cars and forced to wear a yellow star when outside his house. A month before, the German forces had conquered Denmark unopposed and had invaded Norway, swatting aside the British Government's naval campaign to repel them. Chamberlain was forced out of office and replaced by Churchill – the man regarded by many as a belligerent class warrior soon became the saviour of his country and the embodiment of bulldog spirit, a national hero.[20] The Diracs gathered round their radio to listen to his broadcasts and to reports of his speeches. Three days after he entered 10 Downing Street he told the House of Commons in his first speech as Prime Minister that the aim was 'Victory – victory at all costs, victory in spite of all terror; victory, however long and hard the road may be; for without victory, there is no survival.' Manci was star-struck: she sent Churchill a note consisting of just two words – 'God's blessings' – after a broadcast he had made a few days after the Luftwaffe dropped its first bombs on Cambridge on 18 June 1940.[21]

At 11.30 p.m. on that night, the air-raid sirens began to wail, and the Diracs scurried down to the shelter of their cellar. Moments before midnight, they heard a Heinkel bomber dive low overhead and, after a piercing whistle, a huge explosion when the plane dropped two high-explosive bombs about a mile away. Ten people were killed, a dozen were injured, and a row of Victorian houses was laid waste.[22] The following night, the bombers struck Bristol for the first time, targeting the British Aeroplane Company's factory in Filton. Dirac's mother was desperate to speak to her son but, with no telephone, the best she could do was to write to him:

The awful raiders pay a midnight call every night. The first was a downright shock on Monday. I flew down with all my dressing gowns, collected all the green cushions from the big chairs & made myself warm & comfortable propped against the kitchen door [. . .] To my surprise I got intensely angry at their cheek & impudence in disturbing my night's rest & daring to visit our Island in such a manner.[23]

Choosing not to take drams of whisky and play poker with her neighbours in their cellars, Flo spent most nights alone, crouched in the cupboard under the stairs with cotton wool in her ears, trying to sleep during the hours of 'fireworks'.[24] At five in the morning, when the sirens and steamers in the docks roared their 'all clear', she went up to Betty's room to catch up on her sleep. Flo was lonely, sick with rheumatism and gout, anxious about her family and disappointed that her son was such a poor correspondent: 'I am sure you can spare five minutes for a few lines if you try very hard.'[25]

By August 1940, the 'Battle of Britain' was underway. The Luftwaffe was pummelling London and fighting over the skies of England with the Royal Air Force, helped by the early warnings made possible by the new radar technology. Despite the widespread fear of an imminent Nazi invasion, daily life in Britain continued normally. Food and everyday supplies were in the shops, the trains and buses were running, and there were queues outside cinemas showing *Gone with the Wind*.[26] It was a summer of almost uninterrupted glorious weather, and the more prosperous Britons, including Dirac, saw no need to forgo their annual vacation. Dirac and Gabriel took a four-week break in the Lake District, renting a cottage in Ullswater with Max Born and his family – his wife, their nineteen-year-old son Gustav, their daughter Gritli and her new husband Maurice Pryce, a theoretical physicist at the University of Liverpool.[27] The outdoor life, primitive facilities and the prospect of communal cooking were not for Manci, who remained in Cambridge with Judy, baby Mary and her nurse, after Dirac had assured her that the danger of air raids in Cambridge had been exaggerated ('you should not let the air raid warnings worry you, dear').[28]

While Gabriel stayed in the cottage, his head buried in a book, Dirac and Pryce headed off early to the mountains with a vacuum flask of hot tea and a packed lunch. With Pryce and Gustav Born, Dirac climbed the highest peak in England, Scafell Pike, rowed on the lakes, climbed up several rock faces and followed some of the paths trodden by Wordsworth, who had lived in nearby Grasmere.[29] At night, the party dined on the balcony, overlooking a lake as still as a pond: it scarcely seemed possible that they were in a country fighting for its life until they switched on their radio and heard the news from London.[30]

Barely four days after Dirac's vacation began, Manci was in the cellar with Mary and Judy, following the first of several air raids. 'I am very sorry to be away during these air raids,' Dirac wrote to his wife, though he was not worried enough to return home.[31] Feeling abandoned and dejected, Manci dropped her usual affectionate tone when she wrote to him:

I know very well that you never do or did what people happened to ask you for. So I am not asking you anything; it is but a question. Would you return to Cambridge if I was not here? Because if you would not, then do not come home please.[32]

As usual, her wrath soon abated. Dirac was habituated to her outbursts and fended them off by remaining silent. It was a singular marriage, not one most people could endure, but it was working.

Dirac's climbing partner Maurice Pryce – formerly a colleague of Dirac and Born in Cambridge – was studying isotope separation with the Liverpool team and had recently asked Dirac's advice about his centrifugal jet method.[33] But it seems that Dirac did not think seriously about developing the method until several months later. This delay is surprising, as many of his peers were talking urgently of the need to develop a nuclear weapon ahead of the Nazis. Perhaps part of the explanation for his tardiness is that he was preoccupied with his stepchildren, constantly quarrelling and consuming more of his energy than he would have liked.[34] Gabriel, then an introverted fifteen-year-old, was developing into a talented mathematician. Encouraged by Manci, he revered his stepfather as a hero, looked to him for advice and even copied his handwriting, down the last detail of the curl on the capital D. Judy, two years his junior, was growing into an attractive young woman and quite different from her brother: she was lazy, headstrong and not at all frightened of provoking her mother. Manci's high-handedness sometimes alarmed Dirac, who privately warned Gabriel that he should not take too much notice of her tantrums.[35]

Dirac agonised about his sister and her family, behind enemy lines. She had written to him from Amsterdam via the Red Cross mail service on 3 July to report that she was safe, and the letter took three months to arrive. Shortly after he read it, Dirac heard that Dutch citizens would be fined £15,000 if they were caught listening to British radio transmissions. He was also concerned about his mother, who occasionally visited Cambridge but spent most of her time alone in 6 Julius Road, going out only occasionally to the shops, the cinema and to volunteer for the emergency canteen service. Bristol was the fourth most heavily bombed city in the UK (after London, Liverpool and Birmingham): almost every night, the planes attacked the city and, though Julius Road was two miles from the worst of the attacks, Flo was in fear of her life. She went to bed early and tried to sleep through the seven-hour barrages, until the sirens blasted the 'all clear' signal into the dawn.[36]

These were among the darkest days of the war. Peierls in Birmingham was one of many who believed that the fight against Hitler was then 'hopeless', as he recalled fourteen years later.[37]

Although Germany had failed to win the Battle of Britain, the war was going its way, as Hitler well knew: he told his ally Mussolini in October 1940 that the war had been won.

In mid-December, Dirac's mother was admitted to a nursing home, suffering from concussion, after a stone had fallen on her when she was out walking. Dirac rushed to Bristol and, between visits to Julius Road, walked around the bombed-out heart of the city. At the Merchant Venturers' College, he saw that many of the buildings he had known since he was a child had been pulverised into smouldering piles of rubble. Several of the homes on his route had been bombed out, their once-private spaces now embarrassingly on show for all to see. 'The middle of Bristol is terribly damaged [. . .] most of the best shopping areas are in ruins [. . .] and many beautiful churches have gone,' he wrote to Manci.[38] She was too angered by being left alone to feel much sympathy:

You know that envy is not in me but I am a little revolted that you had to go, and have to stay. After all 60 years ought to have been enough for anybody to make friends [. . .] she is only interested in people as far as what she will be able to talk about them.[39]

Unmoved, Dirac helped his mother to return home and stayed with her until she could resume her routine, returning to Cambridge shortly before the year's end. All over the UK, the New Year celebrations were subdued, for the country was pinned to the wall.

Most scientists in Britain had put themselves at the service of their country but, as usual, Dirac did not swim with the shoal. In peacetime, he was part of the mainstream of physics but always one step from it, so that his individuality was not constrained. He now had the same relationship with the scientists working for the military: he supported them but only to an extent that neither his daily routine nor his intellectual independence was compromised. One of the first invitations to participate in war work that Dirac received had come, surprisingly, from the mathematician G. H. Hardy, who was contemptuous of the applied mathematics involved in war work as unworthy of 'a first-rate man with proper personal ambitions'.[40] He wrote to Dirac in May 1940, asking him to join a team of twelve mathematicians to code and decode messages at the Civil Defence offices in St Regis, in the event of a Nazi invasion.[41] Dirac appears to have declined, probably because he would not consider moving from Cambridge and because teams, to him, were anathema.

The journalist Jim Crowther did not, however, stop trying to involve his retiring friend in public affairs: in mid-November 1940, he tried to persuade Dirac to attend a meeting of the Tots and Quots dining club, an informal gathering of academics who were interested in exploring how their expertise might be useful to society (the name of the club is a reference to the Latin *quot homines, tot sententiae:* 'so many men, so many opinions'). Its twenty-three members in 1940 – including Bernal, Cockroft and Crowther – were often joined by guests, such as Frederick Lindemann, H. G. Wells, the philosopher A. J. Ayer and the art historian Sir Kenneth Clark.[42] The location of the club's political centre of gravity, well to the left, was reflected in the outcome of their debates, most of them held over a few bottles of wine and an indifferent meal in London's Soho. The meeting Crowther wanted Dirac to attend, on Saturday, 23 November 1940, was scheduled to discuss Anglo-American scientific cooperation and was to take place in Christ's College, Cambridge. Crowther knew the best way to encourage Dirac to attend: 'It would be quite unnecessary for you to join in the discussion if you did not wish to.'[43] Crowther succeeded, and Dirac listened to a wide-ranging discussion about ways of promoting scientific cooperation with American scientists, until shortly after midnight. Bernal opposed the suggestion that British research projects should be transferred to the United States, arguing that the best way forward was to promote personal contacts between British and American scientists. It was important, he stressed, not to give up too easily on preserving the independence of British science.[44]

The record of this special Tots and Quots meeting makes no mention of any contribution from Dirac. So far as records show, he attended no other social gathering of scientists during the war.

At about the time of the meeting, Dirac began to think again about his method of separating mixtures of isotopes.[45] Seven years earlier, he had demonstrated that the technique might work; he now turned to a theoretical analysis of the process, to help engineers investigate ways of separating a mixture of ^{235}U and ^{238}U. His original idea was to deflect a gaseous jet of the mixture through a large angle, so that the heavier and therefore slower-moving isotopes would be deflected less than the lighter ones, and the two components would separate.

He tried to find a general theory of all processes that might separate isotopic mixtures in this way, aiming to deduce the conditions that would most effectively separate them. To solve the problem, he had to use all his talents: the mathematician's analytical skills, the theoretician's penchant for generalisation and the engineer's insistence on producing useful results.

He gave his first account of the theory in a confidential, three-page memorandum. Dirac wrote it for Peierls and his colleagues, probably in early 1941, between the incessant bombing raids, and typed it at home. He wrote the paper in his usual spare style but taking care to highlight the most important conclusions so that they would be clear even to engineers allergic to complicated mathematics. The memo does not focus on his own jet separation method but concerns every conceivable way of separating isotopes in a liquid or gaseous mixture by causing a variation in the isotopes' concentration. The separation might be achieved, for example, by subjecting the mixture to a centrifugal force or by carefully arranging for the temperature to change across the container. To make the calculations tractable, he made the reasonable assumptions that the fluid mixture contains only two isotopes (each made of simple atoms) and that the concentration of the lighter one is small compared with the concentration of the other. In a short calculation, he derived a formula for what he called the 'separation power' of the apparatus, a measure of the minimum effort needed to cream off a given amount of the lighter isotope. He found that every part of such an apparatus, irrespective of how it is built, has its own maximum separation power, and he showed how to calculate it.

Dirac often drove to Oxford to talk with the experimenters who were developing ways of separating isotopes, under the impish Francis Simon, another German refugee physicist. Dirac surprised many of the experimenters by participating vigorously in their meetings and by making practical suggestions about the design of their apparatus. During these discussions, he conceived several other ways of separating isotopes, each of them based on his original centrifugal jet stream method.

The Oxford group built one of Dirac's designs, and it worked, but his method was less efficient than the competing technique of gaseous diffusion, which exploits the fact that the atoms of two isotopes in equilibrium and with the same energy have different average speeds: the lighter, swifter atoms are more likely to diffuse through a membrane than heavier ones, enabling the mixture to be separated. Consequently, at this stage in the development of nuclear energy, resources were diverted to gaseous diffusion, and Dirac's idea was set aside.

Late at night on 9 May 1941, a bomb fell opposite the Diracs' home, damaging two houses and causing small fires that Judy helped the fire fighters to extinguish.[46] This was the most frightening moment for the Diracs in the worst year of bombing in Cambridge, and it was relentless where they lived, close to the strategic target of the railway station. But the Diracs' everyday life was much the same as it was before the war. Part of this routine involved welcoming visitors; Dirac was determined not to follow his father's example of virtually barring the family home from others, apart from paying students. One of the most frequent visitors to 7 Cavendish Avenue was Jim Crowther, 'the newspaper man'.[47] A one-man clearing house of information about the activities of leftist scientists, he was a favourite of Manci's, who entertained him and his wife Franciska as royally as rationing allowed: she could stretch to a cup or two of tea, but biscuits and cakes were luxuries. After one get-together, Crowther lent her Somerset Maugham's *On Human Bondage* to help her improve her English and her understanding of British foibles. Still worried that people in Cambridge thought of her as an outsider, she even sensed disquiet that she might be an enemy agent. Suspicions of aliens intensified in the town in the spring of 1941, when an innocent-looking Dutch seller of second-hand books in Sidney Street was unmasked as a spy. When he heard that military intelligence was on to him, he broke into an air-raid shelter on Jesus Green and shot himself.[48]

During the Diracs' conversations with the Crowthers, Dirac heard Crowther's bulletins on the scientists' war work, delivered with his subtle political colouring, though almost certainly without the political edge that he reserved for conversations with more committed colleagues. Crowther knew that this was time well spent: Dirac would never commit himself to the cause of the left, but he was a powerful ally, if only because no other British physicist came close to his intellectual prestige.

Although Dirac spent most of his time on war work, he was still thinking about quantum mechanics. In one project, he collaborated with Peierls and Pryce to refute accusations made by Eddington that experts in relativistic quantum mechanics, including Dirac, were persistently misusing the special theory of relativity. This disagreement had been rumbling for years: in the summer of 1939, Sir Joseph Larmor had heard that 'Eddington has lately come to blows with Dirac.'[49] Dirac, Pryce and Peierls tried to make Eddington see reason but, by the early summer of 1941, their patience had run out, and they prepared what Pryce dubbed 'the anti-Eddington manuscript'.[50] The paper appeared a year later, and Eddington's arguments were crushed to the satisfaction of everyone except Eddington himself, who never accepted defeat.

Having been awarded the Royal Society's Baker Medal, Dirac had to prepare a special lecture, where he presented some of his new thinking on quantum physics. In the early afternoon of 19 June 1941, when Dirac arrived at Burlington House, he saw that central London had suffered surprisingly little in the Blitz; most of the damage had been done in the City and the East End. Giving the lecture was in keeping with the spirit of the hour – Londoners were going about their business as usual, and that included attending lectures about matters of no practical importance.

Dirac rose to the podium at 4.30 p.m. to describe why he was so unhappy with the current state of quantum mechanics: why is it, he wondered, that the first version – set out by Heisenberg and Schrödinger – is so beautiful whereas the relativistic version is so diseased?[51] It might be possible, he showed, to remove one of the pathologies of the relativistic theory – negative-energy photons – using a technical device later dubbed the 'indefinite metric'. Although not a panacea, the technique demonstrated to the standing army of quantum physicists that Dirac was still one of their generals. Even Pauli was impressed and wrote to Dirac to say so.[52]

Dirac's conclusion to the lecture was that the 'present mathematical methods are not final' and that 'very drastic' improvements were needed. He knew, however, that they were unlikely to be made at a time when most of the best scientific brains were working on top-priority projects for the military. Only rarely did the scientists on opposing sides communicate. One such encounter took place in late September 1941, when Heisenberg travelled to Nazi-occupied Denmark to see Bohr (who knew nothing of the Anglo-American project to build a nuclear bomb) in a fraught meeting that was remembered and interpreted quite differently by the two men.[53] The playwright Michael Frayn dramatised their discussions six decades later in *Copenhagen*, a metaphor for the uncertainty principle: the more the intentions of the participants at the meeting are probed, the murkier they appear to be. Although it will never be possible to know precisely what the two men said, one consequence of their meeting is now clear: their friendship was damaged beyond repair.

Dirac, in touch with neither Bohr nor Heisenberg, knew nothing of the meeting. When it took place, he was in Cambridge, preparing for the new term, no doubt anxiously reading the news of the Nazis' invasion of the USSR, which had begun when Hitler unilaterally broke the pact with Stalin three months before. Kapitza was now in Hitler's sights. On 3 July, a few days after the pact collapsed and Stalin joined the Allies, Kapitza sent Dirac a telegram, one of the few communications that Dirac received from him during the war:

In this hour of stress when our two countries fight against a common enemy I want [*sic*] send you a friendly word. The united strength of all men of science will help the victory over the treacherous enemy who by brutal force destroyed the liberty and crushed the freedom of scientific thought in Germany and is trying to do the same in all the world. My greetings to all friends united in their will for fighting to complete victory for the freedom of all people for the freedom of scientific thought so dear to our two countries.[54]

Later during the conflict, Dirac was moved to similarly grand words in a rare letter to Kapitza. After offering his 'hearty congratulations' to Kapitza on his second Stalin Prize, Dirac wrote that he hoped 'that the great Hitler menace which now darkens this world will soon be obliterated'.[55]

Flo was also thinking about Kapitza and his compatriots: 'Those plucky Russians are putting up such a grand fight!', she wrote to her son. By the summer of 1941, Bristol appeared to have seen the worst of the bombing; about 1,200 people had been killed.[56] She was ailing and desperate to stay at 7 Cavendish Avenue, where Manci was struggling to cope after her maid and cook had departed. In early October, Flo arrived with her luggage and hatbox, having declared that she wanted to help with the housework, though her doctor wrote privately to Dirac: 'I want you to see that she does not do extra work' as 'her heart is overstrained and she is rather run down'.[57] She stayed longer than the month she had planned, working under

Manci's direction as a kitchen maid and house cleaner, helping the servants and Mary's nurse. Soon after the Americans entered the war, following the bombing of Pearl Harbor on 7 December 1941, Flo wrote to one of her neighbours: 'Paul says it will take two years to conquer the Japs.' But she was homesick and tired of being Manci's charlady: 'I really am afraid I will be quite ill if I stay on. Manci imposes on me too much.'[58]

Flo never sent the note as, four days before Christmas, she had a fatal stroke. Dirac seems to have taken her death with his usual almost-inhuman stoicism: his sliver-thin vocabulary of emotions did not include conventional expressions of grief. Manci saw no tears. Yet he knew better than anyone the tragedy of her unfulfilled life: the suicide of her first-born; her servitude during a sham marriage and its horrible final years, when she was like a rabbit domiciled with a bear. Dirac knew that his mother had her flaws: she was absent-minded and disorganised, selfishly determined to keep her younger son to herself. But Dirac knew that life had not been generous to his mother and that he had been her greatest love.

Her funeral took place two days after Christmas.[59] Dirac threw away most of her belongings but not the Christmas card on which she had written her feelings about Manci. He kept that among his papers.

Twenty-three

There is no room now for the dilettante, the weakling, for the shirker, or the sluggard. The mine, the factory, the dockyard, the salt sea waves, the fields to till, the home, the hospital, the chair of the scientist, the pulpit of the preacher – from the highest to the humblest tasks, all are of equal honour; all have their part to play.

WINSTON CHURCHILL, speech to the Canadian Parliament, 30 December 1941, later broadcast on the BBC

To Dirac's neighbours, it appeared that the war had little impact on his life: he remained another professor going quietly about his business, his civic duties involving nothing more than an occasional night on fire watch at the Cavendish.[1] But none of his neighbours knew that he spent most of 1942 and 1943 working on nuclear weapons. Even Manci had only a vague idea of what he was doing: she told the people she knew in Cambridge that he was working on 'decoding'.[2]

Most leading scientists did more to support the military than Dirac. Patrick Blackett was one of several of Dirac's friends who took his place at the top table of the Government's scientific advisers and attended dozens of interminable policy meetings. He joined his former Cavendish colleagues Cockcroft and Chadwick on a special committee set up to consider the implications of Frisch and Peierls' prediction of the small amount of uranium needed to make a bomb.[3] They consulted Dirac, but he had no wish to be part of the proceedings.[4]

By August 1941, Churchill authorised the manufacture of a nuclear weapon, following the advice of the committee and approving comments from his friend and chief scientific adviser, Frederick Lindemann.[5] The British Government allocated the resources its scientists requested to begin to build the bomb and set up the 'Tube Alloys' project, a name chosen to be dull enough to escape the attention of prying eyes and ears. Blackett, the one dissenting voice on the committee, believed that the British could not build the bomb alone: the project would be successful only if it were pursued in

319

collaboration with the Americans. He would soon be proved right. Blackett was no happier in his other dealings with the Government. He was one of the pioneers in the use of science to inform decisions about the management of the war; for example, in weighing the risks and benefits of different military strategies.[6] The hard-headed application of this new discipline of 'operational research' brought Blackett and his colleagues, including Bernal, into disagreements with the military and the politicians, who both preferred to take decisions with their hearts as well as their heads. Blackett insisted that Churchill's policy of aerial bombing enemy civilians – supported by the military and the public – was ineffective, the misguided result of a failure to identify the enemy's key industrial and military targets. It would be better to bomb the enemy's fleet of U-boats, he told an unmoved Lindemann. Churchill persevered with his policy and kept his scientific committees at a distance: for him, 'Scientists should be on tap, not on top.'[7]

Like many mathematicians, Dirac was invited to work at the Government's research station in Bletchley Park. In late May 1942, he was approached by the ancient-history scholar Frank Adcock, who had been charged with recruiting the best Cambridge brains. Adcock wrote to Dirac, 'There is some work concerned with the war which is itself important and would, I believe, be of interest to you. I am not free to say just what the work is.'[8] When Dirac asked to know more, a Foreign Office official wrote to clarify: 'The work would be a full-time job [nominally nine hours a day] and would require you to leave Cambridge.'[9] With Manci four months pregnant, this was too much disruption for Dirac to contemplate, so he never did work in the huts of Bletchley Park with Max Newman and Newman's former student Alan Turing.[10] This would have been one of the most intriguing collaborations of the war.

In Cambridge, Dirac supervised graduate students and gave his quantum-mechanics lectures to about fifteen students on Tuesday, Thursday and Saturday mornings. In 1942, his audience included Freeman Dyson, an exceptionally talented student, then nineteen years old.[11] Dyson was disappointed: in his view, the course lacked all sense of historical perspective and made no attempt to help students tackle practical calculations. Not one to suffer in silence, Dyson amused his fellow students by bombarding Dirac with questions, sometimes catching him off-guard and once causing Dirac to end a lecture early so that he could prepare a proper response.[12]

Almost twenty years before, the young Dirac had pressurised Ebenezer Cunningham in one of his lecture courses; now it was Dirac's turn to be shown the drawn sword of youth.

By early 1942, Dirac was thinking more about technology than quantum mechanics. He was a consultant to the Tube Alloys project and worked closely with Rudolf Peierls. One of the first reports that Dirac wrote for him concerned another way of separating a mixture of isotopes, using a simple method that involves injecting the mixture into the base of a hollow cylinder spinning rapidly about its long axis. The centrifugal force generated by the rotation causes the heavier isotope to move towards the outer rim and the lighter one to accumulate closer to the central axis, thus effecting a separation. When Dirac sent his report to Peierls in May 1942, he wrote that he had 'written up [his] old work' and did not mention its provenance.[13] It is clear from the manuscript that Dirac wanted to investigate the motion of the gases in the tube, to find how far up the spinning cylinder the injected gas will reach. Using classical mechanics, he found that the device would be a stable source of separated isotopes and calculated that, if the cylinder had a radius of one centimetre and rotated almost five thousand times a second, its length should be about eighty centimetres. This confidential report, declassified in 1946, proved to be seminal for the designers of centrifuges. Dirac's calculations provided the theoretical underpinning of the counter-current centrifuge, invented three years earlier by the American scientist Harold Urey. This technique was not used during the manufacture of the first nuclear bombs – other methods made less onerous engineering demands – but later became the nuclear engineer's preferred choice as it gives a particularly efficient way of separating uranium isotopes.

Dirac's other work for Peierls and his group in Birmingham consisted of theoretical investigations into the behaviour of a block of ^{235}U if a nuclear chain reaction took place inside it. These calculations probed in detail the energy changes going on inside such a block of material and investigated whether the growth of neutrons would change if the uranium were enclosed in a container. Dirac was happy for his results to be shared with the American scientists who were working on the bomb, including Oppenheimer, who by the end of 1942 had been appointed the Scientific Director of what became known as the Manhattan Project. Oppenheimer excelled at nurturing young theoreticians in Berkeley, but most of his colleagues were surprised when General Leslie Groves – the Project Director, appointed by Roosevelt – asked him to take on responsibility for building the bomb. One of Oppenheimer's Berkeley colleagues chortled, 'He couldn't run a hamburger stand.'[14] Just as surprising was the authorities' decision to appoint someone who, although a brilliant researcher and teacher, was well known to be a fellow traveller of the Communist Party.

Dirac worked mainly in his study, the one room in 7 Cavendish Avenue for which only he had the key, allowing in cleaners on the strict condition that they did not move any of his papers. If he saw any sign at all that his desk had been disturbed, he flew into a wordless rage.

The children were proving to be a handful. Dirac and Manci may well have been alarmed when Gabriel, soon after he began his mathematics degree in Cambridge, joined the Communist Party, though he kept up his membership for only six months.[15] Judy was less academic and more rebellious: when she was sixteen, in 1943, Manci furiously ordered her out of the house and threw her clothes out of her bedroom window.[16] Although she was allowed home a few days later, relations with her parents did not improve. Manci, always trying to enforce strict discipline, was frustrated by the feeble support she was given by Dirac – when she needed him to back her up in some altercation with one of the children, he retired sheepishly to his study or escaped to his garden. He spent hours tending his rhododendrons and gardenias, pruning his apple trees, sewing seeds and digging up asparagus, carrots and potatoes to help fill the larder. In the summer, he would shield his balding head from the sun by wearing a handkerchief knotted at each of its four corners.[17] Friends noticed that he practised horticulture using the same top-down methods that he used in theoretical physics, trying to base every decision on a few fundamental principles.[18] He stressed that the best way of ripening apples was to place them in linear rows, each item of fruit separated from its neighbour by precisely the same distance. In one project, he coated pea seeds with dripping and rolled them in red lead oxide powder to discourage birds from eating the newly emerged seedlings, a practice that would today induce palpitations in any self-respecting health and safety inspector.

Dirac's heart remained in quantum mechanics. In July 1942, he took time off from war work, left his family at home and travelled with Eddington to attend a conference in Dublin organised by Schrödinger, who tried to tempt Dirac to accept a job alongside him. 'There is plenty of food here – ham, butter, eggs, cakes, as much as one wants,' he wrote in one of his fond letters to Manci.[19] The Irish Prime Minister Éamon de Valera, a trained mathematician who had helped bring Schrödinger to Ireland, took the two guests on a joyride around the local countryside, having met them during the conference. Dirac had been amazed to see him there, attending lectures and taking detailed notes.[20]

On 29 September, six weeks after his return to Cambridge – still under attack from Nazi bombers – Manci gave birth to a daughter, Florence, named after Dirac's mother, though she was always called by her second name, Monica. Two days after her birth, Dirac received a letter from Peierls gently enquiring, at the request of the project directorate, if he would be prepared to move from Cambridge to work full-time on the war effort.[21] Predictably, Dirac refused.

His family was now complete. He never had a son of his own, a disappointment Manci later described as one of the saddest of his life.[22]

Dirac saw in Cambridge evidence of the prominent role the USA was now taking in the war. Every day, hundreds of uniformed American servicemen – on leave from the nearby airbases – walked the streets of Cambridge, with plenty of money to spend. They organised baseball games and, in November 1942, were visited by the stately Eleanor Roosevelt.[23] At home, Dirac received intelligence reports of the American-led experiments to build a nuclear bomb and, towards the end of the year, heard that a key experiment in the programme had been completed. In a makeshift laboratory built in a disused squash court in Chicago, Enrico Fermi and his team had built a nuclear reactor, and, in the mid-afternoon of 2 December 1942, they got it working for the first time. They had arranged a self-sustaining nuclear chain reaction, releasing energy at a rate of half a watt.[24] Wigner presented Fermi with a bottle of Chianti, which he shared in silence with his team, who had good cause to celebrate but also to be nervous: for all they knew, Hitler's scientists were ahead of them. A member of Fermi's team, Al Wattenberg, later recalled: 'The thought that the Nazis might get the bomb before us was too terrifying to contemplate.'[25]

Shortly before, Peierls asked Dirac to study a sheaf of technical papers written by Oppenheimer and his Manhattan colleagues describing the explosion of a sample of uranium undergoing fission. Early in January, Dirac pointed out inconsistencies in the papers and discussed how a nuclear bomb might be constructed, including the optimal shapes of the two masses of uranium that could be propelled together to make the bomb. During the next six months of 1943, Dirac investigated theoretically the passage of neutrons in a fissioning block of uranium and presented his results in two reports, one of them in collaboration with Peierls and two of his younger Birmingham colleagues. One of them was Peierls' lodger, Klaus Fuchs, a Bristol-educated refugee from Nazi Germany, an inept but courteous young man in his early twenties. When he and Peierls visited 7 Cavendish Avenue to talk about their secret research with Dirac, they all adjourned to the middle of the lawn in the back garden to ensure that they were out of earshot of everyone near by.[26] Manci, asked to stay inside the house, resented what she knew was the implication: she was a potential eavesdropper. During some of these al fresco discussions, Dirac and Peierls noticed that Fuchs sometimes behaved oddly, complaining that he was unwell and leaving them for surprisingly long periods before returning.[27] It would be another seven years before Dirac and Peierls understood Fuchs' behaviour.

The collaboration between the scientists working on the bomb in the USA and their counterparts in Britain was tense and difficult, but the problems were apparently resolved in the late summer of 1943, after peace-making conversations between Roosevelt and Churchill. It was obvious to most of the British scientists that they should join the Manhattan Project, and about two dozen of them – including Peierls, Chadwick, Frisch and Cockcroft – joined Oppenheimer and his team in their Los Alamos headquarters in the New Mexico desert.[28] Through Chadwick, Oppenheimer asked Dirac to join the Manhattan team, but he declined.[29] About a year later, he stopped working on the project, but never fully explained why. Peierls later suggested, probably correctly: 'I believe this was because he was beginning to feel that atom bombs were not a matter he wanted to be associated with, and who could blame him?'[30]

Dirac may have come to believe that the Nazis could be defeated without nuclear weapons. Or perhaps Dirac was influenced by Blackett, who protested that American scientists on the Manhattan Project were given access to all the research done by their British colleagues but did not reciprocate, except with Chadwick, the only Briton to be given full security clearance. Blackett felt so strongly about this that he tried to persuade his British colleagues to take no part in the Manhattan Project.[31]

On the night of 5 November 1943, the Luftwaffe dropped their bombs on Cambridge for what turned out to be the last time. Since the outbreak of the war, the sirens had sounded 424 times to warn of the bombings that had killed thirty people and destroyed fifty-one homes.[32] As the nights closed in, Dirac and his family were hoping that the blackout would end soon, but the authorities did not lift it until September in the following year.[33] By then, he was worrying constantly about his sister Betty and her family. At Dirac's request, Heisenberg had attested to the occupying Nazis that she was not Jewish, but Joe and their son were still in grave danger.[34] When Dirac last heard from them, in early September 1943, they had recently fled their home in Amsterdam – a short tram ride from Anne Frank's secret annex – after the Nazis told Joe that he could either be sterilised or interned in Poland. He probably knew that internment was tantamount to a death sentence, so the family headed for Budapest, hoping that it would quickly be liberated by the Allies.[35]

Powerless to help Betty, Dirac sat out the end of the war at home. Several of the family photographs taken around this time show him in his back garden, sitting in a deckchair, teaching Mary to read from *The Wizard of Oz*. One of her earliest memories was of her father spelling out the letters D-o-r-o-t-h-y.[36] She and Monica were given a disciplined upbringing, following the motto of English family life, 'Children should be seen and not heard,' but without any exposure to religious ideas.[37] Yet Dirac appears to have had at least some regard for religion as he and Manci followed the convention of having both their daughters christened.[38] Probably as a result of his wife's influence, the hard-line atheist had softened his line.

Try as Dirac might to concentrate on quantum physics when he was in college, the continuing presence of the military reminded him that although victory over Hitler was in sight, it could not be taken for granted. Royal Air Force officials still occupied much of the college, and the military had taken over the Combination Room for purposes they kept secret.[39] Only much later did the Fellows of St John's find out that the room contained a huge plaster model of the stretch of the Normandy coastline on which Allied troops landed on 6 June 1944. Churchill's leading general, Montgomery, believed the end of the war was in sight and didn't believe the Germans could go on much longer. Yet still Dirac could not walk over the Bridge of Sighs without being challenged. When the sentry asked, 'Who goes there?', he was satisfied with only one reply: 'Friend.' Dirac knew the threat still posed by the enemy better than most. Even when victory looked inevitable, from June 1944, Dirac was aware that German scientists, including Heisenberg, might already have developed a nuclear weapon. About a year before, he had heard from the refugee Norwegian chemist Victor Goldschmidt that Heisenberg was working on the Germans' counterpart of the Allies' Tube Alloys project. Dirac knew that the fate of hundreds of potential victims could depend on the scientific success of his closest German friend.[40]

While he waited for the war to end, Dirac began work on another edition of his book. His main innovation this time was to introduce a new notation he had first invented shortly before the war broke out. This system of symbols enabled the formulae of quantum mechanics to be written with a special neatness and concision: just the sort of scheme that Dirac had learned to appreciate in Baker's tea parties.

The centrepiece of the notation was the symbol <q for a quantum state labelled q and the complementary q>; together they can be combined to form mathematical constructions such as <q | q>, a bracket. With his rectilinear logic, Dirac named each part of the 'bracket' after its first and last three letters, *bra* and *ket*, new words that took several years to reach the dictionaries, leaving thousands of non-English-speaking physicists wondering why a mathematical symbol in quantum mechanics had been named after an item of lingerie. They were not the only ones to be flummoxed. A decade later, after an evening meal in St John's, Dirac was listening to dons reflecting on the pleasures of coining a new word, and, during a lull in the conversation, piped up with four words: 'I invented the bra.' There was not a flicker of a smile on his face. The dons looked at one another anxiously, only just managing to suppress a fit of giggling, and one of them asked him to elaborate. But he shook his head and returned to his habitual silence, leaving his colleagues mystified.[41]

The war in Europe ended in anti-climax on 8 May 1945. The relief felt like a national exhalation. In the centre of Cambridge, thousands gathered in Market Square in the blazing heat of the afternoon, dozens of Union Jacks fluttering limply in the breeze. After the Lord Mayor's speech, two bands marched separately round the town, each followed by hundreds of people, with dozens of couples dancing cheek-to-cheek in the streets. The authorities in St John's College abandoned all formalities for the day: the Combination Room swelled not only with Fellows but with dozens of normally excluded undergraduates raising their glasses to the new peace.[42] Dirac and his family celebrated with neighbours at an impromptu tea party in a local street, munching on scones and spam sandwiches served from trestle tables.[43]

If Dirac believed that science would quickly return to normal, he was mistaken. In the spring of 1945, he and seven colleagues – including Blackett and Bernal – applied for visas to enable them to attend the June celebrations of the 220th anniversary of the USSR Academy of Sciences; for Dirac, the trip would give him the opportunity to see Kapitza and other Russian friends again. But Churchill refused to allow visas to be issued on the grounds, it was later revealed, that Dirac and his colleagues might share with Stalin's scientists some of the nuclear secrets kept from the Soviets during the war.[44] During a discussion about the matter at the Admiralty in London, Blackett lost his temper and strutted magnificently out of the building, furious that the Government had dared to impugn his integrity.[45] Dirac was angry, too, but showed his emotion only by withdrawing into complete silence and talking a long, solitary walk.[46]

For several weeks after the end of the war in Europe, news had been seeping out about the Nazi concentration camps. Manci was outraged not only with the Germans but also with 'these dirty Poles' – she was sure they had connived in the atrocities. She wrote to Crowther that she had one of her rare rows with Dirac, apparently because his reaction to the revelations of unconscionable cruelty was too restrained for her taste.[47] The Diracs knew that several of

Manci's relatives had probably been murdered in the camps and that Betty's husband Joe might also be dead. News of him arrived in a telegram delivered to the Diracs' home at the beginning of July, when they were preparing to visit the Schrödingers in Dublin.[48] Joe was alive. In Budapest, he had fallen into the hands of the Nazis, who dispatched him to the Mauthausen-Gusen concentration camp in Austria, where he was one of thousands forced to work in the Wiener Graben quarry, mining granite with a pickaxe and carrying the slabs up the hundred and eighty-six steps to the top.[49] Many of his fellow prisoners perished from the freezing cold, were worked to death or were summarily shot through the neck by SS guards after being injured or collapsing from exhaustion. After the camp was liberated in the summer of 1945, he emerged looking close to death – desperate for a morsel of food and with a broken wrist, a seriously infected kidney and missing a finger.[50] While recuperating in an American military hostel in France, desperate for news of Betty and their son Roger, he wrote to Manci to suggest that Kapitza might help to find her, as the Russians had taken over Hungary. He did not have to wait long to hear the denouement: in early September, he heard from Manci that Betty and Roger were safe.

On 6 August, Dirac heard the news he had been dreading: with the tacit agreement of the British Government, the Americans had dropped a nuclear bomb on Hiroshima, killing about forty thousand Japanese civilians. At nine o'clock that evening, Dirac was in his front room listening to the radio news bulletin: 'Here is the news: it's dominated by the tremendous achievement of Allied scientists – the production of the atomic bomb. One has already been dropped on a Japanese army base. It alone contained as much explosive power as two thousand of our great ten-tonners.'[51]

After reading official statements, including one from Churchill and President Truman, the BBC announcer ended with almost comic bathos: 'At home, it's been a Bank Holiday of sunshine and thunderstorms; a record crowd at Lords has seen Australia make 273 for five wickets.'[52] All was well again – cricket had resumed. The national press rushed to praise the achievement of the leading British scientists, including Cockcroft and Darwin, who had helped to design the bomb. None mentioned Dirac, probably to his relief. One of the few civilians who were not shocked by the destructiveness of 'the atomic bomb' was the seventy-nine-year-old H. G. Wells, who first coined the term in 1914. On 9 August, just as President Truman ordered the dropping of another nuclear bomb on Nagasaki, the *Daily Express* published a weary personal perspective on the age he had foreseen.[53] He died a year later.

On 14 August, when news reached Britain of Japan's surrender, public euphoria resurged, and, in Cambridge, Market Hill swelled with an encore of the VE Day celebrations.[54] In the USA, the press showered Oppenheimer with praise and likened him to Zeus. He was the triumph of physics personified.[55]

Dirac had no idea that, only fifteen miles from Cambridge, Heisenberg had been interned by the British Secret Service with nine other German scientists in Farm Hall, a red-brick Georgian House on the outskirts of the village of Godmanchester.[56] They were treated well – given the run of the house, provided with daily newspapers and allowed to walk freely around the grounds, though they were warned that their liberties would be curtailed if any of them tried to escape. A few days after their arrival, Heisenberg wondered why the authorities were keeping him and his colleagues interned without making it public: 'It may be that the British Government is frightened of the communist professors, Dirac and so on. They say "If we tell Dirac or Blackett where they are, they will report it immediately to their Russian friends, [like] Kapitza".'[57]

When Heisenberg and his colleagues heard about the dropping of the first nuclear bomb, soon after the news was broadcast on BBC radio, they were both perplexed and incredulous. One detainee, Otto Hahn, observed sourly: 'If the Americans have a uranium bomb then you're all second raters. Poor old Heisenberg.'[58] Not knowing that the British were recording their conversations – it was unthinkable, Heisenberg chuckled – the Germans talked freely about their feelings. The British authorities declassified their conversations only in 1992; ever since, historians have pored over the transcripts and have come to a variety of conclusions. Some experts believe that Heisenberg never came close to an understanding of how to make a nuclear bomb; others that he could have made one but slow-pedalled his research in order to prevent the Nazis from acquiring the device. It is, however, indisputable that, during the conversations recorded at Farm Hall, neither Heisenberg nor any of his colleagues expressed any serious qualms about working for the Nazi regime.

*

By October 1945, Dirac's life in Cambridge had almost returned to normal. A few weeks before, he had been surprised by the high number of students attending his quantum-mechanics course, several of them still in uniform. At the beginning of the first lecture he announced to the audience, 'This is a lecture on quantum mechanics,' evidently believing that many of the students were in the wrong room. When none of them got up to leave, he repeated his announcement, this time more loudly. But still no student left.[59]

A few weeks later, Betty and her son Roger – both hungry, traumatised and anxious – returned to stay in 7 Cavendish Avenue before they were reunited with Joe. Betty and her son had almost starved to death in Budapest, and she had seen that the liberation was not as joyous as many journalists reported; in her opinion, the Russian troops who liberated the city were far more brutal than the Nazi army they had ousted. In Betty's later years, her memories of the conflict were too painful to share, though she often remarked that she regarded the survival of her family as a miracle: 'Everything afterwards was a bonus.'[60] Best of all was the birth of her daughter, Christine, just over nine months after Betty and Joe were reunited.

For the sake of tact, Betty may not have mentioned during her stay in Cambridge that she despised most of the Hungarian acquaintances she had met. Her memories of the double-dealing and inhospitable citizens of Budapest were to become a running sore in her relationship with Manci, with Dirac the embarrassed and ineffectual peacemaker.[61]

The university and St John's College were settling back into their clockwork routine. Dirac preferred this way of life, free of distractions, but he had a few other duties to discharge: during the war, Crowther had persuaded him to support their French colleagues behind Nazi lines by taking on the undemanding role of the British presidency of the Anglo-French Society of Sciences, working with an informal committee whose members included Blackett, Cockcroft and Bernal.[62] After the war, Crowther decided to relaunch the Society with a prestigious series of talks about scientific developments during the conflict, and he persuaded Dirac to give the first presentation, on 'Developments in Atomic Theory'.[63] The venue for the occasion – a red-letter day in French science – was Le Palais de la Découverte, a public science centre that stands like a Greek temple on a dark side road in the seventh arrondissement. Soon after sun-down on Tuesday 6 December, hundreds of the city's leading scien-tists made their way to Le Palais to hear Dirac talk. Two thousand people clamoured for a seat in the lecture theatre, expecting to hear the secrets of the atomic bomb.[64]

Minutes after Dirac began to speak, the audience realised that it was not going to hear about the latest in nuclear technology but a presentation on the state of quantum mechanics. Dozens tried to leave, but there was no escape: the exit was jammed with the over-flow crowd of hundreds, listening to the lecture via loudspeakers. For the physicists who were interested, a treat was in store: they heard Dirac coin two of the best-known technical terms that he introduced: 'fermions', quantum particles that obey the laws that he and Fermi had set out in 1926, and 'bosons', the other type of quantum parti-cles, which obey laws set out by Einstein and the Indian theoretician Satyendra Bose. For most of the audience, this was not much conso-lation for a wasted evening: at the end of the lecture, several of them bolted for the door.

At the dinner party afterwards, embarrassment was no doubt still in the air, but Dirac was probably oblivious to it. During six bleak years for science, in which he had contributed more to engineering than to quantum physics, he was relieved that life was returning to normal. But he was now well past thirty, the age he once believed marked the end of the theoretician's productive career: was he now too old to have radically new ideas?

Twenty-four

In September 1946, Dirac was scratched again by the next generation's talons. He was at a conference on 'The Future of Nuclear
Science' at Princeton's Graduate College, half a mile from the campus. Nestled among trees at the top of a grassy hill, the college
looked like a Gothic abbey, its majestic tower dominating the surrounding countryside – a picture of English arcadia. Many visitors
thought the college had been a landmark in Princeton for centuries,
but it had stood there for only thirty-three years.

The conference was the first of a series of international events during the university's bicentennial celebrations – months of ceremonial
glad-handing, sybaritic dinners and colourful parades.[1] The conference organiser Eugene Wigner, fresh from the Manhattan Project,
had put together an impressive guest list, including Blackett, Fermi,
Oppenheimer, Van Vleck and the Joliot-Curies, all ready to put the
war behind them and begin the next chapter of physics.

At 9.30 a.m., at the beginning of the conference's second day, Dirac
was introduced by one of the most exciting scientific talents in America,
Dick Feynman (he called himself Dick rather than Richard). Brought up
in the New York suburb of Far Rockaway, he was a clean-cut twenty-
eight-year-old, brimming over with ideas and sophomoric humour
but still grieving after the death of his first wife fourteen months
before, from tuberculosis. He was afraid he was already burnt out, he
later admitted. When he introduced Dirac, Feynman seemed unbur-
dened by self-doubt but felt 'like a ward-heeler [machine politician] in
the 53rd district introducing the President of the United States'.[2]
Feynman was not expecting to be impressed: a few weeks before, he
had been disappointed by his hero's handwritten script, which
Feynman thought was backward looking, stale and 'unimportant'.

Dirac discussed how elementary particles could be described using
his favourite mathematical device, the Hamiltonian: for Dirac, this

was the only way to proceed, and he did not spare his audience – many of them non-specialists – the technical details. As Feynman feared, the talk fell flat. Worse, Dirac was bereft of new ideas.[3] After the applause, Feynman tried to give lay members of the audience a sense of what Dirac was saying, not hiding his disappointment and remarking that Dirac was 'on the wrong track'. He cracked even more than his usual quota of jokes, prompting Bohr to stand up and ask Feynman to take the proceedings more seriously.

A few hours later, Feynman looked out of the window of the lecture room and saw that Dirac had excused himself from the conference programme and was 'paying no attention to anybody', lying on a patch of grass, leaning on an elbow, gazing lackadaisically at the early-autumn sky. Here was Feynman's opportunity to talk informally with Dirac about a matter that had intrigued him for the past four years. When Feynman was a graduate student, he had studied Dirac's 'little paper' on how the classical least-action principle can be applied in quantum mechanics, demonstrating that it could be used to build another version of quantum mechanics, different from Heisenberg's and Schrödinger's but giving the same results.[4] In his paper, Dirac had cryptically remarked that a critical quantum quantity is 'analogous' to its classical counterpart, but Feynman believed that the correct phrase was 'proportional to' (that is, if the quantum quantity changes, the classical one always changes proportionately). Here, at last, was Feynman's chance to find out what Dirac meant.

Feynman described his problem to Dirac and came to the crunch:

FEYNMAN: Did you know that they were proportional?
DIRAC: Are they?
FEYNMAN: Yes they are.
DIRAC: That's interesting.[5]

Dirac then got up and walked away. Feynman subsequently became famous for his new version of quantum mechanics but thought the credit was undeserved. The more closely he looked at the 'little paper', the more he realised that he had done nothing new. He later said, repeatedly, 'I don't know what all the fuss is about – Dirac did it all before me.'[6]

Feynman knew he had much to do if he was to prove himself a great physicist. When the conference photograph was taken, he appeared to hint at the extent of his ambition by standing behind

Dirac, just as Dirac had done in the 1927 Solvay Conference photograph, when he stood directly behind Einstein. Within a few years, Feynman's power as an analyst and intuitionist made him, in the eyes of many, the finest theoretician in America. Wigner agreed with that judgement: 'Feynman is a second Dirac, only this time human.'[7]

The next five years saw the emergence of a new theory of electrons and photons, in some ways the climax of fifty years of theoretical physics. This was largely an American success, the accomplishment of hungry young scientists who had suspended their academic careers during the war to work on nuclear weapons, radar and other projects.[8] Physicists had worked in lavishly funded, goal-driven international teams, having set aside the elitist traditions of European academia and collaborated in the less formal, can-do social environment of the United States. Now it was time for payback.

On Capitol Hill, the physicists argued that they deserved the support of the government's tax dollars to pursue curiosity-driven research. It is a fair bet Willy Loman and the other struggling breadwinners of middle America would have baulked at the physicists' case if they had been aware of it, but the politicians were persuaded and gave unheard-of levels of federal support for basic physics research and training. The US Government and private institutions funded theoretical physics. At much greater expense, Uncle Sam equipped experimenters with machines that could probe the structure of matter even more finely, using beams of subatomic particles accelerated to within a whisker of the speed of light in a vacuum. The pursuit of 'high-energy physics' had flourished in Europe in similar ways, though there was no doubt that in this branch of science – and many others – America led the world.

The first conference of leading subatomic physicists to take place in the USA after the war, at the beginning of June 1947, set their subject's agenda for the next thirty years.[9] Twenty-three carefully selected scientists – all of them men – gathered at an inn on Shelter Island, a small and secluded spot near the eastern tip of Long Island, to review their subject. The gathering could scarcely have had a more spectacular opening: in the first two presentations, experimenters announced that the Dirac equation made predictions that disagreed with new experimental results. The first speaker, Willis Lamb, had the air of a cowboy who had strayed into a physics laboratory.

But his appearance was deceptive: he was a deep thinker, an accomplished experimentalist who could hold his own with the best theorists. He got the meeting off to a flying start by announcing a serious flaw in Dirac's theory: two energy levels of atomic hydrogen that, according to the theory, should have the same energy turn out to be slightly different. Photons emitted by hydrogen atoms when they jump between the two energy levels had been detected by Lamb and his student Robert Retherford, at the Columbia Radiation Laboratory. In a masterly experiment using microwave technology developed during the war, they studied these photons and showed that each of them has only about a millionth of the energy of a quantum of visible light.

In the next presentation, given by the experimenter Isidor Rabi, of Columbia University in New York, the audience heard yet more unexpected news: the strength of the electron's magnetism appeared to be weaker than the Dirac theory had predicted. The audience was euphoric: here were two observations that heralded the end of the reign of Dirac's beautiful theory and provided crucial tests for any theory that presumed to succeed it. Oppenheimer steered the conference, incisively cross-examining the speakers and interspersing the proceedings with his elegant, if ostentatious, editorial arias. By the end of the meeting, it was clear that the main challenge was to explain Lamb's result. But Dirac knew nothing of all this: he had declined an invitation to attend and read about the wounding of his theory on an autumn Sunday in Princeton, on the front page of the *New York Times*.[10]

Within two years of the Shelter Island Conference, Lamb and Retherford's results had been explained by two of the youngest theorists in the audience. One of them was Feynman, the other was a fellow New Yorker, Julian Schwinger, a loner with the manners of a prince and the self-belief of a boxer. Feynman and Schwinger were both the same age and had read Dirac's book when they were precocious teenagers, and both based their theories on Dirac's 'little paper'. Yet the two versions appeared to be quite different: Schwinger's mathematical approach was hard to understand, but Feynman's approach was intuitive and involved special diagrams that made the underlying science easy to visualise, at least superficially. The two methods gave the same results, and everyone except Schwinger agreed that Feynman's methods were quicker and easier.

It turned out that the same results had been obtained several years earlier by the Japanese theoretician Sin-Itiro Tomonaga, who had based his ideas on Dirac's version of quantum field theory. As a student, Tomonaga had been a fanatical student of Dirac's book and was in the Tokyo audience when Dirac and Heisenberg gave their lectures during their tour of Japan in 1929. This pioneering work had been completed in Tokyo, where Tomonaga was one of the tens of thousands of starving citizens who were trying to rebuild the city after American bombers had flattened it towards the end of the war.[11]

So there were now three versions of quantum electrodynamics that looked quite different and yet seemed to give the same results. It was Freeman Dyson, the student who had snapped at Dirac's heels during his wartime lectures, who first demonstrated that the three theories were versions of the same underlying theory. Now, at last, physicists could claim they understood the interactions of the photon and the electron in terms of a theory that agreed with observation to within a few parts in ten thousand – roughly a human hair's breadth compared with the width of a door. Four decades later, when much more accurate measurements were still in excellent agreement with the theory, Feynman referred to it as 'the jewel of physics'.[12] As he often stressed, its fundamental concepts had been set out by Dirac in his 1927 theory: Feynman, Schwinger, Tomonaga and Dyson had, in essence, introduced a collection of ingenious mathematical tricks and techniques that made the theory viable and showed how to remove the embarrassing infinities.

Thoroughly pleased with himself for becoming 'a big shot with a vengeance' after his triumph, Dyson was keen to hear Dirac's opinion on the new theory. He was expecting a few words of congratulation from his former teacher, but was disappointed:

DYSON: Well, Professor Dirac, what do you think of these new developments in quantum electrodynamics?
DIRAC: I might have thought that the new ideas were correct if they had not been so ugly.[13]

The feature of the new theory that Dirac most loathed was the technique of renormalisation.[14] According to this theory, the observed energy of an electron is the sum of its self-energy – resulting from the interaction between the electron and its field – and the bare energy, defined to be the energy the electron is supposed to have when completely separate from its electromagnetic field. But the bare energy is a meaningless concept because it is actually impossible to switch off the interaction between the electron and its field; only the *observed* energy can be measured.

The virtue of renormalisation is that it enables every mention of bare energies in the theory to be removed and replaced with quantities that depend only on observed energies. Using this technique, theorists could use quantum electrodynamics to calculate – to any degree of accuracy – the value of any quantity the experimenters cared to measure. Despite the success of the technique, Dirac abominated it, partly because he could see no way of visualising its mathematics but mainly because he felt that the process of renormalisation was artificial, an inelegant way of sweeping the fundamental problems of theory under the carpet. In his opinion, a fundamental theory of nature must be beautiful, whereas renormalisation seemed to Dirac's taste to be as devoid of beauty as the dissonances of Arnold Schönberg.[15]

Engineers, schooled to worry more about the reliability of their results and less about the rigour of their mathematics, might be expected to be happy with renormalisation, as the process gives answers that always tally with observations to extremely high accuracy. But, paradoxically, Dirac believed his engineering training was at the root cause of his hostility to the technique.[16] At the Merchant Venturers' College, he had learned the engineer's art of using well-chosen approximations to simplify complicated, real-life problems so that they can be analysed mathematically. Dirac made this the theme of his 1980 lecture 'The Engineer and the Physicist': 'The main problem of the engineer is to decide which approximations to make.'[17] Good engineers make wise choices, often based on physical intuition, about the mathematical terms they can ignore in their equations: 'The terms neglected must be small and their neglect must not have a big influence on the result. He must not neglect terms that are not small.'[18]

Renormalisation entails a practice that no self-respecting engineer would countenance, Dirac pointed out: the neglect of large terms in an equation. To neglect infinitely large quantities in an equation was, for an engineer, anathema. Most physicists had no such compunctions, and leading theorists paid little heed to Dirac's objections. As

Dyson pointed out, although the infinities in the theory had not been eliminated, they were isolated in mathematical expressions that were quite separate from formulae representing the effects experimenters actually observe. Dirac was unconvinced. He, Schrödinger, Heisenberg, Pauli, Born and Bohr – the 'old gang', as Dyson dubbed them – had now joined Einstein in the wings of theoretical physics, while the next generation took centre stage. Of the *ancien régime*, only Pauli kept closely abreast of new developments in their subject; the rest withdrew into their own private worlds. Dyson and his friends were contemptuous of their elder colleagues:

In the history of science there is always a tension between revolutionaries and conservatives, between those who build grand castles in the air and those who prefer to lay one brick at a time on solid ground. The normal state of tension is between young revolutionaries and old conservatives [. . .] in the late 1940s and early 1950s, the revolutionaries were old and the conservatives were young.[19]

In a sense, Dirac was the Trotsky of theoretical physics: he envisioned his subject progressing through one revolution after another, each an improvement on its predecessor. But new quantum electrodynamics did not constitute progress so far as Dirac was concerned: the theory offended the aesthetic sensibilities he had first developed in Bristol, when he was an Eton-collared cherub at junior school, a greasy-aproned engineering student – moonlighting in general relativity – at college, and a budding mathematician at university. Whether this unique aestheticism would be a dependable guide remained to be seen.

When Dirac was a young man, he had been uninterested in human companionship, but he had come to value it. The result was that, after the war, Cambridge seemed to him like a ghost town – Fowler and Eddington had died, and all of Rutherford's former 'boys' had left. Manci also felt the pain of the exodus, complaining to her brother Wigner in Princeton that 'Life here is utterly and completely different.'[20]

With the ascendancy of American physics, Cambridge looked to Dirac to give leadership in the new era, but to no avail. Concerned only with his own research and in doing a modicum of teaching, he did nothing to improve the primitive facilities for students of theoretical physics in Cambridge: there were no offices for them in the department, and they even had to organise the programme of seminars.[21] Dirac now preferred to work at home, as he had done during the war. Manci ensured that the children did not disturb him: woe betide them if they tried to attract his attention by banging on his study door.

By late 1950, Gabriel and Judy had left home. Gabriel was pursuing his career, and Judy – apparently settling down after her tempestuous adolescence – had married, leaving the Diracs to bring up their two youngest daughters. According to Manci, Dirac 'kept himself too aloof' from them, and she had to encourage him to kiss them.[22] Neither Mary nor Monica recalled having any sense that their father was a famous or distinguished man – only that he was exceptionally quiet and good-natured, although unemotional and extremely slow to anger. Monica cannot recall seeing him laugh. But in many ways Dirac was a typical father, taking an interest in their hobbies, helping them do their homework and encouraging them to have pets, though he forbade them to bring dogs into the house because, as Monica recalls, 'he did not like being startled when they barked'.[23] Animal welfare was one of his concerns: when designing a flap for the girls' cat, he measured the span of its whiskers to ensure that the animal would not be incommoded as it passed through the hole.

Among the visitors to the Diracs' home were Esther and Myer Salaman. Esther, born and raised in the Ukraine, had been a student of Einstein's in the early 1920s, joined the Cavendish in 1925 and married Myer, a physiologist, a year later.[24] She was the kind of fine-looking, self-assured woman Dirac admired. He listened carefully to her effusions on the leading nineteenth-century Russian novelists, including her favourite, Tolstoy, whose *War and Peace* took Dirac two years to complete, having digested every word of it. He brought this same attention to detail to Dostoevsky's *Crime and Punishment*, which he thought was 'nice', though he pointed out that 'In one of the chapters the author makes a mistake: he describes the sun as rising twice on the same day.'[25]

Manci was still feeling out of place in Cambridge, contemptuous of its drab provincialism and despondent at the thought that she might have to spend the rest of her life in colourless England. Every day, newsreaders delivered discouraging news of the sluggish economy, continued rationing and product shortages; there was no sign of an end to the austerities of wartime. Manci, feeling the pinch, complained to Monica that 'Uncle Eugene pays his cleaner more every week than your father gives me in housekeeping.'[26] These were grim times, accurately summarised by the worldly-wise senior civil servant Bob Morris as 'a right, tight, screwed-down society walled in in every way'.[27]

The treatment of the dons' wives by the colleges and university was still a sore point with Manci, though she saw a few hopeful signs. In 1948, the authorities symbolically enrolled Queen Elizabeth (later the Queen Mother) as the first woman to take a bona-fide degree, albeit an honorary one.[28] A year later, under this legislation, women students at Cambridge first graduated. Slowly, much more slowly than Manci wanted, women in Cambridge University were making progress towards equality.

To the emerging generation of physicists, Dirac was a cool and wary stranger, but for Heisenberg and other fellow pioneers of quantum mechanics, he was an attentive friend. After the war, Heisenberg knew he had to justify the work he had done for the Nazis, but this was an enervating struggle – several of his former colleagues, including his former friend and student Peierls, wanted nothing to do with him, and Einstein treated him with contempt.[29] In 1948, when Heisenberg returned to Cambridge – at a time when Dirac was absent – he looked haggard and anxious but was excellent company, delighting his hosts one evening with an unrehearsed performance of Beethoven's *Emperor Concerto*. He discreetly explained to everyone who would listen that he was never a Nazi and had stayed in Germany out of loyalty to his colleagues and to mitigate the worst of Hitler's intentions. Determined to leave a good impression in Cambridge, as a gesture of remembrance he bought forty-eight rose bushes from a plant centre in nearby Histon and made it known he would plant them in his garden in Göttingen.[30]

When Dirac first met Heisenberg after the war, he accepted Heisenberg's explanation of his wartime conduct at face value and believed Heisenberg had behaved reasonably in an extremely difficult situation. 'It is easy to be a hero in a democracy,' Dirac would observe, as Manci laughed at his naivety.[31] She scorned Heisenberg as a tricky character: 'That Naaaaazi.'[32]

Dirac was supportive of Heisenberg even when he was working for Hitler. Max Born had been startled when Dirac asked him to support

Heisenberg for foreign membership of the Royal Society. 'Heisenberg's discovery will be remembered when Hitler is long forgotten,' Dirac commented.[33] Dirac also strongly supported Schrödinger's election to a reluctant Royal Society. The consensus among its officials was that 'one hunch, however good and however important [. . .] needed more following up with sustained evidence of ability', an insider told Dirac.[34] Probably incredulous, Dirac took up Schrödinger's cause and helped to ensure his election in 1949. Schrödinger was profuse in his thanks, telling Dirac, 'You really are very nearly a saint.'[35] Dirac showed no such conscientiousness when it came to supporting his former peers for the Nobel Prize: strong candidates for the award – Pauli, Born, Jordan or even Dirac's Cavendish friends Blackett, Chadwick, Cockcroft and Walton – received no support from him.[36] The only physicist Dirac nominated was Kapitza.[37]

Dirac had heard little from Kapitza during the war, though he had read in his copy of *Moscow News* of Kapitza's invention of a method of liquefying oxygen that did much to raise the productivity of the hard-pressed steel manufacturers and several branches of the Soviet chemical industry.[38] Stalin never met Kapitza but showed every sign of having a soft spot for him, telephoning him occasionally and showering him with awards, including the USSR's highest civil title 'Hero of Socialist Labour'.[39] By the end of the war, Kapitza had proved himself the scientist best able to work with the Government and with Stalin, whom he flattered shamelessly: 'The country has always been fortunate to have leaders [such as you and Lenin].'[40]

Two weeks after Americans dropped the bomb on Japan, Kapitza's fortunes took a turn for the worse when Stalin set up a special committee to develop nuclear technology and weapons, headed by his first lieutenant Lavrentiy Beria. Of all Stalin's courtiers, Beria was the most feared – a bully, a serial rapist and a casual murderer – but he was a consummate manager, the kind of man who would have no trouble running an industrial conglomerate. At Stalin's request, Beria took over leadership of the Soviets' nuclear project and soon fell out with Kapitza, who complained to Stalin in the autumn of 1945 about Beria's scientific ignorance and incompetence.[41] When Kapitza realised that he could not oust his boss, he asked to be released from the project. Stalin agreed and, though apparently ensuring that Kapitza's life was not in danger, did nothing when all his responsibilities were removed. By early 1946, Kapitza was in disgrace. Dirac knew nothing of this – he did not know that Kapitza had survived the war until the summer of 1949.[42]

In September 1947, Dirac began his most productive year for a decade. Accompanied by his family, he was on sabbatical at the Institute for Advanced Study, which had relocated eight years before to Fuld Hall, a four-storey red-brick building with a spire like a New England church. It stood, symmetric as a crystal, in almost three hundred acres of meadows, fields, woods and wetlands, about half an hour's walk from the centre of Princeton. This was a realisation of Abraham Flexner's vision of a small academic institution focusing on a few disciplines and with a world-class faculty, all of them unencumbered by administration and unwanted students. The Institute was, for Dirac, a 'paradise'.[43]

Manci felt at home in Princeton and thrived in its prosperous academic milieu and – compared with Cambridge – its liveliness and informality. The community treated her with the respect she wanted, not just as Dirac's wife but as a bright woman in her own right. The institute had become even more attractive to Dirac in 1946, when Oppenheimer became its director and gave him an open invitation to visit. Fresh from the Manhattan Project, Oppenheimer was 'ablaze with power', though ill at ease: 'I feel I have blood on my hands,' he had told President Truman.[44]

It was a relief for Dirac and his family to be far away from the austerities of post-war Britain, and they took away from Princeton an album of memories: their young daughters scurrying around in the empty tea room at the weekend, their yells shattering the institute's chapel-like quiet; Einstein, visiting the Diracs for afternoon tea, signing a portrait of himself for Manci; Oppenheimer showing off his van Gogh; setting off with Veblen at the weekends, axes slung over their shoulders, to clear a path in the local woods.[45] Freeman Dyson recalls meeting the Diracs during their visit to the institute in early September 1948:

Everyone loved Manci: she was a real character, always full of life, always ready to chat. Dirac was more communicative than he had been in Cambridge. He was not terribly difficult to talk to. If you asked him a serious question, he would ponder it and give a reply that was always short and to-the-point.[46]

However, he still had no time for strangers who tried to lure him into small talk. Louise Morse, wife of one of the institute's mathematicians, remembers that when she asked Dirac how he was settling in at Princeton, he looked dumbfounded and leaned sharply away from her, as if she were a leak in a sewer. She remembers: 'Without saying a word, his whole body seemed to ask "Why on earth are you talking to me?"'[47]

At the Institute, Dirac worked in a modest office on the third floor of Fuld Hall, next door to Niels Bohr. One of Dirac's main projects in his 1947–8 stay was to develop the theory of the magnetic monopole he had conceived sixteen years before. During the war, he heard reports of the particle's discovery and, although they turned out to be false, they probably rekindled his interest in the idea.[48] He produced an exquisitely crafted theory predicting how monopoles might interact with electrically charged particles, but the theory failed to make a splash. One of the few who followed it closely was Pauli, who was prompted to give one of his more polite nicknames to Dirac: 'Monopoleon'.[49]

In another project, he returned to the roots of quantum field theory. Unhappy with the new theory of electrons and photons, he looked afresh at the application of quantum theory to quantities such as electric and magnetic fields that describe physical conditions at each point in space-time. This was another piece of research that failed to strike a chord at the time but was appreciated later. The same is true of the review he wrote in 1949 about how Einstein's special theory of relativity could be combined with Hamilton's description of motion. Its deceptively straightforward presentation led most physicists to pay no attention to it, a mistake several of them would rue.

Dirac still believed that modern quantum electrodynamics was wrong because it was based on a classical theory of electrons that was fundamentally flawed. So, in 1951, he produced a new theory, quite different from the one he had developed thirteen years before. This time, his classical theory described a continuous stream of electricity, flowing like a liquid – individual electrons emerged only when the classical theory was quantised.[50] The theory was the dampest of squibs. No one disputed Dirac's technical ingenuity but it seemed that he had lost his intuition for productive lines of research. He demonstrated this yet again when, as a by-product of his new theory of electrons, he reintroduced a concept that most scientists believed Einstein had slain: the ether.

Dirac's ether was quite different from the nineteenth-century version: in his view, all velocities of the ether are equally likely at every point in space-time.[51] Because this ether does not have a definite velocity with respect to other matter, it does not contradict Einstein's theory of relativity. Dirac's imagination slipped through this loophole and reinvented the ether as a background quantum agitation in the vacuum; later, he went further and speculated that it might be 'a very light and tenuous form of matter'.[52] The press were more interested than scientists in the idea, which appeared to go nowhere: the logic was impeccable but it seemed to have no connection with nature.[53]

By the time Dirac reached his fiftieth birthday, he seemed to be following the path Einstein had taken, towards isolation from mainstream physicists. In Princeton, Einstein was a lonely figure, uninterested in the latest research headlines and absorbed by his quixotic project to find a unified field theory without introducing quantum mechanics from the outset. He was still active in politics and annoyed J. Edgar Hoover, Director of the Federal Bureau of Investigation (FBI), by supporting several leftist and anti-racist organisations. In 1950, Hoover ordered a secret campaign to 'get Einstein', aiming to have him deported.[54] Unaware that he was being watched, Einstein strolled to his office in the institute from his nearby home on Mercer Street, his briefcase under his arm, pausing only to pick up and sniff discarded cigarette butts. On his favourite route, he walked down the straight section of Battle Road, towering sycamores lining each side, their overarching branches entangled like the swords of a guard of honour.[55]

At the Institute for Advanced Study, he was free to work and ignore the day-to-day trivia of politics. But this tranquillity was about to be disturbed by the FBI agents and journalists who were sniffing around the past of the institute's director. Oppenheimer's former Communist sympathies – and Dirac's – were about to return to haunt them.

Twenty-five

The former Communist was guilty because he had in fact believed the Soviets were developing the system of the future, without human exploitation and irrational waste. Even his naiveté [. . .] was now a source of guilt and shame.

ARTHUR MILLER, *Time Bends*, 1987

'What happened to daddy's brother?' Dirac's daughters would ask their mother. 'Shhh! Don't talk about it,' was Manci's stock reply. Dirac spoke about Felix's suicide only with her and even then he could not bring himself to go into any details. She knew that he still had not come to terms with it. On one occasion, when Mary and Monica persisted, Dirac took out from a drawer a small tin and prised it open to reveal some photographs of his late brother, before hurriedly snapping the tin closed and putting it back. More than twenty-five years after his brother's death, a brief look at Felix's face was all he could bear.[1]

From Dirac's behaviour at home, it appears that he tried to avoid what he regarded as the worst mistakes his father had made in bringing up his children. Unlike Charles, Paul encouraged his daughters to bring their friends home; he did not lean on them to study science or any other subject, nor did he offer them any career advice. They knew that there is more to life than work. The family always ate together, but the mealtimes were not what most people would regard as normal: Dirac would sit at the head of the table, eating slowly, sipping regularly from his glass of water and making it clear that he preferred to eat in silence. If one of his daughters pressed him to speak, he would point to his mouth and mutter irritably, 'I'm eating.' He was quite fussy about food – for example, refusing to eat pickles on the grounds that they were always bad for digestion – and would not allow Manci to use a drop of alcohol in any food, especially if it might be eaten by the girls. There was trouble in the kitchen if he sniffed or tasted in the Christmas pudding so much as a drop of brandy.

Mary and Monica were growing into sharply contrasting personalities that, as Dirac noticed, resembled those of their parents. Mary was rather like him – quiet, trusting and literal-minded – while Monica bore a resemblance to her mother – confident, questioning and assertive. The girls did not get on well: Mary was intimidated by Monica and their mother, while Monica felt psychologically manipulated by Mary. Dirac and Manci, perhaps trying to atone for Mary's vulnerability, treated her as their favourite and often left Monica feeling angry and resentful. Monica still recalls that her parents organised only two birthday parties for her when she was a child, while they gave one to Mary every year.

Worried that these tensions were getting out of hand, Dirac and Manci separated their daughters using the classic English institution of boarding school, sending Mary to a strict and devoutly religious school near Cromer, in East Anglia.[2] On the first weekend she was away, Dirac went on a Sunday morning cycle ride with Monica, who was hoping to begin a new stage in her relationship with her father. But this time he did not stop and chat as he had always done when Mary was with them: during the three-hour ride, he said not a word to her. She was devastated.

No one in Cambridge counted Dirac and Manci as among the most attentive parents: as soon as the Cambridge term was over, they usually headed off on a foreign trip, leaving their children with friends. But the family did take vacations together. In the summer, Dirac would take two days to motor to their favourite destination, Cornwall, driving like a caricature vicar. During the Christmas vacation, shortly after the New Year, the family would stay for a few days in the pea soup of London fog.[3] While Manci lunched with friends or went shopping, Dirac took the girls to South Kensington and walked them round the Science Museum, where they pushed the buttons on the interactive displays and filed past the relics of the Industrial Revolution. In the evening, the family headed to the West End for entertainment – Mary recalled that her father's favourites included the musical *The Pajama Game* and Tchaikovsky's ballet *The Sleeping Beauty*.[4]

Dirac's taste in the arts defies conventional classification, ranging from high culture to catchpenny trivia. On Saturday mornings, he raced his daughters to the front door to pick up the latest edition of their favourite comics, the *Dandy* and the *Beano*, which he would study as if they were works of literature. Mostly, he pursued his leisure interests alone, reading a Sherlock Holmes story, listening to a classical concert at full blast on the radio or sitting impassively watching the television he had first rented so that the family could watch the Queen's coronation. But pageantry was not for him: he preferred the new variety shows and, with millions of other male viewers, sat agog as lines of feathered young women high-kicked their way through their risqué dance routines. This was rather unbecoming, Manci thought, though she happily accompanied him on at least one discreet trip to a London production of the Folies Bergère.[5]

Like Einstein, Dirac was a modernist in science but not in art. His favourite music was the classical canon of Mozart, Beethoven and Schubert, and he had no time for the experiments of contemporary composers. He also had no taste for the extremes of abstract art: the nearest he came to liking a modern artist was a fondness for the surrealism of Salvador Dalí. When he visited his sister Betty and her family in Amsterdam, two minutes' walk from where Ehrenfest shot himself and his son, Dirac would set off in the morning with a compass – but not a map – on the six-mile walk to the Rembrandts of the Rijksmuseum.

If Cambridge colleagues knew anything of these interests, Dirac would have been more engaging than the desiccated figure he cut in the early 1950s, rather like a prototype for Bertrand Russell's fictional don, Professor Driuzdustades.[6] Dirac no longer seemed at home in the mathematics department, though he remained a loyal Fellow of St John's, observing all its rituals without complaint. Every Tuesday night during term, he would don his gown and eat at High Table, while Manci – not allowed to eat with him – ate at a cheap Indian restaurant with Monica on St John's Street, Manci grumbling over her curry and samosas that the college made her feel like an impostor.[7]

Sensing that the university no longer held her husband in the highest regard, she blamed him for not insisting on the respect that was due to him. But he was too self-effacing to assert himself: he had no interest in status for its own sake and was indifferent to the baubles handed down by the establishment. In the early 1930s, he declined an honorary degree from Bristol University because he believed degrees should be qualifications, not gifts, and later declined honorary degrees, replying to offers with 'regretfully, no'.[8] In 1953, he refused a knighthood, infuriating Manci, mainly because his decision deprived her of the chance to become Lady Dirac.[9] He did not want people outside the university to call him Sir Paul but to address him by the name he used on the rare occasions he answered the telephone at home: 'Mr Dirac'.

He did not oppose honours on principle, but he believed that they should be awarded on merit, and not be awarded to athletes and show-business celebrities. When the jockey Gordon Richards was awarded a knighthood by the Queen, Dirac shook his head: 'Whatever next?'[10]

Fundamental physics appeared to be in a mess, just as bad as the one in the early 1920s when Bohr's theory was the creaky framework for atomic physics. Having seen theory swept aside by quantum mechanics, he believed that nothing less than a similar revolution was needed now to replace quantum electrodynamics. Dirac wanted the initiative to come from theorists: since he was a boy, they had been setting the agenda of physics, but now experimenters were ensconced in the driving seat.

Results from cosmic-ray projects and from the new high-energy particle accelerators had shown that the subatomic world was much more complicated than any theoretician had imagined. By the mid-1950s, it was plain that there were many more than two subatomic particles – there were dozens or even hundreds, most of them living for no longer than a billionth of a second, before they fall apart into stable particles. All these decay processes obeyed the laws of quantum mechanics and relativity, but no one knew how to apply them. Fermi had set out the first theory of the weak interaction, which acts only over very short distances, within the ambit of a nucleus, about a ten-thousandth of the distance across an atom. By then, another fundamental type of interaction had emerged, the strong interaction, which also extends only over distances on the scale of the atomic nucleus. Much stronger than the electromagnetic force, the strong force binds the protons and neutrons in the atomic nucleus and prevents the protons from repelling each other. Without this force, stable atomic nuclei could never have formed, and ordinary matter would not exist.

Nature seemed unwilling to disclose its deepest secrets: when experimenters probed strong interaction, they found it all but incomprehensible. But, like Einstein, Dirac did not trouble himself with the complications introduced by the new interaction. In his opinion, there was no point in paying much attention to them until electrons and photons had been properly understood in the context of a mathematically defensible theory. While most others moved on, he remained – in their view – transfixed by an obsolete view of physics, hidebound.

Oppenheimer had also retreated from the front line of research. He was a prominent adviser to the Eisenhower administration on nuclear policy, uneasy that so many aspects of the research were kept secret under the pretext of national security; he preferred Bohr's view that superpowers should, like scientists, share their knowledge as a matter of principle. In a perceptive speech in February 1953, Oppenheimer startled a closed meeting of the Council on Foreign Relations by likening the USA and the USSR to 'two scorpions in a bottle, each capable of killing the other, but only at the risk of his own life'.[11] He believed that, despite the superpowers' posturing and bluster, reason would prevail.

Shortly before midnight on 14 April 1954, Dirac arrived home in Cambridge after spending a month with his stepson Gabriel in Vienna. Dirac had visited him every afternoon at the Viktor Frankl Institute, where he was being treated for psychiatric disorders, including a persecution complex and schizophrenia. Dirac had written to tell Manci of the doctors' assessment: Gabriel had been 'badly brought up'.[12] Soon after he arrived home that night, Dirac would have told his wife of her son's progress, and they may well have discussed the news that had broken in European newspapers that day: the American Government had withdrawn Oppenheimer's security clearance.

The Oppenheimer case was the climax of the anti-Communist paranoia in 1950s America. It had begun with the start of the Cold War and intensified in the late summer of 1949, when the Soviet Union tested its first nuclear weapon at least two years earlier than the Central Intelligence Agency (CIA) expected from its intelligence reports.[13] The USA, terrified that its technological primacy would be eclipsed by the Soviet Union, feared that Communists held important positions in public life. An early victim was Oppenheimer's popular brother Frank, an experimental physicist who had been fired in 1949 by the University of Minnesota when it found out that he was a card-carrying Communist (a few weeks afterwards, Dirac tried to find him a post at the University of Bristol).[14] In early February 1950, there was a national outcry when Klaus Fuchs – Dirac and Peierls' collaborator during the war, later a member of the Manhattan team – confessed to having passed critical secrets to the Soviet Union, an act of espionage that had been responsible for the unexpectedly early detonation of the Soviet nuclear weapon. J. Edgar Hoover called Fuchs' treachery 'the crime of the century'.[15] After the revelation, Dirac and Peierls came up with an explanation of Fuchs' peculiar behaviour during his conversations with them in the back garden of 7 Cavendish Avenue – he had been passing notes on the conversation to a Soviet intermediary. Eighteen days after Fuchs had been unmasked, the Wisconsin Republican Joseph McCarthy stoked up the febrile anti-Soviet rhetoric in the press when he claimed, in a six-hour speech on the Senate floor, that Communists infested the entire government apparatus. When Bohr complained about the apparently unending deluge of insults in the newspapers, Dirac told him not to worry as it would end in a few weeks because, by then, the reporters would have used up all the invective in the English language. Bohr shook his head, incredulous.[16]

In June 1952, the Senate passed an Immigration Act that obliged applicants for US visas to list all their past and current memberships of organisations, clubs and societies. Decisions about whether to grant visas were usually left to consuls, most of them nervous of being seen as 'soft on Commies'. No record of Dirac's submission survives. It is most likely that he would have been open with the American authorities about his relatives behind the Iron Curtain in Hungary and his association with left-leaning organisations before the war. He may also have mentioned that he signed a petition two years before to deplore Bernal's expulsion from the Council of the British Association for the Advancement of Science, after Bernal had made a scathingly anti-Western speech in Moscow.[17] That signature had been noted by MI5.[18]

Soon after Oppenheimer's hearing began, on the rainy Monday morning of 12 April in Washington DC, he realised that he was being subjected not to an enquiry but to a kangaroo court. The FBI had illegally tapped his and his attorneys' phones, forwarding transcripts to the prosecuting lawyers to help them prepare for the next day's proceedings.[19] During the second weekend break in the hearing, Oppenheimer read a pessimistic note from Dirac, who was planning to visit the institute for a year, beginning in the following summer. There was, Dirac believed, little chance that the US Government would grant him a visa.[20]

The enquiry closed on 5 May, and Oppenheimer returned to Princeton tired, depressed and irritable. He knew that it had gone badly: under ferocious cross-examination he had been evasive, mendacious and sometimes even disloyal to his friends. One of the most damning testimonies had been delivered by Edward Teller, who had been angry with Oppenheimer for not making him head of the Manhattan Project's theory group and, in his opinion, for delaying his pet programme to build the first hydrogen bomb. Teller declared that, 'if it is a question of wisdom and judgement, as demonstrated by actions since 1945, then I would say that it would be wiser not to grant [Oppenheimer] security clearance'. Immediately after Teller left the witness stand, he offered his hand to a stunned Oppenheimer, who took it. 'I'm sorry,' Teller said.[21]

When Oppenheimer was waiting for the board's verdict, he received a letter from Dirac: 'I regret to have to tell you that my application for a US visa has been refused.'[22] On both sides of the Atlantic, news of the refusal broke on 27 May 1955, most of the articles declaring or hinting that Dirac's Russian connections had been the cause. Among the journalists who called at 7 Cavendish Avenue was Chapman Pincher, the well-connected *Daily Express* security correspondent. Manci told him, with more pith than accuracy, 'My husband has no political interests,' a phrase that Pincher included in a brief article in the *Express* ('US-Barred Scientist "Not Red"').[23] A reporter from the *New York Times* somehow managed to interview Dirac and was told that his application had been 'turned down flat': the American Consul had told him he was ineligible for a visa under Regulation 212A, without specifying which of the points specified in its five pages he had transgressed.[24] Dirac was uncharacteristically decisive: he asked the British Government to release him from all defence work and started to make arrangements to change the location of his sabbatical to the Soviet Union.[25] He appeared to be cocking a snook at the American authorities, but he made no comment.

A month later, Oppenheimer heard the outcome of his 'hearing': the Board voted two to one that he was a loyal American, though nevertheless a security risk. To ram home their victory, his enemies in the Atomic Energy Commission withdrew his security clearance a day before it was due to expire. Oppenheimer was shattered, and he considered emigrating to England to take up a professorship in physics at Cambridge University, an offer that he discussed with Dirac.[26] His fiercely loyal wife, who had given one of the powerfully supportive testimonies during the hearing, became an alcoholic and remained one for the rest of her life. After a family vacation in the Caribbean, where he was watched by FBI agents suspicious that a Soviet submarine might whisk him back to Russia, he returned to the institute. His eloquence and appetite for his work were undiminished, though many of his colleagues thought his spirit was broken. He looked less like the blazingly confident scientist, an American hero after the Manhattan Project's success, than a scientific martyr, the Galileo of the McCarthy era.

Three days after the *New York Times* announced the Oppenheimer verdict as the lead story on its front page, it printed a short report on Dirac's case, featuring quotes from an interview with Dirac, printed below a photograph that made him look like a criminal. Embarrassed and angry, senior American physicists seized on this latest of many rejected visa applications from top scientists, and it became a cause célèbre. Two days after the report was published, John Wheeler and two Princeton colleagues fired off a letter to the newspaper, deploring the Government's action: '[we] believe this action is exceedingly unfortunate for science and this country', adding that the Act that led to the refusal of Dirac's visa 'seems to us a form of organized cultural suicide'.[27] Dozens of other physicists turned the screws on the State Department and the American Consulate in London, who blamed each other for the outcome of the decision, which had been 'close', they told journalists. Within two weeks, the *New York Times* reported that the State Department was reviewing the ban; a humiliating climb-down looked certain and was duly announced on 10 August. But it was too late: Dirac had made other arrangements.

Dirac's plans for a sabbatical in Russia fell through, so he accepted a long-standing invitation to visit India. At the end of September 1954, Dirac and his wife set sail for Bombay, the first stage of their round-the-world trip, scheduled to last almost a year. The Diracs arranged for their friends Sol and Dorothy Adler to stay in 7 Cavendish Avenue to look after Mary and Monica, both anxious and dreading their parents' long absence. Monica, then twelve years old, cannily observed one important reason why her parents were going far away: Manci believed that Dirac had a female admirer who was showing him rather too much affection, so she wanted him away from Cambridge for as long as possible.[28] Dirac may well have wanted to see something of the country described to him in the fireside reminiscences of his confidante Isabel Whitehead, who had died in the previous year, six years after her husband.

The Diracs' four-month stay in India was organised by the physicist Homi Bhabha, Dirac's former colleague in Cambridge and founding director of the Tata Institute in Bombay.[29] He was exceptionally cultured, an exhibited artist and a connoisseur of poetry in several languages. Bhabha made sure that the Diracs were treated like royalty from the moment they arrived on 13 October, though he could do nothing about Bombay's unbearable heat and humidity, which quickly drove them to depart for the comparative cool of the Mahabaleshwar Hills nearby.[30] Manci disliked much more than the climate: she hated the spicy food and the chauffeur-driven rides through vast, stinking vistas of destitution and squalor; nor did she appreciate being treated as a second-class celebrity, her husband's consort. The experience did, however, give her a glimpse of the respect and reverence that she would later expect, and a little of this taste for glamour later appeared to have rubbed off on Dirac.[31] For the first time in his life, he felt the adulation of a mass crowd when he gave a public lecture during the evening of 5 January 1955 as part of the Indian Science Congress in Baroda, near Calcutta. In a special enclosure at Baroda cricket ground, he delivered his talk to thousands of wide-eyed spectators, many of them watching the presentation on a cinema screen outside the ground.[32]

Perhaps having learned from the debacle at Le Palais in Paris, Dirac had found a way of talking to people who wanted to learn about quantum physics but who knew nothing about it. Shedding his dislike of metaphor and visual imagery in descriptions of the subatomic domain, he spoke in simple, equation-free language and introduced a simile, later given wide currency, to link subatomic particles with his favourite game:

When you ask what are electrons and protons I ought to answer that this question is not a profitable one to ask and does not really have a meaning. The important thing about electrons and protons is not what they are but

how they behave – how they move. I can describe the situation by comparing it to the game of chess. In chess, we have various chessmen, kings,
knights, pawns and so on. If you ask what a chessman is, the answer would
be [that] it is a piece of wood, or a piece of ivory, or perhaps just a sign written on paper, [or anything whatever]. It does not matter. Each chessman has
a characteristic way of moving and this is all that matters about it. The
whole game of chess follows from this way of moving the various chessmen
[. . .][33]

The physicists in the front row as well as the non-experts in the audience gave a warm reception to Dirac's forty-minute summary of the
fundamentals of quantum mechanics. Though he had none of
Eddington's verve as a populariser, it was clear that he had somehow
acquired the skill vital to scientists who detest administration and
who are well past their peak as researchers: the ability to share his
work with the public.

Most eminent among the politicians Dirac met in India was its
charismatic Prime Minister, Jawaharlal Nehru, who had led India
since its independence from Britain in 1947. Although he had the
politician's talent for casting broad-brush thinking in colourful, pop-
ulist language, Nehru was also a cultured thinker who would lighten
a quarrel by quoting the poetry of Robert Frost. During the meeting
in Delhi with Dirac on 12 January 1955, Nehru asked him if he had
any recommendations for the future of the new republic of India.
After his usual reflective pause, Dirac replied: 'A common language,
preferably English. Peace with Pakistan. The metric system.'[34] The
men apparently did not discuss nuclear weapons, though the subject
was on their minds. Eleven days before, at the Science Congress in
Baroda, Dirac heard Nehru lecture scientists about the imperative to
help with the reality of the new weapons, commenting that 'We are
not playing with atomic bombs at present.'[35] With Nehru's support,
Bhabha would later spearhead plans for India's programme and
become his country's Oppenheimer.[36]

Two weeks after the Diracs sailed from Bombay on 21 February
1955, the trip turned unpleasant. After contracting jaundice, Dirac
spent eight days in hospital in Hong Kong, where his doctor agreed
to allow him to sail on to Vancouver, though with a litany of health
warnings and dietary instructions.[37] Manci thought he should not
travel, but he insisted and paid dearly for his obstinacy by spending
most of the voyage in bed, sick with jaundice, vomiting every few

hours, plagued by itches, sometimes unable to sleep through the night.[38] When the Diracs sailed into Vancouver in mid-April, he was exhausted and dispirited, his skin a pale shade of yellow.[39] The University of British Columbia accommodated them on one storey of a finely appointed mansion, where he immediately took to his bed.

Two days later, he heard the news from Princeton that broke his heart: Einstein had died. For the first time, Manci saw him weep – a sight she had never seen before and would never see again.[40] It was for a hero, not a friend, that Dirac shed those tears. During those first hours of grief, he may have recalled his student days in Bristol when he first became acquainted with relativity theory, which inspired him to be a theoretician. What mattered most to Dirac were Einstein's science, his individualism, his indifference to orthodoxy and the ability he demonstrated later in life to ignore his critics' catcalls, muted only by timidity and cowardice. After Einstein's ashes had been scattered into the New Jersey winds, Dirac succeeded him as the most famous loner in theoretical physics, an elderly rebel with a cause that no one else could quite understand.

Sick, depressed and believing he was dying, Dirac told Manci that he had just one request: to see Oppenheimer. She quickly succeeded in bringing together the two friends in the Vancouver apartment, each of them broken, each at their nadirs, each looking fifteen years older than when they last met. No record of their conversation remains, but it is likely that Dirac's main wish was to commiserate with Oppenheimer over the outcome of the trial and, perhaps, over the conduct of Teller and the prosecutors. Teller, a pariah to many of his former friends, had become one of the few physicists Dirac disliked and would criticise, if only to those close to him.[41] Oppenheimer was at his considerate best: he advised Dirac to get treated in the USA and to recuperate for a few weeks in one of the apartments at the Institute for Advanced Study.

Colleagues at the institute noticed the change in Dirac's gait. No longer lissom, he walked slowly and deliberately, as if recovering from surgery, but his vigour was returning. He spent the mornings preparing lectures for a forthcoming meeting in Ottawa, the afternoons sleeping, the early evenings on long, restorative walks round the grounds of the institute, alone except for the squirrels, rabbits and the occasional deer.[42] But misfortune struck: during a visit by Judy and her baby girl, he fractured a metatarsal bone in his right foot – he was an invalid again.[43] In Ottawa, for the first time in his life, he gave his lectures sitting down and looked, as he approached his fifty-third birthday, like an old man.[44]

When the Diracs arrived home in Cambridge at the end of August 1955, to see their daughters for the first time in almost a year, Manci wrote a gushing thank-you note to Oppenheimer, passing on from Dirac a suggestion to help him come to terms with his tormentors. Dirac recommended Oppenheimer read the new Somerset Maugham novel, *Then and Now*, set in fifteenth-century Florence, about the intrigues and deceptions in the relationship between Cesare Borgia and Niccolò Machiavelli.[45]

In the first seminar Dirac gave in Cambridge at the beginning of the next term, he announced to his students: 'I have just done this work. It could be important. I want you to learn it.' This was an extremely rare instance of Dirac publicly pointing the way ahead.[46] His enthusiasm for research had been rekindled.

Dirac's new theory suggested that the universe might not fundamentally consist of point-like particles but of tiny, one-dimensional things that he called 'strings'. The theory, first outlined in his Ottawa lectures, was a new approach to quantum electrodynamics that dispensed with one of the foundations of renormalisation theory that Dirac most disliked – the 'bare electron', the idea that the theory could be built from the fictional notion of an electron that had no surrounding field. In his new approach, he concentrated on one of the theory's underlying symmetries, known as gauge invariance. Long familiar to theorists, this symmetry implies that the theory makes identical predictions if a quantity known as the electromagnetic potential, closely related to the electromagnetic field, is changed at every point in space-time, but only if the changes across the whole of space-time are orchestrated by a governing formula known as a gauge transformation. Dirac found a way of rebuilding quantum electrodynamics in terms of gauge-invariant quantities so that, whenever the electron features in a calculation, it is inseparable from its field. The result was a theory that gave the same results as the renormalised version but that was, for him, superior.

Dirac disliked the concept of bare electrons so much that he wanted 'to set up a theory in which [they] are not merely *forbidden* but *inconceivable*'.[47] He found a way of doing that using the equations of his theory, by applying them to the lines of force describing the electric field of the electron, which resemble the field lines of a magnet. In the classical picture of the electron, the particle is surrounded by continuously varying lines of force: each set of lines of force is, in a sense, infinitesimally close to the next. This led Dirac to imagine a quantum version of the field and to picture the electron not as a particle but as a string:

We may assume [that] when we pass over to the quantum theory the lines of force become all discrete and separate from one another. Each line of force is now associated with a certain amount of electric charge. This charge will appear at each end of the line of force (if it has ends) with a positive sign at one end and a negative sign at the other. A natural assumption to make is that the amount of charge is the same for every line of force and is just the [size of the charge of the electron]. We now have a model in which the basic physical entity is the line of force, a thing like a string, instead of a particle. The strings will move about and interact with one another according to quantum laws.[48]

Dirac had found what he was seeking: 'a model in which a bare electron is inconceivable, because the end of a piece of string is inconceivable without the string'. But it was only the germ of an idea, not a complete new theory. Several of his students examined it but soon set it aside, as Dirac did soon afterwards. Years later, it would transpire that he had once again been ahead of his time.

Dirac was about to reach the low point of his career: apart from wartime, 1956 was the first year since he had begun research that he had published nothing at all.[49] Now semi-detached from the physics community, he had lost touch with many of his closest friends, including Kapitza – they had not been together for almost twenty years.[50] Dirac will have wanted to know how Kapitza was faring in Nikita Khrushchev's regime, which began soon after Stalin's death in March 1953. British newspapers had reported a new mood in the country after the Soviet public heard that Khrushchev had, in a speech to stony-faced party bosses in February 1956, denounced the personality cult of Stalin and the cruelty of his regime.[51]

In the early autumn, Dirac arrived in Moscow to find it very different from the city he and Manci had seen in 1937: it was now focusing on consolidation, not revolution, and the paranoid, inwardly focused nationalism of the late 1930s had been superseded by a dread of a pre-emptive nuclear strike by the USA. Dirac found Kapitza as self-confident as he had ever been and just as full of colourful stories: in one, he told Dirac of how his arch-enemy Beria had sidelined him after he had refused to work on nuclear weapons. Kapitza believed that 'It is a horrible thing for scientists to engage in secret war work,' and he probably mentioned this to Dirac, who may have flinched, at least inwardly.[52] While most other leading Soviet physicists had given their services to the nuclear project, Kapitza worked on ways to destroy incoming nuclear weapons using intense beams, apparently a precursor to the American Strategic Defence ('Star Wars') Initiative. Stalin's good opinion had saved him from execution by one of Beria's henchmen, Kapitza was sure. When Stalin died, Lev Landau danced for joy, but Kapitza knew his own life was in danger if Beria was the country's next leader.[53] Khrushchev outmanoeuvred Beria, but Kapitza's life was still in peril: on what seemed to be an ordinary summer morning, towards the end of the official discussions about Stalin's succession, Kapitza told Dirac, two state officials visited him in his small laboratory and asked for a guided tour. Their questions revealed that they knew little about science and cared even less, yet they insisted on prolonging their visit beyond its natural duration, until their departure on the stroke of noon. According to Kapitza's account of the story, the two men had been deputed – probably by Khrushchev or his associates – to protect him from a last-minute reprisal while Beria was being arrested and taken into custody.[54] A few weeks later, Beria and six of his accomplices were tried and sentenced to death; he was executed by one of Khrushchev's three-star generals, who fired a bullet into his forehead.[55] Kapitza heard the news on Christmas Eve, a joyous moment for him.

Dirac never tired of praising Kapitza's refusal to work on the nuclear-bomb project. This was the story Kapitza told Dirac and everyone else, but it is almost certainly untrue. Kapitza's letters to Stalin – published several years after Dirac's death – make it plain that Kapitza wanted to work on the project, and he shows no hint of any moral scruples; he declined to work on the bomb only because he would not work under Beria's heel. It is also possible that he did not command support from his colleagues, as some of them believed he was contemptuous of scientists outside his cosmopolitan circle.[56] A much stronger case for Kapitza's heroism can be made by pointing to the case of Landau, Stalin's outspoken enemy, whom Kapitza repeatedly defended, often putting his life in grave danger.[57] Hundreds of thousands of Russians were executed for showing only a fraction of Kapitza's insubordination.

Dirac spent most of his visit to Moscow in October 1956 sightseeing – he saw that Lenin was then sharing his tomb with Stalin – as well as reacquainting himself with his old Russian friends, including Tamm, Fock and Landau. It is surprising that Dirac was allowed to meet Tamm, as he was leading the secret project to build the hydrogen bomb (Tamm's participation in this work may have been one reason why his friendship with Dirac fizzled out in the next decade).[58] Landau, the permanent juvenile, was by then in the front rank of theoreticians and still flaunting his irreverence: he replaced the toilet roll in his bathroom with pages from Stalin's autobiography.[59]

Landau was in the audience of Dirac's lectures at Moscow University, where Dirac responded to the request made to some of their guests to summarise their philosophy of physics. He wrote on the blackboard: PHYSICAL LAWS SHOULD HAVE MATHEMATICAL BEAUTY.[60] In public, Landau was respectful of Dirac's aestheticism, but in private he was cutting, once remarking to the physicist Brian Pippard, 'Dirac is the greatest living physicist and he has done nothing of importance since 1930.'[61] Overstated to the point of cruelty, this was typical Landau. He was, however, only giving voice to what many leading physicists in the mid-1950s thought but dared not say in public. Yet, as events were about to prove, Dirac's detractors had been too hasty in writing him off.

Twenty-six

How some they have died, and some they have left me,
And some are taken from me; all are departed;
All, all are gone, the old familiar faces.
    CHARLES LAMB, 'The Old Familiar Faces', 1798

In early December 1958, when Pauli was approaching his fifty-eighth birthday, he was looking sallow and unwell. He complained of stomach pains during a lecture at his university in Zurich in the afternoon of Friday 5 December and took a taxi home. On the following day, he went to the city's Red Cross Hospital, where he was admitted for tests which proved inconclusive, so doctors decided there was no alternative but to operate. A week later, a surgeon cut into the hillock of his midriff and found a pancreatic tumour so large and advanced as to be inoperable. Within forty-eight hours of the operation, he was dead.[1]

The final year of Pauli's life had not been among his happiest – a quarrel with his friend Heisenberg over an ambitious theory they were developing had turned nasty and had suppurated. But the end of Pauli's career had also seen the seal put on one of his finest contributions to physics: during an early summer morning in 1956, he received a telegram from two experimenters in the Los Alamos laboratory to confirm that they had discovered the neutrino, the particle that Pauli had predicted, though Dirac and others had doubted that his arguments held water. Just as Pauli had foreseen, the neutrino has no electrical charge, the same spin as an electron and apparently no mass. The newly discovered particle interacts with matter primarily through the weak interaction, which is extremely feeble: of the ten thousand trillion trillion neutrinos zipping through planet Earth every second, all but a few pass straight through without deflection.

The discovery was a triumph for Pauli but, two years later, nature put him firmly in his place when his intuition about the weak interaction was shown to be quite wrong. The story began at the Brookhaven National Laboratory in 1956, when a duo of young Chinese theoreticians – C. N. 'Frank' Yang and T. D. Lee (usually known as 'TD') – suggested what Pauli and almost all other theorists regarded as ridiculous: when particles interact weakly, nature might choose to break the perfect symmetry between left and right, the so-called parity symmetry. At a fundamental level, gravity and electromagnetism are ambidextrous: every experiment that investigates this type of interaction would give the same result if the configuration of the particles involved were swapped left to right, in their mirror image. At Columbia University in New York, experiments (suggested by Lee and Yang) to investigate whether weak interactions are left–right symmetric were carried out by two groups, one led by the aggressively confident Chien-Shiung Wu, born in Shangai, the other by Leon Lederman, a wisecracking New Yorker. The experiments each came to a climax in the bitter cold of New York in mid-January 1957, when they confirmed that Pauli had been wrong and that the suspicions of Lee and Yang were right: in weak interactions, nature *does* distinguish between left and right.

The result was a sensation, and not only among physicists – it even featured prominently on the front page of the *New York Times*. But the observation was no surprise to Dirac.[2] He had foreseen the possibility that parity symmetry might be broken, in the introduction to the review of relativity he wrote in 1949. There, he considered whether quantum descriptions of nature would remain the same if the positions of the particles are reversed in a mirror (a left–right swap) and, separately, if time runs backwards instead of forwards. In his conclusion, he took the unusual step in a technical article of using a personal pronoun: 'I do not believe that there is any need for physical laws to be invariant under these reflections [in space and in time], although all the exact physical laws of nature so far known do have this invariance.'

Dirac had realised that although the laws of gravity and electromagnetism had left–right symmetry and time-reversal symmetry, the laws of other fundamental interactions may not have this property. No leading physicist had remembered reading these words, and even Dirac himself forgot that he had written them.[3] After 1949, he was aware of the possibility of quantum asymmetries in space and time but apparently said nothing about it, except once during a cross-examination of a Ph.D. student.[4] A few years later, when he heard colleagues talk of the shock of parity violation, he would calmly draw attention to this passage in his paper.[5] To students who asked him about it, he said simply, 'I never said anything about it in my book.'[6] He knew, however, that he could not expect many plaudits for his contribution: the winners-take-all rule of scientific conduct entitled Lee and Yang to take the credit for fully appreciating the importance of the breaking of parity symmetry.[7] Theirs was one of the great discoveries of the modern era.

The death of Pauli had removed from the fraternity of senior theoreticians the one member Dirac disliked. Although they did not overtly compete with one another, undercurrents of rivalry swirled beneath their superficial rapport. Their approaches to theoretical physics were different, as Pauli was a conservative analyst, while Dirac was a revolutionary intuitionist. But that need not have divided them. Most of Pauli's peers thought that his scabrous insults were a small price to pay for the high quality of his insights. But Dirac demurred; he often went out of his way to remind lecture audiences that Pauli 'very often bet on the wrong horse when a new idea was introduced', including the time he 'completely crushed' the idea of spin when it first hatched.[8] Nor, it appears, could Dirac forgive Pauli's pitiless strafings. When Pauli stood over him, damning hole theory, demanding that he recant, perhaps Dirac could see the ghost of his father?

Dirac's daughters never saw him show much interest in politics except perhaps when he watched the television news, with the inscrutability of a sphinx. Manci was quite different: she closely followed international events and had strong opinions about many of them, which she spent afternoons discussing on the telephone with friends. In November 1956, she and her family – including her brother Wigner – looked on sadly when Soviet tanks and troops crushed the uprising in Hungary against its government, a puppet of Moscow, and killing some twenty thousand Hungarians. Landau condemned Khrushchev and his Politburo as 'vile butchers'.[9] In the UK, the *New Statesman*, usually a moderate critic of the Soviet Union, denounced the invasion as 'loathsome', 'indefensible' and 'unforgivable'.[10] Soon, the Communist Party haemorrhaged, and the hard-left core of Cambridge academics was reduced to an ineffectual rump, including Bernal, one of the few whose loyalty to the cause was undiminished. Dirac appears to have said nothing about the

Hungarian invasion even to his closest friends: by the mid-1950s, he appears to have lost every vestige of his youthful idealism. He took the rare step of giving vent to this distaste when he first met Tam Dalyell, an Eton-educated Tory who switched allegiance to the Labour Party in 1956 after the disastrous British invasion of Egypt, following the nationalisation of the Suez Canal. Dirac indicated that he welcomed the maverick Dalyell's change of political heart, but added pointedly, 'I don't *like* politicians.'[11]

Yet Dirac was still following reports from the Soviet Union. 'We're all very excited by the sputniks,' he wrote to Kapitza at the end of November 1957.[12] Dirac had first heard about the launch of the artificial satellite, apparently to mark the fortieth anniversary of the Bolshevik Revolution, on the morning of 5 October.[13] That evening, he and Monica went to the back garden of 7 Cavendish Avenue shortly after dusk hoping to see the twinkling satellite pass over in the night sky.[14] Newspaper reports of the orbiting 'Red Moon', a beach-ball sized sphere girdling the Earth in ninety-five minutes, made front-page headlines for a week, and Dirac wolfed the reports down.[15] Sputnik's success transformed the West's view of Soviet technology from condescension to fearful admiration. For Americans, the Sputniks were frightening wake-up calls, even more disturbing after the attempt to launch their own satellite in early December ended in fiasco, when it exploded a few seconds after lift-off (one jeering journalist suggested that it should have been called 'Stayputnik').[16] The Sputnik missions demonstrated that the Soviets were well on the way to developing intercontinental ballistic missiles and to launching a human being into space. The missions panicked the media and politicians into believing that the Soviet Union – which many Americans believed was a backward, agrarian country – was way ahead of the USA in science education. Edward Teller went on television to pronounce that 'The United States has lost a battle more important and greater than Pearl Harbor.'[17] *Life* magazine pointed out that three in four American high-school students studied no physics at all. As a result of all this pressure, President Eisenhower ordered a renaissance in school science and, between 1957 and 1961, Congress doubled federal expenditure on research and development, to $9 billion. An unlikely beneficiary of this largesse was high-energy physics: a new generation of subatomic particle accelerators were, in a sense, the Sputnik's progeny.

Dirac was as interested in the technology of space flight as in any scientific benefits it might bring. He watched television footage of the launches with the same enthusiasm that he had shown when observing from the back garden of 6 Julius Road the launches of some of the first aeroplanes. But he was puzzled: why were the space rockets launched vertically rather than horizontally? So far as he could see, the challenge of propelling a rocket into space is much the same as that of launching a heavily loaded aeroplane, and vertical take-off is extremely inefficient as much of the fuel is used before the rocket is clear of the launch pad; it would therefore be best to launch the rocket horizontally, at high speed. Dirac was fascinated by this question. In May 1961, soon after the Americans put an astronaut into space – less than a month after the Soviets had beaten them to it – Dirac took aback his two fellow diners over lunch at St John's College by sitting not in his habitual silence but, instead, talking about rocketry non-stop for almost an hour.[18]

In the coming decades, he followed reports of the Soviet and American space programmes and attended specialist meetings on them at the Royal Society. Even after talking with several experts, he remained unconvinced that the rockets were being launched in the most economical way, so he took the unusual step of asking NASA for an explanation.[19] Its officials informed Dirac that he was wrong because he was underestimating the importance of the 'drag' effect of the atmosphere on a space rocket and the performance of the rocket's engine, which improves with altitude.[20] Such rockets are launched vertically so that they can climb quickly, enabling them to reach altitudes where the inhibiting aerodynamic pressures on the rocket are much lower than they are at ground level. As the air thins with height, the engine's exhaust can impart greater thrust. These advantages together make it much more economical to launch the rockets vertically, as several experts explained to Dirac, though it seems that he never quite believed them.

Since Dirac's arrival in Cambridge in 1923, his working environment had hardly changed. But, towards the end of the 1950s, there was a concerted drive in the Cambridge science departments to manage themselves more efficiently, partly so that they could compete more successfully with other international centres of science and, indeed, with other parts of the university. In Dirac's bailiwick, the leader of the drive was George Batchelor, an Australian-born mathematician with an uncompromising manner that made clear the extent of his ambition to anyone who doubted it. Then in his late thirties, Batchelor was an expert in fluid mechanics, the branch of applied mathematics concerned with the flow of gases and liquids, a subject for which Dirac had little time – he regarded it as the small fry of theoretical physics. Nor did he like Batchelor, one of the few people who could bring out the snob in him; their colleague John Polkinghorne recalls that Dirac once offended the rhino-skinned Batchelor by dismissing George Stokes, one of the pioneers of fluid mechanics, as 'a second-rate Lucasian professor'.[21]

From the beginning of the autumn term in 1959, Dirac officially worked in the Department of Applied Mathematics and Theoretical Physics, headed by Batchelor. Polkinghorne admired Batchelor as an effective, congenial leader, but Dirac and his colleague Fred Hoyle – now a top-flight cosmologist and a popular broadcaster – both declined offices in the new department and disliked virtually every change he wanted to make. One of the proposed changes was to adopt a more communal approach to research, a notion that could not have been more inimical to Dirac, who looked like a refugee from another age on the rare occasions he attended the new social gatherings. In seminars, he often appeared to be catching up on his sleep but would sometimes give the lie to that by asking a pertinent question. But he would also embarrass senior colleagues by showing how little he knew about the latest research discoveries, even about new particles familiar to greenhorn students.[22]

Although Dirac was not one to stand on his dignity, he was stung when Batchelor ejected him from the office he had occupied for some twenty-five years and 'volunteered' him to give additional lectures. Having been wounded by a series of such slights, he snapped when an officious parking attendant in the Cavendish told him that he had no right to leave his car there. John Polkinghorne recalls Dirac's response: 'He was furious. He told the attendant that he had parked there for twenty years.'[23] He accepted Batchelor's executive decision, but Manci was less compliant and wrote a scathing letter to the Vice Chancellor, who wrote back soothingly and then forgot about her.[24] The authorities no longer felt obliged to keep Dirac happy, and he knew it.

Perhaps in part because of his unhappiness at work, Dirac's marriage was for the first time under strain. The wife of one of the Fellows at St John's briefly caught sight of this when Manci light-heartedly accosted her outside Woolworth's: 'Let's go for a coffee – he hasn't spoken to me for a week and I'm *so* bored.'[25] Stories like this did not surprise the Diracs' acquaintances in Cambridge as most of them had never fully understood how such different people could be happy together. But this happiness was partly an act. Behind their front door, her attitude towards him swung from one extreme to another: one day, she would throw her arms round him and enquire coquettishly whether he loved her; the next, she would tell him angrily: 'I'd leave if I had somewhere to go.'[26] Such threats left Dirac unmoved. According to one story, she once snapped at him when he was eating his dinner, 'What would you do if I left you?' only for him to reply – after a half-minute pause – 'I'd say "Goodbye dear".'[27]

Although he sometimes gave the impression that his research had dried up, Dirac was still thinking hard about his physics. When he gave Manci the signal that he was at work, she ordered the girls to be quiet: Monica would retire to her room, while Mary switched off the gramophone, endlessly blaring out the soundtrack of *Oklahoma!* Now in their teens, the girls had realised that their father was a dis-tinguished scientist and that he was exceptionally quiet and self-effacing.[28] 'I was lucky,' he told Monica. 'I went to good schools, I had excellent teachers. I was in the right place at the right time.'[29]

Gabriel, recovered from his illness, was acutely aware of his step-father's status: his surname drew amused comments from his mathe-matical colleagues and did him no harm at all. Dirac was close to Gabriel and went out of his way to promote his career, often exchanging letters with him to chew over chess problems they had read in newspapers (G. H. Hardy had described such problems as 'the hymn tunes of pure mathematics'[30]). Judy and her family – by the summer of 1960, she had three children – were more distant, and she was in one long fight with her mother, who had all but lost patience with her. As many family friends confirm, Manci was a much better wife than a mother, always supportive and loyal to her husband but often insensitive to her children. It seems that Mary suf-fered most from her mother's tongue: Manci repeatedly browbeat her, told her she was 'ugly' and also 'lazy', a word she used to describe everyone in the family who did not earn a wage, including

Dirac's sister Betty.[31] No one, least of all Dirac, dared to remind Manci that she had yet to do a day's paid work.

By the late 1950s, Mary was back at home and working in Cambridge, contemplating emigration; Monica was preparing to study geology at university. The girls were rapidly becoming independent, and the Diracs wanted to make the most of their new freedom by travelling even more. For someone so friendly, Manci had surprisingly few friends in Cambridge – she was close only to Sir John Cockcroft's wife Elizabeth – and she was continually planning trips to see her family and friends abroad, the further from Cambridge the better. Dirac felt much the same way: an outsider in his own department and resentful of Batchelor's machinations, he preferred to be where he was appreciated. The result was that, in the dozen years before his retirement in 1969, the Diracs were away from Cambridge almost as much as they were there.

Soon after the neutrino was discovered, Dirac had the idea that the particle's existence might be explained by Einstein's general theory of relativity.[32] This was at the back of his mind in September 1958, when he began another sabbatical at the Institute for Advanced Study in Princeton, intending to develop a new version of Einstein's theory based on his favourite way of setting out fundamental theories, using Hamiltonians to describe the interactions. His aim was to find a general classical description of every basic type of field – electromagnetic, gravitational and so on – preparing the ground for their quantisation.

Although his project failed, his method of analysing the general theory of relativity gave new insights into gravity. He described some of them in his lecture at the annual meeting of the American Physical Society, held in New York in the grip of a bitterly cold spell, at the end of January 1959. Always averse to large gatherings, Dirac was probably not looking forward to his stay as he walked the two blocks from Penn Station to the huge, overheated New Yorker hotel, to join the five thousand delegates, most of them in a starched white shirt and tie, sleeves rolled up. Without Dirac's scientific celebrity, he would have been just another of the meeting's invisible men, but his renown made his attendance one of the talking points in the bars and lounges. Many of the audience arrived early after lunch to secure a seat in the huge ballroom, between the imitation Ionic columns reaching to the ceiling, and below the three giant chandeliers decorating the room like cheap jewellery.

Dirac began his talk by making it clear that he was not going to comment on the particle physics in fashion but about the electromagnetic and gravitational interactions, both known for centuries but still not fully understood. Everyone in the audience knew that Maxwell's field theory of electromagnetism predicted the existence of electromagnetic waves, including visible light, and that the energy of the field comes in quanta, known as photons. By a similar token, Einstein had shown that the general theory of relativity predicts the existence of gravitational waves. Dirac announced that his study of the gravitational field's energy indicated that it is delivered in separate quanta, which he called 'gravitons', a long-neglected term first introduced a quarter of a century before in the journal *Under the Banner of Marxism*.[33] After Dirac reintroduced the name, it stuck. These particles will be much harder to detect than photons, he pointed out, but experimenters should lose no time in beginning the hunt for them. He gave the impression to the *New York Times* journalist Robert Plumb that this was an important prediction; the next day, Plumb's report appeared on the front page: '[Dirac] believed that his postulation at this time was in the same category as his postulation of positive electrons a quarter of a century ago.'[34]

Dirac did not succeed in quantising the general theory of relativity, but his Hamiltonian method turned out to be his most influential contribution to the theory.[35] His approach, and similar techniques developed independently by other physicists, enabled Einstein's equations to be conveniently set out in a comparatively simple form, especially in situations when gravitational fields change rapidly. This excursion by Dirac into relativity theory looked odd to most physicists. In the late 1950s, the development of the general theory of relativity was a cottage industry by comparison with the industrial scale of particle physics. Relativity was an unfashionable subject for theorists, and Dirac was one of the few who thought it important to develop it and to find a single theoretical framework to understand gravity and electromagnetism. The main topic at the conference was the strong interaction and the particles that feel it, including the newly discovered mesons. One of the leaders in the field was Feynman, who met Dirac again in the autumn of 1961 at the Solvay meeting, where they had another of their Pinteresque exchanges:

FEYNMAN: I am Feynman.
DIRAC: I am Dirac. [*Silence*]
FEYNMAN (*admiringly*): It must have been wonderful to be the discoverer of that equation.
DIRAC: That was a long time ago. [*Pause*]
DIRAC: What are you working on?
FEYNMAN: Mesons.
DIRAC: Are you trying to discover an equation for them?
FEYNMAN: It is very hard.
DIRAC (*concluding*): One must try.[36]

Dirac's reticence had surprised even his former student Abdus Salam, sitting next to him: from the conversation, Salam concluded that Feynman and Dirac had not previously met. One explanation for Dirac's behaviour, strange even by his standards, is that he did not recognise Feynman: Dirac had an unusually poor memory for faces, which is why he rarely remembered physicists he had met only once, even if their characters were as memorable as Feynman's.

Dirac was convinced that the best way to understand strongly interacting particles was to describe their behaviour with equations, just as he had done when he discovered the electron equation. But most theoreticians were not now thinking along those lines: some were exploring new types of field theory; others gave up all hope of finding equations to describe the particles' motion and sought only to describe in broad terms what can happen when they interact. In this approach, a 'scattering matrix' gives, for every possible initial state of the particles, the likelihood that it will lead to each of the possible final outcomes. Dirac rejected it as 'a façade'.[37]

Apart from the strongly interacting particles, experimenters had also discovered another family in the subatomic zoo. The first hint had arrived from experiments on cosmic rays in 1946, when Carl Anderson identified a particle later to be called the muon. It was some two hundred times as heavy as the electron and unstable, but in other respects it bore a close resemblance to the electron: it had the same spin and did not feel the strong interaction. But there was one crucial difference: in 1962, experimenters showed that the muon is associated with its own variety of neutrino, different from the familiar neutrino linked with the electron. All four particles – the electron, the muon and their neutrinos – appeared to have no constituents and to be part of a family, later known as leptons, following Leon Lederman's introduction of the term, taken from the Greek word for something small and delicate, *leptos*.

The arrival of new particles normally did nothing to excite Dirac – he still had not come to terms with the photon and electron. But in late 1961, Dirac broke his rule of not working on new problems until he had solved the ones already on his plate: he tried to understand the muon, which he believed might simply be an excitation of the electron. He abandoned the usual image of the electron as a point particle and pictured it as a spherical bubble in an electromagnetic field: 'One can look upon the muon as an electron excited by radial oscillations,' he suggested. Dirac described the bubble using a relativistic theory whose equations described its motion in space-time. It was a sublime piece of applied mathematics but most physicists ignored it, apparently because its account of the electron was so unconventional: it gave a geometric account of a particle usually assumed to have no size and paid no attention to its spin. Nor did the theory's predictions do much to win over doubters – Dirac calculated that the mass of the first quantum excitation of his electron accounted for only a quarter of the measured mass of the muon.

Dirac first presented his theory of 'the extended electron' to his colleagues at the Institute for Advanced Study in Princeton on the warm autumn afternoon of 16 October 1962. Oppenheimer was sitting in the front row, his deep-blue eyes still alert and penetrating, his complexion as fragile as an eggshell.[38] Still a master inquisitor, after making one of his smart comments, usually at the speaker's expense, he would sometimes turn round and survey the audience, to check that everyone had appreciated it. When Dirac was the speaker, however, Oppenheimer was on his best behaviour.

An hour after Dirac's audience had dispersed, at 6.30 p.m., President Kennedy met his officials in the White House to discuss urgent intelligence reports: the Soviets were building secret missile bases in Cuba, ninety miles from Florida and therefore potentially a threat to the USA.[39] Six days later, Kennedy went public with the intelligence, announcing a naval blockade of Cuba and demanding that the Soviets remove the missiles. Khrushchev angrily refused to back down. Oppenheimer's scorpions were staring straight into each other's eyes.

The tension dropped on 28 October, when the Soviets agreed to remove the missiles in return for concessions from the Americans; it seemed to many – including Dirac, watching the crisis unfold on his television in Princeton and possibly wondering whether he was about to see his third world war – that humanity had been lucky to survive. The planet seemed to be at the mercy of its Dr Strangeloves.

Bohr lived just long enough to see the Cuban missile crisis. Three weeks later, after Sunday lunch at home with his wife Margrethe, he went upstairs for a nap and died of heart failure. In a letter of condolence to Margrethe, Dirac said that he was 'excessively sorry' to hear of 'the loss of one of my closest friends' and recalled his first stay with the Bohrs in Copenhagen in 1926: 'I was greatly impressed by the wisdom that Niels showed, not only in physics but in all branches of human thought. He was the wisest man I knew, and I did my best to absorb some of the wisdom he imparted.'[40]

This was the latest of a series of blows to Dirac, who was seeing his closest colleagues die off one by one. In Princeton, von Neumann had died in 1957, followed by Veblen in 1960. And only eleven months before Bohr's death, Dirac had written the obituary in *Nature* for Schrödinger, who had died in his Vienna home of heart disease. In his article, Dirac went out of his way to defend Schrödinger's apparent welcoming of Nazism in May 1938: 'He was forced to express his approval of the Nazi regime, and he did this in as ambiguous a way as he could.'[41] Many of those who had read Schrödinger's article joyfully pledging support for 'the will of the Führer' will not previously have noticed that it contained many ambiguities. But, as Heisenberg and Kapitza had seen, Dirac could not be faulted on his loyalty.

Until 1962, Dirac had shown no interest in publicly discussing his recollections of the beginnings of quantum mechanics. But that year, when he turned sixty, he changed his mind. He agreed to be interviewed by the American philosopher of science Thomas Kuhn, a former student of Van Vleck. Kuhn persuaded Dirac to help compile the archive for the history of quantum physics. Kuhn knew that Dirac was nervous of talking to strangers in unusual environments, so he held the first interview in Wigner's home in Princeton, with Wigner present and often chipping in with tactfully phrased questions to draw him out. During the forty-minute session, Dirac spoke quietly and clearly, often sounding tentative and mildly amused that anyone would be interested in what he would have to say.

For almost forty years, Dirac had hardly spoken a word to his physicist colleagues about his upbringing, but Kuhn and Wigner heard childhood memories pour out of him, including a torrent of domestic detail.[42] About ten minutes into the interview, Dirac began to talk about his brother. It is clear from Wigner's delicately phrased questions and from his mild incredulity at Dirac's responses that the two men had scarcely broached the subject in the thirty-five years they had known each other. During this part of the interview, Dirac speaks as gently as usual, but each of his carefully articulated words seems to bear a heavy burden of sadness and regret, especially when he responds to Wigner's question about why Felix took his own life:

I suppose he was just very depressed. And, well . . . that kind of life where we were brought up without any social contacts at all must have been very depressing to him as well as to me and having a younger brother who was brighter than he was must have depressed him also quite a lot.[43]

Dirac left much unsaid, but Kuhn and Wigner were wise not to press him; if they had, he would almost certainly have clammed up and perhaps even refused further interviews.

Privately, Dirac was in no doubt why his brother killed himself. Dirac told Kurt Hofer that he was sure his father was primarily responsible for the tragedy: Charles had denied Felix a normal upbringing, forced him to speak French against his will and crushed his ambition to be a medical doctor.[44] But, even after decades of reflection, Dirac could not understand the depth of his father's grief after Felix's suicide: his father was still a mystery to him and still, as he told his closest friends, the only person he had ever 'loathed'.[45]

Three months after the interview, Kuhn wrote to thank Dirac for his participation and informed him that his taped disclosures about Felix's death would be removed from the published version and 'filed separately for future use'.[46] The material was made public only after Dirac's death.

In 1962, Dirac was about to enter the final stage of his career in Cambridge. His family circumstances were changing rapidly: his daughter Mary was preparing to emigrate to the USA; Monica had gone off to university 'to discover the Beatles'. Shortly before leaving, Monica had been thrown out of the house by her mother, just as she ejected Judy in her teenage years.[47] Now Judy and her family were settled in the USA and Gabriel was pursuing his academic career in Europe.

Dirac imagined that he would spend the rest of his life at home in Cambridge, tending his garden and working in his study. But Manci had other plans.

Twenty-seven

[Some critics] act as if Flaubert, or Milton, or Wordsworth were some
tedious old aunt in a rocking chair, who smelt of stale powder, was
only interested in the past, and hadn't said anything new for years. Of
course, it's her house, and everybody's living in it rent free; but even
so, surely it is, well, you know . . . time?
JULIAN BARNES, *Flaubert's Parrot*, 1984

By the mid-1960s, Dirac was spending most of the week working at
home. At the department he looked increasingly out of place: 'He
was irrelevant,' his young colleague and former student John
Polkinghorne remembers.[1] Other Cambridge physicists thought the
same but followed the scientists' unwritten code of chivalry: when
great researchers go to seed and speak out against modern trends in
their subject, they should be ignored and even mocked in private, but
be heartily praised in public for their past achievements.

Outside the university, too, Dirac cut the lonely figure of a misfit
from another age, uncomfortable with the new popular culture and
its irreverence. It was unthinkable to him that serious critics could
treat a painting of a soup tin as a mainstream work of art and that
many of the defining songs of a generation were written by cheeky,
working-class Liverpudlians who could not read music. What, Dirac
wondered, was he to make of a group whose lead vocalist claimed to
be a walrus?[2]

Dirac was beginning to fear old age and the prospect of being
effectively abandoned by his colleagues: all the signs were that
Batchelor was going to bundle him out of his Lucasian Chair at the
statutory retirement age of sixty-seven. The threat led Dirac to make
a brief venture into the poisonous netherworld of university politics
in the spring of 1964, when he joined Hoyle and a few others to seek
Batchelor's removal after his first five-year stint as head of their
department. Outmanoeuvred, they failed miserably.[3] With no wish
to be part of Batchelor's empire, and with his child-rearing responsibilities behind him, Dirac – encouraged by Manci – resumed his travels and spent even more time in his garden, trimming his immaculate

lawn, pruning his roses and growing far more vegetables than Manci needed for her larder. His bookshelves heaved with horticultural magazines and books, making his study look as if it belonged not to a research physicist but to a landscape gardener.[4] He still did research but knew that he had next to no chance of coming up with a radically new idea. He was enduring the fate of all ageing theoretical physicists: his spirit was outliving his imagination.

Though marginalised in Cambridge, he was treated kindly at his favourite academic address in the USA. In the spring of 1963, Dirac heard from Oppenheimer that he had arranged for a framed photograph of him to be mounted on a wall at the Institute for Advanced Study, next to a snapshot of Einstein: 'You two are alone on that wall.'[5] This simple gesture symbolised the generosity of the American academic system, much more willing than British universities to find room for leading scholars to spend their unproductive twilight years in dignity. Mainly for this reason, Dirac spent more time in the USA. From 1962 to his retirement in 1969, Dirac visited the United States every year, for at least a couple of months, twice for almost an entire academic year (1962–3 and 1964–5).[6] For much of the rest of the time, he and Manci were visiting conferences or on vacation in Europe and Israel (the USSR was no longer on their itinerary, apparently because even they could not get a visa). During these seven years, Stephen Hawking – a colleague of Dirac's and a rising star – did not see him in the department.[7]

Manci had set her heart on escaping from Cambridge. Dirac disliked change and wanted to be loyal to his university but eventually agreed that it was time to emigrate, preferably to the USA. He did not have the initiative to secure a new position: that task fell to Manci, who assumed a new role as the pushy manager of a tongue-tied talent, chasing royalties and upgrades, insisting on sea-facing cabins and the room with the finest view. He was her Elvis, and she was his Colonel Parker.

Lecturing had become Dirac's forte. Although his voice was weakening, he could be relied on to keep his audience hooked, not through wit and humour but through clarity and humility. At the podium, he looked and sounded like an elderly preacher from Bristol but had the innocence of a young lad reading an essay on Prize Day, clipping his vowels, emphasising his consonants with the force of a stab. It was often a surprise to people in the audience that such a taciturn man was so fluent, hardly ever hesitating with an 'er' or an 'um' and rarely showing a sign of even approaching a grammatical tangle. His most unnerving idiosyncrasy was a propensity to go silent in mid-sentence: when he needed to think or find the right words, he would suddenly stop talking, typically for ten seconds but sometimes for over a minute, before resuming without comment.

He presented fewer specialist talks but occasionally gave guest lectures, including a series on quantum field theory at Yeshiva University in New York in the spring of 1964. In these lectures, later recognised as classics, he developed the theory logically from its beginnings and, unusually for him, spelt out in detail the calculations that led to the prediction of the energy shift of the hydrogen atom, measured by Lamb in 1946. Although the theory and experiment agree to within experimental uncertainties, Dirac left his audience in no doubt that the theory of quantum electrodynamics is profoundly flawed: 'If one is a research worker, one mustn't believe in anything too strongly; one must always be prepared that various beliefs one has had for a long time may be overthrown.'[8]

A year earlier at Yeshiva, he gave his lecture 'The Evolution of the Physicist's Picture of Nature', which he adapted into an article for the May 1963 edition of *Scientific American*, the only article he ever wrote for a popular-science magazine. The style and content of the talk foreshadowed dozens of similar presentations: he explained in plain, stripped-down language why fundamental physics was in crisis, drawing lessons from an often simplistic overview of the history of physics. In the article, he dwelt on one of his favourite anecdotes: Schrödinger claimed that he had discovered a mathematically beautiful relativistic version of his equation a few months before the famous non-relativistic version but did not publish the relativistic equation because it failed to account for observations on the hydrogen atom (the disagreement arose because it was not known at that time that the electron has spin). Schrödinger published his non-relativistic version only when he was sure it was in good agreement with the data, but if he had been bolder he would have been the first to publish a relativistic quantum theory. For Dirac, this story had a moral: 'It is more important to have beauty in one's equations than to have them fit experiment.'

Dirac suggested to his readers that 'God is a mathematician of a very high order, and He used very advanced mathematics in con-structing the universe,' having apparently forgotten that he first encountered the God–beauty link forty years before in the writings of his colleague Sir James Jeans.[9] In his positivist youth, Dirac would have regarded the link as unverifiable and therefore meaningless, but he had changed his tune: after spending decades on the *terra firma* of experiment-based science, he was ready to take pleasure trips on the seas of metaphysical philosophy.

The physicist in Dirac now seemed to prefer the past to the present. Uncomfortable in the company of the leading young physicists, he was most at ease when he was reminiscing with his old friends. He missed none of the triennial meetings of Nobel Laureates at Lindau, a relaxed town in southern Germany, where he talked with physicists and, with rather more reserve, to the students invited to join them. *Horizon*, the flagship science series of the new British television chan-nel BBC2, made a film at the 1965 meeting, produced by Peter Loïzos. He saw that the two Nobelists most lionised by the students were Dirac and Heisenberg, who attracted swarms of admirers like Hollywood stars, and that, away from the mêlée, Dirac followed Heisenberg like a butler.

Loïzos knew it was not going to be easy to persuade Dirac to talk, as several BBC radio and television producers had asked him for interviews but had been turned down firmly.[10] But Dirac agreed to be filmed in conversation with Heisenberg and the result is a unique recording of Dirac in relaxed conversation.[11] Always with an agree-able smile, Heisenberg was as smartly dressed and easy-going as he had been thirty years before, but Dirac had changed rather more. His comically ill-combed hair helped to maintain his reputation for peer-less dishevelment, but he was more relaxed than he had been as a young man, constantly smiling with his eyes and his mouth, speaking with a surprising assertiveness. Most striking about the encounter is that Dirac led the discussion, especially after he steered the subject towards beauty, via his anecdote about Schrödinger's premature shelving of the relativistic version of his equation. When Heisenberg gently remarked that beauty is less important than agreement with experiment – the conventional view – Dirac took up the cudgels for aestheticism, forcing Heisenberg on to the defensive:

HEISENBERG: I do agree that the beauty of an equation is a very important point and that one can get already a lot of confidence from the beauty of an equation. On the other hand, you have to check whether it fits or whether it doesn't. It's only physics when it really fits with nature. But that may turn out much later.

DIRAC: And if it doesn't fit you'd hold up publication would you? Just like Schrödinger?

HEISENBERG: I'm not sure whether I would. In at least one case I have not done so.

Smiling beatifically, Heisenberg appeared to concede the point: thirty years before, he would have persisted with the tenacity of a terrier, but his appetite for competition had been weakened by years of post-war humiliation. Delighted to have won the argument, Dirac's face lit up with the broadest of smiles, revealing two rows of rotting teeth.

Dirac still had faith in the large numbers hypothesis, though he knew most physicists regarded it as a blot on his CV after Edward Teller had published an apparently damning refutation of it in 1948. Teller pointed out that the hypothesis implied that because the universe is expanding, gravitational forces were greater millions of years ago than they are today. Teller showed that Dirac's idea implied that the Earth's oceans would have boiled and evaporated away 200–300 million years ago, contrary to the geological evidence that life had existed on the planet for at least 500 million years.[12] Interest in the hypothesis had flickered again in 1957, when the American cosmologist Robert Dicke demonstrated that the large numbers hypothesis is a consequence of the fact that human life occurs after stars were formed and before they die.[13] If the hypothesis were wrong, astronomers, and all other life forms, would not exist. Dirac was unimpressed with Dicke's reasoning and would not budge: he believed in the importance of the hypothesis 'more than ever'.[14] In November 1961, Dirac wrote his first public comment on cosmology in twenty-two years:

On Dicke's assumption habitable planets could exist only for a limited period of time. With my assumption they could exist indefinitely in the future and life need never end. There is no decisive argument for deciding between these assumptions. I prefer the one that allows the possibility of endless life.[15]

Dirac's vision of the fate of the universe was consonant with one of the articles of faith he wrote in his philosophical jottings of January 1933: 'the human race will continue to live for ever', a subjective assumption he had to make 'for his own peace of mind'.[16] Evidently, this most detached of theoreticians could not bear to think of a universe without human beings.

One of the few cosmologists who still believed that it was worth spending time on Dirac's hypothesis was the vodka-swilling giant George Gamow. In 1965, he took a sabbatical in Cambridge, accompanied by his new wife Barbara, whom he had married shortly after his divorce from Rho in 1956 'on mental grounds'.[17] The Gamows stayed at the new Churchill College, whose first Master, Sir John Cockcroft, had been chosen by the Prime Minister after whom it was named.[18]

One topic of discussion between Dirac and Gamow was the beauty of the 'steady state' theory of the universe, which says that the universe has no beginning or end, but goes on for ever like a film with an endlessly repeated plot. That summer, this was a topical question because the steady-state theory seemed to have been discredited by one of the most telling astronomical observations to have been made in decades. Two astronomers at the Bell Laboratories in New Jersey had detected an all-pervading background bath of low-energy radiation. It was only after the astronomers made their observations that they heard that just such a bath of radiation had been predicted long before by Gamow and others, using the Big Bang theory. For most cosmologists, the theory afforded a beautifully simple description of the development of the universe, compatible with the general theory of relativity and all the other great theories of science. Fred Hoyle, who had given the Big Bang theory its name in 1949 during one of his BBC radio broadcasts, was the most vocal of the diminishing number who did not give up on the steady-state theory.[19] Hoyle found the idea of the Big Bang distasteful and compared the notion of the universe emerging out of nothing to a 'party girl' jumping out of a cake: 'it just wasn't dignified or elegant'.[20]

After one of his discussions with Dirac, Gamow wrote to ask if he had heard of a tongue-in-cheek summary of the role of aesthetics that appears to have dated from their days in Copenhagen (Gamow uses the word 'elegant' where Dirac would use 'beautiful'):

Case I Trivial statement
If an elegant theory agrees with experiment, there is nothing to worry about.

Case II Heisenberg's postulate
If an elegant theory does not agree with experiment, the experiment must be wrong.

Case III Bohr's amendment
If an inelegant theory disagrees with experiment, the case is not lost because [by] improving the theory one can make it agree with experiment.

Case IV My opinion
If an inelegant theory agrees with experiment, the case is hopeless.[21]

Dirac believed that if observations agree with an ugly theory – such as quantum electrodynamics – it is little more than a coincidence. He had a fundamentalist belief in beauty, as Heisenberg found when he produced a new theory of particle physics and pressed Dirac for 'specific criticism', only for Dirac to give the thumbs down to the theory because its basic equation had 'insufficient mathematical beauty'.[22]

Kapitza was one of the few who understood Dirac's passion for beauty, perhaps because he had helped to foster it in their early conversations in the Cavendish and in Trinity College. Dirac may have feared that he would never again feel the thrill of Kapitza's company in Cambridge, but he heard in the spring of 1966 that both Kapitza and his wife had secured exit visas to enable them to return for a short stay. In late April, as the Kapitzas' arrival drew near, Dirac and Manci were like children on the eve of a royal visit, so excited that they could barely concentrate on the preparations.

By 1966, Kapitza was the Soviet Union's most famous scientist, in the address books of most of the country's leading artists and a licensed critic of the Government. The British Ambassador wrote in advance to Cockcroft to warn him that Kapitza was still 'a bit of a rebel' and suggested that 'the public relations aspect of the visit will require rather careful watching'.[23] But the Ambassador need not have worried; Kapitza was on his best behaviour, having learnt from Rutherford how to balance irreverence and propriety so that he could be seen as both close to the establishment and fiercely independent. In his interviews he was always careful to stress that he had played no part in the development of nuclear weapons and that he was as patriotic as ever, as he demonstrated in his lecture 'The Training of the Young Scientist in the USSR' in the Hall of Trinity College.[24]

When the Kapitzas visited the Diracs for lunch, Manci made a special effort in the kitchen, poaching a salmon and serving it with home-made mayonnaise and a chilled Burgundy: Mary recalled that it was the closest her parents ever came to giving a banquet.[25] For just that one afternoon, the front room had the warmth of a jacuzzi – their reminiscences darted around from the summer they spent in the Kapitzas' *dacha* to their days in the Cavendish, with Kapitza telling wedding-night jokes so blue that Anna left the room, leaving Dirac and Manci to giggle their way to the punchline.[26]

They will also have talked about Kapitza's Club, which had ceased to exist in the spring of 1958, superseded by programmes of seminars. The Club was, however, reconvened on 10 May for its 676th meeting, so that some of its surviving members – including Dirac and Cockcroft – could meet one last time and so that Kapitza could close it.[27] The venue was a smart common room in Gonville and Caius College, where the participants sipped fine dessert wines, in contrast to the meetings forty years before, when they would drink dishwater coffee. A photograph of the occasion shows Kapitza and a forlorn-looking Dirac, his left elbow leaning on the table, his left hand supporting his head. He gives the impression of being bored out of his mind.

The highlight of the meeting was a joint presentation by Dirac and Kapitza on the effect they had identified in 1933, a year before Kapitza had been detained in the Soviet Union: the possibility that electrons could be bent (diffracted) by light. When they first predicted the effect, it was impossible to observe because the available sources of light were too weak and the electron-detectors were too insensitive. But now detection looked possible, following improvements to the sensitivity of the detectors and the invention of lasers, devices that had become familiar to the public since they featured in the 1964 James Bond film *Goldfinger*. The barrel-chested Kapitza, standing by a blackboard and easel, pointed out that it was now odds-on that experimenters would soon observe the effect; the question was: would Dirac and Kapitza be alive to see it?[28]

A few days after the Kapitzas left Cambridge, Dirac switched his attention from the past to the future. He attended an entire course of lectures on modern particle physics given by the American theoretician Murray Gell-Mann, a source of many of the most productive new ideas in particle physics since the early 1950s. Then thirty-six and still at the height of his powers, he was admired for his imagination and technical brilliance but feared for his waspish tongue and disliked for his egoism, not least by Dirac.[29] In the 1960s, Gell-Mann and others suggested that strongly interacting particles could be classified in mathematical patterns, and he used one of them in 1963 to predict the existence of a new particle. When experimenters detected it in the following year, it was a signal success for theoretical physics. Gell-Mann and his colleague George Zweig, working independently, also proposed that strongly interacting particles might consist of different combinations of three varieties of a new type of fundamental particle that Gell-Mann called quarks (he took the word from James Joyce's *Finnegans Wake*: 'Three quarks for muster mark!') But Gell-Mann himself was sceptical: he remarked in his lectures that quarks were probably not real particles but mathematical artefacts that help to explain the symmetries among the properties of the strongly interacting particles.[30] A year later, Gell-Mann recalled that he was surprised that Dirac 'loved' quarks, despite their having – in Gell-Mann's opinion – 'many annoying properties', including their apparently permanent confinement inside strongly interacting particles, such as protons and neutrons.[31] When Gell-Mann asked Dirac why he thought quarks are so 'marvellous', Dirac replied that they have the same spin as the electron, the muon and the neutrino. Perhaps Dirac had seen that it was possible that all fundamental constituents of matter have the same spin – the spin of the electron. And perhaps he had sensed that it might soon be possible to set out a description of strong interactions in terms of a field theory, as he had hoped.

Gell-Mann's lectures taught Dirac a lesson: the bottom-up way of doing theoretical physics – drawing inspiration from experimental observations – was proving much more productive than the top-down style – taking cues from beautiful mathematics – that Dirac practised and preached. Dirac privately admitted this, though he had no intention of changing his approach.[32]

In mid-September 1967, the Diracs heard that Sir John Cockcroft, one of their closest friends, had died suddenly of a heart attack in the Master's Lodge of Churchill College. Several of his friends believed that his death had been hastened by his anxiety over a classic Cold War melodrama that had taken place two days before: Soviet Embassy officials abducted his colleague Vladimir Tkachenko – a student protégé of Kapitza – on the Bayswater Road in London and had whisked him off to Heathrow, where they put him on a plane bound for Moscow. But, just as his plane was setting off, it was surrounded by squad cars of airport police and MI5 agents, who boarded the plane and found him looking sick and bleary-eyed, apparently under sedation. They forcibly removed him, outraging Soviet authorities, who protested that he was leaving Britain of his own volition, having been blackmailed and intimidated by British agents. Cockcroft died on the morning after the incident became public, when the story was on the front page of *The Times*.[33]

His wife Elizabeth knew she would soon have to leave the Lodge to make way for the next Master, and the College assisted her in making the move. In the opinion of the Cockcrofts' children, the authorities treated her sensitively and with a good deal of generosity, but Manci disagreed: she told everyone who would listen that the College was shooing Lady Cockcroft out of the Lodge with despicable haste.[34] Manci's patience with Cambridge finally ran out, and she made up her mind that Dirac must move to an institution that behaved better towards its senior academics. She also vowed to take her revenge on Churchill College.

Dirac and Manci began making plans to settle in the USA. Some of its universities were certain to offer Dirac a professorship, and Mary and Monica, both married by the summer of 1968, now lived there. Manci's brother Eugene Wigner was also in the USA and was one of the elder statesmen of American science, an adviser to the Government, and – to Manci's irritation – moving politically further to the right each year. From his letters to the Diracs, it is plain that Wigner was a thoughtful and caring member of his family but, in the public eye, his humility had become something of an affectation: he was now so self-deprecating that many of his acquaintances thought he was using it as a subtle form of mockery. Ideally, the Diracs would have liked to have settled in Princeton, but that was no longer an option: after Oppenheimer's retirement in June 1966 – seven months before he died of throat cancer – the Institute for Advanced Study was unlikely to offer Dirac an academic home, nor could Princeton University be expected to accommodate a physicist so far past his best.

Two branches of Dirac's family remained in Europe. Betty was a contented housewife in Amsterdam, doing the chores to the soundtrack of the BBC Home Service (now Radio 4) and going regularly to the highest Catholic mass she could find. In 1965, Gabriel was appointed to the mathematics faculty at the University of Swansea soon after the US Government rejected his application for a visa, apparently because of his brief membership of the Communist Party in Cambridge.[35] Two years later, he and his family moved to the University of Aarhus in Denmark, and Dirac and Manci visited them during their summer vacations.

Of all their children, Dirac and Manci were most concerned about Judy, who had lost custody of her children after an acrimonious divorce in 1965. Soon afterwards, she moved to Vermont and spent several lonely months each year in the Wigners' summer cottage on the shore of Lake Elmore. Wigner feared for her mental health. He wrote to Manci, telling her that Judy was desperate for her mother's affection and pleading with her to support her troubled daughter: 'You must not abandon her,' he told Manci in September 1965.[36] Two and a half years later, Judy was holed up in a motel near Lake Elmore, lonely, penniless and delusional. She desperately needed psychiatric help, Wigner believed, and he begged his sister to intervene, but Manci told him that she would have nothing to do with Judy until she got a job and that he should stop interfering.[37] Manci felt no responsibility for her daughter's plight, she wrote to Wigner:

Why should I in the name of heaven feel guilty? . . . I DID my duty, and who can throw a stone at me? J is an expert in hurting deeply, and may be she does this to those she loves. In that case she must seek a remedy.[38]

Manci's indignation was suddenly punctured on 17 September 1968, when she read a telegram from her brother: 'JUDYS CAR FOUND ABANDONED DO YOU KNOW WHEREABOUTS LOVE.' This was the worst day of Manci's life, she later said.[39] Manci had no idea where Judy was, as they were no longer in touch. In the following days, the Diracs heard nothing from Vermont or from the Wigners. Manci was distraught, lurching between wildly different accounts of Judy's disappearance, always refusing to believe that her depression had led her to take her life. It was most likely, Manci believed, that Judy had been murdered.[40] Dirac's reactions to all this were known only to Manci, who appears to have shared them with no one.

The Diracs decided not to travel to Vermont but to stay in Britain and monitor events from there: they left it to the Wigners to deal with the authorities in Vermont. In early October, after visiting the site where Judy's car was found – a country lane near Morrisville, Vermont – Wigner and his wife wrote to the Diracs with details of the police hunt for her in the surrounding countryside and ponds.[41] The search parties found nothing. Gradually, the Wigners, tearful and depressed, came to believe that Judy would never be seen again, but the Diracs clung to every last hope. For three years, they tried to imagine scenarios in which Judy might suddenly reappear, but the weight of probability gradually crushed what remained of their optimism. They accepted that it was practically certain that Judy was dead.[42]

Mary later recalled that her mother was inconsolable, 'insane with grief'.[43] The Diracs kept the pain of their loss private, but two of his later acquaintances, the sculptor Helaine Blumenfeld and her husband Yorrick, the *Newsweek* journalist, glimpsed deeper feelings.[44] The Blumenfelds recall that, two years after Judy went missing, Dirac and Manci were still losing sleep over her fate and talked about it endlessly. From Dirac's comments about her, the Blumenfelds assumed that he was her biological father – he was as sad and bereft as if he had lost his own daughter.

In the early weeks of 1969, the Diracs were in Miami, pondering life after Cambridge. Of the American universities wanting to employ Dirac, one of the most tempting offers had been made by his former student Behram Kurşunoğlu at the University of Miami. A wheeler-dealer Turkish theoretician – always smart in his Stetson hat, jacket and tie – Kurşunoğlu had spent his career searching for a unified theory of fundamental interactions, following Einstein's agenda.[45] Kurşunoğlu had founded the annual Coral Gables conferences, which gave several leading theorists a good reason to leave their home cities in the depths of January and spend a few days in the bright, warm sun of south Florida. Kurşunoğlu employed Dirac at the university on a temporary contract and tried hard to persuade him to accept a permanent post, making him and Manci as welcome as family, taking them out on trips round the area and giving Dirac a taste for coconuts, alligators and exotic birds.[46] Manci was embarrassed by the time Dirac took to weigh Kurşunoğlu's offer, but he was not to be hurried – he disliked Miami's oppressive heat and felt uncomfortable in a place where recreational walkers are regarded as perverse.[47]

The most memorable of Kurşunoğlu's outings was a trip to the cinema on New Year's Day. Kurşunoğlu and his wife asked Dirac to go with them to see Stanley Kubrick's *2001: A Space Odyssey*. The film had divided critics and audiences since its release eight months before: it inspired Steven Spielberg and a new generation of filmmakers, but it left John Updike's Rabbit Angstrom bemused and sent his wife to sleep.[48] Firmly on Spielberg's side, Dirac was enraptured: he had seen hundreds of movies, but had never imagined it was possible for a film to have such a powerful impact and enable him 'to see his dreams', as he told Mary's husband Tony Colleraine. Dirac disliked opaque and open-ended narratives, so his love of *2001* was not predictable. It is easy, however, to imagine him being moved by Kubrick's use of Johann Strauss's 'Blue Danube' and the rest of the classical soundtrack and by the appeal of a story told mainly through visual images rather than words. Dirac's opinion that a good deal of quantum mechanics can be expressed accurately only through mathematics, not words, is echoed by a comment Kubrick made about *2001*: 'I don't like to talk about [it] much, because it's essentially a non-verbal experience.'[49]

Still excited two days later, Dirac saw the film again at a matinee with Tony Colleraine and also with Manci and Mary, who spent most of the two and a half hours in the theatre whispering to each other. Dirac suggested to Tony that they see it again 'without the running commentary'. Without telling Manci, they stayed to watch the next two screenings and returned home to find their hot dinner left to get cold on the table. But Dirac was too excited to care about food: he was like a child after three consecutive rides on a roller coaster. Several of the scenes had possessed him, especially the Star Gate sequence and the emergence of the grizzled astronaut into the eighteenth-century bedroom: 'I would not be able to sit alone through that scene,' he later told Colleraine.[50] Manci was not interested in Dirac's observations on 'that weird film'; her idea of a good movie was the romantic epic *Dr Zhivago*, not one whose most memorable character was a talking computer.

2001 stoked Dirac's interest in the Apollo space programme. During the evening of 20 July 1969, he sat open-mouthed in front of the television in the Kurşunoğlus' front room when Neil Armstrong prepared to set foot on the moon. He sat up all night watching the coverage. Kubrick's images were sharper and his soundtrack was clearer, but the grainy television pictures and the muffled sound of that first moon landing had a compelling reality of their own. And for Dirac, the former engineer, reality mattered most: the first moon-walk was the culmination of aeronautic technology, whose beginnings he had seen as a boy and which now enabled human beings to set foot on a landscape a quarter of a million miles away. The Apollo team, having achieved the most impressive technological feat Dirac had seen in his lifetime, may well have given him a twinge of regret that he had chosen science rather than engineering: he had been a leader of a scientific revolution that, in his opinion, had led to a dead end, whereas the Apollo engineers could declare 'Mission Accomplished' and move on.

In the summer of 1969, Dirac prepared to leave his post and say his goodbyes to the few friends left in Cambridge, including Charlie Broad, the philosopher who gave him his first proper introduction to the theory of relativity. Broad, aged eighty-one, still lived in Trinity College, where he died two years later.

On Tuesday 30 September, Dirac spent his final day in Cambridge as its Lucasian Professor, the most distinguished holder of the Chair since Sir Isaac Newton. Dirac's retirement passed without ceremony, probably because the university authorities assumed that Dirac would feel uncomfortable if he was the cynosure of a leaving party. This was an error, though an understandable one: Dirac would have liked his contribution to the university to be marked officially as his sense of propriety was, contrary to the impression he gave, stronger than his aversion to ceremony.[51] Manci was disgusted. But she was gratified by the sensitivity of St John's College, which extended Dirac's fellowship for life so that he could return there whenever he wished. Batchelor wanted to be generous, too, and offered Dirac the use of a room in the department whenever he was passing through the town, but he declined. His true home in the university was his college, not his department.

For two years, the Diracs divided their time between the UK and the United States, and, by March 1971, Manci could hardly wait another day to leave Britain, 'that lazy impossible island'.[52] Labour unrest, steadily increasing since the war, had become critical: in the first year of Edward Heath's government, more working days had been lost to withdrawals of labour than in any year since the General Strike. Postal workers had gone on strike and slowed down communications in the country for seven weeks. Even Rolls Royce had gone bankrupt.

The Diracs were about to move to a country that was no less troubled. The USA's prosecution of the war in Vietnam was as controversial in the extended Wigner family as it was in thousands of others: the doveish Manci seethed over 'young American lives mutilated fighting for a bastard government' and argued with her hawkish brother Eugene, who believed that the war was essential to stem the spread of Communism.[53] She did not know that the FBI had opened a file on her and was seeking evidence that she was a subversive.[54] Dirac knew that his past political sympathies would raise eyebrows in some American institutions, as he noted when he declined an invitation to the University of Texas at Austin because he was technically ineligible: 'I do not have strong political views, but [. . .] I am a member of the Soviet Academy of Sciences and this makes me, according to [the university's] definition, a member of the communist party.'[55]

Whenever he left the USA in the late 1960s and 1970s, he was nervous that the authorities might forbid him to re-enter it. As he probably suspected, the FBI was still watching him.[56]

Dirac, a dissenter from America's foreign policy in south-east Asia, followed the fierce opposition to the war in American universities through the newspapers and the television news. Although Miami University was one of the less volatile campuses, its students harassed the authorities almost every day, condemning the Vietnam War, demanding free contraception and more support for civil rights. The protestors would talk only to the university's President, Henry King Stanford, who stood on 'The Rock' – a stage-like stone structure in the centre of the campus – making conciliatory speeches to the students and trying to avoid further trouble.[57] On the periphery of these crowds, Stanford often saw the slender, inquisitive figure of Dirac.

On Wednesday, 6 May 1971, the students were especially angry. It was two days after the Ohio State Guard had opened fire on student demonstrators at Kent State University, during a protest triggered by the American invasion of Cambodia.[58] Thirteen seconds of gunfire had killed four students, wounded nine others and brutally curtailed the flower-power hedonism that had flourished only briefly since the

Sgt Pepper summer of love in 1967. The mood of America turned ugly. Even the usually sober campus of Princeton University was unstable: Wigner thought many of the students were 'selfish and nihilistic', behaving 'like the Hitler Youth'.[59] Miami University teetered on the edge of anarchy, when its students – supported by many staff – began a four-day strike, joining two hundred and fifty other campuses across the country. After lunch, at the beginning of a warm afternoon, Stanford made his way to The Rock to address a volatile rally of over a thousand students, many of them with their arms folded aggressively or holding banners with messages such as 'U$ out of S.E. Asia'. Earlier, the crowd had made an effigy of President Nixon out of newspapers, old clothes and firecrackers and then set fire to it. Dirac had seen nothing remotely like it since the Cambridge demonstrations in the 1930s.

During Stanford's walk towards the crowd, he saw an elderly man on the periphery and was quite taken aback to be approached by him. It was Dirac, who asked gently, 'Are you afraid?' Stanford, his heart pumping hard in his chest, replied, his tongue firmly in his cheek, that he was quite looking forward to addressing the students. It seems that Dirac saw that the President was anxious and could use a little reassurance, as he took what was, for him, the unusual step of offering him advice: 'Tell them what you think and listen to what they have to say.' The tone of Dirac's voice gave the impression that he had a 'spiritual kinship' with the protestors, as Stanford later wrote, perhaps identifying a faint echo from the days when Dirac was on the fringe of left-of-centre radicalism. In his emollient address, Stanford described the Kent State incident as 'One of the saddest chapters in the history of higher education', adding that the students' deaths 'dramatise the deterioration of reason' in the USA.[60] Shortly after the speech, the protest ended peacefully, though the university remained on edge for weeks. Dirac probably wondered what future lay ahead of him.

A few weeks later, the Diracs took a break and drove up to Florida's state capital of Tallahassee. Compared with tense, crime-ridden Miami, it was as friendly and safe as a village.[61] Dirac knew that he was being wooed by Florida State, known best not for its physics department but for its student parties and the high quality of its football team. Joe Lannutti, the physics department's ambitious leader, saw an opportunity to persuade the dithering Dirac to become a 'professor at large' at the university, a mascot for the physics department's aspiration to be a 'centre of excellence'.[62] Lannutti had already invited the Diracs to Tallahassee in March 1969, when the Holiday Inn welcomed them with banners fluttering over the entrance, and the physics department had given tenure to Mary's husband Tony a few months later.[63] For the Diracs, the prospect of spending their final years near Mary was attractive, and the warm climate would be good for the worsening arthritis in Manci's hands, but Dirac wanted to delay his decision until he could see how he coped with the fiercest of Tallahassee's heat and humidity and with the barking dogs that ruined his walks.[64] Swimming was now his favourite form of exercise so, in his spare hours, he visited the local lakes and sinkholes, usually taking a thermometer to check the temperature of the water. If it was above precisely sixty degrees Fahrenheit, he would dive in; if not, he would return home.[65]

In early January 1971, Florida State University formally offered Dirac the post of Visiting Eminent Professor, to be renewed annually.[66] The FBI had found no evidence that either Manci or Dirac was a subversive, so there was to be no official barrier to their emigration. After reflecting on the offer for five months, Dirac accepted and shortly afterwards returned briefly to Cambridge with Manci to pack up their belongings. During one of their conversations with the Blumenfelds, Helaine asked Dirac whether he was excited about moving to Tallahassee; he replied, gesturing to Manci, 'She is, that's why we're going. I would like to stay here.'[67]

Twenty-eight

Old men have a weakness for generality and a desire to see structures
whole. That is why old scientists so often become philosophers [. . .].
EUGENE WIGNER, *The Recollections of Eugene P. Wigner,* 1992

The advice Barbara Walters, doyenne of celebrity interviewers, gave
in her 1971 book *How to Talk to Practically Anybody about
Practically Anything* did not quite extend to making conversation
with Dirac. Yet the Director of Publicity at the Miami Museum of
Science, Dorothy Holcomb, wished she had read the book when she
was trying to wrest a few words from him during a buffet reception
in his honour on the evening of 8 March 1971.[1] After he replied to
her 'Hi!' with a blank 'Hello,' she realised that the only way to get
him to speak more than a few words at a time was to ask him to pick
the topic of conversation. He chose comic strips. For several minutes,
he talked with surprising fluency about the merits of two strips he
had been reading since the 1930s: the fifth-century adventurer
'Prince Valiant' and 'Blondie', a carefree flapper girl who settled
down to family life in suburbia. Holcomb was charmed. When Dirac
admitted that he could not make head or tail of the quirkier humour
of 'Peanuts', she suggested he should try a little harder to understand
American humour; he agreed. Afterwards, Holcomb made up her
mind to buy a copy of *The Principles of Quantum Mechanics* and
also of *How to Talk to Practically Anybody about Practically
Anything.* As Holcomb will have seen, if she got to the end of
Walters' book, it concludes with good advice for everyone who had
tried vainly to draw Dirac into conversation: 'You can't win 'em all.'[2]

Before this conversation, Dirac had given a lecture entitled
'Evolution of Our Understanding of Nature', which ranged well
beyond physics. Still haunted by the early scenes in *2001: A Space
Odyssey,* he began by discussing how early humans understood the
mechanics of growing grain, graduating from beliefs based on super-
stition to ideas based on theories grounded in observations. He
opposed critics of the Apollo space programme who believed that the
money should be spent instead on social programmes: 'People who

equate all the different kinds of human activity to money are taking too primitive a view of things.' The solution to social problems was not, he argued, to be cheese-paring with the space programme and fundamental research but to avoid 'the great waste that we see around us', especially the unemployment of people who want to work. Look at the hippies in California, he said: they welcome the challenge to help fight forest fires rather than just laze around.[3]

Dirac's reputation as a speaker enabled him and Manci to sate their appetite for international tourism.[4] Florida State gave him the freedom to travel and everything else he needed, in addition to a modest income: an office, companionship, financial support for his research and – most important – respect. The university officials treated him with a reverence that often cloyed into obsequiousness, and they regarded Manci as his queen. She whiled away hours chatting and exchanging risqué jokes with the university's clubbable President, Bernie Sliger, knowing that he would always take her phone calls and be sympathetic to her every request. In return, the university asked only that Dirac be available when they wanted to display their most illustrious professor to visiting dignitaries; he played along and had some success in disguising his boredom. Only once, when his compliance was taken for granted, did his patience run out: he locked himself in his house and Kurt Hofer had to persuade him to come out, just in time to meet an important visitor.[5]

Beyond the light supervision of a few graduate students, Dirac had no teaching responsibilities. But in 1973, he agreed to present a series of lectures on the general theory of relativity, aiming to develop the theory from its fundamental principles and to lay bare its logical structure. One of the physics students in the audience, Pam Houmère, recalls:

The first lecture was 'standing room only'. He began so simply that the office cleaners could have understood it: what is meant by position, what we mean by time, and so on. Later, he built on these foundations brick by brick, making every step of the construction look inevitable. The funny thing was, he never compared the theory with experiment, he just kept stressing how beautiful it was. Only a few students made it to the end of the course, but for those who did, it was an unforgettable experience.[6]

Dirac presented the lectures most years until 1980 and used them as the basis of his short book *General Theory of Relativity*, a minor classic of exposition, describing the theory in sixty-nine pages, without a single diagram.

In Tallahassee, the Diracs' home was about twenty minutes' leisurely walk from Dirac's office on the third floor of the university's Keen Building, in the heart of the campus. Each weekday morning after breakfast, he would link his hands behind his back and walk slowly to his office across a local field, the route that ensured minimum contact with the neighbourhood dogs. In summer, when he wore his baseball cap, he looked like an all-American retiree, but on the coldest days of winter, when he put on the heavy overcoat he had bought almost fifty years before in Lord and Taylor, he looked every inch the venerable English professor. He often carried a forty-year-old umbrella: 'It was my father's,' he told colleagues.[7]

In his office, he worked at his desk for three hours, pausing occasionally to visit the library. To unexpected visitors who knocked on his office door, he had a simple message: 'Go away.'[8] When the phone rang, he would often lift the receiver off the hook and immediately drop it, without bothering to listen to the caller's voice.[9] At noon, he would join a few colleagues for a brown-bag lunch. Dirac usually said nothing but would occasionally interject with a comment, perhaps on the impenetrability of American football or about the wisdom of trying to educate so many undergraduates in science when so few of them had an aptitude for the subject or even took much pleasure from studying it. He was fond of jokes, especially ones dependent on the interpretation of a single word and ones with a slight sexual edge. This was one of his favourites:

In a small village, a newly appointed priest decided to call on his parishioners. In one modest home, teeming with children, he was greeted by the lady of the house. He asked her how many children she and her husband had. 'Ten,' she replied. 'Five pairs of twins.' The priest asked, 'You always had twins?' to which the woman replied, 'No, Father, sometimes we had nothing.'[10]

After lunch, he would return to his office for a nap on his sofa and sometimes attend a seminar, often appearing to sleep through most of it, before returning home for late afternoon tea with Manci. After dinner, he would relax. He and Manci might go to a classical concert, or he might read a novel – Edgar Allen Poe mysteries, Le Carré spy thrillers and Hoyle's science-fiction stories were among his favourites

– or watch television with Manci in the family room, dominated by a painting of Judy when she was a child.[11] Dirac watched most of the *Nova* science documentaries, but the programmes that he and Manci regarded as unmissable were period dramas: *The Forsyte Saga* – Dirac was spellbound by the leading lady, Nyree Dawn Porter – and *Upstairs, Downstairs*, dramatising the class divisions between the servants and their masters in an Edwardian household. On the night an episode of the programme was broadcast, the Diracs would accept dinner invitations with friends only if their hosts agreed in advance to watch it with them in silence. One dispute about the evening television schedule threatened to get out of hand, when there was a clash between Cher's Sunday-night television show – a highlight of Dirac's week – and the live broadcast of the Oscar ceremony, which Manci was desperate to see. The dispute was resolved several days later, but at a price: they bought a second television.[12]

The couple did not always resolve their differences so amicably. In August 1972, they had what may have been the worst row of their marriage, when they were visiting the recently widowed Betty at her apartment in Alicante, on the south-east coast of Spain. The relationship between the sisters-in-law had long been brittle: part of the problem was that Manci made no secret that she found Betty dull and idle, while Betty was vexed by Manci's unrelenting bossiness. Tempers flared during a conversation on the apartment balcony when Dirac backed up his sister after she made a sly comment about the behaviour of Hungarians in Budapest at the end of the war. Manci stormed out of town and wrote to Dirac in a rage:

You looked at me, then did all you could to hurt, scare & humiliate me, & embarrass me greatly [. . .] It is a fact that most mental inmates have been driven there by their families. On that 5th floor balcony I felt your presence whenever I was there alone, urging me to jump [. . .] You cruelly, unjustly uncaringly completely identified yourself with my tormentor, and this I did not earn or deserve. I do not feel you are a husband as it is understood by millions. Yes, keep your loyalties to the one so similar to you in lacking human emotions, & I learn not to care or want to die.[13]

A few days later, she wrote to him again, in a rather different tone:

Thank you for your loving care. For your love, warm & affectionate. For your taking notice when sick or in pain. For heeding for needs I have. For allowing me to read your wishes from unspoken words. For allowing me near you when ill or depressed. For forgiving my ills and extravagances. For never making me anxious and panicky. For treating me as an equal: always justly & fairly. For trying your best to make us around you happy and cheerful. I thank you.[14]

In Trieste a month later, at a symposium organised by Abdus Salam to mark Dirac's seventieth birthday, Heisenberg and all the other guests saw the Diracs on their best form, the model of the contented elderly couple. But Dirac apparently did not want to put the unpleasantness of the previous few weeks completely behind him: he clipped Manci's two notes together and filed them among the papers in his office. He appeared to regard all her attacks – and the makings-up that always followed – with an equanimity bordering on indifference; whether he suffered more deeply than others saw we shall probably never know, as he appears not to have discussed her behaviour – still less to have complained about her – with anyone.

To the Diracs' acquaintances in their later years, Manci was a controversial figure. No one questioned that her gift for friendship hugely enriched his social life and that she was devoted to her husband, 'my little Mickey Mouse'. Many colleagues attest to the care she took to look after him and make him look presentable; one visitor was touched to see her adjusting his clothes when he returned home one evening looking like a scarecrow. 'She takes such *good* care of me,' Dirac beamed as Manci adjusted his tie.[15] Without her, he would probably have spent almost his entire adult life living alone in college, like Charlie Broad.

Yet many friends could not help flinching when she shouted at him, 'Are you listening to me?' and wondered how he felt when he silently bore her tirades against 'nigger' doctors and Jews (that Manci was both Jewish and occasionally anti-Semitic was one of the most baffling paradoxes of her personality).[16] Yorrick Blumenfeld gives a bleak summary of the state of their thirty-four-year-old marriage: 'She was tired of hen-pecking him, and he just wanted to live in his dream world.' Helaine Blumenfeld is surprised that he could tolerate her: 'He was a lovely man. She was simply an awful person.'[17] But Lily Harish-Chandra, a frequent visitor to the Diracs' home and a family friend, disagrees: 'Manci was extremely warm and loyal, a great listener and a very caring woman. Paul cannot have been easy to live with. Their marriage worked because they gave each other what they wanted: he gave her status and she gave him a life.'[18]

*

In the early 1970s, Dirac was briefly optimistic about his research on particle physics. He had happened on a way of describing isolated elementary particles with a spin equal to a whole number, using an equation that he believed had a special mathematical beauty. Better yet, it described only *positive* energies – the mathematics yielded no embarrassing negative-energy solutions. But his excitement waned after he found it impossible to use the equation to describe how a particle interacts with other particles or with a field – the real-world case. Mathematical beauty had again proved a treacherous beacon.

Dirac then wound down his work on the theory of fundamental particles and returned to general relativity and his still-unproven large numbers hypothesis. He knew that Einstein's theory and the hypothesis were incompatible because general relativity requires – in the language of Newtonian mechanics – that the strength of the gravitational force between two identical masses separated by the same distance has always had the same value, contrary to the hypothesis. So he tried to reconcile them using ideas set out by a former colleague at the Institute for Advanced Study, the German mathematician Hermann Weyl, whose approach to theoretical physics resembled Dirac's. Weyl once said: 'My work always tried to unite the truth with the beautiful, but when I had to choose one or the other, I usually chose the beautiful.'[19] In 1922, Weyl had produced a prototype theory that gave a tantalising glimpse of how a mathematical account of gravity and electromagnetism might be given with a unified set of equations. Enthralled by their beauty, Dirac believed Weyl's approach might furnish a link between the general theory of relativity and the large numbers hypothesis, in a way that involved a gradual weakening of gravity over time.[20]

Dirac was assisted in the project by Leopold Halpern, a general relativity specialist who arrived in Tallahassee in 1974, a year short of his fiftieth birthday. Born and raised in Austria, he and his family had fled, on Hitler's invasion in 1938, when he was thirteen years old. He spent twenty-seven years working in several European research institutions, including a spell with Schrödinger, and Dirac had first met him at a conference in 1962. Halpern was a homoeopath and a certified African medicine man, a twenty-four-carat eccentric who slept outdoors all year round, sliced baked potatoes with karate chops and refused to wash with soap. He was not always popular in elevators. Colleagues with conventional manners were often disconcerted by the prickliness that disguised his shyness: when his phone rang, he would answer with a rasping, impatient 'Hello', his voice softening into a lilt the moment he realised that he was talking to a friend.

The oddities and coarse manners of Halpern grated on Manci, but they endeared him to Dirac, and the two men became close friends. At least once a week, they went swimming in Silver Lake and Lost Lake, two of Dirac's favourite spots near Tallahassee, mainly because the waters there were so quiet. Dirac did not like to swim anywhere near motorboats, but on one trip, when he was seventy-six years old, he hailed one and asked the owner if he could have a go at water-skiing. The owner obliged. When Halpern told Manci, she was horrified: 'Paul is still *very* immature!'[21]

Most weekends, the two men headed off in Halpern's Volkswagen Super Beetle – his sixteen-foot canoe and a pair of paddles tied to the roof rack – on the hour-long drive to the Wakulla river.[22] Minutes after setting off from the shore, they were alone in one of Florida's most pleasant microclimates, a near wilderness. They would row for some two hours upstream on the slow-flowing river, through forests of sassafras and American beech trees, draped with Spanish moss. The alligators made scarcely a sound: the silence was broken only by the rhythmic sloshing of the paddles, the cry of a circling osprey, the occasional shuffling of wind passing through shoreline gaps in the forest. After a snack lunch at Snake Point, Dirac and Halpern would strip off and go for a swim, before they rowed back to their starting point, scarcely exchanging a word. These were idyllic, private hours. Occasionally, they would invite a visitor to join them – but it had to be someone who could be relied on to stay silent most of the time. One of the visitors was Kurşunoğlu, who went along dressed in his three-piece suit, tie and Stetson. Halfway through the trip, he stood up in the canoe to admire the scenery only for Dirac to dump him in the river and then collapse in a fit of laughter.

Dirac and Halpern often arrived home several hours late, half-heartedly suppressing shame – like a pair of errant schoolboys – when they were explaining themselves to a frantic Manci. Halpern assured her week after week that the Wakulla wildlife posed no danger at all: 'If you leave the snakes and alligators alone, they will do

397

nothing to harm you.' Halpern could not understand what she was so worried about.[23]

In the 1970s, particle physics underwent what amounted to a revolution. After decades of uncertainty, physicists achieved a new clarity about the workings of the universe at the finest level: everything in the universe is made of a few basic particles – a handful of leptons and quarks and a small number of particles that mediate their interactions – and described by a quantum field theory simple enough to be spelt out on a T-shirt. The Dirac equation describes the electromagnetic interactions of all the leptons and quarks, each with the same spin as the electron.[24]

In the past fifty years, physicists had come up with quite a few attention-grabbing labels for their new concepts, but they allowed this description of weak, electromagnetic and strong interactions – one of the supreme syntheses of twentieth-century thought – to be given the most prosaic of names: the Standard Model. One of the first important steps towards the consensus was taken by Dirac's former student Abdus Salam and by the American theorist Steven Weinberg, who independently suggested in 1967 that the weak and electromagnetic interactions might be understood in a unified way, by describing them in terms of a special type of gauge theory whose underlying mathematical symmetry is broken.[25] For several years, the Weinberg–Salam theory was not taken seriously as it appeared to suffer an even more serious infestation of unwanted infinities than quantum electrodynamics, the theory of photons and electrons. All this changed in the early 1970s. After the Dutch theoreticians Gerard 't Hooft and Martin Veltman proved that the infinities in the theory – and in all other gauge theories – could be removed by renormalisation, the Weinberg–Salam theory quickly commanded wide interest and support.[26] Also at around this time, theorists improved their understanding of renormalisation so that it was much more rigorous than the 'sweeping under the carpet' dodge that Dirac deplored. Renormalisation was now widely accepted as a rigorous branch of mathematical physics, with no sleights of hand; Dirac vehemently disagreed.

Soon physicists found a gauge theory of strong interactions, called quantum chromodynamics, with the same underpinnings as the Weinberg–Salam theory. It turned out that it was possible to describe

the strong interaction between quarks, mediated by massless particles which Gell-Mann named gluons. Quarks are never observed in
isolation, the theory says, because the strong force prevents them
being separated, though when quarks are close together they behave
as if they were free. So the neutron, first observed by Chadwick just
over thirty years before, could be re-envisaged as a compassionate
prison for quarks – they cannot escape their confinement, but they
are free when inside.

Rutherford's vision of a typical atom as electrons orbiting a tiny
nucleus of protons and neutrons ('a gnat in the Albert Hall') had
been superseded. Now, the most fundamental way of imagining an
atom was in terms of relativistic quantum field theory: the quarks in
the nucleus were quantum excitations of the field associated with the
strong interaction, just as the orbiting electrons were the quantum
excitations of the electromagnetic field. Everything in an atom can be
described in terms of such fields. Rutherford would have choked on
such abstractions, yet they were the apparently inevitable consequence of a century of labour by his fellow experimenters and their
theoretical colleagues.

Although the Standard Model left many questions unanswered –
no one fully understood the particles' masses, for example – its setting out in the 1970s was a high point in the history of science. But
Dirac was unmoved: ensconced with Halpern in their Tallahassee
redoubt, the new discoveries left him cold, and he appeared to take
no great pleasure to see other theoreticians find a way of describing
strong interactions using a field theory of the type he had pioneered,
as scattering matrices fell into disuse. He no longer kept up to date
with the latest physics journals and was beginning to make errors in
his science, though no one was ungracious enough to say so in public.[27] By the mid-1970s, Dirac had lost interest in particle physics,
and Halpern noticed that he was less interested in news about field
theory than the renewed public debate about the origin of the Turin
Shroud, believed by some to be the burial shroud of Jesus Christ.[28]

Although Dirac was impressed by the best young particle theoreticians, he thought they were deluded. Through his talks and occasional publications, he urged them to devote all their time to clearing
and disinfecting the Augean stables of renormalisation, a job almost
all physicists believed had already been done.[29] By contrast,
Heisenberg in Munich kept an open mind about new theoretical

developments until liver cancer took his life in February 1976, six years after his former teacher and friend Max Born had died in Göttingen.[30] All Dirac's friends among the pioneers of quantum mechanics were now dead.

At one time, the historical perspective on atomic physics was not important to him, but now he was keen to put his side of the story to historians and other physicists. In these talks, he always took pains to emphasise the excitement of the early years of quantum mechanics – an emotion that, by all accounts, he rarely showed when he was living through them. He even included a reference to his feelings in the account that was the nearest he ever came to writing a scientific memoir: *Recollections of an Exciting Era.*[31]

In May 1980, while suffering from a bad dose of flu, Dirac travelled to Chicago to attend a conference on the history of particle physics at the Fermi National Accelerator Laboratory (Fermilab), where he spoke about the origins of quantum field theory. In a round-table session, he went out of his way to criticise the destructiveness of Pauli's opposition when the idea of spin and the positron were first conceived.[32] In another session, he presented his versions of the history of anti-matter in a talk that Leon Lederman recalled as 'quintessential Dirac' – clear, fluent and modest: 'the content poured out of him like heavy cream'.[33] When he had finished speaking, Vicki Weisskopf commented that Einstein had suggested the existence of a positive electron in 1925, some six years before Dirac's prediction.[34] But Dirac was unperturbed; he waved a hand dismissively, remarked, 'He was lucky,' and moved on. Even for Dirac, modesty had its limits.

Manci was a generous hostess, able to make everyone in the room feel special and at ease. She often threw dinner parties, attentively filling her guests' glasses, serving generous portions of her favourite dishes, ensuring that the conversation kept moving. Dirac, sitting at the head of the table, would apparently spend most of the evening asleep. He could, however, be drawn into conversation if he were approached by a young woman, especially if she was friendly and attractive.[35] His advice was often sought but he usually declined to offer any; however, when pressed, he would sometimes offer a few words. One of his favourite replies was: 'Think about yourself first. If nobody gets hurt, do it' – a slightly egoistic summary of the view of the individual's moral responsibility in the opinion of John Stuart Mill.[36]

Manci would point out to guests a favourite photograph of Dirac warmly shaking hands with Pope John Paul II in the Vatican. 'Paul and the Pope get on so well,' Manci would beam, as if the two men met every weekend for a round of golf.[37] The photograph was taken at one of several meetings between Dirac and the Pope at the Papal Academy, a group of distinguished scientists that offers the Pope disinterested scientific advice. Dirac had been elected to the Academy in 1961, the year after his friend the cosmologist George Lemaître became President. The Diracs' friend Kurt Hofer recalls Manci's pride in her husband: 'After showing guests the papal photograph, she unpacked a collection of postal stamps from all over the world, each bearing a portrait of Paul. He pretended to be embarrassed, but he never did anything to prevent her.'[38]

It was during one of Hofer's weekly visits to 223 Chapel Drive that Dirac unexpectedly disgorged his recollections of his father. Dirac trusted only his closest friends with these unexpurgated memories, although the circumstances of Felix's death were still too distressing for him to share with anyone, even with Manci.[39] Dirac did, however, speak of his happiest memories of Felix's life to Betty in October 1969 when she was in an Amsterdam hospital, lying in a coma after a stroke and a seven-hour brain operation.[40] Alone at her bedside, he tried to coax her back into consciousness by telling her stories of their childhood – playing on the Downs with Felix, the three of them bathing on Portishead beach, sharing each other's books and comics. She regained consciousness a few weeks later and gradually made a partial recovery.

Hofer recalls that Dirac thought organised religions were primitive and socially manipulative 'myths'. Once, as he walked past a local Mormon church with a huge satellite dish, he scoffed that the church needed such a large dish 'so that it can communicate directly with God'.[41] Yet Dirac was now much more willing to introduce the concept of God into discussions about science. In June 1971, he had startled his audience at the Lindau meeting by considering 'Is there a God?' to be one of the five most important questions in contemporary physics. He said it would be useful to approach the question scientifically:

A physicist would need to make this question precise by understanding what is meant by a universe with a God and what is a universe without a God, having a clear distinction between the two types of universes, and then looking at the actual universe and seeing which class it belongs to.[42]

The audience laughed nervously and went quiet when he suggested a way of detecting the presence of a God. If future scientists demonstrated that the creation of life is overwhelmingly unlikely, then, in his opinion, this would be evidence for the existence of God. Until that time, the hypothesis must be regarded as unproven.[43] Dirac was taken to task by the press for these speculations but he was not to be deflected and often returned to the topic, in public and private. He took a dim view of any religion declaring itself to offer the only hope of salvation, Hofer remembers: 'Paul believed it was the height of arrogance for any group of people to claim that they alone know the truth. He often pointed out there are hundreds of religions on this planet and that it is impossible to know which one, if any, is correct.'[44]

There was 'no trace of religiosity' in Dirac, Halpern later wrote. He remembered that Dirac was especially critical of Catholicism and other religions that acknowledged miracles, because, in his view, the existence of a miracle implies a temporary breaking of the underlying laws of nature, whose beauty he regarded as sacred.[45] Like Einstein, and largely following the philosopher Spinoza, Dirac appeared to take the pantheist view that the universe is either identical with God or in some way an expression of God's nature, a view that – though vague almost to the point of tautology – appears to rule out the notion of a God that can influence human affairs. Dirac's pantheism was an aesthetic faith: that observations on nature at the most fundamental level will be described perfectly by theories whose mathematical beauty is also perfect. If he had a religion, this was it.

Dirac's modesty was genuine, but he was not above a little vanity. The Danish sculptor Harald Isenstein, a specialist at portraying leading physicists, made two busts of Dirac, and both are good likenesses, if lacking in character: the first in 1939, which Dirac displayed in his home, the next thirty-two years later.[46] He offered the first Isenstein bust to St John's College, who accepted it and displayed it in their library, where it stands today. The college also wanted a painting of Dirac in oils to be displayed in their Hall, and Dirac went out of his way to oblige.[47] In the early summer of 1978, he sat several times for Michael Noakes, portrait painter of the British royal family and, the year before, of Frank Sinatra.[48] In the first session, Noakes tried to help Dirac relax by drawing him into conversation:

NOAKES: Can you put into layman's terms what you're working on, Professor?
DIRAC: Yes. Creation.
NOAKES: Wow! Tell me more.
DIRAC: Creation was one vast bang. Talk of a steady state is nonsense.
NOAKES: But if nothing existed beforehand what was there to bang?
DIRAC: That is not a meaningful question.

Dirac would say no more. Though unsettled by Dirac's reticence and apparent lack of interest, Noakes captured his abstracted gaze to infinity, Dirac looking as innocent as a five-year-old, as detached as an oracle.[49] A comparison between this portrait and the first to be painted – by his friend Yakov Frenkel in 1933, shortly after they heard of Ehrenfest's suicide – shows how much Dirac's confidence had drained away in the ensuing forty-five years. His personality is perhaps best caught in the drawing made in 1963 by Robert Tollast, whose portrait expertly catches Dirac's childlike innocence. Less accomplished, but nevertheless competent, is the drawing of Dirac made two years later by Feynman, whose portrait shows signs of reverence ('I'm no Dirac,' Feynman often said).[50] Dirac kept his drawing in his filing cabinet.

Twenty years after Dirac declined a knighthood, he accepted the most prestigious honour of all, membership of the Order of Merit, which did not oblige him to call himself anything other than 'Mr Dirac'.[51] The order is limited to twenty-four members of the British Commonwealth judged by the sovereign to have given exceptional service (previous members had included Florence Nightingale, Winston Churchill and William Walton). Manci deplored that her husband was the last of his generation of Cambridge scientists to be honoured – J. J. Thomson, Eddington, Rutherford, Cockcroft and Blackett had been admitted long before.[52]

In June 1973, the Diracs returned to the UK so that he could collect his award. A chauffeur drove them in a Rolls Royce to

Buckingham Palace, where he received the award in private from the Queen for a few minutes, while Manci waited in an ante-room. A few weeks later, he shared with Esther and Myer Salaman his discussions with the Queen about the challenges faced by a female scientist who is also the mother of young children:

I said it was difficult for a woman who had to choose between her career and her family and there could not be real equality between the sexes. The Queen said she did not press for equality of the sexes.[53]

On his return to the USA, Tallahassee colleagues quizzed Dirac about his impression of the Queen, but he would say very little. His description of her consisted of two words: 'Very small.'[54]

That summer, Dirac visited CERN in Geneva to see its newest particle accelerator, capable of increasing the energy of protons to some fifty thousand times the energy reached by Cockcroft and Walton's device. During his visit, he walked to the rue Winkelried, a side street near the lake and close to the main railway station, to see the apartment owned until the mid-1920s by his paternal grandmother, where he and his family stayed in 1905. As he strolled around the nearby statue of Rousseau, Dirac may have thought of the time he spent running around in the lakeside park with Felix, watched by his father and mother, baby Betty in her arms. Dirac had not visited Switzerland since then, despite many invitations. The pain of the country's association with his father had been so deep that Dirac had not been able to bring himself to visit it until he was seventy years old.

In 1979, the centenary of Einstein's birth, Dirac was feeling weak and listless. But he was determined to speak at as many of the celebratory meetings as he could, so that he could 'make clear what a great scientist Einstein was', as Halpern recalled.[55] During that year, Dirac achieved one of his ambitions – of flying across the Atlantic on Concorde, the first supersonic passenger aircraft. The aircraft, developed by an Anglo-French collaboration in the 1960s, was noisy, a prodigious guzzler of fuel and hopelessly uneconomic, but it symbolised the best and most exciting in contemporary engineering. It was also the apogee of the aviation industry in Dirac's native city: the Bristol Aeroplane Company had led the first British design team to work on the aircraft and build the first British prototype in Filton, a few miles from Julius Road.[56]

Somehow, Manci persuaded UNESCO to fund transatlantic flights on the aircraft for Dirac and herself as a condition of his attending the organisation's Einstein celebration in Paris, as guest of honour.[57] He and Manci took the flight on 5 May 1979, cruising at almost 60,000 feet – the nearest he would ever get to outer space. During the flight, he probably read on the front page of the *New York Times* the news from Britain that Margaret Thatcher had just become Prime Minister.[58] He may have wondered whether his mother's fears about the notion of a woman prime minister would be realised, whether Mrs Thatcher would, in Flo's words, 'vacillate in her feminine way' so that 'her supporters would fall off right and left'.[59]

By spring 1982, when Dirac and Kapitza were tired of travel, three opportunities to meet that summer arose, and they seized them.[60] Accompanied by their wives, they met first at the Lindau meeting at the end of June. Kapitza had been eligible to attend the meeting only since he received the Nobel Prize for physics in 1978, after Dirac had lobbied for him for almost forty years. During that time, Dirac had seen the honour awarded to almost all of Rutherford's most able 'boys' – Blackett, Chadwick, Cockcroft and Walton – and virtually all the pioneers of quantum mechanics from the 1920s and 1930s had received the prize, including Born, Fermi, Landau, Pauli, Tamm and Van Vleck, but not Jordan, whose Nazi past probably cost him the honour.

At the Lindau meeting, Dirac mounted one of his last attacks on renormalisation in front of an audience of some two hundred students and Nobel laureates.[61] Looking as fragile as a cut-glass figurine, Dirac stood at the rostrum giving a speech almost identical to ones he had been giving for almost fifty years; he had no praise for the Standard Model or any other successes of particle physics. A microphone amplified his trembling voice, each letter 's' accompanied by a whistle from his ill-fitting dentures. Current theories were 'just a set of working rules', he said; physicists should go back to basics and find a Hamiltonian description of nature free of infinities. '*Some* day', he said with a gentle and weary defiance, 'people will find the correct Hamiltonian.' But he was preaching a lost cause: physicists no longer based their descriptions of fundamental particles on Hamiltonians, as other methods were much more convenient. But the audience listened respectfully to Dirac's twenty-five-minute speech, partly, perhaps, in anticipated sadness that his lone voice would soon be silent. Here was someone, like Einstein, who was unafraid of bucking contemporary trends and taking the consequences, to be his own man.

The Diracs and Kapitzas met again a few days later in Göttingen. Kapitza had pleasant recollections of the town, as did Dirac – it was, in his opinion, the birthplace of quantum mechanics, where he had first become acquainted with Born and his group, where he became friends with Oppenheimer and probably where he first saw a Nazi in uniform. The Diracs stayed in Gebhard's Hotel overlooking Göttingen railway station, where Dirac had first arrived in the town from Copenhagen fifty-five years before.[62] Then, his journey from the station to his room in the Carios' home was a luggage-laden walk; now, he and Manci were met by a welcoming party that whisked them in a taxi to the town's most luxurious accommodation.

There are photographs of Kapitza and Dirac sitting at a table in the garden of the hotel, looking exhausted and a little dispirited. Physics, once one of their main topics of conversation, was now much less important than international affairs, the preoccupation of Kapitza. He will almost certainly have spoken with Dirac about the recently ended Falklands War between Argentina, led by General Galtieri, and the United Kingdom, led by Mrs Thatcher, over the disputed island territory in the South Atlantic. Dirac was in two minds about Thatcher: he feared the impact of her radicalism on British education and science but sympathised with her determination to protect the Falkland Islanders' wish to remain British. He thought, however, that the dispute should have been resolved through negotiation: at the beginning of the war, it had seemed absurd to him that the number of people likely to die would exceed the number whose British citizenship would be protected.[63] In politics, if not in physics, Dirac was now a pragmatist.

The Falklands War was a trivial matter compared with nuclear proliferation, a subject Dirac and Kapitza talked about at length when they met again a few weeks later, at the Erice summer school in Sicily, organised by the physicist Antonino Zichichi. Dirac took risks in the subject matter he addressed there: during the previous summer, he had given a presentation on 'The Futility of War', an uncomplicated statement of an argument that few would oppose.[64] In the summer of 1982, he collaborated with Kapitza and Zichichi to write the one-page 'Erice Statement', which urged governments to be less secretive in defence matters (one of Bohr's favourite themes), to prevent the spread of nuclear weapons and to help non-nuclear powers feel more secure.[65] The well-intended phrasing of the document, later signed by ten thousand scientists, was so bland that its first signatories at the Erice meeting included not only opponents of nuclear weapons but also the right-wing Eugene Wigner and the obdurately pro-nuclear Edward Teller, who had done more than almost any other American to fuel the arms race.

On the last stages of the Diracs' 1982 European tour, they visited Betty in Amsterdam and Gabriel in Aarhus, before travelling to Cambridge. Dirac returned to St John's College, which, as he was to tell the Master soon afterwards, 'has been the central point of my life and a home to me'.[66] That summer, the talk of the Combination Room was the imminent arrival of the college's first women undergraduates: another all-male bastion of Cambridge was about to fall. Earlier, the theoretical physicist Peter Goddard asked Dirac whether he thought women students should be admitted to the college, and, after a long pause, Dirac replied, 'Yes, provided we don't admit fewer men.'[67]

Before he left St John's, Dirac left his gown at the Porters' Lodge, where he had first registered as a student almost sixty-nine years before. He wrote a label: 'Professor Dirac's Gown. Please take it to the Master and ask him to keep it until the next time I come to Cambridge.' But he would not see the town again.

Twenty-nine

I bade, because the wick and oil are spent
And frozen are the channels of the blood,
My discontented heart to draw content
From beauty that is cast out of a mould
In bronze, or that in dazzling marble appears,
Appears, but when we have gone is gone again,
Being more indifferent to our solitude
Then 'twere an apparition. O heart, we are old;
The living beauty is for younger men:
We cannot pay its tribute of wild tears.
W. B. YEATS, 'The Living Beauty', 1919

The confidence Dirac displayed when he spoke about physics hid a despair that he apparently revealed only once, to someone he hardly knew – Pierre Ramond, a theoretical physicist at the University of Florida in Gainesville.[1] A courteous and articulate man, Ramond is an American who speaks in a richly musical voice whose accent is a constant reminder to his listeners that he was born and raised in France. After lunch one Wednesday in the early spring of 1983, he drove from Gainesville to give a colloquium at Florida State University, hoping that his 'hero and guiding light' Dirac would be in the audience. Sure enough, when Ramond arrived in the seventh-floor seminar room, overlooking the campus, he saw in his audience the daydreaming figure of Dirac, slight as a pixie.

In his presentation, speculative but assured, Ramond discussed the possibility of setting out fundamental theories not in the usual four dimensions of conventional space-time but in a higher number of dimensions.[2] Throughout, Dirac appeared to be snoozing, and, during the questions afterwards, he said nothing. But when the seminar broke up, he – unusually – lingered until he was with the speaker, alone, and the door was shut.

Ramond had met Dirac twice before, but had not been able to draw him into anything resembling a normal conversation. 'I had heard that the only way to persuade Dirac to talk was to ask him a non-trivial question that required a direct answer,' Ramond recalls.

So he asked Dirac directly whether it would be a good idea to explore high-dimensional field theories, like the ones he had presented in his lecture. Ramond braced himself for a long pause, but Dirac shot back with an emphatic 'No!' and stared anxiously into the distance. Neither man moved, neither sought eye contact; they both froze in a silent stand-off. It lasted several minutes. Dirac broke it when he volunteered a concession: 'It *might* be useful to study higher dimensions if we're led to them by beautiful mathematics.' Encouraged, Ramond saw an opportunity: doing his best to sound understanding, he invited Dirac to give a talk on his ideas at Gainesville any time he liked, adding that he would be glad to drive him there and back. Dirac responded instantly: 'No! I have nothing to talk about. My life has been a failure!'

Ramond would have been less stunned if Dirac had smashed him over the head with a baseball bat. Dirac explained himself without emotion: quantum mechanics, once so promising to him, had ended up unable even to give a proper account of something as simple as an electron interacting with a photon – the calculations ended up with meaningless results, full of infinities. Apparently on autopilot, he continued with the same polemic against renormalisation he had been delivering for some forty years. Ramond was too shocked to listen with any concentration. He waited until Dirac had finished and gone quiet before pointing out that there already existed crude versions of theories that appeared to be free of infinities. But Dirac was not interested: disillusion had crushed his pride and spirit.

Dirac said goodbye and walked off, looking impassive, but Ramond was shattered. He took the elevator to the ground floor and walked alone in the fading light of the afternoon back to his car. Twenty-five years later, he could still recall how upset he was: 'I could hardly believe that such a great man could look back on his life as a failure. What did that say about the rest of us?'

Ramond cannot recall whether he had explicitly mentioned to Dirac the idea that nature is fundamentally built not of point-like particles but of tiny pieces of string. In the late 1970s and early 1980s, Ramond was one of the small band working on the idea, then a backwater of theoretical physics. Dirac had tentatively suggested in 1955 that electrons and other quanta might be pictured as lines rather than points, but the mathematical form of Dirac's idea was completely different to that of the modern string theory, itself still only embryonic. The theory had, however, used contributions Dirac had made in the late 1950s and early 1960s, including his methods of describing two- and three-dimensional objects in ways consistent with both quantum mechanics and the special theory of relativity. The mathematics he used to describe a small sphere – his model of a muon – resurfaced in a different context, to describe the motion of a string moving through space and time.

Among the encouraging features of the new string theory was a pleasing absence of the infinities of conventional field theories, such as quantum electrodynamics, the best-available description of electrons and photons. Most impressive was that string theory made the existence of gravity inevitable: if the theory is correct, gravity *must* exist. Although there was no experimental evidence to favour string theory over other field theories, to its supporters it looked too beautiful to be entirely wrong. Dirac will have heard about the theory in seminars at Florida State but he gave it no credence – his curiosity was spent. A few months after his eightieth birthday, the local journalist Andy Lindstrom had found him 'a painfully spare man [. . .] stoop-shouldered and frail'. His once-black hair had 'retreated to a wispy cowl at the very fringes of his forehead, as though worn away by the great thoughts fermenting below . . . A web of wrinkles etches his gentle, lonely face, outlining eyes that seem to be forever questing.'[3]

Since overcoming his digestive trouble in late 1980, Dirac had become more relaxed about his health, but his anxieties returned three years later when he started to suffer from apparently unrelated problems – night sweats and occasional fevers. He consulted Hansell Watt, a local doctor and lay preacher whose calm, comforting words were all the more reassuring for being spoken with a rich southern drawl. Dirac took to him, and, for Manci, he could do no wrong. Watt diagnosed the source of Dirac's medical problems to be his right kidney, which X-ray photographs showed to have been infected by tuberculosis, probably when he was a child. This was a surprise to Dirac, who had never suspected that he had been infected, having been assured by his mother: 'T.B. runs in families and it is absolutely not in ours.'[4]

When Dr Watt advised Dirac that his tubercular kidney should be removed, Halpern was outraged.[5] Wary of surgical cures and wanting only to try herbal remedies, Halpern opposed Watt's strategy and – to Manci's anger – did all he could to undermine it. Manci, fighting Halpern's influence over Dirac like a tigress guarding her wounded cub, did not tell him when she arranged the operation at the Tallahassee Memorial Hospital on 29 June 1983, a month after what would be his final talk.[6] The surgeon found that Dirac had only the last remains of a right kidney with a cyst the size of a hockey ball.[7]

The operation was technically successful but it left Dirac an invalid. Weak and dispirited, he spent the summer recuperating at home, watching television and playing Wei Chi and other board games but unable to do serious work. After several weeks, he could walk a few steps but did not have the strength to venture out of his air-conditioned home into the heat and humidity outside. For the first time in decades, he could not spend the summer walking in the countryside – especially cruel for someone who had trodden a distance comparable with Wordsworth's total of about 180,000 miles.[8] One of Dirac's most frequent visitors was Halpern, who sat at his bedside several times a week, chatting about their work and anything else that took their fancy, including politics. Dirac said that he could not help liking President Reagan, though he disagreed with most of his policies; at heart, Dirac remained a liberal, though with no loyalty to the Democrats or any other political grouping.

Halpern's relationship with Manci became more fraught by the week. Upset by what he saw as her unending nagging, he often found himself leaving the Diracs' home red-faced and purse-lipped with anger. Whenever Dirac mentioned his discomfort at Tallahassee's oppressive summer climate, she would shoot back with her favourite rejoinder, 'It's better than Cambridge,' Halpern recalled.[9] For her part, Manci thought Halpern was a rude, interfering busybody who was shamelessly taking advantage of his helpless friend by foisting quack medicine on him. Aware of her hostility, Halpern decided that subterfuge was the only hope. When Manci was out shopping, he instituted a secret programme of homoeopathic treatment, furtively dropping herbal essences into Dirac's drinking water when the nurse was not looking.[10] According to Halpern, Dirac's energy resurged like Popeye's after he had downed a can of spinach. As soon as Manci found out about 'the herbal conspiracy', she returned Dirac to his usual diet, whereupon he slipped back into lethargy and indifference, if Halpern's testimony is correct.

Dirac spent most of his waking hours in a wheelchair, talking to visitors, including his daughter Mary and her dashing new husband, Peter Tilley. After a few months, Dirac was fit enough to return occasionally to his office in Florida State University, to supervise his final graduate student Bruce Hellman and to oversee what would be his final publication. Halpern drafted the text of 'The Inadequacies of Quantum Field Theory' for Dirac, who wanted his final published words to execrate renormalisation, the technique born of one of his most profound contributions to science.[11] For the last time, he refused to accept that, as Feynman had advised him in 1946, he was on 'the wrong track'. Feynman might as well have counselled a train to depart from its rails.

Early in April 1984, Dirac heard that Kapitza was dead. The Soviet Union knew it had lost one of its most loyal subjects: the entire Politburo and many of the country's scientific leaders signed *Pravda*'s announcement of his death. Dirac had lost his dearest friend, his surrogate brother, but he showed only resignation. More sad news followed a few weeks later: the Diracs' son Gabriel had a skin cancer so aggressive that his doctors gave him only a few months to live. In June, Manci flew to Europe to see her son, leaving Dirac in the care of friends. A few weeks after her return, Gabriel died on 20 July, aged fifty-nine. Three days later, Dirac was too ill to put himself to bed.[12] Halpern was away in Europe, so Manci had her husband to herself and had to cope with his sinking morale and hardening stubbornness.[13] Dirac's spirits rallied during a visit by Gabriel's daughter Barbara, a radiantly attractive young woman and a special favourite of the Diracs. ('You look like Cher,' he told her a few years before.[14]) In sharp contrast to Halpern, Barbara's view of Manci was that she was a sensitive and humane nurse – there were occasional quarrels between her and Dirac but they would dissolve swiftly into an affectionate holding of hands. Dirac's energy had all but ebbed away, Barbara observed, but his love of physics still flickered: he returned to his papers and whispered resolutely, 'I have work to do.'[15] His greatest fear, of losing his mind, was never realised.

At the beginning of October 1984, after Barbara had returned to Europe, Manci hired nurses to be with Dirac round the clock: he was hanging on to the last thread of life. But he still received the occasional visitor, including Mary's husband Peter Tilley, who sat for hours at Dirac's beside, mostly in silence. During his final visit, Tilley recalls, Dirac leant over to him and said firmly, in a matter-of-fact tone: 'The biggest mistake of my life was marrying a woman who wanted to get out of the house.'[16] Dirac sounded neither bitter nor regretful, Tilley remembers, but was making a factual statement in a way that invited no further discussion. Perhaps Dirac was thinking of what Manci had said to him soon after they met – that she had married her first husband only to get out of her parents' home – and of the veiled warnings his mother had given him about marrying Manci forty-seven years before.

The battle of wills between Manci and Halpern resumed. When he knew she was out, Halpern sneaked into the house and stirred his fortifying herbs into Dirac's drinking water. The nurse had almost given up trying to interest him in food, and it was left to Halpern to feed his friend, who took his food like a baby. All Dirac wanted to do was to talk about Kapitza. Dirac spent many of his last conscious hours recounting favourite stories about his friend's colourful life – over and over again, Dirac told the story of how Kapitza refused to work on the bomb, standing alone among lesser mortals who did not have the moral courage to make a stand. It was a tape loop of delusion.

On Thursday 18 October, as Halpern was leaving the Diracs' home, he bumped into Manci. He was expecting a telling off for visiting his friend, but Manci did not mention it; she told him calmly that she had just been to the mortician to reserve Dirac's grave. But the next day Halpern received the phone call he had feared for weeks: Manci forbade him from setting foot in the house again – Dr Watt had told her, she said, that Dirac was too weak to see anyone except close family. Angry, bitter and tearful, Halpern heard nothing until four days later, when he read on the front page of the *Tallahassee Democrat*: 'FSU physicist is dead at 82'. On the Saturday evening, with Manci and his nurse at his side, Dirac's heart had failed and stopped beating at five minutes before eleven.[17]

'I want to be put down like a horse,' Manci told Dr Watt. But in public she showed her usual spirit and fortitude, informing friends and relatives of Dirac's passing with business-like calm and attending to every detail of the funeral arrangements.[18] She took great pains to ensure that Dirac was remembered as she wanted: the day after his death, she told friends that he was 'a very religious man' and that he would have wanted a high-Episcopalian funeral.[19]

The ceremony took place in the open air at the Roselawn cemetery in Tallahassee, on 24 October, under an overcast sky, rain threatening. When the guests arrived, shortly before 11 a.m., they saw Dirac's coffin was on a plinth beside his freshly dug grave, under a bright blue marquee-like roof mounted on four wooden poles, in the shade of a group of conifers swaying slightly in the breeze. Among the mourners was Dirac's one-time confidant, Pierre Ramond, who was surprised when he saw the congregation: 'Considering how famous he was, there were very few people there.'[20] There were about ninety mourners, including dozens from Florida State University but – as Manci bitterly noted – no one from Cambridge. Several in the congregation were uneasy to see that they were not alone: they had been joined by scribbling journalists and a flotilla of television crews. Manci had decided that her husband should be buried under the encircling gaze of TV cameras.[21]

The rector Dr W. Robert Abstein read slowly from the oldest-surviving version of the Anglican Bible, the text Manci had insisted on. She had forbidden Halpern to speak, and there was no eulogy. After half an hour, as the sky brightened, Abstein crumbled soil on the coffin and traced the sign of the cross in the dirt. The place of Dirac's burial was marked a few weeks later with a neat white-marble stone, engraved with words he had used, chosen by Manci: 'because God said it should be so'.

A few days after Dirac's funeral, Manci had to take another blow. She heard from the police in Vermont that they now presumed that Judy was dead and that they had called off the hunt for her.[22] The pain for Manci was terrible: in just four months, she had suffered the grief of losing her best friend in Russia, two of her children and her husband. Life seemed to hold little for her – but she was a fighter.

'Dirac was a militant atheist,' objected the Dean of Westminster, Edward Carpenter, when he was asked if Dirac might be commemorated in the Abbey's science corner.[23] The Oxford physicist Dick Dalitz led a group of scientists that began to press for Dirac to be remembered alongside Newton and Rutherford. For someone to be worthy of a place in such company, the Abbey authorities had to be sure that he or she was a Christian – or at least not inimical to religion – and was judged, after a decade's reflection, to be of 'millennial significance'.[24] Carpenter was easily persuaded of Dirac's status, but Dalitz found it hard to demonstrate that Dirac passed the religious test, especially after the Dean found out about Pauli's comment 'There is no God and Dirac is his prophet.' Pauli could make things difficult for Dirac even when they were both dead.

During the stalemate, Dalitz found an unanswerable way to counter the objection: if Dirac's parents had christened him, then – regardless of any derisive comments he had made about religion – he was officially a Christian.[25] Dirac would have been amused by the absurdity. In the late 1980s, Dalitz spent weeks trawling through parish records in Bristol but could find no evidence that Charles and Flo Dirac had christened their children, and this line of investigation drew a blank. However, the church authorities were impressed to hear that Dirac was a member of the Papal Academy and that he had made no antireligious comments during its meetings. Dalitz and his colleagues kept up their pressure on the authorities, and, in early 1990, after six years of lobbying, the new Dean of Westminster declared himself 'very sympathetic' to their cause. It was finally won in early 1995.[26]

The commemoration took place in Westminster Abbey on Monday, 13 November 1995, beginning with Evensong at 5 p.m. Though much less well publicised, the ceremony was on a scale as grand as Rutherford's fifty-eight years before: the Abbey looked gorgeous, the choir sounded magnificent, and the congregation was in good voice. After tributes to Dirac's scientific work had been read, the mathematician Sir Michael Atiyah, President of the Royal Society, unveiled the commemorative stone in the nave of the Abbey, next to Newton's gravestone and just a few paces from Darwin's. Stonemasons in Cambridge had used a piece of Burlington Green slate quarried from the Lake District to produce a two-foot square slab of stone and etch into it the inscription 'P. A. M. Dirac OM physicist 1902–84', with a statement of his equation.[27]

Stephen Hawking gave the final address, using his voice synthesiser to speak through the Abbey's antiquated public-address system.[28] He began with his usual arresting clarity and humour:

It has taken eleven years for the nation to recognise that he was probably the greatest British theoretical physicist since Newton, and belatedly to erect a plaque to him in Westminster Abbey. It is my task to explain why. That is, why he was so great, not why it took so long.[29]

His final words consisted of another barb: 'It is just a scandal that it has taken so long.' Dalitz threw anxious glances at his fellow organisers; evidently, Hawking did not know that at least a decade had to elapse after a subject's death before he or she could be commemorated – Dirac's ceremony was at most only a year late.[30] Afterwards, Dalitz sought out the Abbey authorities and apologised.[31]

After the organist had played Bach's *Prelude and Fugue in A Major*, Dirac's daughter Monica and her two children laid flowers on the memorial plaque, before the congregation sang the hymn 'Lord of Beauty, Thine the Splendour'. The music had been well chosen.

Angry that Westminster Abbey had questioned Dirac's suitability for commemoration, Manci did not attend the ceremony: 'The English are hypocrites,' she fumed. 'Lord Byron is buried in the Abbey, [and] he was the greatest rogue of the century.'[32] After Dirac's death, Manci become the keeper of his flame, firing off affronted notes to obituarists and chroniclers of her husband's life who cast any doubt on her view that he was a scientific saint.[33] Abraham Pais was startled when he received a letter from her, insisting that Dirac was not an atheist. 'Many times did we kneel side by side in Chapel, praying. We all know, he was no hypocrite.'[34] Friends of Dirac, certain that he was agnostic, were puzzled: did he join her at prayer out of politeness? Or had Dirac privately practised a religion he had mocked among friends? Or was Manci fantasising?

After she had come to terms with Dirac's death, Manci remained lively and active for ten years, travelling in Europe and the USA, and entertaining an almost unbroken stream of guests, including Lily Harish-Chandra, Leon Lederman and his wife Ellen, and Wigner's daughter Erika Zimmermann.[35] When she was alone, Manci's idea of a perfect day was to spend it shopping, playing with her dog, hobnobbing with Florida State officials, adjusting her investments and driving out with her pals for lunch at a local Marriott hotel, where she traded gossip while munching on cheese blintzes.[36] She was in close touch with her daughters, constantly worrying about Mary, who lived nearby and was often in poor mental health. In the evening, Manci would settle down in front of the television with a glass of sherry to watch public-service documentaries and her favourite game shows, *Jeopardy!* and *The Price Is Right*. Through letters and endless phone calls, she kept in touch with friends and family all over America and Europe, though not with her sister-in-law Betty, who died in 1991.

Still angry with Churchill College for what she regarded as their terrible treatment of Elizabeth Cockcroft, Manci took her revenge when she withdrew Dirac's archive from the college. She arranged for it to be transferred to Florida State University, where the archive is now stored in the Dirac Science Library, which Manci formally opened in December 1989.[37] Outside the library, she unveiled a statue of Dirac by the Hungarian sculptor Gabriella Bollobas, showing him in old age, reading *The Principles of Quantum Mechanics*. The statue is peculiarly lifeless, with no sign of the energy and imagination that propelled him to greatness.

Manci never mellowed: she would still switch in an instant between mean-spiritedness and generosity. After railing at Halpern for an entire morning, she would spend the afternoon trying to sweet-talk Florida State officials into giving him a permanent position in the physics department.[38] She behaved no more consistently towards her brother Eugene, suffering from Alzheimer's disease: in public, she adored him but in private she described him witheringly as 'a third-rate physicist'.[39] On the telephone, she argued with him for hours about family matters, haranguing him for his politics and for associating with 'the Moonies'. On New Year's Day 1995, she called Leon and Ellen Lederman hours after Wigner's death, and said to each of them in turn: 'Thank God the monster is dead.'[40]

Even in her ninth and tenth decades, Manci kept abreast of the news. In late 1989, she was jubilant when, following the fall of the Berlin Wall, the Soviet-backed Hungarian Socialist Workers' Party abdicated its monopoly power and agreed to free elections. Soon afterwards, during the presidency of George Bush Senior, she considered applying for American citizenship so that she could vote against him if he stood for re-election. Delighted when Bill Clinton first won the presidency, in late 1995 she wrote supportively to Hillary Rodham Clinton, who sent a courteous reply on White House notepaper ('Dear Ms Dirac [. . .]').[41] No letter ever gave Manci more pleasure.

She spent her last few years in pain with arthritis and suffering grievously from asthma. Friends and family urged her to move into a care home, but she would hear nothing of it: she was going to live out her days at home, no matter what the cost of round-the-clock home assistance. Early in 2002, after she tripped over her dog and broke her hip, she had no choice but to be admitted to hospital, where she died a few days later. Mary and Monica arranged for her to be buried with Dirac under a joint gravestone; his epitaph was unchanged, hers was 'Let her generous soul rest in peace.'

Thirty

Then she showed me this picture ☺ and I knew that it meant 'happy', like when I'm reading about Apollo space missions, or when I am still awake at 3 a.m. or 4 a.m. in the morning and I can walk up and down the street and pretend that I am the only person in the whole world.

CHRISTOPHER BOONE, narrator in Mark Haddon's *The Curious Incident of the Dog in the Night Time*, 2003

Bristol has never taken Dirac to its heart. Today, the few reminders in the city of its association with Dirac include a little-noticed abstract sculpture, the name on a grimly functional building and a few plaques. During my many visits to Bristol over the past five years, I have met scarcely half a dozen people outside the university who have heard of him. A few minutes after I first walked through the front door of the Bristol Records Office, in May 2003, I enquired of the bracingly confident assistant if she had any material on Paul Dirac; she looked at me quizzically and asked, 'Who's he?'

In the Records Office, the best way of finding out about Dirac's early school years is to ask to see the well-fingered documents about his fellow pupil at Bishop Road School, Cary Grant. Local journalists and television crews were always ready to record Grant's sojourns in the city, a prospect that would have frightened off Dirac; his visits were always anonymous. In the 1970s, however, he welcomed the campaign led by the local Member of Parliament William Waldegrave to celebrate the city's association with him, an initiative that led to the founding of a mathematics prize in local secondary schools.[1] Waldegrave had noticed that while Dirac is not well known by the people of Bristol, they were proud of their association with the charismatic engineer Isambard Kingdom Brunel, though he had not been born in the city or even lived there.

In 2006, Bristol's veneration of Brunel was clear during a five-month celebration of the bicentenary of his birth. Local businesses and cultural organisations collaborated to present 'Brunel 200', an eight-month festival of exhibitions, theatrical events, concerts, art installations and poetry readings.[2] Some forty thousand people –

most of them from Bristol and the surrounding towns – attended the opening weekend in April. Four years before, the centenary of Dirac's birth was marked in Bristol rather more modestly. The main event, organised by the University's physics department, was an afternoon of lectures to celebrate Dirac's life and legacy, followed by a formal dinner on Brunel's SS *Great Britain*. Following an interview about the Dirac equation on Radio 4's *Start the Week*, I was called by one of the organisers who asked me to give a lecture on Dirac's life and work. This was a special moment for me as I had been fascinated by Dirac since I was a teenager.

I first heard his name on a suburban doorstep, when I was hawking subscriptions for a weekly raffle in aid of the Liberal Party in Orpington, a suburb in south-east London. When I was closing a sale on a spring evening in 1968, my new customer – a distracted, oddly engaging man by the name of John Bendall – mentioned perfunctorily that he was a theoretical physicist. We became friends, and, during several Sunday-morning chats in his front room, I realised that he was a Dirac fanatic: Bendall would find an excuse to introduce his hero's name in every conversation lasting longer than a few minutes.[3] I found out that it was no coincidence that the younger Bendall daughter, playing with her dolls at our feet, had been named Paula. Every Christmas, he would take a plate of mince pies from the kitchen, sit back in his armchair with a glass of sherry and read *The Principles of Quantum Mechanics*, savouring every sentence. Minutes after I first browsed through his copy, I knew I too wanted to be a theoretical physicist.

A few months later, it dawned on me that, when Dirac was a boy, he lived just a few miles from my Bristol-born paternal grandmother Amelia ('Mill') Jones. She was fond of telling me about that time in her life, when she worked in a corset factory. At weekends, she and her fiancé Charley – a docker, later my grandfather – would promenade arm in arm around the centre of the city, her expansive skirt almost touching the ground, his moustache daringly trimmed. 'I wonder if we ever saw Cary Grant before he 'opped it to America?' I heard her ask. She may well have set eyes on him around the city, perhaps around the Hippodrome, one of her haunts. It is also possible that she and my grandfather knew the high reputation of Charles Dirac and almost certain that they saw at least some members of the Dirac family, perhaps the two French-speaking brothers walking together.

In middle age, Dirac made several trips back to the city. In 1956, after a summer holiday in his mother's home county of Cornwall, he returned through Bristol with his family and stopped outside 6 Julius Road to point out to his daughters Mary and Monica where he had lived since he was ten.[4] But he said nothing about his memories of the twenty-five years he spent there. During my visits to Bristol, I lurked several times outside this unremarkable home, trying unsuccessfully to imagine my way into it. My problem was solved during a visit in the early summer of 2004, when the owner of the property generously invited me inside, allowing me to enter the theatre of Dirac's most traumatic memories.[5]

Overlooking the front garden is Charles's tiny study, where he taught his private students, away from the gaze of the tax inspectors. Under the stairs is the tiny cupboard where Flo crouched during the German bombing raids, cotton wool in her ears. Above is the little bedroom where, a few months after Felix killed himself, Dirac first read Heisenberg's path-breaking paper and realised that it contained the key to quantum physics. Felix's bedroom, for many years a shrine, is now scattered with the toys and games of the children who occupy the room. Flo's tiny kitchen overlooks the back garden, where Dirac had looked up at the stars and had watched some of the first British-made aeroplanes take off, and where he had begun to learn gardening during the Great War. It seemed barely possible that this suburban home had seen events that had left Dirac, as Manci had described him, 'an emotional cripple'.[6]

Her words might sound cruel, but Dirac would probably have agreed that they were accurate. He always attributed his extreme taciturnity and stunted emotions to his father's disciplinarian regime; but there is another, quite different explanation, namely that he was autistic. Two of Dirac's younger colleagues confided in me that they had concluded this, each of them making their disclosure *sotto voce*, as if they were imparting a shameful secret. Both refused to be quoted. Yet one should be extremely careful about making this diagnosis: rather too often, people are labelled autistic on the flimsiest of evidence except that they are exceptionally reserved, focused and unsociable. Besides, it is not easy to psychoanalyse someone who is dead.

Before one can say whether there is a strong case that Dirac was a person with autism, it is important to be clear about the nature of the condition. For someone to be diagnosed as autistic, he or she must have all three of the following characteristics since early childhood:

1. Social skills are poorly developed compared with the development of other 'classroom' skills, such as reading and arithmetic.
2. The development of verbal and non-verbal communication is impaired compared with the development of other 'classroom' skills. Behavioural signs of repetitive or stereotyped movements, a delay in the acquisition of language and a lack of varied, spontaneous make-believe play.
3. An unusually narrow repertoire of activities and interests that are abnormally intense.[7]

A few days before the Nobel Prize ceremony in 1933, Flo told journalists that Dirac was a precocious, industrious and unusually quiet child.[8] There is not nearly enough detail in her comments or in reports of Dirac's behaviour at school to justify a diagnosis that he was then autistic. His behaviour as an adult, however, had all the characteristics that almost every autistic person has to some degree – reticence, passivity, aloofness, literal-mindedness, rigid patterns of activity, physical ineptitude, self-centredness and, above all, a narrow range of interests and a marked inability to empathise with other human beings. Extremes of these characteristics are at the root of the humour in almost all the tales about Dirac that physicists have been telling each other for decades: almost all of these 'Dirac stories' might also be called 'autism stories'.

The word 'autism', derived from the Greek word *autos* for self, covers a wide spectrum of conditions, spanning people with mental retardation through to those like Dirac who are gifted in their specialist fields and often described as 'high functioning'. An unusual case was dramatised in the Hollywood film *Rain Man*, where Dustin Hoffman portrays the autistic character Raymond Babbitt, who also has the much more rare Savant Syndrome, manifested in his prodigious arithmetic skills and in his amazing memory for baseball statistics and telephone numbers.

Clinicians believe just over half a million people in the UK are autistic to some degree, almost one in a hundred, and it is clear that it is predominantly a male condition. Statistical studies also show that depression is especially common among people with autism and that about 20 per cent of children with the condition speak fewer than five words a day.[9] About one person with autism in ten has a special talent – for example, in drawing, working with computers or rote-memory learning. Another characteristic, yet to be properly quantified, is that young people with autism are exceptionally fussy about the food they are prepared to eat.[10]

There is currently a good deal of speculation of a modern-day epidemic of autism, especially in the USA, where, as *Nature* put it in 2007, the condition is the 'golden child of the fundraising circuit'.[11] But talk of a sudden rise in the number of people with autism is probably ill founded because diagnoses often differ from one doctor to another, with the result that the data have large uncertainties.[12] Reliable information has been available only since the mid-1960s, when high-quality empirical studies began, long after Leo Kanner, an Austrian-born child psychiatrist at Johns Hopkins University in Baltimore, first identified and named the condition in 1943. A year later, the Viennese psychiatrist Hans Asperger independently described a condition now known as Asperger's Syndrome, part of the spectrum of autistic behaviour.[13]

Although the study of autism is developing rapidly, it is still in its infancy: like atomic physics in the early 1920s, there is a huge amount of observational information about the condition, but the experts know that their understanding of the data is only fragmentary. But some firm conclusions have emerged. A few decades ago, scientists believed that people with autism had some disorder of the mind, but it is now plain that this is incorrect: there is now overwhelming evidence that the condition is a disorder of the tissue in the *brain*.[14] Using modern brain-imaging techniques – including positron emission tomography – clinicians have demonstrated that the regions linked with the process of 'reading other people's minds' in the brains of people with autism are noticeably less active than in most other people.

Some of the most productive research into autism is now being done in Cambridge at the Autism Research Centre. Its director, Simon Baron-Cohen, is a pioneer of the idea that autism is a manifestation of the extreme male brain – comparatively weak in the typically female characteristic of empathy but strong in the typically male characteristic of systemising, such as working out how mechanical devices function, solving mathematical puzzles, poring over league tables and filing CDs.[15] In one of Baron-Cohen's research projects, he and his colleagues are studying the behaviour of leading mathematicians and scientists, many of whom – including Newton and Einstein, some believe – exhibit at least some of the traits of autism.[16] The great majority of top mathematicians and physical scientists are undoubtedly male; this may indicate a predisposition of the male brain, though critics point out that it may also be a consequence of rearing children in ways that perpetuate sexual stereotypes.

When I visited Baron-Cohen in his rooms in Trinity College, I was struck by two remarks that seemed especially relevant to Dirac. First, he said that he had noticed the high proportion of autistic men who were in a stable marriage with a foreign wife, perhaps because the women were more tolerant of unusual behaviour in foreign men than in men from their own culture. Baron-Cohen had no idea that Dirac was married for almost fifty years to a Hungarian. That, of course, could be a coincidence. I was taken aback again by another remark he made a few minutes later, however, when he pointed out that although people with strongly autistic personalities appear to be detached from most other people, when they believe that a friend has suffered an injustice, they are often so indignant that they will disrupt or abandon their almost invariable daily routines to rectify it.[17] Baron-Cohen knew nothing of Dirac's one venture into international politics when he spent a few months concentrating on the campaign to free Kapitza from his detention in the Soviet Union. Heisenberg, pilloried by many of his former colleagues after the war, had cause to regard Dirac as one of his most loyal friends. Again, these may be coincidences.

But Baron-Cohen argues that it is not happenstance that the young Dirac bloomed in 1920s Cambridge:

Cambridge was a niche where his eccentricity would have been tolerated and his skills valued. College life provided him with a regular daily routine and everything he needed. His bed was made for him, food was provided for him. High Table in College would have provided social contact if he wanted it, with its own rules and routines to render it highly predictable. In the mathematics department, he would have been free to do as he wished, he was surrounded by like-minded people, with no pressure to socialise. An environment like this would have been optimal for someone like Dirac.[18]

A fruitful source of insights into autism is the American business executive and teacher Temple Grandin, who describes herself to be 'a high-functioning person with autism'.[19] In her books and articles, Grandin stresses two particular aspects of her personality that she shares with most other autistic people; both are characteristics that Dirac shared. First, she is hypersensitive to sudden sounds, bringing to mind the great care Dirac always took to ensure that he would not be disturbed by chiming bells or by the sudden barks of neighbourhood dogs. Second, she points out that she thinks visually and that, in several respects, her brain does not function like those of most people she has met.

Here's how my brain works: It's like the search engine Google for images. If you say the word 'love' to me, I'll surf the Internet inside my brain. Then, a series of images pops into my head. What I'll see, for example, is a picture of a mother horse with a foal, or I think of *Herbie the Lovebug*, scenes from the movie *Love Story* or the Beatles song . . . 'All you need is Love'.[20]

Like Temple Grandin, Dirac was certain that his mind was 'essentially a geometrical one'.[21] He was always uneasy with algebraic approaches to physics and with any mathematical process he could not picture – one of the reasons why he was so uncomfortable with renormalisation.

Yet again, it is possible that this correlation between autistic characteristics and Dirac's behaviour is a coincidence, but, in the light of other such correlations, this seems unlikely. I believe it to be all but certain that Dirac's behavioural traits as a person with autism were crucial to his success as a theoretical physicist: his ability to order information about mathematics and physics in a systematic way, his visual imagination, his self-centredness, his concentration and determination. These traits certainly do not explain his talent but they give some insight into his unique way of looking at the world.

One of the strongest clues about the true nature of autism is that the condition has a genetic component – it runs in families. The theory, although powerful, cannot predict with the precision of a theory in physics how most characteristics are passed down the generations, especially for conditions such as autism, associated with several genes. Observational studies show that it is rare for families to have more than one child with autism, though the probability that a second child will be autistic is about one in twenty, almost eight times the usual likelihood. This raises the question of whether Felix Dirac was autistic. Again, it is impossible to say one way or the other because too little information about his personality survives. I was, however, given pause for thought one evening during my visit to the family's genealogist, Gisela Dirac. As she surveyed the family tree, she remarked, 'It's amazing how many people in the family had acute depression. And how many killed themselves.' At my request, she later sent me a family tree annotated with such instances: in the previous century, there had been at least six.

Charles Dirac also showed signs of autistic behaviour. Most of the descriptions of him by his colleagues and students refer to his self-centredness, his dedication to work and his rigid teaching methods. Like his son Paul, Charles appears to have had only a modest ability to understand other people's feelings, but whereas lack of empathy in Paul was manifest in his reserve, in Charles it seems to have appeared as a tendency to behave like a human bulldozer. Neither man was ever going to be the easiest of husbands to live with: Flo's teenage infatuation with the charming Swiss man she met in the library had led to a wretchedly unhappy union, whereas Manci somehow found ways of living stably with a man few women would contemplate as an acceptable partner for a second.

Dirac was aware that he was in some ways similar to his father. Three months after Charles died in June 1936, Manci suggested to Paul that he thought too much about these similarities and that he might unconsciously be seeking to emulate some of his father's habits.[22] Shortly afterwards, Paul had pondered on his father's biological inheritance when he attended Bohr's conference on genetics and heard in detail about genetic characteristics and how they are passed from one generation to the next. Sitting on one of the wooden benches in the lecture theatre of Bohr's institute in Copenhagen, listening to the lectures, Dirac may well have wondered which of these heritable characteristics were written into his own genes.

Whatever their genetic profiles, there is no doubt that Dirac and his father were incompatible. Having heard so much about the harrowing mealtimes together, I found myself shuddering when I first walked into the dark dining room of 6 Julius Road overlooking the back garden. The original fireplace is still there. It was easy to imagine Flo passing bowls of steaming porridge from the kitchen through the hatch in the dividing wall and urging the worryingly thin Paul not to leave a morsel uneaten. Although he had a weak appetite, one of the symptoms of tuberculosis, his parents seem not to have suspected that he had the disease and so had no reservations about putting him under pressure to consume much more food than he wanted to eat.[23]

The elderly Dirac remembered this dining room as a torture chamber. It was here, he said many times, that his father drove him into a life of silence and inhibition – the young Dirac, forced to speak French, found it easier to say nothing than to make errors that his father would punish unmercifully. No one else in the family left an account of these mealtimes, so we shall probably never know if he was exaggerating. Nor are we ever likely to know what his parents felt about the problems of bringing up a child who was both precociously clever and emotionally withdrawn.[24] From a modern perspective, Charles and Flo were coping with a challenge they did not know they faced, one that may well have made their marital problems even worse. If they were living in Bristol today, the city council would – like most local authorities in the UK – give them support and enable their son to go to a special school.

I for one accept the testimony of Paul Dirac and his mother that Charles Dirac was a domineering and insensitive father, though I don't believe he bullied his younger son into taciturnity. Much more likely, it seems to me, is that the relationship between Paul and Charles was doomed by nature rather than nurture: the young Dirac was born to be a child of few words and was pitiably unable to empathise with others, including his closest family. He laid all the blame for this at the feet of his father, though he disliked him for other reasons, too, with a bitterness that surprised the few people – including Kurt Hofer – who saw the extent of it. 'Why was Paul so bitter, so obsessed with his father?' Hofer wondered after hearing his outburst. Perhaps the main reason was that Dirac knew in his heart he was not just his own man but, inescapably, his father's.

Thirty-one

Dirac told physics students they should not worry about the meaning of equations, only about their beauty. This advice was good only for physicists whose sense of purely mathematical beauty is so keen that they can rely on it to see the way ahead. There have not been many such physicists – perhaps only Dirac himself.

STEVEN WEINBERG, Dirac Centenary Meeting, University of Bristol, 8 August 2002[1]

All scientists, even the most eminent, are dispensable to science. Although inspired individuals influence it in the short term, the absence of any of them would be unlikely to make much difference to it in the long run. If Marie Curie and Alexander Fleming had never been born, radium and penicillin would have been discovered soon after the dates now in the textbooks.

Every scientist can hope, however, that posterity will judge him or her to have revealed more than a typical share of nature's secrets. By this criterion, there is no doubt that Dirac was a great scientist, one of the few who deserves a place just below Einstein in the pantheon of modern physicists. Along with Heisenberg, Jordan, Pauli, Schrödinger and Born, Dirac was one of the group of theoreticians who discovered quantum mechanics. Yet his contribution was special. In his heyday, between 1925 and 1933, he brought a uniquely clear vision to the development of a new branch of science: the book of nature often seemed to be open in front of him. Freeman Dyson sums up what made Dirac's work so unusual:

The great papers of the other quantum pioneers were more ragged, less perfectly formed than Dirac's. His great discoveries were like exquisitely carved marble statues falling out of the sky, one after another. He seemed to be able to conjure laws of nature from pure thought – it was this purity that made him unique.[2]

Dirac's book *The Principles of Quantum Mechanics* was one of these statues, Dyson points out: 'He presents quantum mechanics as a work of art, finished and polished.' Never out of print, it remains the most insightful and stylish introduction to quantum mechanics and is still a powerful source of inspiration for the most able young theoretical physicists. Of all the textbooks they use, none presents the theory with such elegance and with such relentless logic, a quality of Dirac's highlighted by Rudolf Peierls in 1972: 'The thing about Dirac is that he has a way of thinking logically [. . .] in a straight line, where we'd all tend to go off in a curve. It's this absolutely straight thinking in unexpected ways that makes his works so characteristic.'[3]

Most young physicists, however, are concerned not with the internal logic of quantum mechanics but with using the theory as a way of getting quick and reliable results. In effect, it gives scientists a completely dependable set of practical tools for describing the atomic and molecular world. Every day, tens of thousands of researchers in the microelectronics industry routinely employ the techniques developed by Dirac and his colleagues: ideas that took years to clarify are now used without a thought for the headaches they once caused their creators.

The modern trend to miniaturisation is making quantum mechanics even more important. In the growing field of ultra-miniature technology – usually called nanotechnology (from the Greek word for dwarf, *nanos*) – quantum mechanics is as indispensable as classical mechanics was to Brunel. In one branch of this new technology, spintronics (short for spin-based electronics), engineers are trying to develop new devices that rely not only on controlling the flow of the charge of electrons – the way conventional devices work – but also the flow of the electrons' *spins*. Because these can be flipped from one state to another much more quickly than charge can be moved around, spintronic devices should operate faster than conventional ones and produce less heat. If, as engineers hope, they can produce a spin-based transistor to replace conventional transistors in memory and logic circuits, it may be possible to continue the trend towards ever-more compact computers beyond the currently feasible limits.

It could be that, just over a century after Dirac first brought electron spin into the logical structure of quantum mechanics, his equation – once seen as mathematical hieroglyphics with no relevance to everyday life – becomes the theoretical basis of a multi-billion dollar industry.

Great thinkers are always posthumously productive. By this criterion, Dirac can be counted as one of the greatest of all scientists –

many of the concepts he introduced are still being developed, still instrumental in modern thinking. The Dirac equation, for example, is still a fecund source of ideas for mathematicians, who have long been fascinated by spinors, mathematical objects that first appeared in the equation. In Sir Michael Atiyah's opinion:

No one fully understands spinors. Their algebra is formally understood but their geometrical significance is mysterious. In some sense they describe the 'square-root' of geometry and, just as understanding the concept of the square root of −1 took centuries, the same might be true of spinors.[4]

Dirac's influence is felt most strongly by scientists studying the universe's tiniest constituents. Experimenters can now smash particles together with energies so high that even Rutherford would have been impressed: at the Large Hadron Collider, the huge particle accelerator at CERN, they can recreate the conditions of the universe to within a millionth of a millionth of a second of the beginning of time. During the subatomic collisions produced in this and other accelerators, experimenters routinely see subatomic particles created and destroyed, processes that can be explained only using relativistic quantum field theory. Dirac's hand is all over this theory – he was one of its co-discoverers and the author of the action–principle formulation of quantum mechanics, now a crucial part of modern thinking about fields.

Over the past twenty-five years or so, the gap between the energies accessible by particle accelerators and the energies needed to test the latest theories has widened alarmingly. The building of the accelerators is increasingly difficult and expensive for the international collaborations needed to fund and operate them, so new devices come on stream only slowly. One consequence has been that the theory of subatomic particles has run ahead of the supply of data from experiment, producing a scenario of the type Dirac envisaged in his landmark paper of 1931 where he set out an agenda for theoretical physics led by mathematics rather than experiment. One physicist who believed this was prescient was C. N. Yang: at a Princeton meeting they both attended in 1979, Yang suggested that when Dirac set out this idea, he had hit on a 'great truth'.[5] In the same 1931 paper, Dirac suggested the existence of the anti-electron and the anti-proton and developed a quantum theory of magnetic monopoles using a geometric approach that has influenced generations of theoreticians. As experimenters were unable to detect monopoles, Dirac regarded this project as another disappointment, and he died believing it unlikely that monopoles occur in nature.[6] But, today, many physicists disagree, as monopoles are predicted by some simple generalisations of the Standard Model (the 'modern' monopole is a mathematically better-defined relative of Dirac's).[7] Moreover, according to cosmologists, monopoles should have been created during the Big Bang in vast quantities and should now be detectable; that they are not is known as 'the monopole problem'.[8]

The detection of Dirac's monopole would raise a question in virtual history: what would have been the effect on his reputation if the monopole had been detected around the time the positron was first observed? Such a pair of successes would have further bolstered his reputation among his colleagues and may well have made him much better known to the public. But there was never any chance that he would become a media celebrity like his most recent Lucasian successor, Stephen Hawking: it seemed not to have occurred to Dirac to write a popular book, nor would he have contemplated making the kind of forays into the media spotlight undertaken by Hawking, such as his appearances on *Star Trek*, *The Simpsons* and on the dance floor of a London nightclub.[8] Yet Dirac admired such boldness more than most of his colleagues knew.

Dirac left his mark on several other fields besides quantum mechanics. One of his least typical contributions was his invention of a new way of separating different isotopes of a chemical element. He developed the method during the Second World War but it seemed that the idea was impracticable; it was soon forgotten, only to be independently rediscovered thirty years later by engineers in Germany and South Africa.[9] His method still does not appear to be economically viable, but the development of new, ultra-strong materials still leaves open the possibility that the method could be used in the nuclear industry.

Another of Dirac's less characteristic pieces of work was his exploration of the wave and particle nature of electrons with Kapitza in 1933. Modern improvements in laser technology provided fresh opportunities to verify the existence of the Kapitza–Dirac effect, the diffraction (bending) of a thin beam of electrons by a standing wave of light. Kapitza and Dirac had themselves discussed the new possibilities at the final meeting of the Kapitza Club in 1966. Several groups attempted to demonstrate the effect, but none was successful until, in the early spring of 2001, a team at the University of Nebraska observed it with a high-powered laser and a fine beam of electrons, using apparatus that would have fitted on a dining-room table.[10] The Kapitza–Dirac effect is now used as a subtle probe of the wave-like and particle-like behaviours of both electrons and light.

Dirac also left a legacy in general relativity, if one not quite equal to his talent. It is a mystery that he showed so little interest in following up the discovery – made by Oppenheimer and his colleagues in 1939 – that Einstein's theory predicted the existence of black holes, objects with such a strong gravitational field that not even light can escape them. In Dirac's most important contribution to the theory of relativity, he set it out in analogy to his favourite Hamiltonian version of quantum mechanics and devised a set of complementary mathematical techniques (other physicists did similar work at about the same time).[11] These methods have proved useful to astronomers studying closely spaced pairs of rotating neutron stars (usually called pulsars), orbiting each other, slowly losing energy. This gradual loss of energy can easily be explained by Einstein's general theory of relativity, especially if it is interpreted using the methods Dirac co-invented: the pulsars emit gravitational radiation, in much the same way as accelerating electrons emit electromagnetic radiation. The study of gravitational waves is now one of the most promising areas of astronomy.

Dirac's intuition for the workings of the universe on the largest scale was not nearly as strong as it was when he was focusing on atoms. There is, however, no denying the far-sightedness he showed when he reviewed the state of cosmology in the Scott Lecture he delivered shortly before the outbreak of the Second World War, when the subject was in its infancy. In one of a string of astute remarks he makes in passing, he hazarded an inspired guess that the complex structure of everything around us has its seeds in a quantum fluctuation in the initial state of the universe. 'The new cosmology', Dirac suggested, 'will probably turn out to be philosophically even more revolutionary than relativity or the quantum theory,' perhaps looking forward to the current bonanza in cosmology, where precise observations on some of the most distant objects in the universe are shedding light on the nature of reality, on the nature of matter and on the most advanced quantum theories. In the view of Nathan Seiberg, Dyson's colleague at the Institute for Advanced Study, 'The lecture would look just as impressive if the date on the front were not 1939 but 1999.'[12]

Although Dirac was, towards the end of his life, often defensive about his large numbers hypothesis, he always had faith in its truth.[13] The modern view about the large numbers that fascinated him for decades is that only one of them is a mystery: the ratio between the strength of the electrical force and that of the gravitational force between an electron and a proton (10^{39}). The fundamental problem is to understand why the gravitational force is so feeble compared with the other fundamental forces.[14] All the other huge numbers that puzzled Dirac now follow from the standard theory of cosmology, so there is no need to guess links between them – the coincidences he spotted are illusory.[15]

Dirac was convinced that the strength of the gravitational force had fallen since the beginning of time, and he invested many of his later years in trying to prove it, though observations made by astronomers on nearby planets in the solar system have now all but ruled it out. Although it is still just possible that Dirac's intuition was correct, the subject is currently low on today's research agenda. One scientist who always believed in his bones that Dirac was right was Leopold Halpern, who left Florida State University in 2004 and became a resident theoretician with a satellite-based experimental programme run by NASA and Stanford University, aiming to check some of the unverified predictions of Einstein's general theory of relativity.[16] Halpern hoped to compare the predictions of his theory with the satellite's observations but he was unable to complete his work before he died of cancer in June 2006.[17]

Regardless of how Dirac's predictions about the gravitational force fare in the future, his name will always be associated with the role of anti-matter at the beginning of the universe. According to modern Big Bang theory, matter and anti-matter were created in exactly equal amounts, at the very beginning of the universe, about 13.7 billion years ago. Soon afterwards, the decay of some of the heavy particles formed from the quarks and anti-quarks led to a small but crucial surfeit of matter over anti-matter, by just one part in a billion. The first scientist to analyse this difference in detail was Tamm's student

Andrei Sakharov – later a courageous human-rights activist in the Soviet Union – who discussed in 1967 how this excess came about and why the universe was left with an overwhelming preponderance of matter. Without that imbalance, the matter and anti-matter formed at the beginning of time would have annihilated each other immediately, so that the entire universe would only ever have amounted to a brief bath of high-energy light. Matter would, in that case, never have had the opportunity to discover anti-matter.[18]

The surplus of matter over anti-matter at the beginning of the universe is still not understood, and thousands of physicists are working to understand it. Their main sources of experimental information are particle accelerators, where anti-matter is produced by smashing ordinary particles into each other and then quickly 'separating off' the anti-matter, before it is annihilated by matter. By comparing the decays of particles with those of their anti-particles, experimenters hope to get to the bottom of the matter–antimatter imbalance.

Every day, particle accelerators now generate about a hundred thousand billion positrons and five thousand billion anti-protons – a total of roughly a billionth of a gram. Although this quantity is only tiny, the ability to produce it at will demonstrates that *Homo sapiens* now – a million years after our species evolved – uses anti-matter as a tool. Today, positrons are routinely generated in mass-produced equipment all over the world: doctors use positron emission tomography (PET) to see inside their patients' brains and hearts, without the need for surgery. It is a simple technique: the patient is injected with a tiny amount of a special radioactive chemical that spontaneously emits positrons, which interact with electrons in the tissue where the chemical settles. The photograph is a record of the radiation given off in the electron–positron annihilations.

Within just a few decades, positrons changed in the eyes of scientists from appearing outlandish novelties to being just another type of subatomic quantum; the public has become more familiar with anti-matter, too, from the fictional treatments of it in, for example, *Star Trek* and Dan Brown's *Angels and Demons*. But what is most remarkable about the story of anti-matter is that human beings first understood and perceived it not through sight, smell, taste and touch but through purely theoretical reasoning inside Dirac's head.

*

Like Einstein, Dirac was fundamentally in search of generalisations – theories that explain more and more about the universe, in terms of fewer and fewer principles. Both men believed, too, that the best way of achieving this was through theories expressed in terms of beautiful equations.[19] As a physicist, Dirac had been well served by mathematics, as he wrote in an unusually candid passage in 1975:

If you are receptive and humble, mathematics will lead you by the hand. Again and again, when I have been at a loss how to proceed, I have just had to wait until [this happened]. It has led me along an unexpected path, a path where new vistas open up, a path leading to new territory, where one can set up a base of operations, from which one can survey the surroundings and plan future progress.[20]

Although he never acknowledged it in public, the guiding hand of beauty had led Dirac not only to some rich new pastures of research but also into the deserts that yielded no fruit at all. In his talks, he was an ambassador of mathematical beauty, repeatedly underlining the triumphs of theories with this quality but not mentioning the years he had spent trying in vain to use sensually appealing mathematics to describe nature. It is striking that he put forward the principle of mathematical beauty several years after he had done his best work, and we have to suspect that some of his accounts of his greatest discoveries – usually portrayed as successes for his type of aestheticism – were reinterpreted in the light of his faith in the principle. In his pioneering papers on quantum mechanics, he never explicitly says that beauty was his guide; he recalled its value only in the tranquillity of his least productive years.[21]

Dirac first made it clear that he was using the principle of mathematical beauty in the late 1940s, when he dismissed the renormalised theory of photons and electrons on the grounds that it was too ugly. He was, however, unable to use his principle constructively, to build new theories. It could therefore be argued that Dirac's passion for beauty was to some extent destructive, but he knew no other way: he was temperamentally unable to focus on any other subject in particle physics until he had found a truly beautiful theory of electrons and photons, without the disfiguring infinities.

A way out of this alleged flaw in quantum field theory arrived, tragically, just too late for him: a particularly promising, infinity-free theory of electrons and photons began to circulate among theoreticians in the autumn of 1984, as he lay dying. Michael Green, of the University of London, and John Schwarz, of Caltech, had written a crucial paper showing that string theory might be able to form the basis of a unified theory of fundamental interactions.[22] Previously, the theory appeared to say that the weak interactions must have perfect left–right mirror symmetry, contrary to experimental evidence. By proving that the theory can naturally describe the *breaking* of this symmetry, and by resolving other embarrassing anomalies in the theory, Green and Schwarz began a revolution. Within weeks, string theory was the hottest topic in theoretical physics. Although the theory was far from complete – it was really a collection of inchoate concepts, all in need of development – there were strong signs that it contained the seeds of an exciting new framework for giving a unified account of all the fundamental interactions, encompassing the Standard Model and Einstein's general relativity.

The new theory describes nature not in terms of point-like particles but of pieces of string, so small that if they could be aligned end to end, it would take a billion billion of them to span a single atomic nucleus. In this picture of the fundamental constituents of the universe, there is only one fundamental entity – the string – and every type of particle, including the electron and the photon, is simply an excitation of the string, analogous to a mode of vibration of a tuning fork.[23] The mathematics of the theory is fearsome, but underneath the complexities is a modern version of John Stuart Mill's desideratum of fundamental physics: a unified description of all the fundamental interactions.

What would surely have impressed Dirac is that modern string theory has none of the infinities he abhorred. He would have revelled in the mathematical beauty of the theory, which delights not only the physicists who use it but also many mathematicians who have mined it for new concepts. It has turned out that string theory, much like the Dirac equation, is a fertile source of purely mathematical ideas that have a value for their own sake, not just as tools to understand nature. Dirac often said that he was interested in theories only as ways of accounting for nature, but he would probably have been intrigued to see, at the heart of string theory, mathematics known as complex projective geometry, a generalisation of his favourite branch of geometry.[24]

No one has done more to shed light on string theory than the mathematical physicist Edward Witten, at the Institute of Advanced Study. In 1981, when he was a lecturer at the Erice summer school and thirty years old, he met Dirac briefly and heard his familiar condemnation of renormalisation but chose not to follow his advice. Dirac followed Witten's work and, in 1982, wrote – in his trembling hand – to the Papal Academy, supporting Witten's nomination for a special award and describing his mathematical work as 'brilliant'.[25] From the early 1980s, Witten's reputation among string theorists has been comparable to Dirac's among quantum theorists half a century before.

Witten believes that string theory seems to be the kind of theory that Dirac had in mind when he argued that a revolution was needed to produce a new theory free of infinities so that renormalisation was not needed:

In some ways Dirac's reaction to renormalization was vindicated because the better theories he said he wanted were eventually developed, with the advent of string theory. But by far the most progress towards the new theory was made by physicists who used and studied renormalization. So you'd have to look at the outcome for Dirac as bittersweet: he was partly right, but his approach was not entirely pragmatic.[26]

It is hard to disagree with this tactfully expressed judgement about Dirac's principled but counterproductive attitude to renormalisation. If he could have shed some of the insistence on rigour that he learned as a student of pure mathematics and been able to retain some of the pragmatism he learned when training to be an engineer, his achievement would, in all likelihood, have been even greater. Perhaps, if he had been more active in quantum field theory, it would have advanced more quickly, and modern string theory would have arrived sooner.

Although string theory is the only strong candidate for a unified theory of the fundamental interactions, by no means all theoreticians are convinced of its value. A substantial number of physicists worry that the theory makes sense only in more than four dimensions of space-time (it is easiest to formulate in ten or even eleven dimensions). More worrying, it has received little support from experiment: string theory has yet to make a clear-cut prediction that experimenters have been able to test. These are among the key signals, several physicists have argued, that the theory is absurdly overvalued and that it would be better to pursue other avenues. One of the most vocal sceptics is the Standard-Model pioneer Martin Veltman: 'String theory is mumbo jumbo. It has nothing to do with experiment.'[27]

But it is clear from the comments Dirac repeatedly made in his lectures on the way theoretical physics should be done that he would have disagreed with these criticisms: he would have counselled string theorists to let the theory's beauty lead them by the hand, not to worry about the lack of experimental support and not to be deterred if a few observations appear to refute it. But he would have cautioned string theorists to be modest, to keep an open mind and never to assume that they are within sight of the end of fundamental physics. If past experience is anything to go by, another revolution will follow eventually.

Such was the advice this extraordinarily unemotional man offered to his colleagues: be guided, above all, by your emotions.

Abbreviations in Notes

AHQP Archives for the History of Quantum Physics, multiple locations, provided by Niels Bohr Library & Archives, American Institute of Physics, College Park, Maryland., USA (http://www.amphilsoc.org/library/guides/ahqp/).

AIP American Institute of Physics, Center for the History of Physics, Niels Bohr Library, Maryland, USA.

APS Archive of the American Philosophical Society, Philadelphia, USA.

BOD Bodleian Library, University of Oxford, UK.

BRISTU Bristol University archive, UK.

BRISTRO Bristol Records Office, UK.

CALTECH California Institute of Technology, archive, USA.

CHRIST'S Old Library, Christ's College, Cambridge University, UK.

CHURCHILL Churchill Archives Centre, Churchill College, Cambridge University, UK.

DDOCS Dirac letters and papers, property of Monica Dirac.

EANGLIA Tots and Quots archive, University of East Anglia, Norwich, UK.

FSU Paul A. M. Dirac Papers, Florida State University Libraries, Tallahassee, Florida, USA. All of the letters Dirac's mother wrote to him are in this archive.

IAS Institute for Advanced Study, archive, USA.

KING'S King's College, Cambridge; unpublished writings of J. M. Keynes.

LC Library of Congress, Collections of the Manuscript Division.

LINDAU Archive of Lindau meetings, Germany.

NBA Niels Bohr Archive, at the Niels Bohr Institute, Copenhagen.

PRINCETON Eugene Wigner Papers, Manuscripts Division, Department of Rare Books and Special Collections, Princeton University Library, USA.

ROYSOC Archives of the Royal Society, London, UK.

RSAS Royal Swedish Academy of Sciences, Center for History of Science, Stockholm.

SOLVAY Archives of the Solvay Conferences, Free University of Brussels, Belgium.

STJOHN St John's College archive, Cambridge, UK.

SUSSEX Crowther archive, Special Collections at the University of Sussex, UK (the university holds the copyright of the archive).

TALLA Dirac archive at the Dirac Library, Florida State University, USA, http://pepper.cpb.fsu.edu/dirac/diracFA(HTML)_011.htm.

UCAM University of Cambridge archive, UK.

UKNATARCHI National Archives of the UK, Kew.

WISC University of Madison, Wisconsin, archives, USA.

1851COMM Archives of the Royal Commission of 1851, Imperial College, London, UK.

Notes

Prologue

1 A version of the 'more people who prefer to speak than to listen' remark, one of Dirac's favourites, is cited by Eugene Wigner in Mehra (1973: 819).
2 Dirac made the 'God is a mathematician' remark in his *Scientific American* article in May 1963.
3 The quote from Darwin is taken from Part VII of his autobiography. The words were written on 1 May 1881.
4 The author of the quote relating to Shakespeare was the late Joe Lannutti, a leading member of the Physics Department at Florida State University when Dirac arrived. The source of the quote is Peggy Lannutti, interview 25 February 2004. Lannutti also tells the story in J. Lannutti (1987) 'Eulogy of Paul A. M. Dirac' in Taylor (1987: 44–5).
5 This account is taken from interviews with Kurt Hofer on 21 February 2004 and 25 February 2006, and many subsequent e-mails. The account was checked in detail via e-mails on 22 September 2007. Hofer's recollections are consistent in every detail with the account given by Dirac in Salaman and Salaman (1986), in his interview, AHQP, 1 April 1962 (pp. 5–6), and in the account he gave of his early life to his friends Leopold Halpern and Nandor Balázs. I spoke to these former colleagues of Dirac on 18 February 2003 and 24 July 2002, respectively. Dirac's wife gives her recollections of his experiences at the dining table in her letter to Rudolf Peierls, 8 July 1986, Peierls archive, additional papers, D23 (BOD).

Chapter one

1 Letter from André Mercier to Dirac and his wife, 27 August 1963, Dirac Papers 2/5/10 (FSU).
2 Interview with Dirac, AHQP, 1 April 1962, p. 5.
3 Dirac Papers 1/1/5 (FSU), see also the records of the Merchant Venturers' School in BRISTRO.
4 See, for example Jones (2000: Chapter 5).
5 Pratten (1991: 8–14).
6 Although Flo lived in Cornwall only briefly, she would later insist that she was not English but Cornish. Source: interview with Christine Teszler, 22 January 2004.
7 Flo Dirac mentions this in an undated letter to Manci Dirac, written in early February 1940 (DDOCS). By 1889, when Richard Holten was fifty, he was captain of the 547-ton *Augusta*.
8 Richard Holten was aware that official documents often name his wife as the head of the family. His sailing record is in *They Sailed Out of the "Mouth"'* by Ken and Megan Edwards, microfiche 2001, BRISTRO, FCI/CL/2/3. See also Holten's Master's certificates, stored in the archives at the National Maritime Museum, Greenwich, London, UK.
9 The details of Charles and Flo's early life together are in Charles's documents in Dirac Papers 1/1/8 (FSU).
10 Louis Dirac was the illegitimate son of the recently widowed Annette Vieux, who

gave him her maiden name Giroud. Only later, when the baby's parents settled down together, did he take the surname of his father, Dirac; otherwise, his physicist grandson would have been called not Paul Dirac but Paul Giroud. Source: civil records in St Maurice, Switzerland. Louis Dirac's paeans to the beauty of the Alpine countryside are still in print, though rarely read. His poetry is published in Bioley (1903).

11 Dalitz and Peierls (1986: 140).

12 The pine cones are against a blue background; the leopard and clover are against a silver background (http://www.dirac.ch/diracwappen.html). After the first member of the Dirac family obtained citizenship in the town of Saint Maurice, Swiss law accorded the same rights of citizenship to succeeding generations.

13 This letter was written from Flo to Charles on 27 August 1897. This and the other extant letters from their correspondence are in Dirac Papers 1/1/8 (FSU). I am taking the arrival of e-mail for the UK public to be c. 1995.

14 Felix's full name was Reginald Charles Félix. His mother always anglicised his name, so I shall use that version of it here.

15 The Diracs' address was 15 Monk Road, Bishopston, Bristol. The house still stands. The date of the Diracs' move are in UKNATARCHI HO/144/1509/374920.

16 The details of Dirac's birth are given in a letter from Flo to Paul and Manci, 18 December 1939, Dirac Papers, 1/5/1 (FSU). The description of Dirac as 'rather small' and the colour of his eyes is given in the poem 'Paul', Dirac Papers, 1/2/12 (FSU). Charles gave his children names used in his mother's family, the Pottiers. The origins of his children's names are as follows: Reginald Charles Felix was named after himself and after his grandfather Felix Jean Adrien Pottier; Paul Adrien Maurice's second name was that of Charles's maternal grandfather Pottier, and Maurice is probably in memory of his native town, Saint Maurice; Beatrice Isabelle Marguerite Walla's last name came from Charles's mother Julie Antoinette Walla Pottier, and she was probably named after Flo's sister Beatrice.

17 Letter to Dirac from his mother, 18 December 1939, Dirac Papers, 1/4/9 (FSU).

18 Sunday Dispatch, 19 November 1933 (p. 17).

19 On 16 May 1856, the Bristol Times and Mirror called the area 'the people's park' soon after the council had taken the popular step in the early 1860s of acquiring it from its owners, who included the Merchant Venturers' Society.

20 Mehra and Rechenberg (1982: 7n). The authors point out that Dirac checked the information they included about his early life.

21 Dirac Papers, 1/1/12 (FSU).

22 Dirac Papers, 1/1/9 (FSU).

23 In the Dirac family archive, there is a copy of one of these postcards, marked by Charles Dirac on the back with the date 3 September 1907, presumably the date on which the photograph was taken (DDOCS).

24 The friends were Esther and Myer Salaman, see Salaman and Salaman (1986: 69). The Salamans comment that Dirac read their account of his memories and verified them. For the earlier interview with AHQP on 4 April 1962, see p. 6.

25 Interview with Dirac, AHQP, 4 April 1962; Salaman and Salaman (1986).

26 Dirac told his daughter Mary that his parents always denied him a glass of water at the dinner table: interview with Mary Dirac, 21 February 2003.

27 Letter from Dirac to Manci Balázs, 7 March 1936 (DDOCS).

28 Letter from Dirac to Manci Balázs, 9 April 1935 (DDOCS).

29 The school-starting age of five was introduced in the 1870 Education Act. Dirac's mother was in the first generation to benefit from compulsory education in England. Woodhead (1989: 5).

30 Detail about the late serving of breakfast from Manci Dirac to Gisela Dirac in August 1988 in Caslano, Ticino. Interview with Mary Dirac, 21 February 2003.

31 Details of the Bishop Road School in this period are available in the Head Teacher's report, in the BRISTRO archive: 'Bishop Road School Log Book' (21131/SC/BIR/L/2/1).

32 The source of these comments is family photos of the Dirac brothers and data on the boys' heights obtained when they were at school (see Felix's records in Dirac Papers, 1/6/1, FSU). In November 1914, Felix's height was five feet four inches, and his weight was one hundred and ten pounds, whereas Paul's height was four feet ten inches and his weight was sixty-six and a half pounds. Two years earlier, when Felix had the same age as Paul in late 1914, he was about the same height as his brother but was some twenty pounds heavier.

33 Felix's school reports (1908–12) are in Dirac Papers, 1/6/1 (FSU).

34 The description of Dirac as 'a cheerful little schoolboy' is given in his mother's poem 'Paul' in Dirac Papers, 1/2/12 (FSU).

35 See 'Report cards' in Dirac Papers, 1/10/2 (FSU).

36 Quoted in Wells (1982: 344). As an adult, Dirac did not add a letter L to the ends of words that end in the letter A, but he did have the characteristic practice among Bristolians of warmly accentuating the letter R; for example, in his pronunciation of 'universe'.

37 Dirac's school reports are in Dirac Papers, 1/10/2 (FSU).

38 Interview with Mary Dirac, 21 February 2003.

39 Interview with Flo Dirac, *Svenska Dagbladet*, 10 December 1933.

40 The technique, applied to engineering, became popular in Renaissance Florence. The architect 'Pipo' Brunelleschi used such drawings to help his clients visualise the buildings and artefacts and to give his assistants a set of instructions so that they could do their work in his absence.

41 In 1853, the first report of Sir Henry Cole's Department of Practical Art urged teachers to give the students exercises that 'contain some of the choicest elements of beauty, such as elegance of line, proportion and symmetry' (minutes of the Committee of the Council of Education [1852–3], HMSO, pp. 24–6). Aesthetic recommendations like this continued unabated in reports and guides to teaching for decades. In 1905, the Government's Board of Education stressed to junior schoolteachers that 'the scholar should be taught to perceive and appreciate beauty of form and colour. The feeling for beauty should be cherished, and treated as a serious school matter.' See Board of Education (1905).

42 Gaunt (1945: Chapters 1 and 2). The Aesthetic Movement was not the first flowering of the importance of beauty in British cultural life. For example, in the eighteenth century, it was important for people of taste to refer to the concept of beauty to demonstrate that they were cultured and intellectually distinguished. See Jones (1998). In 1835, Gautier defined the essence of aestheticism in the preface to one of his novels: 'Nothing is beautiful unless it is useless; everything useful is ugly, for it expresses a need and the needs of a man are ignoble and disgusting, like his poor and weak nature. The most useful place in the house is the lavatory.' Quoted in Lambourne (1996: 10).

43 Hayward (1909: 226–7).

44 Examples of Dirac's early technical drawings are in Dirac Papers, 1/10/2 (FSU). In one drawing, he gives an idealised image of a small building, showing two of its four vertical sides, this time taking full account of the perspective. Dirac underlines his understanding of perspective by showing that parallel lines on each side all meet at a single point in the far distance.

45 The Government's Board of Education had recommended: 'No angular system of hand-writing should be taught and all systems which sacrifice legibility and a reasonable degree of speed to supposed beauty should be eschewed,' Board of Education (1905: 69).

46 Government report on inspection on 10–12 February 1914, reported in the log book of Bishop Road School, stored in BRISTRO: 'Bishop Road School Log Book' (21131/SC/BIR/L/2/1).

47 Westfall (1993: 13).

48 Betty refers to her skating at the Coliseum rink in her letter to Dirac, 29 January 1937 (DDOCS).

49 'Paul', a poem by his mother, Dirac Papers, 1/2/12 (FSU). The relevant lines are: 'At eight years old in quiet nook / Alone, he stays, conning a book / On table high, voice strong and sweet / Poems of length he would repeat.'

50 Interview with Flo Dirac in *Svenska Dagbladet*, 10 December 1933.

51 'Recollections of the Merchant Venturers', 5 November 1980, Dirac Papers 2/16/4 (FSU).

52 Salaman and Salaman (1986: 69).

53 Dirac's scholarship covered his expenses at his next school, rising from £8 in the first year (1914–15) to £15 in the final year (1917–18). BRISTRO, records of the Bishop Road School, 21131/EC/Mgt/Sch/1/1.

54 Winstone (1972) contains dozens of photographs of Bristol during the period 1900–14.

55 Interview with Mary Dirac, 14 February 2004.

56 Dirac Papers, 1/10/6 (FSU). The lectures were held at the Merchant Venturers' Technical College, where Dirac would later study.

57 Testimony of H. C. Pratt, who attended Bishop Road School from 1907 to 1912, to Richard Dalitz in the mid-1980s.

Chapter two

1 Words by H. D. Hamilton (School Captain, 1911–13). This is the second verse of the song.

2 Lyes (n.d.: 5).

3 Pratten (1991: 13).

4 The following recollections were given to Richard Dalitz. Leslie Phillips attended the Merchant Venturer's School from 1915 to 1919. Some of Charles's codes are extant in Dirac Papers, 1/1/5 (FSU). In 1980, Dirac described his father's reputation in Dirac Papers, 2/16/4 (FSU).

5 Interview with Mary Dirac, 7 February 2003.

6 These comics, named after the 'penny stinker' (a cheap and nasty cigar), first became popular in the 1860s and were still popular in Dirac's youth. They were widely frowned upon for their lack of seriousness.

7 Interview with Mary Dirac, 21 February 2003.

8 Bryder (1988: 1 and 23). See also Bryder (1992: 73).

9 Interview with Mary Dirac, 26 February 2004.

10 Dirac's reports when he was at the Merchant Venturers' School are in Dirac Papers, 1/10/7 (FSU).

11 See, for example, the reports of the Government's Department of Science and Art, from 1854, London: Her Majesty's Stationery Office.

12 Stone and Wells (1920: 335–6).

13 Stone and Wells (1920: 357).

14 Stone and Wells (1920: 151).

15 Interview with Dirac, AHQP, 6 May 1963, p. 1.

16 Testimony to Richard Dalitz of J. L. Griffin, one of Dirac's fellow students in the chemistry class.

17 *Daily Herald*, 17 February 1933, p. 1.

18 Interview with Dirac, AHQP, 6 May 1963, p. 2.

19 Interview with Dirac, AHQP, 6 May 1963, p. 2.

20 Dirac remarked that he 'was very interested in the fundamental problems of nature. I would spend much time just thinking about them'. See Interview with Dirac, AHQP, 1 April 1962, p. 2.

21 Dirac (1977: 11); interview with Dirac, AHQP, 1 April 1962, pp. 2–3.

22 Wells (1895: 4).

23 See, for example, Monica Dirac, 'My Father', in Baer and Belyaev (2003).

24 Pratten (1991: 24).

25 Dirac (1977: 112).

26 Testimony of Leslie Roy Phillips (fellow pupil with Dirac at Merchant Venturers' School, 1915–19) given to Richard Dalitz in the 1980s.

27 Dirac Papers, 2/16/4 (FSU).

28 Interview with Dirac, AHQP, 6 May 1963, p. 2.

29 Later, Dirac received more books as prizes at the Merchant Venturers' School, including *Decisive Battles of the World* and Jules Verne's *Michael Strogoff*, an adventure story set in tsarist Russia. Some of the books Dirac won for school prizes at the Merchant Venturers' School are stored in the Dirac Library at Florida State University. Other information about Dirac's reading choices is from his niece Christine Teszler.

30 Letter from Edith Williams to Dirac, 15 November 1952, Dirac Papers, 2/4/8 (FSU).

31 From Merchant Venturers' School yearbooks 1919, BRISTRO 40659, 1.

32 Stone and Wells (1920: 360).

33 In the spring of 1921, Dirac planned the planting of vegetables on what looks like a geometric drawing of the garden in 6 Julius Road, with some annotations by his father. The plan, dated 24 April 1921, is in Dirac Papers, 1/8/24 (FSU).

34 The Bishopston local Norman Jones told Richard Dalitz in the mid-1980s that his most vivid memory of Charles was 'seeing him always carrying an umbrella, struggling up the hill, often with his daughter, of whom he was very fond', interviews with Richard Dalitz, private communication.

35 Interview with Dirac, AHQP, 1 April 1962. Felix's reports when he was at the Merchant Venturers' School are in Dirac Papers, 1/6/4 (FSU).

36 Quoted in Holroyd (1988: 81–3).

37 Interview with Monica Dirac, 7 February 2003; interview with Leopold Halpern, 18 February 2003.

38 The Merchant Venturers' School used the facilities during the day, and the college used them during the evening.

39 See Felix's university papers in Dirac Papers, 1/6/8 (FSU); the scholarships are recorded in BRISTRO 21131/EC/Mgt/sch/1/1.

40 Dirac took the qualifying examinations for the University of Bristol in 1917, three years earlier than most other applicants. He then spent a year studying advanced mathematics and finally qualified in 'physics, chemistry, mechanics, geometrical and mechanical drawing and additional mathematics', enabling him to take a degree in any technical subject. See Dirac Papers, 1/10/13 (FSU); details of Dirac's matriculation are also in a letter to him from his friend Herbert Wiltshire, 10 February 1952, Dirac Papers, 2/4/7 (FSU).

41 Interview with Dirac, AHQP, 6 May 1963, p. 7.

42 Interview with Flo Dirac, *Svenska Dagbladet*, 10 December 1933.

Chapter three

1 Stone and Wells (1920: 371–2).
2 *Bristol Times and Mirror*, 12 November 1918, p. 3.
3 'Recollections of Bristol University', Dirac Papers, 2/16/3 (FSU).
4 Lyes (n.d.: 29). At the Dolphin Street picture house, for example, Fatty Arbuckle starred in *The Butcher Boy*.
5 Quoted in Sinclair (1986).
6 Dirac Papers, 2/16/3 (FSU).
7 The list of textbooks that Dirac studied as an engineering student is in Dirac Papers, 1/10/13 and 1/12/1 (FSU).
8 BRISTU, papers of Charles Frank. '[N]ot the faintest idea' is the testimony of Mr S. Holmes, a lecturer in electrical engineering, given to G. H. Rawcliffe, who, in turn, passed it to Charles Frank on 3 May 1973.
9 Papers of Sir Charles Frank, BRISTU. 'Even as an engineering student, he spent much time reading in the Physics Library,' wrote Frank in a note in 1973.
10 The college had classes on Saturday mornings as well as during weekdays (as was traditional, Wednesday afternoons were usually free for sporting activities). Information on Dirac at the Merchant Venturers' College is in the college's Year Books (BRISTRO 40659/1). Dirac's student number was 1429.
11 Letter to Dirac from Wiltshire, 4 May 1952, Dirac Papers, 2/4/7 (FSU). The first two names of Wiltshire, known to most people as Charlie, were Herbert Charles.
12 Dirac Papers, 2/16/3 (FSU).
13 Dirac Papers, 2/16/3 (FSU).
14 Interview with Leslie Warne, 30 November 2004.
15 Records of the Merchant Venturers' Technical College, BRISTRO.
16 The photograph shows the visit of the University Engineering Society's visit to Messrs. Douglas' Works, Kingswood, 11 March 1919, Dirac Papers, 1/10/13 (FSU).
17 'Miscellaneous collection, FH Dirac', September 1915, Dirac Papers, 1/2/2 (FSU).
18 Testimony to Richard Dalitz by E. B. Cook, who taught with Charles from 1918 to 1925.
19 Testimony to Richard Dalitz by W. H. Bullock, who joined the Cotham Road School staff in 1925 and was later Charles's successor as Head of the French Department.
20 Charles Dirac's letter is reproduced in Michelet (1988: 93).
21 See Charles Dirac's Certificate of Naturalization, Dirac Papers, 1/1/3 (FSU). The papers concerning Charles Dirac's application for British citizenship are in UKNATARCHI HO/144/1509/374920.
22 Interview with Mary Dirac, 21 February 2003.
23 Interview with Dirac, AHQP, 1 April 1962, p. 6.
24 Letter to Dirac from Wiltshire, 10 February 1952, Dirac Papers, 2/4/7 (FSU).
25 Dirac (1977: 110).
26 Sponsel (2002: 463).
27 Dirac (1977: 110).
28 Five shillings (25 pence) secured a copy of *Easy Lessons in Einstein* by Dr E. L. Slosson, a guinea (£1.05) *The Reign of Relativity* by Viscount Haldane.
29 Eddington (1918: 35–9).
30 Dirac Papers, 1/10/14 (FSU).
31 Testimonies of Dr J. L. Griffin, Dr Leslie Roy Phillips and E. G. Armstead, provided to Richard Dalitz.
32 Letter to Dirac from his mother, undated but written at the beginning of his sojourn in Rugby, c. 1 August 1920, Dirac Papers, 1/3/1 (FSU).

33 *Rugby and Kineton Advertiser*, 20 August 1920.

34 Letters to Dirac from his mother, August and September 1920, especially 30 August and 15 September (FSU).

35 Interview with Dirac, AHQP, 1 April 1962, p. 7.

36 Letter from G. H. Rawcliffe, Professor of Electrical Engineering at Bristol to Professor Frank on 3 May 1973. BRISTU, archive of Charles Frank.

37 Broad (1923: 3).

38 Interview with Dirac, AHQP, 1 April 1962, p. 4 and 7.

39 Schilpp (1959: 54–5). I have replaced Broad's archaic term 'latches' with 'laces'.

40 Broad (1923: 154). This book is based on the course of lectures that Broad gave to Dirac and his colleagues. Broad prepared all his lectures meticulously and wrote them out in advance, making it easy for him to publish them. What we read in this book is therefore likely to be the material that Broad presented to Dirac.

41 Broad (1923: 486).

42 Broad (1923: 31).

43 Dirac (1977: 120).

44 Dirac (1977: 111).

45 Schultz (2003: Chapters 18 and 19).

46 Galison (2003: 238).

47 Skorupski (1988).

48 Mill (1892). His most important comments about the nature of science are in Book 2 and in Book 3 (Chapter 21).

49 Dirac (1977: 111).

50 Interview with Dirac, AHQP, 6 May 1963, p. 6.

51 See http://www.uh.edu/engines/epi426.htm (accessed 27 May 2008).

52 Nahin (1987: 27, n. 23). Heaviside never completed his autobiography.

53 Interview with Dirac, AHQP, 6 May 1963, p. 4. Another example of the kind of neat tricks that engineers use and that Dirac read about as an engineering student is featured in the appendix to one of his set textbooks (Thomälen, 1907).

54 The two books that Dirac used to study stress diagrams were Popplewell (1907) (see especially Chapter 5) and Morley (1919) (see especially Chapter 6).

55 Dirac (1977: 113).

56 The 'spoilsport' taught Dirac in the autumn of 1920. Dirac's reports are in Dirac Papers, 1/10/16 (FSU).

57 Interview with Dirac, AHQP, 6 May 1963, p. 13. Dirac's lack of a qualification in Latin was not a bar to his admission to postgraduate study at Cambridge, but it would have made him ineligible to study there as an undergraduate.

58 Warwick (2003: 406 n.); Vint (1956).

59 Letter from Charles Dirac, 7 February 1921, STJOHN.

60 Dirac took the examination on 16 June 1921. The examination papers are in Dirac Papers, 1/10/11 (FSU).

61 Letter from Dirac to the authorities at St John's College, 13 August 1921, STJOHN.

62 Boys Smith (1983: 23). A much higher estimate of the amount needed to live as a student in Cambridge at the time is given in Howarth (1978: 66): about £300.

63 Letter from Charles Dirac, 22 September 1921, STJOHN.

64 Unsigned letter from St John's College to Charles Dirac, 27 September 1921, STJOHN. The signatory concludes his letter: 'Perhaps before deciding [what to do] you would be so kind as to let me know the sum total of means that he would have at his disposal, I could then better advise what he can do.'

Chapter four

1 Interview with Dirac, AHQP, 6 May 1963, p. 9.
2 Recollections of Dirac's first term in the mathematics class are from the testimony of E. G. Armstead in a letter to Richard Dalitz. The lecturer concerned was Horace Todd.
3 Dirac (1977: 113); interview with Dirac, AHQP, 6 May 1963, p. 10.
4 Interview with Dirac, AHQP, 1 April 1962, p. 3.
5 It is likely that Dirac learned this subject from *Projective Geometry* by G. B. Matthews (1914), published by Longmans, Green and Co. This book apparently meant a lot to him as it was one of the few books from his youth that he kept until his death. His copy is kept in his private library, stored in the Dirac Library, Florida State University.
6 Dirac studied four courses in pure mathematics: 'Geometry of Conics; Differential Geometry of Plane Curves', 'Algebra and Trigonometry; Differential and Integral Calculus', 'Analytical Projective Geometry of Conics' and 'Differential Equations, Solid Geometry'. See Bristol University's prospectus for 1922–3, BRISTU.
7 Dirac studied four courses in applied mathematics: 'Elementary Dynamics of a Particle and of Rigid Bodies', 'Graphical and Analytical Statics; Hydrostatics', 'Dynamics of a Particle and of Rigid Bodies' and 'Elementary Theory of Potential with Applications to Electricity and Magnetism'. See Bristol University's prospectus for 1922–3, BRISTU.
8 Testimony of Norman Jones (who attended the Merchant Venturers' School from 1921 to 1925) to Richard Dalitz in the 1980s. Private communication from Dalitz.
9 Interview with Dirac, AHQP, 1 April 1962, p. 8, and 6 May 1963, p. 10.
10 The inclusion of the lectures on special relativity can be deduced from the presence of examination questions on the subject. See Dirac Papers, 1/10/15 and 1/10/15A (FSU).
11 The term 'non-commuting' was introduced by Dirac later in the 1920s.
12 Cahan (1989: 10–24); Farmelo (2002a: 7–12).
13 Letter from Hassé to Cunningham, 22 March 1923, STJOHN.
14 Interview with Dirac, AHQP, 6 May 1963, p. 14. During Dirac's first visit to Cambridge, he had met Cunningham.
15 Warwick (2003: 466, 467, 468, 493 and 495).
16 Stanley (2007: 148); see also Cunningham (1970: 70), STJOHN.
17 Letter from Ebenezer Cunningham to Ronald Hassé, 16 May 1923, and letter from Dirac to James Wordie, 21 July 1923, STJOHN. The grant from the Department of Science and Industrial Research was technically a maintenance allowance for research. Wordie became Dirac's tutor in his early years in Cambridge. Postcard from Dirac to his parents, 25 October 1926 (DDOCS).
18 Dirac often spoke to close friends of the significance of this gesture by his father. Among those to attest to this: Kurt Hofer in an interview on 21 February 2004, Leopold Halpern in an interview in February 2006 and Nandor Balázs in an interview on 24 July 2002.

Chapter five

1 Gray (1925: 184–5).
2 Boys Smith (1983: 10).
3 See contemporary issues of the Cambridge students' magazine *The Granta*; for example, the poem 'The Proctor on the Granta', 19 October 1923.

4 Boys Smith (1983: 20).

5 Dirac kept the lodging accounts for the digs where he stayed as a student. See Dirac Papers, 1/9/10 (FSU). Dirac's landlady at 7 Victoria Road was Miss Josephine Brown, and he resided with her from October 1923 to March 1924. From April to June 1924, he stayed at 1 Milton Road. In his final postgraduate year, he lived at 55 Alpha Road.

6 College records attest that he took his meals there: his bill for food in college during his first term was £8 17s 0d, about the same as other students who ate there (STJOHN). The bill from Miss Brown includes no charges at all for either 'cooking' or 'food supplied'.

7 From documents in STJOHN. A typical example of a menu that Dirac would have been offered is the following, served on 18 December 1920: 'Hare soup / Boiled mutton / Potatoes, mashed turnips, carrots au beurre / Pancakes / Ginger mould / Hot and cold pie / Anchovy eggs'. He will not have gone hungry.

8 Interview with Monica Dirac, 7 February 2003.

9 Interview with Mary Dirac, 21 February 2003. Dirac's words were 'give myself courage'.

10 Interview with John Crook, 1 May 2003.

11 Boys Smith (1983: 7).

12 See contemporary issues of the Cambridge students' magazine *The Granta*.

13 Werskey (1978: 23).

14 Snow (1960: 245). See also Dirac (1977: 117).

15 Needham (1976: 34).

16 Stanley (2007: Chapter 3), especially pp. 121–3; Earman and Glymour (1980: 84–5).

17 Hoyle (1994: 146).

18 de Bruyne, N. in Hendry (1984: 87).

19 This description is taken mainly from Snow (1960), and from Cathcart (2004: 223).

20 Wilson (1983: 573).

21 Oliphant (1972: 38).

22 Mott (1986: 20–2); Hendry (1984: 126).

23 Oliphant (1972: 52–3).

24 Carl Gustav Jung introduced the words 'extrovert' and 'introvert' into the English language in 1923.

25 'Naval diary, 1914–18. Midshipman', by Patrick Blackett, pp. 80–1. Text kindly supplied by Giovanna Blackett.

26 Nye (2004: 18, 24–5).

27 Boag et al. (1990: 36–7); Shoenberg (1985: 328–9).

28 Boag et al. (1990: 34).

29 Chukovsky's first book, *Crocodile*, was published in 1917. I am indebted to Alexei Kojevnikov for this information. Chadwick later recalled Kapitza's first explanation of the nickname: when discussing his work with Rutherford, Kapitza was always afraid of having his head bitten off. (Chadwick papers, II 2/1 CHURCHILL). Chadwick dismissed other explanations (e.g. Boag et al. 1990:11).

30 Letter from Keynes to his wife Lydia, 31 October 1925, Keynes archive, JMK/PP/45/190/3/14 to JMK/PP/45/190/3/16 (KING'S © 2008).

31 Spruch (1979: 37–8); Gardiner (1988: 240). See also *The Cambridge Review*, 7 March 1942; Boag et al. (1990: 30–7).

32 Parry (1968: 113).

33 Letter from Kapitza to V. M. Molotov, 7 May 1935, translated in Boag et al. (1990: 322).

34 See Hughes (2003), Section 1.

35 Childs, W., Scotland Yard, to Chief Constable, Cambridge, 18 May 1923, KV 2/777, UKNATARCHI.

36 Werskey (1978: 92); Brown (2005: 26, 40).

37 I am grateful to Maurice Goldhaber for his recollections of the meetings of the Kapitza Club, moderated by Kapitza, in 1933 and the first two terms of 1934.

38 Blackett (1955).

39 Postcard from Dirac, 16 August 1925 (DDOCS).

40 See, for example, letters to Dirac from his mother, 26 October and 16 November 1925, 2 June 1926, 7 April 1927: Dirac Papers, 1/3/5 and 1/3/6 (FSU).

41 Ramsay MacDonald's Labour Government was a minority one, whose survival depended on support from at least one of the other two parties. This partly explains the Government's moderate agenda.

42 Letter to Dirac from his mother, 9 February 1924, Dirac Papers, 1/3/3 (FSU).

43 In one letter, c. 1924, Felix requests a weekly wage of £2 10s 0d. Dirac Papers, 1/6/3 (FSU).

44 The spelling of the Reverend's name is not completely clear. His letters to Felix, including one dated 25 September 1923 and another dated 21 September, are in Dirac Papers, 1/6/6 (FSU). I am grateful to Peter Harvey for his advice on the theosophy of Felix's correspondent and to Russell Webb for pointing out the tone of the Reverend's letters, from the point of view of a follower of Eastern philosophy.

45 Interview with Dirac, AHQP, 1 April 1962, pp. 5–6.

46 Cunningham (1970: 65–6).

47 Description of Compton is from the article 'Compton Sees a New Epoch in Science', *New York Times*, 13 March 1932.

48 Einstein (1949), in Schilpp (1949: 47).

49 Hodge (1956: 53). Details of Dirac's early mathematical and scientific influences in Cambridge are in the final section of Darrigol (1992).

50 Cunningham, E., 'Obituary of Henry Baker', *The Eagle*, 57: 81. Dirac (1977: 115–16).

51 *Edinburgh Mathematical Notes*, 41, May 1957.

52 Quoted in Darrigol (1992: 299–300).

53 Moore (1903: 201); Baldwin (1990: 129–30). Moore's conception of the role of art in relation to morality is prefigured in Hegel and thence by his successors. Moore adapts this position to the utilitarian scheme that he took over from the Victorian thinker Henry Sidgwick. John Stuart Mill anticipates Moore through the conception of the great value of the 'higher' pleasures.

54 As Budd describes Kant's conception of the experience of beauty, it was 'the facilitated play of imagination and understanding, mutually quickened (and so made pleasurable) by their reciprocal harmony' (2002: 32).

55 Boag et al. (1990: 133).

56 Letter from Einstein to Heinrich Zangger, 26 November 1915.

57 This, and all of Dirac's publications until the end of 1948, is reproduced in Dalitz (1995).

58 Interview with Dirac, AHQP, 7 May 1963, p. 7.

59 Orwell (1946: 10).

Chapter six

1 Reference for Dirac by Cunningham, April 1925, provided for Dirac's application for a Senior Studentship, 1851COMM.

2 Undated to Dirac from his mother, c. May 1924.

3 Dirac was in room H7 on the first floor of New Court in Michaelmas (autumn) term. Later, he moved into other rooms: in Lent (winter) and Easter term 1925, he was in New Court room E12; from Michaelmas term in 1927 to Easter term 1930, he was in New Court room A4; in Michaelmas term in 1930, he was in Second Court room C4; from Michaelmas term 1936 to Michaelmas 1937, he was in New Court room I10.

4 Letter from Dirac to Max Newman, 13 January 1935, Newman archive in STJOHN.

5 Letter to Dirac from his mother, undated, c. November 1924, Dirac Papers, 1/3/3 (FSU).

6 Letter from 'Technical Manager' (unnamed) at W & T Avery Ltd, 10 January 1925, Dirac Papers, 1/6/3 (FSU).

7 Interview with Dirac, AHQP, 1 April 1962, p. 5; Salaman and Salaman (1986: 69). I am assuming that the date of Felix's death on his gravestone, 5 March 1925, is correct; on his death certificate, the date of his death is given as the day after.

8 Letter to Dirac from his Auntie Nell, 9 March 1925, Dirac Papers, 2/1/1 (FSU).

9 *Express and Star* (local paper in Much Wenlock), 9 March 1925; *Bristol Evening News*, 27 March 1925.

10 Interview with Mary Dirac, 21 February 2003; interview with Monica Dirac, 7 February 2003. In an interview with Leopold Halpern, 18 February 2003, Halpern commented that Dirac found the suicide of Felix too painful to talk about.

11 *Bristol Evening News*, 9 March 1925.

12 *Bristol Evening News*, 10 March 1925.

13 Dirac often remarked on this. His feelings are recorded in Salaman and Salaman (1986: 69). His close friend Leopold Halpern also mentioned that Dirac had mentioned this to him, quite independently (interview on 18 February 2002).

14 Letter to Dirac from his mother, 4 May 1925, Dirac Papers, 1/3/4 (FSU). Dirac always mentioned this when he opened his heart to friends and even mentioned it to his children.

15 Flo wrote her poem 'In Memoriam. To Felix' on 5 March 1938. The poem is in Dirac Papers, 1/2/12 (FSU).

16 Letter to Dirac from his mother, 22 March 1925, Dirac Papers, 1/3/4 (FSU).

17 Death cerificate of Felix Dirac, registered 30 March 1925.

18 Interview with Leopold Halpern, 18 February 2003.

19 Interview with Christine Teszler, 22 January 2004.

20 The problem that Dirac addressed was: if light consists of photons, as Compton had argued, how would these particles be affected by collisions with electrons swirling around on the surface of the Sun?

21 Mehra and Rechenberg (1982: 96).

22 Dirac (1977: 118).

23 C. F. Weizsächer, in French and Kennedy (1985: 183–4).

24 Pais (1967: 222). Pais gives a vivid description of Bohr's strange oratory, noting 'Bohr's precept never to speak more clearly than one thinks.'

25 Letters from Bohr to Rutherford, 24 March 1924 and 12 July 1924, UCAM Rutherford archive.

26 Elsasser (1978: 40–1).

27 In his AHQP interview on 1 April 1962 (p. 9) and in an interview on 26 June 1961 (Van der Waerden 1968: 41), Dirac says he was not present, whereas elsewhere he says he was there (Dirac 1977: 119).

28 Heisenberg recalls his experience at the Kapitza Club, and of staying with the Fowlers, in the BBC *Horizon* programme 'Lindau', reference 72/2/5/6025. The recording was made on 28 June 1965, in Dirac's presence.

29 The application is held by the 1851COMM.

30 Letter to Dirac from his mother, with a contribution from his father, June 1925, in Dirac Papers, 1/3/4 (FSU). The application was advertised in the *Times Higher Education Supplement*, his mother says.

31 This proof copy is in Dirac Papers, 2/14/1 (FSU).

32 An English translation of this paper, together with other key papers in the early history of quantum mechanics, are reprinted in Van der Waerden (1967).

33 Dirac (1977: 119).

34 Interview with Flo Dirac, *Stockholms Dagblad*, 10 December 1933.

35 Darrigol (1992: 291–7).

36 Dirac (1977: 121).

37 Letter from Albert Einstein to Paul Ehrenfest, 20 September 1925, in Mehra and Rechenberg (1982: 276).

38 Dirac (1977: 121–5).

39 Dirac (1977: 122).

40 Here, X and Y are mathematical expressions of a type known as partial differentials. What is important is the superficial similarity between the form of the Poisson bracket and the difference AB – BA.

41 Eddington (1928: 210).

42 Elsasser (1978: 41).

43 Reference for Dirac, written by Fowler in April 1925, for the Royal Commission of the Exhibition of 1851, 1851COMM.

44 Dalitz and Peierls (1986: 147). The student was Robert Schlapp, who was studying under the veteran Sir Joseph Larmor.

45 Van der Waerden (1960).

46 Letters from Oppenheimer to Francis Fergusson, 1 November and 15 November 1925; in Smith and Weiner (1980: 86–9).

47 Bird and Sherwin (2005: 44).

48 Letter to Dirac from his mother, 16 November 1925 (she repeats the image of 'the block of ice' in another letter to Dirac, written on 24 November), Dirac Papers, 1/3/4 (FSU).

49 Heisenberg later remarked that when he read Dirac's first paper on quantum mechanics, he assumed that its author was a leading mathematician (BBC *Horizon* programme, 'Lindau', reference 72/2/5/6025).

50 Frenkel (1966: 93).

51 Born (1978: 226).

52 Letter to Dirac from Heisenberg, 23 November 1925, Dirac Papers, 2/1/1 (FSU).

53 All these letters from Heisenberg to Dirac at this time are in Dirac Papers, 2/1/1 (FSU).

54 Beller (1999: Chapter 1); see also Farmelo (2002a: 25–6).

Chapter seven

1 Letter from Einstein to Michel Besso, 25 December 1925, quoted in Mehra and Rechenberg (1982: 276).

2 Letter from Einstein to Ehrenfest, 12 February 1926, quoted in Mehra and Rechenberg (1982: 276).

3 Bokulich (2004).

4 Dirac (1977: 129).

5 Slater (1975: 42).

6 Jeffreys (1987).

7 Bird and Sherwin (2005: 46).

8 Interview with Oppenheimer, AHQP, 18 November 1963, p. 18.

9 'The Cambridge Review', 'Topics of the Week' on 14 March and 12 May 1926.

10 Letters to Dirac from his mother, 16 March 1926 and 5 May 1926, Dirac Papers, 1/3/5 (FSU).

11 Morgan et al. (2007: 83); Annan (1992: 179–80); Brown (2005: 40 and Chapter 6); Werskey (1978: 93–5).

12 Quoted in Brown (2005: 75).

13 Wilson (1983: 564–5).

14 Morgan et al. (2007: 84).

15 Morgan et al. (2007: 80–90).

16 Dirac Papers, 2/1/2 (FSU).

17 This description follows the one given by Kapitza of his Ph.D. graduation ceremony three years before, when the proceedings were the same. See Boag et al. (1990: 168–9).

18 Letter to Dirac from his mother, 28 June 1926, Dirac Papers, 1/3/5 (FSU).

19 The Cambridge newspapers reported a wave of heat deaths in July. See the *Cambridge Daily News*, 15 August 1926, the hottest day in the town for three years.

20 Dirac had carefully studied a derivation of the radiation spectrum produced by the previously unknown Satyendra Bose, a student in Calcutta. No one had understood quite why his derivation worked. Einstein developed Bose's ideas to produce a theory that is now named after both men.

21 Postcard from Dirac to his parents, 27 July 1926, DDOCS.

22 Letter to Dirac from Fermi, Dirac Papers 2/1/3 (FSU).

23 Greenspan (2005: 135); Schücking (1999: 26).

24 Letter to Dirac from his mother, 2 October 1926, Dirac Papers 1/3/6.

25 Mott (1986: 42).

Chapter eight

1 Wheeler (1998: 128–9). On 24 April 1932, Jim Crowther wrote of hearing a similar anecdote from Bohr over afternoon tea (Book I of Crowther's notes from his meeting with Bohr, pp. 99–100 [SUSSEX]).

2 Book I of Crowther's notes from his meeting with Bohr, 24 April 1932, pp. 96–101, SUSSEX. See also the article on Dirac by John Charap in *The Listener*, 14 September 1972, pp. 331–2.

3 Book I of Crowther's notes from his meeting with Bohr, p. 99, SUSSEX.

4 Dirac (1977: 134).

5 Bohr's words (*Nicht um zu kritisieren aber nur um zu lernen*) are quoted in Dirac (1977: 136).

6 Postcard from Dirac to his parents, 1 October 1926 (DDOCS).

7 Letter from Dirac to James Wordie, 10 December 1926, STJOHN; Dirac (1977: 139).

8 The phrase 'liked the sound of his own voice' is taken from the letter John Slater wrote to John Van Vleck on 27 July 1924, John Clarke Slater papers APS. See also Cassidy (1992: 109).

9 Crowther notes, p. 99, SUSSEX.

10 The wave is what is known mathematically as a complex function, which means that the wave at any point has two parts: one real, the other imaginary. The 'size' of the wave at any point, related to both parts, is called its modulus. According to Born, the probability of detecting the quantum in a tiny region near a point is related to the *square* of the modulus of the wave.

11 Pais (1986: 260–1).

12 Heisenberg (1967: 103–4).

13 Interview with Oppenheimer, AHQP, 20 November 1963.

14 Weisskopf (1990: 71).

15 Interview with Dirac, AHQP, 14 May 1963, p. 9.

16 Garff (2005: 308–16, 428–31).

17 Interview with Monica Dirac, 3 May 2006.

18 Quoted in Garff (2005: 311); interview with Dirac, AHQP, 14 May 1963, p. 9.

19 Møller (1963).

20 Dirac had also seen the need for the function when he was studying Eddington's *The Mathematical Theory of Relativity* (1923). On page 190, Eddington uses non-rigorous mathematics, and he drew attention to this in a footnote, which Dirac read. This was an example of the case where the delta function is needed to make some sense of a scientific equation which would otherwise be mathematically unintelligible. See interview with Dirac, AHQP, 14 May 1963, p. 4.

21 Interview with Dirac, AHQP, 6 May 1963, p. 4.

22 Heaviside (1899: Sections 238–42).

23 Lützen (2003: 473, 479–81).

24 Interview with Heisenberg, AHQP, 19 February 1963, p. 9.

25 Dirac (1962), report of the Hungarian Academy of Sciences, KFKI-1977-62.

26 Letter from Einstein to Conrad Habicht, 24 December 1907, see Pais (1982: 441).

27 Dirac mentioned this in a press release issued by Florida State University on 24 November 1970; Dirac Papers, 2/6/9 (FSU).

28 Letters to Dirac from his mother, 19 November, 26 November, 2 December, 9 December 1926, Dirac Papers, 1/3/6 (FSU).

29 It is possible that Charles wrote other letters to Dirac. If so, Dirac did not keep them – uncharacteristically, as he appears to have kept most of his family correspondence. Moreover, the frequent letters from Dirac's mother often send messages from his father, indicating that his father was communicating to his son via her, a common arrangement in family correspondence of this type.

30 Letter to Dirac from his father, 22 December 1926, Dirac Papers, 1/1/7 (FSU).

31 Letter to Dirac from his mother, 25 December 1926, Dirac Papers, 1/3/6 (FSU).

32 Mehra (1973: 428–9).

33 Postcard from Dirac to his parents, 10 January 1927, DDOCS.

34 Slater (1975: 135).

35 Elsasser (1978: 91).

36 Born (2005: 88).

37 'The deepest thinker': Dirac (1977: 134).

38 'The most remarkable scientific mind . . .': Crowther notes, p. 21, SUSSEX. The 'logical genius' comment is in the interview with Bohr, AHQP, 17 November 1962, p. 10.

39 Both quotes from the Crowther notes, p. 97, SUSSEX.

40 'PAM Dirac and the Discovery of Quantum Mechanics', Cornell colloquium, 20 January 2003, available at http://arxiv.org/PS_cache/quant-ph/pdf/0302/0302041v1.pdf (accessed 24 September 2007).

Chapter nine

1 Bird and Sherwin (2005: 62).

2 Bernstein (2004: 23).

3 Bird and Sherwin (2005: 65).

4 The address of the Carios' home was Giesmarlandstrasse 1. See interview with Oppenheimer, AHQP, 20 November 1963, p. 4.

5 Michalka and Niedhart (1980: 118).

6 Frenkel (1966: 93).

7 Interview with Gustav Born, 6 April 2005.

8 Frenkel (1966: 93).

9 Weisskopf (1990: 40).

10 Bird and Sherwin (2005: 56, 58).

11 See Frenkel (1966: 94) for a reference to the practice of Mensur in Göttingen. See also Peierls (1985: 148).

12 Interview with Oppenheimer, AHQP, 20 November 1963, p. 6.

13 Interview with Oppenheimer, AHQP, 20 November 1963, p. 11.

14 Delbrück, M. (1972) 'Homo Scientificus According to Beckett', available at http://www.ini.unizh.ch/~tobi/fun/max/delbruckHomoScientificusBecket1972.pdf, p. 135 (accessed 13 May 2008).

15 Greenspan (2005: 144–6).

16 Elsasser (1978: 71–2).

17 Letter from Raymond Birge to John Van Vleck, 10 March 1927, APS.

18 Elsasser (1978: 51).

19 Frenkel (1966: 96).

20 Delbrück (1972: 135).

21 Wigner (1992: 88).

22 Mill's comment is in Mill (1873: Chapter 2).

23 Interview with Oppenheimer, AHQP, 20 November 1963, p. 11.

24 During his time in Göttingen, Dirac successfully applied his theory to the light emitted by atoms when they make quantum jumps, apparently after discussions with Bohr. See Weisskopf (1990: 42–4).

25 Letter from Pauli to Heisenberg, 19 October 1926, reprinted in Hermann et al. (1979). See also Beller (1999: 65–6); Cassidy (1992: 226–46).

26 Heisenberg (1971: 62–3).

27 Heisenberg demonstrated that the principle also applied to energy and time and to other pairs of quantities known technically as 'canonically conjugate variables'.

28 This was a popular walk with students. See, for example, Frenkel (1966: 92). On 5 April 1927, Dirac referred to the walk in a postcard of the path to his parents (DDOCS).

29 Lecture by Dirac, 20 October 1976, 'Heisenberg's Influence on Physics': Dirac Papers, 2/29/19 (FSU); see also the interview with Dirac, AHQP, 14 May 1963, p. 10.

30 See the article on complementarity in French and Kennedy (1985), e.g. Jones, R.V. 'Complementarity as a Way of Life', pp. 320–4; see also the illustration of Bohr's coat of arms, p. 224.

31 Interview with Dirac, AHQP, 10 May 1969, p. 9.

32 Eddington (1928: 211). This book is an overview of the latest ideas in physics based on a series of lectures he gave between January and March 1927.

33 Eddington (1928: 209–10).

34 Dirac (1977: 114).

35 Dirac Papers, 2/28/35 (FSU). The seminar took place on 30 October 1972. See Farmelo (2005: 323).

Chapter ten

1 Interview with Oppenheimer, AHQP, 20 November 1963, p. 5.
2 Greenspan (2005: 137).
3 Goodchild (1985: 20). Even if Dirac did not write these words, he agreed with their sentiment; see interview with von Weizsächer, AHQP, 9 June 1963, p. 19.
4 Dirac (1977: 139); Greenspan (2005: 141).
5 Greenspan (2005: 142), and von Meyenn and Schücking (2001: 46). The student was Otto Heckmann. Boys Smith's comment is from a conversation with his former colleague at St John's College, Cambridge, Peter Goddard, 5 July 2006.
6 Information on scholarship from Angela Kenny, archivist, Royal Commission for the Exhibition of 1851 (e-mail, 10 December 2007).
7 Letter from Dirac to James Wordie, 28 February 1927, STJOHN.
8 Letter to Dirac from his mother, 28 June 1928, Dirac Papers, 1/3/8 (FSU).
9 Greenspan (2005: 145).
10 Greenspan (2005: 146).
11 Letter to Dirac from his mother, 7 April 1927, Dirac Papers, 1/3/7 (FSU).
12 Letter to Dirac from his mother, 20 May 1927, Dirac Papers, 1/3/7 (FSU).
13 Letter to Dirac from his mother, 6 January 1927, Dirac Papers, 1/3/7 (FSU).
14 Letter to Dirac from his mother, 10 February 1927, Dirac Papers, 1/3/7 (FSU).
15 Letter to Dirac from his mother, 20 May 1927, Dirac Papers, 1/3/7 (FSU).
16 Letter to Dirac from his mother, c. 26 March 1927, Dirac Papers, 1/3/7 (FSU).
17 Flo enjoyed the company of several men in her classes and even put Dirac in touch with one of them, a German-speaking insurance clerk Mr Montgomery ('Monty'). Letter to Dirac from his mother, 18 March 1927, Dirac Papers, 1/3/7 (FSU).
18 These recollections were given to Richard Dalitz in the 1980s.
19 Letter from Dirac to Manci Balázs, 7 April 1935, DDOCS.
20 Letter from Dirac to Manci Balázs, 17 June 1936, DDOCS.
21 Their address was 173 Huntingdon Road. Fen (1976: 161); Boag et al. (1990: 78).
22 The conference was held at L'Institut de Physiology Solvay au Parc Léopold, from 24 to 29 October 1927.
23 Letter from John Lennard-Jones (of Bristol University) to Charles Léfubure (Solvay official), 9 March 1928, SOLVAY.
24 See http://www.maxborn.net/index.php?page=filmnews (accessed 13 May 2008).
25 Heisenberg (1971: 82–8); interview with Heisenberg, AHQP, 27 February 1963, p. 9. The location of the hotel is specified in a letter to Dirac from the conference administrator on 3 October 1927: Dirac Papers, 2/1/5 (FSU).
26 Dirac (1982a: 84).
27 Interview with Heisenberg, AHQP, 27 February 1963, p. 9.
28 Heisenberg (1971: 85–6).
29 In the early 1850s, the *Punch* humorist Douglas Jerrold quipped about the controversial feminist writer Harriet Martineau, 'There is no God, and Harriet Martineau is her prophet.' See A. N. Wilson (2002), *The Victorians*, London: Hutchinson, p. 167.
30 Dirac Papers, 2/26/3 (FSU).
31 Dirac (1977: 140).
32 Dirac (1977: 141).

Chapter eleven

1 Menu from College records, STJOHN.
2 Crowther (1970: 39) and Charap (1972).
3 Interviews with Dirac, AHQP, 1 April 1962, p. 15; 7 May 1963, pp. 7–8.
4 Dirac gave contradictory accounts of the goal he was pursuing at that time. In one account, he stated that he was seeking the answer to the question 'How could one get a satisfactory relativistic theory of the electron?' (Dirac 1977: 141). In another account, he says that 'my dominating interest was to get a satisfactory relativistic theory of a particle, of the simplest possible kind, which was presumably a spinless particle.' Dirac wrote the latter words on a single sheet of paper headed 'Sommerfeld Atombau un Spektralinen II 539.18' in Dirac Papers, 2/22/15 (FSU). I prefer to use the 1977 account as it is the nearest thing we have to a carefully prepared history of Dirac's thinking in his own hand.
5 Farmelo (2002a: 133).
6 See the notes for Dirac's lectures in the 1970s and 1980s: 2/28/18–2/29/52 (FSU).
7 Huxley's 1870 Presidential Address to the British Association for the Advancement of Science in Huxley (1894). Dirac uses similar words: 'The originator of a new idea is always rather scared that some development may happen which will kill it' (1977: 143).
8 Interview with Dirac, AHQP, 7 May 1963, p. 14; Dirac (1977: 143).
9 Letter from Darwin to Bohr, 26 December 1927 (AHQP).
10 Interview with Rosenfeld, AHQP, 1 July 1963, pp. 22–3.
11 Mehra (1973: 320).
12 The talented young physicist Rudolf Peierls remarked that, even after a few days studying the equation, 'I have begun to have an inkling of what it deals with, but I haven't understood a single word.' Letter from Peierls to Hans Bethe, 4 May 1924, quoted in Lee (2007b: 33–4).
13 *Florida State University Bulletin*, 3 (3), 1 February 1978.
14 Slater (1975: 145).
15 Postcard from Darwin to Dirac, 30 October 1929, Dirac Papers, 2/1/9 (FSU).
16 Dirac gave courses on quantum mechanics in the Michaelmas and Lent terms of 1927–8 and was paid £100 for the pair: see the letter from the Secretary to the Faculty of Mathematics, 16 June 1927, Dirac Papers, 2/1/4 (FSU).
17 Crowther later affirmed that he had left the Communist Party by 1950, but it is not clear when he left it. I thank Allan Jones for this information.
18 Clipping, annotated by Charles Dirac, in Dirac Papers, 1/12/5 (FSU).
19 *The Times*, 5 October 1931, p. 21. This well-briefed article was written by a journalist who appears to have succeeded in persuading Dirac to speak about his work.
20 'Mulling over the Universe with Paul Dirac', interview by Andy Lindstrom, *Tallahassee Democrat*, 15 May 1983.
21 Letter to Dirac from his mother, 26 January 1928, Dirac Papers, 1/3/8 (FSU). See also postcard from Dirac to his parents, 1 February 1928 (DDOCS).
22 See the entry for Bishop Whitehead in *Crockford's Clerical Dictionary*, 1947, p. 1,416. See also Billington Harper (2000: 115–26, 129–33, 293–5). The quoted description of Mrs Whitehead is on p. 145. I thank Oliver Whitehead and the late David Whitehead, grandsons of Isabel Whitehead, for the information in the description of Isabel Whitehead's home.

Chapter twelve

1 Kojevnikov (1993: 7–8).

2 Peierls (1985: 62–3).

3 Kojevnikov (2004: 64–5).

4 Letter from Tamm to his wife, 4 March 1928, in Kojevnikov (1993: 7).

5 'The tulip fields are all in flower now': postcard from Dirac to his parents, 29 April 1928 (DDOCS). '[Leiden] is below sea level and there are nearly as many canals as streets': postcard from Dirac to his parents, 29 June 1927 (DDOCS).

6 Letter from Tamm to his wife, undated, Kojevnikov (1993: 8).

7 Casimir (1983: 72–3).

8 Brown and Rechenberg (1987: 128).

9 Letter from Heisenberg to Pauli, 31 July 1928, in Kronig and Weisskopf (1964).

10 Peierls (1987: 35). In this account, Peierls remembers going to the theatre, but it seems from his letter to Dirac of 14 September 1928 (Lee [2007: 50]) that they went to the opera. I am grateful to Professor Olaf Breidbach for his comments on early twentieth-century Prussian politesse.

11 Born (1978: 240) and Greenspan (2005: 151–3).

12 Schücking (1999: 27).

13 Bohr nicknamed Gamow 'Joe' after the standard name for cowboys in western movies, which Bohr especially liked (interview with Igor Gamow, 3 May 2004). See also Reines (1972: 289–99; see pp. 280); Mott (1986: 28).

14 The only exception is the paper that Dirac co-authored with Rutherford's student J. W. Harding, 'Photoelectric Absorption in Hydrogen-Like Atoms', in January 1932.

15 Gamow (1970: 14).

16 Wigner (1992: 9–15).

17 Letter from Gabriel Dirac to Manci Dirac, 5 September 1940: 'It may interest you to know that everybody (Prof [Max] Born, Morris [Pryce] and Daddy [Paul Dirac]) says that Johnny von Neumann is the world's best mathematician' (DDOCS).

18 Fermi (1968: 53–9).

19 Wigner (1992: 37–43).

20 Interview with Pat Wigner, 12 July 2005.

21 Dirac wrote to his parents on 18 July 1928: 'The woods here are full of fireflies in the evening. I have been to the top of the Harz mountains' (DDOCS).

22 Dirac's wife would later write to him: 'It seems the beautiful scenery has the same effect on you as a beautiful book has on me', 12 August 1938 (DDOCS).

23 Letter to Dirac from his mother, 12 July 1928, Dirac Papers, 1/3/8 (FSU).

24 Sinclair (1986: 32–3).

25 Letter from Dirac to Tamm, 4 October 1928, Kojevnikov (1993: 10). The conference lasted from 5 August to 20 August.

26 Brendon (2000: 241).

27 Salaman and Salaman (1986: 69). In this article, Dirac is quoted as giving 1927 as the date of the experience; this is impossible as he did not visit Russia that year.

28 He first took a boat to Constantinople (renamed Istanbul in the following year), then sailed on to Marseilles via Athens and Naples, before travelling across France and then home. He planned to arrive in Bristol on Monday, 10 September (letter from Dirac to his parents, 8 September 1928, DDOCS).

29 Letter to Dirac from his mother, 28 October 1928, Dirac Papers, 1/3/8 (FSU). A copy of the speech is in this file of the archive.

30 In mid-December, Dirac read a paper by Klein showing that the Dirac equation pre-

dicted that if a beam of electrons is fired at a barrier, more electrons will be reflected than were present in the original beam. It was as if a tennis ball struck a player's racket and not one but several balls flew off it.

31 Howarth (1978: 156).

32 *Cambridge Review*, 29 November 1929, pp. 153–4. See also the rhapsodic review in the *Times Literary Supplement*, 24 October 1929.

33 Draft letter to Dirac from L. J. Mordell, 4 July 1928, Dirac Papers, 2/1/7 (FSU).

34 Mott (1986: 42–3).

35 Letter from Jeffreys to Dirac, 14 March 1929, Dirac Papers, 2/1/8 (FSU).

36 St John's awarded Dirac a praelectorship in mathematical physics, which enabled him to devote himself entirely to research, apart from the presentation of his lecture course.

Chapter thirteen

1 Letter from Dirac to Oswald Veblen, 21 March 1929, LC, Veblen archive.

2 Scott Fitzgerald (1931: 459).

3 Letter from Dirac to Veblen, 21 March 1929, LC (Veblen archive).

4 Diaries of Dirac (DDOCS).

5 Fellows (1985); see the introduction (p. 4) and the conclusion.

6 Comment made by Bohr to Crowther, recorded by Crowther on 24 April 1932 in the Crowther archive, SUSSEX, Book II of his notebooks, pp. 96–7. For one of many retellings of this anecdote, see Infeld (1941: 171).

7 See the article on Roundy in the *Wisconsin State Journal* on the day after his death, on 10 December 1971.

8 The article is reproduced in its entirety in Kragh (1990: 72–3). The original is in Dirac Papers, 2/30/1 (FSU).

9 A check of the microfilm records of the *Wisconsin State Journal* reveals that the article was not published between 1 April and 29 May 1929 (the microfilm for 30 May is missing).

10 Van Vleck (1972: 7–16; see pp. 10–11).

11 Record of Dirac's payment as 'Lecturer in physics April and May 1929' is in WISC. Early in his stay, from 10–16 April, Dirac had spent almost a week based at the University of Iowa.

12 Dirac left Madison on 27 May and travelled to the Grand Canyon via Minneapolis, Kansas City and Winslow, Arizona.

13 Quoted in Brown and Rechenberg (1987: 134). This article gives much detail about Dirac and Heisenberg's preparations for their 1929 trip and the trip itself.

14 Mehra (1973: 816).

15 Brown and Rechenberg (1987: 136–7).

16 Interview with Leopold Halpern, 18 February 2003.

17 Brown and Rechenberg (1987: 139–41).

18 Heisenberg returned from the 1929 trip to be the best ping-pong player in the quantum community: interview with von Weiszächer, AHQP, 9 July 1963, p. 11.

19 Mehra (1973: 816).

20 Mehra (1972: 17–59).

21 The *jako* was commonly used to scent clothes in Japan at that time. Hearn (1896: 31n).

22 Dirac gives his timetable in his letter to Tamm on 12 September 1929, Kojevnikov (1993: 29); Brendon (2000: 234).

23 Letter to Dirac from his mother, 6 July 1929, Dirac Papers, 1/3/11 (FSU).

24 Letter to Dirac from his mother, 6 May 1929, Dirac Papers, 1/3/10 (FSU).
25 Postcards from Dirac to his parents, autumn 1929, DDOCS.
26 Interview with Oppenheimer, 20 November 1963, p. 23 (AHQP).
27 Fitzgerald (1931: 459).
28 Dirac (1977: 144).
29 Kojevnikov (2004: 56–9).
30 Pais, A. (1998: 36).
31 Letter from Dirac to Bohr, 9 December 1929, NBA.
32 Letter to Dirac from his mother, 11 October 1929, Dirac Papers, 1/3/10 (FSU). The spelling is the one used by Flo Dirac. Dirac expected to arrive home on 19 December (postcard from Dirac to his parents, 27 November 1929, DDOCS).
33 Letter from Dirac to Manci, 26 February 1936 (DDOCS).

Chapter fourteen

1 Cavendish Laboratory Archive, UCAM. The poem was apparently written as a Valentine's card to the electron.
2 Dirac, 'Symmetry in the Atomic World', January 1955. The draft, which features this analogy, is in Dirac Papers, 2/27/13 (FSU).
3 Cited in Kragh (1990: 101).
4 Gamow (1970: 70); letter from Dirac to Tamm, 20 March 1930, in Kojevnikov (1993: 39).
5 On Saturday, 16 February 1935, Van Vleck took D to 'A Disney Day' at a cinema in Boston. The documents, marked with Van Vleck's comment 'Dirac loved Mickey Mouse', are in the Van Vleck papers at AMS.
6 Dirac's formula is $n = -\log_2 [\log_2 (^2\sqrt{(\sqrt{\ldots \sqrt{2}})})]$, where the ellipsis (...) denotes the taking of n square roots. The story is related in Casimir (1984: 74–5), where the author asserts that Dirac killed the game using only three 2s. Each symbol in the formula is very common in mathematics, so Dirac's solution is within the rules of the game.
7 Postcard from Dirac to his parents, 20 February 1930 (DDOCS).
8 Telegram to Dirac from his mother, 22 February 1930, Dirac Papers, 1/3/12 (FSU).
9 Letter to Dirac from his mother, 24 February 1930, Dirac Papers, 1/3/12 (FSU).
10 The certificate of Dirac's election to the Fellowship of the Royal Society is available on the Society's website. The names of the 447 Fellows of the Society on 31 December 1929 are given in the Yearbook of the Royal Society 1931.
11 Letter to Dirac from his mother, 24 February 1930, Dirac Papers, 1/3/12 (FSU).
12 Letter from Hassé to Dirac, 28 February 1930, Dirac Papers, 2/2/1 (FSU).
13 Letter from Arnold Hitchings to the Bristol Evening Post, 14 December 1979.
14 In 1935, Dirac traded in this car. Dirac Papers, 1/8/2 (FSU).
15 Interview with John Crook, 1 May 2003.
16 Mott (1986: 42).
17 Dirac was well known for this practice. It is described explicitly by his climbing tutor Tamm in the course of the letter to his wife on 27 May 1931, Kojevnikov (1993: 55). See also Mott (1972: 2).
18 Interview with Monica Dirac, 7 February 2003; see also M. Dirac (2003: 42).
19 Letter from Taylor Sen (1986: 80). Howarth (1978: 104).
20 See, for example, Daily Telegraph, 12 February 1930, Manchester Guardian, 12–18 February 1930.
21 Peierls (1987: 36).
22 Letter to Dirac from his mother, 12 June 1930, Dirac Papers, 1/3/12 (FSU).

23 Kojevnikov (1993: 40), note on letter from Dirac to Tamm, 6 July 1930.

24 The *Guardian*, 'World Conference of Scientists', 3 September 1930. Crowther was probably the author of this report.

25 Ross (1962).

26 The venue and the time of the talk are in the records of the British Association for the Advancement of Science, BOD.

27 Delbrück (1972: 280–1).

28 The report of the Science News Service is in Dirac Papers, 2/26/8 (FSU).

29 *New York Times*, 10 September 1932.

30 I have translated the German word *quatsch* as 'crap'. Another, similar version of this anecdote is given in the interview with Guido Beck, AHQP, 22 April 1967, p. 23.

31 Among the most able students who were dissatisfied by Dirac's talks was Freeman Dyson, who recalls: 'I read Dirac's book hoping to learn quantum mechanics from it, and found it totally unsatisfactory.' E-mail from Dyson, 19 August 2006.

32 *Nature*, Vol. 127, 9 May 1931, p. 699.

33 Pauli's review is in Kronig and Weisskopf (1964: 1,397–8).

34 Einstein (1931: 73).

35 Leisure reading anecdote: Woolf (1980: 261); 'Where's my Dirac?' anecdote is from *Tallahasse Democrat*, 29 November 1970.

36 Hoyle (1994: 238).

37 Freeman (1991: 136–7).

38 Quoted in Charap (1972: 331).

39 Letter from Tamm to Dirac, 13 September 1930, in Kojevnikov (1993: 43).

40 Einstein (1931: 73).

41 Comment made by Einstein on his arrival in New York on 11 December 1930, reported in the *LA Times*, 12 December 1930, p. 1.

42 Letter to Dirac from Tamm, 29 December 1930, Kojevnikov (1993: 48–9).

43 Letter from Kemble to Garrett Birkhoff, 3 March 1933 (AHQP).

44 Dirac attended the dinner on 17 December 1932, Dirac Papers, 2/79/6 (FSU).

45 Letter from Kapitza to his mother, 16 December 1921, in Boag et al. (1990: 138–9).

46 Da Costa Andrade (1964: 48).

47 Da Costa Andrade (1964: 162).

48 Records of the Cavendish dinners (CAV 7/1) 1930, p. 10 (UCAM).

49 Records of the Cavendish dinners (CAV 7/1) 1930, p. 10 (UCAM).

50 Snow (1931).

51 Snow (1934). Dirac features in the book, and some of his opinions also appear, unattributed. See Snow (1934: 97–8 and 178–83).

52 Letter from Chandrasekhar to his father, 10 October 1930, quoted in Miller (2005: 96).

53 Letter to Dirac from his mother, 8 November 1930, Dirac Papers, 1/3/13 (FSU).

Chapter fifteen

1 Letter to Dirac from his mother, 27 April 1931, Dirac Papers, 1/4/1 (FSU). Dirac appears to have left Bristol on 15 April (postcard from Dirac to his parents, 15 April 1931, DDOCS).

2 Letter from Dirac to Van Vleck, 24 April 1931, AHQP.

3 Kapitza Club, 21 July 1931. See the Kapitza Club notebook in CHURCHILL.

4 Dirac (1982: 604); Dirac (1978).

5 The size of the force between two attracting monopoles separated by a millionth of a millimetre – roughly the distance between the electron and the proton in a hydrogen atom – is about a ten-thousandth of the weight of a medium-sized apple.

6 Heilbron (1979: 87–96).

7 Sherlock Holmes used these words in the novel *The Adventure of the Blanched Soldier* (1926), and used extremely similar words in several other stories.

8 The phrase 'theorist's theorist' is often applied to Dirac. See, for example, Galison (2000).

9 Tamm arrived in Cambridge on 9 May and left on 25 June.

10 Fen (1976: 181).

11 Crowther (1970: 103).

12 Letter from Tamm to his wife, undated c. May 1931, in Kojevnikov (1993: 54).

13 Letter to Dirac from Tamm, 18 May 1931, in Kojevnikov (1993: 54–5).

14 Werskey (1978: 92).

15 Annan (1992: 181).

16 James Bell (1896–1975) was one of Scotland's leading climbers and was fascinated by the Soviet Union. He stayed in contact with Dirac for decades.

17 Wersley (1978: 138–49).

18 Bukharin (1931).

19 Brown (2005: 107).

20 Letter to Dirac from Tamm, 11 July 1931, Dirac Papers, 2/2/4 (FSU).

21 Home Office Warrant 4081, 27 January 1931, KV 2/777, UKNATARCHI.

22 Postcard from Dirac to his parents, 13 July 1931 (DDOCS).

23 Letter to Dirac from his mother, 8 July 1931, Dirac Papers, 2/2/4 (FSU).

24 The most direct comment on this from Dirac was reported by his mother in her letter to Betty from Stockholm in December 1933: '[Dirac] says it is awful and time we made an improvement.' In her letters to Dirac, she often mentions the disrepair of the family home.

25 Letter to Dirac from his mother, 19 July 1931, Dirac Papers, 2/2/4 (FSU).

26 Letter to Dirac from his mother, 20 July 1931, Dirac Papers, 2/2/4 (FSU).

27 Postcard from Flo to Betty Dirac, 1 August 1931: 'Having a sea voyage with Paul. The weather is fine and it is lovely. Back 6.35am Sunday. Hope you are both looking after each other' (DDOCS).

28 The area was officially named the Glacier National Park only in the following year.

29 Robertson (1985).

30 The furniture budget was $26,000; the budget for rugs was nearly $8,000. Batterson (2007: 612). Fine Hall is now called Jones Hall.

31 Jacobson, N., 'Recollections of Princeton' in Robertson (1985).

32 Letter from Pauli to Peierls, 29 September 1931, in Hermann et al. (1979).

33 Enz (2002: 224–5).

34 *New York Times*, 17 June 1931.

35 Pais (1986: 313–17).

36 Brown (1978).

37 Enz (2002: 211).

38 'Lectures on Quantum Mechanics', Princeton University, October 1931, Dirac Papers, 2/26/15 (FSU). These notes were transcribed by Banesh Hoffman and checked by Dirac.

39 'Dr Millikan Gets Medal', *New York Times*, 5 September 1928.

40 Kevles (1971: 180); Galison (1987: Chapter 3, pp. 86–7).

41 Interview with Robert Oppenheimer, AHQP, 18 November 1963, p. 16.

42 De Maria and Russo (1985: 247, 251–6).

43 Letter from Anderson to Millikan, 3 November 1931, quoted in De Maria and Russo (1985: 243). In this letter, Anderson describes data taken over the previous 'very few days'.

44 Interview with Carl Anderson, 11 January 1979, p. 34, available at http://oralhistories.library.caltech.edu/89 (accessed 13 May 2008), p. 34.

45 De Maria and Russo (1985: 243).

46 Letter to Dirac from Martin Charlesworth, 16 October 1931, Dirac Papers, 2/2/4 (FSU). Charlesworth was Dirac's personal tutor during his postgraduate years and was evidently fond of him. Later, on 19 March 1935, he wrote a letter to Dirac 'to send my [i.e. his] love' – a remarkably forward phrase in that cultural milieu, Dirac Papers, 2/3/1 (FSU).

47 Batterson (2006: Chapter 5).

48 Brendon (2000: Chapter 4).

49 *New York Times*, 14 June 1931.

50 Letter from Gamow to Dirac, written in June 1965, Dirac Papers, 2/5/13 (FSU). See also Gamow (1970: 99).

51 Gorelik and Frenkel (1994: 20–2). See also Kojevnikov (2004: 76).

52 Gorelik and Frenkel (1994: 50–1). Gamow gives a partially inaccurate account of this incident in his autobiography (1970).

53 The first Soviet edition is discussed in detail in Dalitz (1995), which includes a translation of the prefaces to the book.

54 Ivanenko had ensured that the book had been translated with no changes, but the Russian edition does include an additional chapter on applying quantum mechanics to practical problems. It is not clear whether Dirac added the section as a result of ideological pressure.

55 Greenspan (2005: 161).

56 Letter from Dirac to Tamm, 21 January 1932, in Kojevnikov (1993: 60). Dirac was learning the branches of mathematics known as group theory and differential geometry.

57 Interview with Oppenheimer, AHQP, 20 November 1963, p. 1.

58 Letter to Dirac from his mother, 9 October 1931, Dirac Papers, 2/2/4 (FSU).

59 Letter to Dirac from his mother, dated 28/31 September 1931, Dirac Papers, 2/2/4 (FSU).

60 Letter to Dirac from his mother, 22 December 1931, Dirac Papers, 2/2/4 (FSU).

61 Brown (1997: Chapter 6).

62 Cathcart (2004: 210–12); Chadwick (1984: 42–5).

63 Brown (1997: 106).

Chapter sixteen

1 Eddington made this remark in Leicester, at the annual meeting of the British Association for the Advancement of Science: 'Star Birth Sudden Lemaître Asserts', *New York Times*, 12 September 1933.

2 An English translation of the play, by Gamow's wife Barbara, is given in Gamow (1966: 165–218). For comments on the production: von Meyenn (1985: 308–13).

3 Wheeler (1985: 224).

4 Crowther (1970: 100).

5 Letter from Darwin to Goudsmit, 12 December 1932, APS.

6 Interview with Beck, AHQP, 22 April 1967, p. 23.

7 Interview with Klein, AHQP, 28 February 1963, p. 18. Klein recalled that 'Heisenberg once told me that when Dirac got the Nobel Prize some years later – in 1933 – he asked Dirac if he believed in his own theory. Dirac answered, in his very precise way, that a year before the positive electron was discovered he had ceased to believe in the theory' (interview with Klein, AHQP, 28 February 1963, p. 18).

8 Cathcart (2004: Chapters 12 and 13).

9 *Reynolds's Illustrated News*, 1 May 1932.

10 *Daily Mirror*, 3 May 1932.

11 Cathcart (2004: 252). Einstein's lecture took place on 6 May; see the *Cambridge Review*, 13 May 1932, p. 382.

12 Howarth (1978: 187).

13 Howarth (1978: 224).

14 Report in *Sunday Dispatch* on 19 November 1933.

15 Interview with von Weizsächer, AHQP, 9 June 1963, p. 19.

16 Note from P. H. Winfield to Dirac, Dirac Papers, 2/2/5 (FSU).

17 Letter from Sir Joseph Larmor to Terrot Reaveley Glover (1869–1943), the classical scholar and historian, 20 February 1934, STJOHN.

18 Infeld (1941: 170).

19 Letter to Dirac from his mother, 27 July 1932, Dirac Papers, 2/2/6 (FSU).

20 Letter to Dirac from his sister, 14 October 1932, Dirac Papers, 2/2/6 (FSU).

21 Letter to Dirac from his sister, 11 July 1932, Dirac Papers, 2/2/6 (FSU).

22 Letter to Dirac from his sister, 15 October 1932, Dirac Papers, 2/2/6 (FSU).

23 Letter to Dirac from his mother, 21 April 1932, Dirac Papers, 2/2/6 (FSU). See also the letter of 1 June 1932.

24 Letter to Dirac from his father, Dirac Papers, 1/1/10 (FSU).

25 The paper combined several parts, one mostly from Dirac, the other mostly from Fock and Podolsky, and also a part that developed during the process of writing in correspondence between the three authors. One snapshot of the collaboration is in the letter written to Dirac by Podolsky in Kharkov on 16 November 1932, Dirac Papers, 2/2/6 (FSU). I thank Alexei Kojevnikov for this information.

26 Weisskopf (1990: 72–3).

27 Infeld (1941: 172).

28 Article in the *Los Angeles Times* by Harry Carr, 30 July 1932.

29 For more detail on the discovery of the anti-electron, see Anderson (1983: 139–40), and Darrow (1934).

30 Interview with Louis Alvarez by Charles Weiner, 14–15 February 1967, American Institute of Physics, p. 10.

31 Von Kármán (1967: 150).

32 Von Kármán (1967: 150).

33 Interview with Carl Anderson, 11 January 1979, available online at http://oralhistories.library.caltech.edu/89 (accessed 13 May 2008).

34 Galison (1987: 90).

35 *New York Times*, 2 October 1932.

36 Letter from Robert Oppenheimer to Frank Oppenheimer, autumn 1932, in Smith and Weiner (1980: 159).

37 Nye (2004: 54). The incident, recalled by Blackett's student Frank Champion, probably took place during the 1931–2 academic year. I am grateful to Mary Jo Nye for this information.

38 See http://www.aps-pub.com/proceedings/1462/207.pdf (accessed 13 May 2008).

39 De Maria and Russo (1985: 254).

40 Contribution of Occhialini to the Memorial Meeting for Lord Blackett, *Notes and Records of the Royal Society*, 29 (2) (1975).

41 Dalitz and Peierls (1986: 167). The anecdote is due to Maurice Pryce.

42 Dirac's notes on Fowler's lectures on 'Analytic Dynamics' are in Dirac Papers, 2/32/1 (FSU).

43 Letter from Dirac to Fock, 11 November 1932, passed to me by Alexei Kojevnikov.

44 Greenspan (2005: 170).

45 *Bristol Evening Post*, 28 October 1932.

46 Letter to Dirac from his mother, 26 October 1932, Dirac Papers, 2/2/7 (FSU).

47 Letter to Dirac from his mother, 9 January 1933, Dirac Papers, 2/2/8 (FSU).

Chapter seventeen

1 IAS Archives Faculty Series, Box 32, Folder: 'Veblen, 1933'.

2 De Maria and Russo (1985: 266 and 266 n.). Anderson's paper had been available in the university library from the mid-autumn of 1932.

3 Archie Clow, contributing to Radio 3 programme *Science and Society in the Thirties* (1965). Script stored in the library of Trinity College, Cambridge.

4 Schücking (1999: 27).

5 Interview with Léon Rosenfeld, AHQP, 22 July 1963, p. 8.

6 Halpern (1988: 467).

7 Letter to Dirac from Isabel Whitehead, 20 July 1932, Dirac Papers, 2/2/6 (FSU).

8 Taylor Sen (1986).

9 Dirac, book review in the *Cambridge Review*, 6 February 1931.

10 Interview with von Weizsächer, AHQP, 9 June 1963, p. 19.

11 Private papers of Mary Dirac. Dirac wrote the notes on 17 January 1933.

12 Letter from Dirac to Isabel Whitehead, 6 December 1936, STJOHN.

13 Compte remarked that 'The greatest problem, then, is to raise social feeling by artificial effort to the position which in the natural condition is held by selfish feeling.' See http://www.blupete.com/Literature/Biographies/Philosophy/Comte.htm (accessed 14 May 2008).

14 The headquarters of the Royal Society were then at Burlington House.

15 Bertha Swirles, Dirac's former student colleague, described the talk as 'sensational' in her letter of 20 February 1933 to Dirac's colleague Douglas Hartree. Hartree archive, 157, CHRIST'S.

16 Dirac was giving a technical talk at the London Mathematical Society on his favourite topic, 'The Relation Between Classical and Quantum Mechanics', at the Royal Astronomical Society in Burlington House, Dirac Papers, 2/26/18 (FSU).

17 The word was used in the 15 March issue of the *Physical Review*.

18 Quoted in Pais (1986: 363).

19 Interview with von Weizsächer, AHQP, 9 July 1963, p. 14.

20 Letter from Tamm to Dirac, 5 June 1933, in Kojevnikov (1996: 64–5).

21 Interview with Dirac, AHQP, 14 May 1963, p. 31.

22 Letter from Pauli to Dirac, 1 May 1933, see Pais (1986: 360).

23 Galison (1994: 96).

24 Darrow (1934: 14).

25 Roqué (1997: 89–91).

26 Brown and Hoddeson (1983: 141).

27 Blackett (1955: 16).

28 Gell-Mann (1994: 179).

29 See the lecture Dirac gave in Leningrad on 27 September 1933 (Dalitz 1995: 721), Dirac's Nobel Prize lecture in December 1933 and most of Dirac's subsequent lectures on the positron.

30 Blackett (1969: xxxvii).

31 Gottfried (2002: 117).

32 Bohr's support was sought by Kapitza. See the correspondence quoted in Kedrov (1984: 63–7).

33 The quote from Rutherford is from Kapitza's letter to Bohr of 10 March 1933, quoted in Kedrov (1984: 63–4).

34 Anon., 'Conservatism and the Young', *Cambridge Review*, 28 April 1933, pp. 353–4.

35 The debate was held on 21 February 1933 and was reported in the *Cambridge Evening News* on the following day. See also Howarth (1978: 224–5).

36 Anon (1935); essay by Blackett (based on a radio broadcast in March 1934), pp. 129–44, see p. 130.

37 Werskey (1978: 168).

38 Werskey (1978: 148).

39 The *Cambridge Review*, 20 January 1933. The article alerted the Cambridge University community to the reservations expressed by the translators of Dirac's book into Russian.

40 Anon. (1933) 'The End of a Political Delusion', *Cambridge Left*, 1 (1): 10–15; p. 12.

41 *Daily Herald*, 15 September 1933, p. 10. McGucken (1984: 40–1).

42 Letters to Dirac from his mother, 20 July and 22 July 1933, Dirac Papers, 1/4/3 (FSU).

43 Letter to Dirac from his mother, 8 August 1933, Dirac Papers 1/4/3 (FSU).

44 Postcards from Dirac to his mother, from September 1933 (DDOCS).

45 Letter from Dirac to Tamm, 19 June 1933, in Kojevnikov (1993: 67); see also the letter from Tamm to Dirac on 5 June 1933 (Kojevnikov 1993: 64).

46 Interview with Beck, AHQP, 22 April 1967, APS, p. 23.

47 The mansion was awarded to Bohr in December 1931, whereupon Bohr and his family moved in during the summer of 1932. The Bohrs' first sleeping-over guests were Ernest Rutherford and his wife, who stayed there from 12 September to 22 September 1932. I thank Finn Aaserud and Felicity Pors for this information.

48 Parry (1968: 117).

49 Casimir (1983: 73–4). Letter from Dirac to Margrethe Bohr, 24 September 1933, NBA.

50 Letter from Dirac to Margrethe Bohr, 24 September 1933, NBA.

51 Letter from Dirac to Bohr, 20 August 1933, NBA.

52 Fitzpatrick (1999: 40–1).

53 Conquest (1986: Epilogue).

54 M. Dirac (1987: 4).

55 Anne Kox, 'Een kwikkolom in de Westertoren: De Amsterdamse natuurkunde in de jaren dertig', available online at http://soliton.science.uva.nl/~kox/jaarboek.html (14 May 2008).

56 Letter from Dirac to Bohr, 28 September 1933, NBA.

57 Letter from Margrethe Bohr to Dirac, 3 October 1933, NBA.

58 Letter from Ehrenfest to Bohr, Einstein and the physicists James Franck, Gustave Herglotz, Abram Joffé, Philipp Kohnstamm and Richard Tolman, 14 August 1933, NBA. Another suicide note, written on the day before Ehrenfest killed himself was unearthed in 2008: see *Physics Today*, June 2008, p. 26–7.

59 Roqué (1997: 101–2).

60 Letter from Heisenberg to Pauli, 6 February 1934, in Hermann et al. (1979).
61 Dirac mentioned his surprise to a reporter from the *Daily Mirror*. See the article on 13 November 1933.
62 Taylor (1987: 37).
63 The youngest experimenter to win the prize was, and remains, Lawrence Bragg, who won it when he was twenty-five. Dirac's record as the youngest theoretician to win the prize was broken (by a margin of three months) in 1957 by T. D. Lee.
64 Reports on 10 November 1933 included the *Daily Mail*, *Daily Telegraph*, *Manchester Guardian*; the *Daily Mirror* reported on the following day.
65 *Sunday Dispatch*, 19 November 1933.
66 Letter from Dirac to Bohr, 28 November 1933, NBA.
67 Greenspan (2005: 242). Maurice Goldhaber remembers that when he remarked that Dirac's award was 'great news', Born scowled. Interview with Maurice Goldhaber, 5 July 2006.
68 *Cambridge Review*, 17 November 1933; Brown (2005: 120). See also Stansky and Abrahams (1966: 210–13). A few days before the march, a few socialists and pacifists clashed with audiences leaving the Cambridge cinema Tivoli, after an evening showing of the patriotic movie *Our Fighting Navy*. The fight was the talk of the town and therefore guaranteed interest in the Armistice Day march.

Chapter eighteen

1 Dalitz and Peierls (1986: 146).
2 Information from RSAS, 14 September 2004.
3 The main sources of the material in this chapter are in the Dirac Papers (FSU): Letter to Dirac from his mother, 21 November 1933 (2/2/9). Florence Dirac's account of her trip is in 'My visit to Stockholm' (1/2/9) and in a long, descriptive letter to Betty (2/2/9).
4 Reports in *Svenska Dagbladet* and *Dagens Nyheter*, both on 9 December 1933.
5 This was one of Dirac's favourite stories about his absent-minded mother. It is well recounted in Kurşunoğlu (1987: 18).
6 Reports in the Stockholm newspapers *Nya Dagligt Allehanda*, 9 December 1933, *Stockholms Dagblad*, 10 December 1933.
7 Reports in the Stockholm newspapers *Nya Dagligt Allehanda*, 9 December 1933, *Stockholms Dagblad*, 10 December 1933.
8 Report in *Dagens Nyheter*, 11 December 1933.
9 *Dagens Nyheter*, 11 December 1933; *Svenska Dagbladet*, 11 December 1933.
10 Women guests were first invited to the banquet in 1909, when the female Swedish wirter Selma Lagerlöf won the Nobel Prize for Literature.
11 *Dagens Nyheter*, 11 December 1933; *Svenska Dagbladet*, 11 December 1933; *Stockholms Tidningen*, 11 December 1933.
12 See http://nobelprize.org/physics/laureates/1933/dirac-speech.html (accessed 14 May 2008).
13 Annemarie Schrödinger notes 'Stockholm 1933', AHQP. Letter from Schrödinger to Dirac, 24 December 1933.
14 I thank Professor Sir Partha Dasgupta for identifying this error and clarifying its nature.
15 Flo Dirac, Dirac Papers, 1/2/9 (FSU) and 2/2/9 (FSU).
16 See http://nobelprize.org/nobel_prizes/physics/laureates/1933/dirac-lecture.html (accessed 14 May 2008).

17 Schuster (1898a: 367); see also Schuster's follow-up article (1898b).
18 Born (1978: 270). See also 'Eamon de Valera, Erwin Schrödinger and the Dublin Institute' (McCrea 1987).
19 Flo Dirac, Dirac Papers, 1/2/9 (FSU) and 2/2/9 (FSU).
20 Dirac read Abraham Pais's book *Subtle is the Lord*, and remarked 'Most interesting for its revelation of the working of Nobel Committee', Dirac Papers, 2/32/12 (FSU). The book mentions that Einstein did not nominate Dirac for a Nobel Prize.
21 Nobel Committee papers, 1929 RSAS.
22 Apart from Bragg, only the comparatively little-known Polish physicist Czeslaw Bialobrzeski nominated Dirac in 1933. No other leading theorist had nominated him.

Chapter nineteen

1 Letter from Pauli to Heisenberg, 14 June 1934, reprinted in Hermann et al. (1979).
2 Schweber (1994: 128–9).
3 Letters from Oppenheimer to George Uhlenbeck, March 1934 and to Frank Oppenheimer, 4 June 1934, in Kimball Smith and Weiner (1980: 175, 181).
4 Interview with Dirac, AHQP, 6 May 1963, p. 8, Salam and Wigner (1972: 3–4). See also Peierls (1985: 112–13).
5 Letter from Rutherford to Fermi, AHQP, 23 April 1934.
6 'Peter Kapitza', 22 June 34, KV 2/777, UKNATARCHI.
7 'Note on interview between Captain Liddell and Sir Frank Smith of the Department of Scientific and Industrial Research, Old Queen Street', 26 September 1934, KV 2/777. Jeffrey Hughes speculates that 'VSO' might be the Russian émigré I. P. Shirov (Hughes 2003).
8 Born (1978: 269–70).
9 I am grateful to Igor Gamow for making available home movies, shot in the 1920s, which show his mother dressed in this way.
10 The correspondence between Dirac and Rho Gamow is in Dirac Papers, 2/13/6 (FSU).
11 Letter from Dirac to Manci, 9 April 1935 (DDOCS).
12 Letter from Dirac to Rho Gamow, Dirac Papers, 2/2/10 (FSU).
13 Conversation with Lydia Jackson's literary executor Rosemary Davidson, 8 January 2006.
14 Letter to Dirac from Lydia Jackson, 20 March 1934, Dirac Papers, 2/2/10 (FSU).
15 Fen (1976: 182).
16 Letter to Dirac from Lydia Jackson, 25 June 1934, Dirac Papers, 2/2/10 (FSU).
17 Letter to Dirac from Lydia Jackson, 5 February 1936, Dirac Papers, 2/3/3 (FSU).
18 Van Vleck (1972: 12–14).
19 The visitor was his sister Manci. M. Dirac (1987: 3–8; see p. 3).
20 The account of Dirac's early courtship of Manci is taken mainly from M. Dirac (1987).
21 Letter to Dirac from Van Vleck, June 1936, Dirac Papers, 2/2/11 (FSU).
22 Dirac was living at 8 Morven Street. See the Dirac archive in IAS (1935).
23 Quoted in Jerome and Taylor (2005: 11).
24 Jerome and Taylor (2005: Chapters 2 and 5).
25 Blackwood (1997: 11).
26 Testimonies of Malcolm Robertson and Robert Walker, 'The Princeton Mathematics Community in the 1930s', available at http://www34.homepage.villanova.edu/robert.jantzen/princeton_math/pm02.htm (accessed 14 May 2008).

27 The *Physical Review* received the paper on 25 March 1935: Pais (1982: 454–7).

28 Blackwood (1997: 15–16).

29 Infeld (1941: 170).

30 See 'The Princeton Mathematics Community in the 1930s', in particular the interviews of Merrill Flood, of Robert Walker and of William Duren, Nathan Jacobson and Edward McShane.

31 Letter from Dirac to Max Newman, 17 March 1935, Newman archive STJOHN.

32 Dirac alludes to his memories of ice-cream sodas and lobster dinners with Manci in his letters to her of 2 May and 25 May 1935 respectively (DDOCS).

33 Manci was divorced from Richard Balázs on 20 September 1932. See Budapest's archive of marriages, microfilm repository no A555, Inventory no 9643, Roll no 155. These papers tell us that Manci married Balázs on 27 February 1924.

34 Manci told her friend Lily Harish-Chandra of these relationships. Interview with Lily Harish-Chandra, 4 August 2006.

35 Wigner (1992: 34, 38–9).

36 Letter to Dirac from Manci, 2 September 1936 (DDOCS).

37 M. Dirac (1987: 4–5).

38 Letter to Dirac from Anna Kapitza, dated beginning December 1937, copy held by Alexei Kojevnikov.

39 Hendry (1984: 130).

40 A detailed account of Kapitza's detention is in: Internal MI5 memo, signed GML, 11 October 3KV 2/777 (UKNATARCHI). See also the letters from Kapitza to his wife in Boag et al. (1990: Chapter 4).

41 For a full account of Rutherford's campaign to secure Kapitza's release, see Badash (1985), notably Chapter 2. See also Kojevnikov (2004: Chapter 5).

42 Letter from Dirac to Anna Kapitza, 19 December 1934, copy held by Alexei Kojevnikov.

43 Dirac wrote of his vacation, without mentioning Manci, to Max Newman in a letter written on 13 January 1935 (Newman archive, STJOHN). The story of the alligator, which Gamow named Ni-Nilich, is related in letters from Dirac to Manci on 2 February, 29 March, 22 April and 2 May 1935 and in the letter from Manci to Dirac on 5 April 1935 (DDOCS). See also the letter from Gamow to Dirac, 25 March 1935, Dirac Papers, 2/3/1 (FSU).

44 Letter from Dirac to Anna Kapitza, 14 March 1935, copy held by Alexei Kojevnikov.

45 Letter from Rutherford to Bohr, 28 January 1935, Rutherford archive, UCAM.

46 Gardiner (1988: 240–8).

47 Gardiner (1988: 241).

48 Gardiner (1988: 242).

49 Kragh (1996: Chapter 2).

50 'Lemaître Follows Two Paths to Truth', *New York Times*, 19 February 1933.

51 Letter from Dirac to Manci, 2 February 1935 (DDOCS).

52 Dirac had heard Lemaître speak at the Kapitza Club in about 1930. Dirac commented on this in a note he wrote on 1 September 1971: 'There was much discussion about the indeterminacy of quantum mechanics. Lemaître emphasised his opinion that he did not believe God influenced directly the cause of atomic events': Dirac Papers, 2/79/2 (FSU).

53 Letter from Dirac to Manci, 2 March 1935 (DDOCS).

54 Letter from Dirac to Manci, 2 May 1935 (DDOCS). Schnabel gave the concert on 7 March 1935.

55 Letter from Dirac to Manci, 10 March 1935 (DDOCS).

56 Letter to Dirac from Manci, 28 March 1935 (DDOCS).

57 Letter from Dirac to Manci, 29 March 1935 (DDOCS).

58 Letter from Dirac to Manci, 2 May 1935 (DDOCS).

59 Letter from Dirac to Manci, 9 May 1935 (DDOCS).

60 Letter to Dirac from Manci, 30 May 1935 (DDOCS).

61 Letter to Dirac from Manci, 4 March 1935 (DDOCS).

62 Letter from Dirac to Manci, 9 April 1935 (DDOCS).

63 Badash (1985: 29).

64 Badash (1985: 31).

65 Letter from Kapitza to his wife, 13 April 1935, quoted in Boag et al. (1990: 235).

66 Letter from Kapitza to his wife, 23 February 1935, quoted in Boag et al. (1990: 225).

67 Letter from Kapitza to his wife, 23 February 1935, quoted in Boag et al. (1990: 225, 226).

68 Kojevnikov (2004: 107).

69 Letter from Dirac to Manci, 2 May 1935 (DDOCS).

70 Lanouette (1992: 151); see also letter from Dirac to Anna Kapitza, 31 May 1935, copy held by Alexei Kojevnikov.

71 Letter from K. T. Compton to the Soviet Ambassador, 24 April 1935, copy held by Alexei Kojevnikov.

72 Letter from Dirac to Anna Kapitza, 27 April 1935, copy held by Alexei Kojevnikov.

73 'Embassy Occupied by Troyanovsky', *New York Times*, 7 April 1934.

74 Letter from Dirac to Anna Kapitza, 27 April 1935, copy held by Alexei Kojevnikov.

75 Letter from Dirac to Anna Kapitza, 27 April 1935.

Chapter twenty

1 Letter from Dirac to Anna Kapitza, written from the Institute for Advanced Study, Princeton, 14 May 1935. Copy of letter held by Alexei Kojevnikov.

2 Letter from Dirac to Anna Kapitza, written in Pasadena, 31 May 1935, copy held by Alexei Kojevnikov.

3 Crease and Mann (1986: 106); Serber (1998: 35–6).

4 Letter from Dirac to Manci, 4 June 1935 and 10 June 1935 (DDOCS).

5 Letter from Dirac to Manci, 1 August 1935 (DDOCS).

6 Letter from Dirac to Manci, 22 June 1935 (DDOCS).

7 Quoted in Brendon (2000: 241).

8 Letter from Kapitza to his wife, 30 July 1935, quoted in Boag et al. (1990: 251).

9 Letter from Dirac to Manci, 17 August 1935 (DDOCS).

10 Letter to Dirac from Manci, 30 September 1935 (DDOCS). See also Dirac, M. (1987: 6).

11 Letter from Dirac to Manci, 22 September 1935 (DDOCS).

12 Letter from Dirac to Manci, 23 October 1935 (DDOCS).

13 Letter to Dirac from Manci, 9 October 1935 (DDOCS).

14 Letters from Dirac to Manci, 3 October 1935 and 8 November 1935 (DDOCS).

15 Letter from Dirac to Manci, 17 November 1935 (DDOCS).

16 Letter to Dirac from Manci, 22 November 1935 (DDOCS).

17 Letter from Dirac to Manci, 3 October 1935 (DDOCS).

18 In Dirac's letter to Manci on 6 February 1937, Dirac mentions that his father owned a copy of Shaw's plays.

19 Letter to Dirac from his mother, 15 July 1934, Dirac Papers, 1/4/4 (FSU).

20 Dirac's father's notebook is in Dirac Papers, 1/1/10 (FSU). Charles dates his first entry September 1933. The latest date he referenced was 4 November 1935, so he

probably ceased compiling the notes in early 1936.

21 Dalitz and Peierls (1986: 146).

22 Letter to Dirac from his mother, 4 August 1935, Dirac Papers, 1/4/5 (FSU).

23 Letter to Dirac from his mother, 4 August 1935, Dirac Papers, 1/4/5 (FSU).

24 Dalitz and Peierls (1986: 155–7).

25 Letter from Dirac to Tamm, 6 December 1935, in Kojevnikov (1996: 35–6).

26 One of the physicists who thought that Dirac was over-excited by the Shankland result was Hans Bethe, who wrote 'What has happened to him?' in a letter to Rudolf Peierls on 1 August 1936, in Lee (2007b: 152).

27 Dirac (1936: 804).

28 Letter from Heisenberg to Pauli, 23 May 1936, Vol. II, p. 442.

29 Letter from Einstein to Schrödinger, 23 March 1936, AHQP.

30 Letter from Schrödinger to Dirac, 29 April 1936, Dirac Papers, 2/3/3 (FSU).

31 Letter from Bohr to Kramers, 14 March 1936, NBA.

32 Letter from Dirac to Blackett, 12 February 1937, Blackett archive ROYSOC.

33 Letter from Dirac to Manci, 15 January 1936. Other details in this paragraph are in his letters to Manci of 25 January 1936, 2 February 1936 and 10 February 1936 (DDOCS).

34 Huxley (1928: 91) ('Emotionally, he was a foreigner') and p. 230 ('a mystic, a humanitarian and also a contemptuous misanthrope'). See also Huxley (1928: 90, 92–6).

35 Letter from Dirac to Manci, 2 February 1936 (DDOCS).

36 Letter to Dirac from Manci, 23 February 1936 (DDOCS).

37 Letter from Dirac to Manci, 7 March 1936 (DDOCS).

38 Letter from Dirac to Manci, 7 March 1936 (DDOCS).

39 Letter to Dirac from Manci, 13 March 1936 (DDOCS).

40 Letters from Dirac to Manci, 23 March 1936 and 29 April 1936, and letter to Dirac from Manci, 24 April 1936 (DDOCS).

41 Letter from Dirac to Manci, 5 May 1936 (DDOCS).

42 Dirac had also fibbed to Kapitza in the previous year. Dirac makes this plain to Manci in his letter to her of 23 June 1936 (DDOCS).

43 Letter from Dirac to Manci, 9 June 1936 (DDOCS).

44 Letter from Dirac to Manci, 5 June 1936 (DDOCS).

45 Sinclair (1986: 55).

46 A. Blunt, 'A Gentleman in Russia', and a review of Crowther's *Soviet Science* by Charles Waddington, both in the *Cambridge Review*, 5 June 1936.

47 Letter to Dirac from his mother, 7 June 1936, Dirac Papers, 1/4/6 (FSU).

48 Letters to Dirac from his sister, 6 June, 8 June and 9 June 1936, Dirac Papers, 1/7/1 (FSU).

49 Letter from Dirac to Manci, 17 June 1936 (DDOCS).

50 Letter to Dirac from his mother, 11 June 1936, Dirac Papers, 1/4/6 (FSU).

51 *Daily Mirror*, 21 May 1934, p. 14. The article concluded: 'Dirac. Our great grandchildren may be repeating that name when the Chaplins, Fords, Cowards and Cantors are forgotten.' Cantor is the American writer and entertainer Eddie Cantor.

52 Letter from Dirac to Manci, 17 June 1936 (DDOCS).

53 Letter to Dirac from his mother, July 1936, Dirac Papers, 1/4/6 (FSU).

54 Letter to Dirac from his mother, 27 August 1936, Dirac Papers, 1/4/6 (FSU).

55 Feinberg (1987: 97).

56 Dalitz and Peierls (1986: 151).

57 Letter from Kapitza to Rutherford, 26 April 1936, quoted in Badash (1985: 110).

58 Letter to Dirac from Manci, 2 September 1936 (DDOCS).

59 Pais (1991: 411).

60 Both preceding quotes are from the letter from Dirac to Manci, 7 October 1936 (DDOCS). Dirac commented to an officer of the Rockefeller Foundation, which funded the conference, that he was 'genuinely enthusiastic', quoted in Aaserud (1990: 223).

61 In Dirac, M. (1987), Manci recalls that she was on the *Queen Mary*'s maiden voyage. At that time, however, she was in Budapest.

62 Letter from Dirac to Manci, 19 October 1936 (DDOCS).

63 Letter from Dirac to Manci, 17 November 1936 (DDOCS).

64 Letter to Dirac from Isabel Whitehead, 29 November 1936, Dirac Papers, 2/3/4 (FSU).

65 Letter from Dirac to Isabel Whitehead, 6 December 1936, STJOHN.

66 Letter to Dirac from Isabel Whitehead, 9 December 1936, Dirac Papers, 2/3/4 (FSU).

67 Interview with Monica Dirac, 7 February 2003. Manci often related this story of Dirac's proposal to her. The description of the car is in the letter from Dirac to Manci, 17 November 1935 (DDOCS).

68 Letter to Dirac from Manci, 29 January 1937 (DDOCS).

69 Letter to Dirac from his mother, 24 December 1936, Dirac Papers, 1/4/6 (FSU).

Chapter twenty-one

1 Dirac, M. (1987: 4).

2 Letter from Dirac to Manci, 18 February 1937 (DDOCS).

3 Letter from Dirac to Manci, 6 February 1937 (DDOCS).

4 Letter from Dirac to Manci, 20 February 1937 (DDOCS). Dirac writes 'How soon after the new moon comes will I be alone with my beloved, and have her in my arms [. . .]'.

5 Letter from Dirac to Manci, 19 February 1937 (DDOCS).

6 Letter from Dirac to Manci, 20 February 1937 (DDOCS).

7 Letter to Dirac from Manci, 16 February 1937 (DDOCS).

8 Letters to Dirac from Manci, 25 January and 16 February 1937 (DDOCS).

9 Letter to Dirac from Betty, 29 January 1937 (DDOCS).

10 Letter to Dirac from Manci, 29 January 1937 (DDOCS).

11 One reading of Manci's cryptic comments in her letter to Dirac of 16 February 1937 is that his parents were sexually incompatible (DDOCS): 'Betty told me today the reason why probably your parents did not like each other. Your father could not help it, don't blame him dear, nor do [*sic*] your mother.'

12 Letter to Dirac from Manci, 18 February 1937 (DDOCS).

13 Letter to Dirac from Manci, 28 January 1937 (DDOCS). Dirac's 'unexpected' marriage was noted in the *Cambridge Daily News*, 7 January 1937.

14 Letter from Rutherford to Kapitza, 20 January 1937, in Boag et al. (1990: 300).

15 Letter from Dirac to Kapitza, 29 January 1937, Dirac Papers 2/3/5 (FSU).

16 Letter to Manci from Anna Kapitza, 17 February 1937, Dirac Papers, 2/3/5 (FSU).

17 Dirac's use of 'Wigner's sister' became famous in his community. Both Dirac's daughters confirm that he used this term of introduction.

18 Manci often used this expression. See, for example, Dirac (1987: 7).

19 Interview with Monica Dirac, 7 February 2003.

20 Salaman and Salaman (1996: 66–70); see p. 67.

21 Daniel (1986: 95–6).

22 Letter from Dirac to Manci, 19 February 1937 (DDOCS).

23 Dirac's wish to have children appears obvious from his delighted reaction to the news

of Manci's later pregnancies.

24 Gamow (1967: 767).

25 Christianson (1995: 257).

26 Dingle (1937a).

27 Untitled supplement to *Nature*, Vol. 139, 12 June 1937, pp. 1001–2; p. 1001.

28 Dingle (1937b).

29 Report on Theoretical Physics to the Institute for Advanced Study, 23 October 1937, in the IAS Archives General Series, 52, 'Physics'.

30 Estate of Charles Dirac, prepared by Gwynn, Onslow & Soars, who prepared the document on 7 October 1936 (DDOCS).

31 Letter to Dirac from his mother, 21 January 1937, Dirac Papers, 1/4/7 (FSU). See also the letter of 1 February 1937 in the same file of the archive.

32 Interview with Kurt Hofer, 21 February 2004.

33 Kojevnikov (2004: 119).

34 Postcard from Manci Dirac to the Veblens, 17 June 1937, LC Veblen archive.

35 Telegram from Kapitza to Dirac, 4 June 1937, KV 2/777, UKNATARCHI.

36 Service (2003: 223).

37 Fitzpatrick (1999: 194).

38 Letter from Kapitza to Rutherford, 13 September 1937, in Boag et al. (1990: 305–6).

39 Kojevnikov (2004: 116).

40 Before Landau fled Kharkov, he had worked at the Ukrainian Physico-technical Institute. He was arrested on 28 April 1938 in Moscow, and Kapitza wrote to Stalin seeking his release. His letter is quoted by David Holloway (1994: 43).

41 Letter from Dirac to Kapitza, 27 October 1937, Dirac Papers, 2/3/6 (FSU).

42 Letter to Dirac from Kapitza, 7 November 1937, Dirac Papers, 2/3/6 (FSU).

43 Letter from Fowler to Dirac, 25 January 1939, Dirac Papers, 2/3/8 (FSU).

44 This was one of Dirac's favourite observations. See R. Dalitz, *Nature*, 19 Vol. 278 (April) 1979.

45 Hoyle (1992: 186).

46 Hoyle (1994: 131).

47 Hoyle (1994: 133).

48 Letter from Dirac to Bohr, 5 December 1938, NBA.

49 At least two of Flo's poems were published in newspapers: 'Cambridge' appeared in the *Observer* on Saturday, 23 July 1938, and 'Brandon Hill' was published in the local *Western Daily Press* on Saturday, 12 March 1938.

50 On 2 February 1938, Princeton University sent Dirac a letter offering him tenure with an annual salary of $12,000, beginning 1 October 1938, Dirac Papers, 2/3/7 (FSU).

51 Letter from Anna Kapitza to Manci Dirac, 9 March 1938, Dirac Papers, 1/8/18 (FSU).

52 *Nature*, 21 May 1938, No. 3577, p. 929. Schrödinger's well-publicised letter was published in *Graz Tagepost*, 30 March 1938. See Moore (1989: 337–8).

53 Letters from Dirac to Manci in August 1938 (DDOCS). Wigner married Amelia Frank on 23 December 1936 in Madison, and she died on 16 August 1937. See 'The Einhorn Family', compiled by Margaret Upton (private communication).

54 Bell wrote to Dirac on 15 March 1938: 'I had already and for a year or two reached the conclusion the Soviet trials were probably of the frame-up type. After all, that is not new. The Tom Mooney case in California in 1918 was such and the victim has been in prison ever since [. . .] also the Sacco & Vanzetti case. Moreover, we seem to do it ourselves to a great extent in India. However, the "confession technique" is

peculiarly Russian, on its present scale at least.' Letter to Dirac from J. H. Bell, Dirac Papers, 2/3/7 (FSU).

55 Moore (1989: 347); letter from Schrödinger to Dirac, 27 November 1938, Dirac Papers, 2/3/7 (FSU).

56 Dirac gives these reasons in his obituary of Schrödinger in *Nature*, 4 February 1961, 189, p. 355–6.

57 Letter from Dirac to Kapitza, 22 March 1938, Dirac Papers, 2/3/7 (FSU).

58 Howarth (1978: 234–5).

59 *The Times*, 6 October 1938.

60 'Eddington Predicts Science Will Free Vast Energy from Atom', *New York Times*, 24 June 1930. He was speaking at the World Power Conference. He suggested that such energy could be released by arranging for particles to annihilate or to make hydrogen nuclei fuse to form a helium nucleus.

61 Rhodes (1986: 28).

62 Weart and Weiss Szilard (1978: 53).

63 Weart and Weiss Szilard (1978: Chapter II).

64 Weart and Weiss Szilard (1978: 71–2).

65 The event took place in the Society's house, 24 George Street, beginning at 4.30 p.m. Max Born was present.

66 Mill (1892: Book 2, Chapter 12).

67 This quote is from the text of the lecture, *Proceedings of the Royal Society* (Edinburgh), 59 (1938–9: 122–9); p. 123.

68 *Granta*, 48 (1): 100, 19 April 1939.

Chapter twenty-two

1 Bowyer (1986: 51).

2 This was one of Manci's favourite expressions about how the British treated her. Interview with Mary Dirac, 21 February 2003.

3 Boys Smith (1983: 44).

4 *Cambridge Daily News*, 2 September 1939, p. 5.

5 *Cambridge Daily News*, 1 September 1939, p. 3. I am grateful to my mother, Joyce Farmelo, for her recollections of her time as an unhappy evacuee and her other wartime experiences.

6 E-mail from Mary Dirac, 5 March 2006.

7 'Cambridge During the War; the Town', *Cambridge Review*, 27 October 1945; 'Cambridge During the War; St John's College', *Cambridge Review*, 27 April 1946. See also 'Thoughts Upon War Thought', *Cambridge Review*, 11 October 1940.

8 Barham (1977: 32–3).

9 Letter to Dirac from his mother, 26 January 1940, Dirac Papers, 1/4/10 (FSU).

10 Manci spent the final months of her pregnancy in the Mountfield Nursing Home in London. Information about Mary's birth from her baby book. Further clarification in an e-mail from Mary Dirac, 16 January 2006.

11 Letter to Dirac from Manci, 20 February 1940 (DDOCS). Manci's exact words are ungrammatical: 'I never felt as much that she has nor heart nor feelings whatsoever as yesterday.'

12 Peierls (1985: 150, 155).

13 Rhodes (1986: 323).

14 Facsimiles of the memos are in Hennessy (2007: 24–30).

15 Peierls (1985: 155).

16 The earliest extant letter about this, from Peierls to Dirac, is dated 26 October 1940, AB1/631/257889, UKNATARCHI.

17 Rhodes (1986: 303–7); Fölsing (1997: 710–14).

18 Letter to Aydelotte from Veblen and von Neumann, 23 March 1940, IAS Archives Faculty Series, Box 33, folder: 'Veblen–Aydelotte Correspondence 1932–47'. The words omitted, marked by the ellipsis, are 'There are considerable deposits of uranium available near Joachimsthal, Bohemia, as well as in Canada.'

19 Letter to Adyelotte from Veblen, 15 March 1940: IAS Archives General Series, Box 67, folder: 'Theoretical Physics 1940 Proposals'.

20 Cannadine (1994: 161–2).

21 Letter from Manci to Crowther, 28 June 1941, SUSSEX.

22 Barham (1977: 54); Bowyer (1986: 51).

23 Letter to Dirac from his mother, 27 June 1940, Dirac Papers, 1/4/10 (FSU).

24 Letters to Dirac from his mother, 16 August and 31 August 1940, Dirac Papers, 1/4/10 (FSU).

25 Letter to Dirac from his mother, 12 May 1940, Dirac Papers, 1/4/10 (FSU).

26 Letter to Dirac from his mother, 21 June 1940, Dirac Papers, 1/4/10 (FSU).

27 Letter from Dirac to Manci, 27 August 1940 (DDOCS).

28 Letter from Dirac to Manci, 23 August 1940. Four days later, he wrote to her: 'I am sorry to be away from you these days, but do not think there is any real danger in Cambridge' (DDOCS).

29 Gustav Born later recalled that Dirac on this vacation was 'a twinkling-eyed, kindly, distant man', happiest when on his own. Interview with Gustav Born, 12 February 2005.

30 'The ladies do the cooking, and the men take it in turns to do the washing up,' Dirac told Manci: letter, 23 August 1940 (DDOCS).

31 Letter from Dirac to Manci, 2 September 1940 (DDOCS).

32 Letter to Dirac from Manci, 8 September 1940 (DDOCS).

33 Letter from Pryce to Dirac, 18 July 1940, Dirac Papers, 2/3/10 (FSU).

34 Letter from Dirac to Manci, 21 January 1940 (DDOCS).

35 Letter from Gabriel to Dirac, 30 August 1945, and another undated later in the same month, Dirac Papers, 1/8/12 (FSU).

36 Letter to Dirac from his mother, 31 August 1940, Dirac Papers, 1/4/10 (FSU).

37 Letter from Peierls to Oppenheimer, 16 April 1954, LC, Oppenheimer archive.

38 The first part of this quotation is from the letter Dirac wrote to Manci on 18 December 1940; the second and third parts are from the letter he wrote to her the next day.

39 Letter to Dirac from Manci, 22 December 1940 (DDOCS).

40 Werskey (1978: 23); see also the foreword by C. P. Snow to Hardy (1940: 50–3).

41 Letter to Dirac from Hardy, May 1940, Dirac Papers, 2/3/10 (FSU).

42 Attendance register of Tots and Quots in 1940, Zucherman archive, wartime papers, SZ/TQ, EANGLIA.

43 Letter from Crowther to Dirac, 15 November 1940, Dirac Papers, 2/3/10 (FSU).

44 Brown (2005: Chapter 9).

45 The first letter to Dirac, from Peierls, in connection with war work is dated 26 October 1940, UKNATARCHI.

46 Bowyer (1986: 181). Manci often spoke of Judy's role in the firefighting (e-mail from Mary Dirac, 23 April 2006). Manci refers to an earlier near-miss on 15 February 1941 in her letter to Crowther on 17 February 1941, SUSSEX.

47 Dirac often referred to Crowther as 'the newspaper man'. See, for example, letter from Dirac to Manci, 4 May 1939 (DDOCS).

48 The spy was Jan Willen der Braak. 'The Spy Who Died Out in the Cold', *Cambridge Evening News*, 30 January 1975.
49 Letter from Harold Brindley, 7 August 1939, STJOHN; Dirac refers calmly to discussions with Eddington in a letter to Peierls, 16 July 1939, Peierls archive (BOD).
50 Letter from Pryce to Dirac, 11 June 1941, Dirac Papers, 2/3/11 (FSU).
51 The time of the lecture is recorded in the Royal Society's Meeting Notices. Afternoon tea began at 3.45 p.m.
52 Letter to Dirac from Pauli (then at the Institute for Advanced Study), 6 May 1942, Dirac Papers, 2/3/12 (FSU).
53 Bohr did not find out about the project until he escaped occupied Denmark in autumn 1943: see Bohr (1950).
54 Telegram to Dirac from Kapitza, 3 July 1941, Dirac Papers, 2/3/11 (FSU).
55 Letter from Dirac to Kapitza, 27 April 1943, Dirac Papers, 2/14/12A (FSU).
56 Penny (2006: 'Fatalities in the Greater Bristol Area').
57 Letter to Dirac from Dr Strover, 2 October 1941, Dirac Papers, 2/3/11 (FSU).
58 Letter from Flo Dirac to her neighbour Mrs Adam, written shortly before Christmas 1941, Dirac Papers, 1/2/1 (FSU).
59 Flo was buried in the Borough Cemetery (now the City Cemetery) in grave space 7283.

Chapter twenty-three

1 Article by Lannutti in Taylor (1987: 45).
2 Interview with Monica Dirac, 1 May 2006.
3 The committee was called MAUD, possibly short for Ministry of Aircraft production Uranium Development committee: Gowing (1964: Chapter 2).
4 Gowing (1964: 53n.).
5 Nye (2004: 73–4).
6 Nye (2004: 75–85).
7 The quote is from Churchill (1965: epilogue).
8 Letter to Dirac from F. E. Adcock, 24 May 1942, Dirac Papers, 2/3/12 (FSU).
9 Letter to Dirac from Nigel de Grey of the Foreign Office in London, 1 June 1940, Dirac Papers, 2/3/10 (FSU).
10 Copeland (2006: Chapter 14).
11 Letter from Sir Denys Wilkinson, who was one of Dyson's fellow students in Dirac's lecture course, 15 January 2004; also phone call, 16 January 2004. 'I went to Dirac's lectures in Cambridge in 1942/3. Freeman Dyson, a year junior to us but very precocious, was also in the class. He was very disruptive because he asked questions. Dirac always took a long time to answer them and on one occasion ended a class early so that he could prepare a proper response' (interview, 15 January 2004).
12 Sir Denys Wilkinson, letter, 15 January 2004; phone call, 16 January 2004.
13 Letter from Dirac to Peierls, 11 May 1942, UKNATARCHI.
14 See Thorp and Shapin (2000: 564).
15 Letter from Wigner to the US Office of International Affairs, 1 September 1965, Wigner archive, PRINCETON.
16 Anecdotes from interview with Monica Dirac, 7 February 2003 and 1 May 2006; and with Mary Dirac, 21 February 2003.
17 Hoyle (1987: 187).
18 Dirac, M. (2003: 41).

19 Letter from Dirac to Manci, 13 July 1942 (DDOCS).

20 With his usual understatement, Dirac wrote to Manci, 'It seems a little strange to have a prime minister at these very specialized lectures. I wonder how he can spare the time.' Letter from Dirac to Manci, 17 July 1942, DDOCS.

21 Letter from Peierls to Dirac, 30 September 1942, AB1/631/257889.

22 Letter from Manci to Dirac to 'Anna', 15 October 1986, Wigner archive in PRINCETON.

23 'Mrs Roosevelt's Village Hall Lunch', *Cambridge Daily News*, 5 November 1942.

24 Wattenberg (1984).

25 Interview with Al Wattenburg, 30 October 1992.

26 One of their meetings probably occurred on 31 July 1943, as Dirac proposes this date for a meeting in his letter to Fuchs of 19 August 1943 (BOD). Dirac wrote another letter to Fuchs on 1 September 1943 (BOD).

27 Peierls (1985: 163–4).

28 Szasz (1992: xix and 148–51).

29 Gowing (1964: 261).

30 Peierls, 'Address to Dirac Memorial Meeting, Cambridge', in Taylor (1987: 37).

31 Brown (1997: 250).

32 A further seventy people in Cambridge had been injured and 1,271 homes in the town had been damaged (Barham 1977: 53).

33 'Cambridge Streets Light-Up at Last!', *Cambridge Daily News*, 26 September 1944.

34 Joe wrote of his family's 'threatening situation' to Heisenberg on 25 March 1943 and sought his assistance. Four months later, Heisenberg replied to say that he was not able to offer specific help but hoped to make contact with Joe during a later visit to Holland (this meeting does not seem to have taken place). Joe wrote again to Heisenberg on 2 February 1944 from Budapest urgently requesting confirmation of Betty's Aryan descent. See Brown and Rechenberg (1987: 156).

35 Letter from Betty to Dirac, 20 July 1946, Dirac Papers, 1/7/2A (FSU).

36 Interview with Mary Dirac, 21 February 2003.

37 Gabriel later recalled that Dirac declared that there 'was no God and no Heaven or Hell'. Letter from Gabriel Dirac to the Diracs, 18 January 1972, Dirac Papers, 1/8/14 (FSU).

38 E-mail from Mary Dirac, 17 February 2006. Monica confirms that both daughters were christened.

39 Boys Smith (1983: 44).

40 Letter from Lew Kowarski to James Chadwick, 12 April 1943 (CHURCHILL).

41 Interview with the late John Crook, 1 May 2003. Professor Crook was present when Dirac made this remark.

42 'Happy Crowds Celebrate VE-Day', *Cambridge Daily News*, 9 May 1945.

43 Interview with Monica Dirac, 1 May 2006.

44 Pincher (1948: 111). The account of this event by Chapman Pincher implied that Dirac lied. Pincher remarks, 'Dr PAM Dirac, one of the scientists involved, told me at the time that he was not then engaged on vital war research. But, as the British White Paper on atomic energy states, he had been helping the British atom-bomb project by theoretical investigations on chain reactions.' Pincher had not allowed for Dirac's literal-mindedness.

45 Brown (2005: 266).

46 Interview with Leopold Halpern, 26 February 2006. Dirac told Halpern that he was disappointed with the actions of the British Government and that he went on long solitary walks in order to cool his anger. Dirac heard of the refusal of his application

for an exit visa from the Home Office official C. D. C. Robinson (letter to Dirac, 13 June 1945, Dirac Papers, 2/3/15 [FSU]). Two days later, Nevill Mott wrote to Dirac to inform him of the protests that would be made by the disappointed scientists. Mott makes it plain that he does not expect Dirac to be an active member of the protesting group (letter to Dirac from Mott, Dirac Papers, 2/3/15 [FSU]).

47 Letter from Manci Dirac to Crowther, 18 May 1945, SUSSEX.

48 Telegram from Joe Teszler to the Diracs, 1 July 1945, Dirac Papers, 1/7/5 (FSU).

49 Interview with Christine Teszler, 22 January 2004.

50 Letters from Joe Teszler to Manci, 19 July, 2 August, 23 August, 31 August, 6 September and 27 September 1945, Dirac Papers, 1/7/5 (FSU).

51 Cornwell (2003: 396).

52 The team playing at Lord's was not an official Australian side, but was called 'The Australian Services' team.

53 Smith (1986: 478).

54 'How Cambridge Heard the Great Victory News', *Cambridge Daily News*, 15 August 1945.

55 See, for example, *Time*, 20 August 1945, p. 35.

56 Cornwell (2003: 394–400).

57 Anon. (1993: 36).

58 Anon. (1993: 71).

59 Dalitz (1987a: 69–70). Also, interview with Dalitz, 9 April 2003.

60 Interview with Christine Teszler, 22 January 2004.

61 Letter from Betty to Dirac, 20 July 1946, Dirac Papers, 1/7/2A (FSU).

62 Brown (2005: 173).

63 Crowther (1970: 264).

64 The official report on the lecture is in the UKNATARCHI (Dirac Papers. BW83/2/257889).

Chapter twenty-four

1 Osgood (1951: 149, 208–11).

2 Interview with Feynman by Charles Weiner, 5 March 1966, 27 March 1966, AIP. Interview with Lew Kowarski by Charles Weiner, 3 May 1970, AIP.

3 The typed manuscript of Dirac's talk is in the Mudd Library, PRINCETON.

4 In Feynman's theory, the probability that a quantum such as an electron will make a transition from one point in space-time to another can be calculated from a mathematical expression related to the action involved in moving between the two points, summed over all possible routes between them.

5 Interview by Charles Weiner of Richard Feynman, 27 June 1966 (CALTECH). See also Feynman's Nobel Lecture and Gleick (1992: 226) and its references.

6 Interview with Freeman Dyson, 27 June 2005. Dyson noted that Feynman made the point repeatedly.

7 Quoted by Oppenheimer in Smith and Weiner (1980: 269). Wigner was one of the examiners of Feynman's Ph.D. thesis; the other was Wheeler. The oral examination was held on 3 June 1942, and the examiners' report is held in the Mudd Library, PRINCETON.

8 See Kevles (1971: Chapter 12) and Schweber (1994: Section 3).

9 Schweber (1994: Chapter 4); Pais (1986: 450–1); Dyson (2005).

10 Lamb (1983: 326). 'Radar Waves Find New Force in Atom', *New York Times*, 21 September 1947.

11 Ito (1995: 171–82).

12 Feynman (1985: 8).

13 Dyson (1992: 306). Interview with Dyson, 27 June 2005. Dyson's description of him-self as a 'big shot with a vengeance' is in Schweber (1994: 550).

14 Dyson (2005: 48).

15 Dirac took no pleasure in abstract art or in Schönberg's music and found neither beautiful.

16 'The Engineer and the Physicist', 2 January 1980, Dirac Papers, 2/9/34 (FSU).

17 Dirac Papers, 2/29/34 (FSU).

18 Dirac Papers, 2/29/34 (FSU).

19 Dyson (2006: 216).

20 Letter from Manci to Wigner, 20 February 1949, PRINCETON.

21 Interview with Richard Eden, 14 May 2003.

22 M. Dirac (1987: 6).

23 M. Dirac (2003: 41).

24 I am grateful to the Salamans' daughter Nina Wedderburn for supplying me with biographical information on her parents. Fen (1976: 375).

25 Gamow (1966: 122); Salaman and Salaman (1986: 69).

26 Interview with Monica Dirac, 7 February 2003.

27 Quoted in Hennesey (2006: 5).

28 It took centuries for women students to win equality with males at Cambridge University. The first women's colleges in Cambridge, Girton and Newnham Colleges, were founded in 1869 and 1871 respectively. From 1881 women were allowed to sit tripos exams but they did not receive any formal qualifications from the university for passing them. From 1882, women's results were published with the men's, but on separate lists. In 1921, a report proposing full admission for women was defeated. Statutes allowing the admission of women to full member-ship of the university finally received Royal Assent in May 1948, and the first woman to graduate at Cambridge was the Queen Mother in the following October. Under this legislation, women students at Cambridge first graduated in January 1949.

29 Reasons for Heisenberg's post-war depression are suggested by Cassidy (1992: 528).

30 R. Eden, unpublished memoirs, May 2003, p. 7a.

31 Dirac first met Heisenberg after the war in 1958. 'Hero' quote from interview with Antonio Zichichi, 2 October 2005.

32 Interview with Monica Dirac, 7 February 2003.

33 Greenspan (2005: 253, 263–4). Dirac supported Heisenberg's nomination, having remarked earlier that his election to a foreign membership of the Royal Society should take precedence over that of Pauli. Cockcroft writes to Dirac in his 15 February letter, 'I agree that he [Heisenberg] is more eminent than Pauli,' Dirac Papers, 2/4/7 (FSU).

34 Letter to Dirac from Douglas Hartree, 22 December 1947, Dirac Papers, 2/4/2 (FSU).

35 Letter to Dirac from Schrödinger, 18 May 1949, Dirac Papers, 2/4/4 (FSU).

36 Soon after Blackett won the prize in 1947, Dirac sent to him 'heartiest congratula-tions', remarking, 'You ought to have had it long ago': letter from Dirac to Blackett, 7 November 1948, Blackett archive, ROYSOC. Yet Dirac had not nomi-nated him.

37 Dirac nominated Kapitza twice before 1953, on 16 January 1946 and 25 January 1950. It is clear from Dirac's records that he later nominated Kapitza several times (RSAS).

38 Letter from Dirac to Kapitza, 4 November 1945, Dirac Papers, 2/4/12 (FSU); See also

letter from Kapitza to Stalin, 13 October 1944, reproduced in Boag et al. (1990: 361–3).

39 Boag et al. (1990: 378).

40 Letter from Kapitza to Stalin, 10 March 1945, cited in Kojevnikov, A. (1991) *Historical Studies in the Physical Sciences*, 22, 1, pp. 131–64.

41 Letters from Kapitza to Stalin, 3 October 1945 and 25 November 1945, reprinted in Boag et al. (1990: 368–70, 372–8).

42 Letter to Dirac from Manci, 12 July 1949 (DDOCS).

43 *Tallahassee Democrat*, 29 November 1970.

44 Bird and Sherwin (2005: 332).

45 Sources of anecdotes: 'young daughters scurrying', interview with Freeman Dyson, 27 June 2005; 'welcoming Einstein for Sunday tea', interview with Monica Dirac, 7 February 2003, interview with Mary Dirac, 21 February 2003; the 'early evening drinks parties', one of the social rituals at the institute during Oppenheimer's tenure as Director; 'amateur lumberjacks', interview with Morton White, 24 July 2004.

46 Interview with Freeman Dyson, 27 June 2005. E-mail from Dyson, 23 October 2006.

47 Interview with Louise Morse, 19 July 2006.

48 Dirac received several importunate letters from the maverick Austro-Hungarian experimenter Felix Ehrenhaft, who asserted that he had evidence for the existence of the magnetic monopole, Dirac Papers, 2/13/1 and 2/13/2 (FSU).

49 Letter from Pauli to Hans Bethe, 8 March 1949, Hermann et al. (1979).

50 The new theory made little impact, though it did interest scientists – including Dennis Gabor at Imperial College in London – who were studying electron beams in television sets. The correspondence between Dirac and Gabor (1951) is in the Gabor archive at Imperial College, London.

51 Dirac (1954).

52 Dirac (1954).

53 'The Ghost of the Ether' was published in the *Manchester Guardian* article on 19 January 1952; the *New York Times* published 'Briton Says Space Is Full of Ether', 4 February 1952. In Dirac's talk to the 1971 Lindau meeting (for former Nobel Prize winners), he said that the ether appeared not to be useful to quantum mechanics, though he did not rule out that the concept might one day be useful.

54 Jerome (2002: Chapter 12, 278–82).

55 Interview with Einstein's acquaintance Gillett Griffen on 20 November 2005, and with Louise Morse on 19 July 2006. The anecdote about Einstein picking up cigarette butts and sniffing them is from Kahler, A. (1985), *My Years of Friendship with Albert Einstein*, IX, 4, p. 7.

Chapter twenty-five

1 The information in this section is mainly from interviews with Monica Dirac (7 and 8 February 2002) and Mary Dirac (21 February 2002 and 17 February 2006). See also M. Dirac (2003: 39–42). Information about Dirac and Betty from interview with Christine Teszler, 22 January 2004.

2 The boarding school was Beeston Hall School in West Runton, near Cromer. E-mail from Mary Dirac, 30 October 2006.

3 The Diracs often stayed at the Barkston Gardens Hotel, Kensington, for a week or two.

4 Interview with Mary Dirac, 21 February 2003.

5 Letter to Dirac from Manci, 5 September 1949 (DDOCS): 'We can have a quiet weekend in London where the Folies Bergère is showing the full Paris show.'

6 Professor Driuzdustades appears in Russell's 1954 short story 'Zahatopolk' (see

Russell 1972: 82–110).

7 Manci and Monica often ate at the Koh-I-Noor restaurant in St John's Street. Interview with Monica Dirac, 7 February 2003.

8 Dalitz (1987b: 17).

9 Interview with Monica Dirac, 7 February 2003.

10 Interview with Tony Colleraine, 15 July 2004.

11 Bird and Sherwin (2005: 463–5).

12 Letter from Dirac to Manci, undated, late March 1954 (DDOCS).

13 Szasz (1992: 95).

14 Letter from Dirac to Oppenheimer, 11 November 1949, LC Oppenheimer archive.

15 Szasz (1992: 86, 95).

16 Pais often told this story. See, for example, Pais (2000: 70).

17 It appears that Dirac was excluded from a conference as early as 1951 because of Manci's Hungarian nationality. See interview with Lew Kowarski by Charles Weiner, 3 May 1970, AIP, pp. 203–4.

18 The documents concerning the petition, dated 23 March 1950, are in the Bernal Papers, KV 2/1813, UKNATARCHI.

19 McMillan (2005: 12, 199).

20 This letter, from Dirac to Oppenheimer on 17 April, does not appear to have survived. However, Ruth Barnett, of the Institute for Advanced Study, refers to it in her letter to Dirac of 28 April 1954, Dirac Papers, 2/4/10 (FSU).

21 McMillan (2005: 214).

22 Letter from Dirac to Oppenheimer, 24 April 1954, IAS Dirac archive.

23 'US-Barred Scientist "Not Red"', *Daily Express*, 28 May 1954.

24 'US Study Visa Barred to Nobel Prize Physicist', *New York Times*, 27 May 1954.

25 Letter to Dirac from Christopher Freeman, Secretary of the Society for Cultural Relations with the USSR, 26 April 1954, Dirac Papers, 2/16/9 (FSU).

26 Pais (1998: 33).

27 Letter from Wheeler, Walker Bleakney and Milton White to the *New York Times*, published in the newspaper on 3 June 1954.

28 The name of the woman is not known for certain. Interview with Monica Dirac, 7 February 2003.

29 Dirac Papers, 2/14/5 (FSU).

30 After the Diracs' stay in Mahabaleshwar, they returned to the Tata Institute in Bombay until 15 December. The Diracs then moved on to Madras and, on 20 December, travelled to Bangalore, where they spent Christmas. On New Year's Eve, they returned to Bombay and then travelled to the Indian Science Congress in Baroda on 5 January. Four days later, they travelled to Delhi and saw the Taj Mahal shortly afterwards. The Diracs were in Calcutta from 18 January to 23 January, before returning to Delhi for a few days and then, finally, back to the Tata Institute. They left India, sailing from Bombay, on 21 February 1955.

31 Interview with George Sudarshan, 15 February 2005. In 1955, Sudarshan was a research assistant at the Tata Institute.

32 Dirac's enthusiastic acceptance of the invitation to give this talk in his letter to Dr Basu, 23 June 1954, Dirac Papers, 2/4/10 (FSU).

33 Manuscript of the talk, corrected by Dirac, is in Dirac Papers, 2/14/5 (FSU). In the published version of this presentation, many of Dirac's finest touches are removed (*Journal of Scientific and Industrial Research*, Delhi, A14, pp. 153–65).

34 Salaman and Salaman (1996: 68).

35 *Science and Culture*, Volume 20, Number 8, pp. 380–1, see p. 380.

36 Perkovich (1999: 59). India became a nuclear power in 1974, eight years after Bhabha died in a plane crash.

37 Letter to Oppenheimer from G. M. Shrum, 4 April 1955 (Oppenheimer archive, Dirac Papers, LC). Dirac may have caught this form of jaundice, homologous serum hepatitis, from a contaminated needle during a medical examination in December 1954, Dirac Papers, 1/9/3 (FSU).

38 Note from Manci to Oppenheimer included in Dirac to Oppenheimer, 25 September 1954 (LC, Oppenheimer archive, Dirac Papers).

39 The Diracs sailed into Vancouver on 16 April. Letters from Manci to Oppenheimer, 15 April 1955, 22 April 1955 and other undated letters written at about the same time (LC, Oppenheimer archive).

40 Manci often remarked on the one time she saw her husband cry. See, for example, Science News, 20 June 1981, p. 394.

41 Interview with Tony Colleraine, 22 July 2004.

42 Letter from Manci to Oppenheimer, 29 August 1955, Oppenheimer archive, Dirac Papers, LC.

43 Medical report on 28 March 1955, Dirac Papers, 1/9/3 (FSU).

44 The Diracs were in Princeton from 22 May to 30 June 1955, and they flew to Ottawa on 1 July.

45 Letter from Manci to Oppenheimer, 29 August 1955 (LC, Oppenheimer archive).

46 Interview with Jeffrey Goldstone, 2 May 2006.

47 Talk on 'Electrons and the Vacuum' by Dirac at the Lindau conference. The manuscript, annotated by Dirac (June 1956) is in Dirac Papers, 2/27/14 (FSU).

48 'Electrons and the Vacuum', pp. 7–8.

49 Dirac spent much of this year working on the fourth edition of The Principles of Quantum Mechanics, which was published in the following year, 1957.

50 For an account of Kapitza's activities between 1937–49 see Kojevnikov (2004: Chapters 5–8).

51 Taubman (2003: Chapter 11).

52 The quote is in a letter from Dirac to Bohr, undated, NBI. The lecture was plainly written after this visit.

53 Dorozynski (1965: 61).

54 Boag et al. (1990: 368). See also Knight (1993: Chapters 9 and 10).

55 Taubman (2003: 256).

56 Fitzpatrick (2005: 227).

57 Dorozynski (1965: 60–1).

58 Feinberg (1987: 185 and 197).

59 Weisskopf (1990: 194).

60 Dirac's writing is still preserved on the blackboard.

61 Landau made this remark at a conference in Moscow in 1957. Interview with Sir Brian Pippard, 29 April 2004.

Chapter twenty-six

1 Enz (2002: 533).

2 Dirac probably heard the news through the grapevine in Cambridge before the news was published. One of the first accounts of the experiment was published in the Guardian on 17 January 1957.

3 Shanmugadhasan (1987: 56).

4 Dirac raised the issue of left–right symmetry in quantum mechanics in the Ph.D.

examination of K. J. Le Couteur in 1948, see Dalitz and Peierls (1986: 159).

5 On 25 August 1970, Dirac gave a piece of paper to the physicist Ivan Waller bearing the message: 'The statement that I do not believe there is any need for P and T invariance occurs in Rev Mod Phys vol 21 p 393 (1949). I never followed it up. PAM Dirac.' Waller archive, RSAS. See also Pais (1986: 25–6).

6 Polkinghorne (1987: 229).

7 Seven years later, in 1964, when two experimenters at Princeton University confirmed that some quantum processes that involve the weak interaction are not symmetric when time is reversed, most physicists were once again shocked. But not Dirac: he had also foreseen that possibility in the two paragraphs of his 1949 relativity paper.

8 The 'wrong horse' quote is from a round-table discussion at the Fermilab Symposium in May 1980, Brown and Hoddeson (1983: 268). The 'complete crushing' quote is from Dirac's talk at the Argonne Symposium on Spin, 26 July 1974, see 'An Historical Perspective on Spin' Lecture notes, pp. 3, Dirac Papers, 2/29/3 (FSU).

9 Taubman (2003: 302).

10 'The Soviet Crime in Hungary', *New Statesman*, 10 November 1956, p. 574.

11 Interview with Tam Dalyell, 9 January 2005. Dalyell recalls that his meeting with Dirac took place in either 1971 or 1972.

12 Letter from Dirac to Kapitza, 29 November 1957, Dirac Papers, 2/4/12 (FSU).

13 The connection with the anniversary was pointed out in the *New Statesman* in 26 October and 9 November 1957.

14 Interview with Monica Dirac, 1 May 2006.

15 Dirac often told his daughter Mary that he would like to travel to the Moon. Interview with Mary Dirac, 10 April 2006.

16 Newhouse (1989: 118).

17 Newhouse (1989: 118).

18 The other two physicists at lunch with Dirac were Peter Landshoff and John Nuttall. Interview with Peter Landshoff, 6 April 2006.

19 Letter from Dirac to Walter Kapryan, 19 July 1974, Dirac Papers, 2/7/6 (FSU).

20 I thank Bob Parkinson and Doug Millard for their advice on the reasons why space rockets were launched vertically rather than horizontally.

21 Interview with the Revd. Sir John Polkinghorne, 11 July 2003.

22 Interview with the Revd. Sir John Polkinghorne, 11 July 2003. Dirac once asked 'What is a rho meson?', a particle then well known to almost all particle physics researchers.

23 Interview with the Revd. Sir John Polkinghorne, 11 July 2003.

24 Interview with Monica Dirac, 7 February 2003. In 1967, Dirac's parking rights were further constrained, and, again, Manci was outraged. Letter from R. E. Macpherson to Dirac, 2 November 1967, Dirac Papers, 2/6/3 (FSU).

25 Interview with John Crook, 1 May 2003.

26 After the Christmas vacation of 1959, Gabriel urged his mother to stop telling Dirac 'I will leave you' in front of them. Letter from Gabriel to the Diracs, 13 January 1960, Dirac Papers, 1/8/12 (FSU).

27 Interview with Stanley Deser, 5 July 2006.

28 Letter to Dirac from Manci, 10 April 1954 (DDOCS).

29 Interview with Monica Dirac, 7 February 2003.

30 Hardy (1940: 87). See, for example, letters to Dirac from Gabriel, 22 September 1957 and 8 October 1957, property of Barbara Dirac-Svejstrup.

31 Interview with Mary Dirac, 21 February 2003.

32 Dirac told Gamow in 1961 that he began his work on general relativity in the hope of finding a connection between the theory and neutrinos, but that the project had failed. Letter from Dirac to Gamow, 10 January 1961, LC, Gamow archive.

33 The word 'graviton' appears to have been used for the first time in print by the Soviet physicist D. I. Blokhintsev in the journal *Under the Banner of Marxism* (*Pod znamenem marxisma*): Blokhintsev (1934). See Gorelik and Frenkel (1994: 96).

34 'Physicists Offer New Theories on Gravity Waves and Atomic Particles', *New York Times*, 31 January 1959.

35 Deser (2003). I am grateful to Sir Roger Penrose (interview 20 June 2006) and Stanley Deser (interview 5 July 2006) for advice on Dirac's contribution to general relativity.

36 Pais (1986: 23) and Salam (1987: 92).

37 Dirac describes the theory in this way in the notes for the talk he gave on 8 October 1970, 'Relativity Against Quantum Mechanics', Dirac Papers, 2/28/19 (FSU). See also Dirac (1970).

38 This description of Oppenheimer is based on the one given by Stephen Spender in *Journals 1939–83*. See also Bernstein (2004: 194).

39 Anon. (2001: 109–34).

40 Letter from Dirac to Margrethe Bohr, 20 November 1962, NBA. Margrethe's reply, dated 19 December 1962, is in Dirac Papers, 2/5/9 (FSU).

41 *Nature*, 4 February 1961, pp. 355–6; see p. 356.

42 Interview with Dirac, AHQP, 1 April 1962, pp. 5–7.

43 Interview with Dirac, AHQP, 1 April 1962, p. 5 (text from the original tape).

44 Interview with Kurt Hofer, 21 February 2004.

45 In my interviews with Leopold Halpern and Nandor Balázs, respectively on 18 February 2003 and 24 July 2002, they both noted that Dirac said he had 'loathed' his father – an extremely strong word for him to use.

46 Letter from Kuhn to Dirac, 3 July 1962, Dirac Papers, 2/5/9 (FSU). Dirac subsequently gave four more interviews with Kuhn in 7 Cavendish Avenue, Cambridge, on 6, 7, 10 and 14 May 1963.

47 Interview with Monica Dirac, 30 April 2006.

Chapter twenty-seven

1 Interview with the Revd. Sir John Polkinghorne, 11 July 2003.

2 Interview with Mary Dirac, 21 February 2003.

3 Dirac co-signed a letter, dated 27 April 1964, to Professor H. Davenport as part of a campaign to oust Batchelor from the headship of the Department of Applied Mathematics and Theoretical Physics, UCAM, Hoyle archive.

4 Interview with Yorrick and Helaine Blumenfeld, 10 January 2004.

5 Letter to Dirac from Oppenheimer, 21 April 1963, Dirac Papers, 2/5/10 (FSU).

6 The Diracs were in the USA in 1962 and 1963 (based at the Institute for Advanced Study in Princeton until late April 1962 and from late September 1962 to early April 1963); in 1964 and 1965, based mainly at the Institute for Advanced Study, from September 1964 to spring 1965; in 1966 in March and April, based in Stony Brook, New York; in 1967, based in the spring at Stony Brook and November and December at the University of Texas at Austin; in 1968 and 1969, in December 1968 based in Stony Brook until after Christmas, when they moved on to the University of Miami, where they stayed until spring 1969.

7 Goddard (1998: xiv).

8 Dirac (1966: 8). One of the themes of these lectures is Dirac's conclusion that the Schrödinger picture of quantum mechanics is untenable when it is applied to field theory and that only the Heisenberg picture is satisfactory.

9 Dirac (1963:53).

10 Several instances of Dirac's declining to appear on BBC radio and television programmes are documented in Dirac's archive at Florida State University, notably when he refused to be interviewed in connection with his *Scientific American* article (letter to Dirac from BBC radio producer David Edge, on 11 June 1963, Dirac Papers, 2/5/10 [FSU]).

11 BBC *Horizon* programme 'Lindau', reference 72/2/5/6025. The recording was made on 28 June 1965 and broadcast on 11 August 1965.

12 Barrow (2002: 105–12). Teller noted, however, that the experimental uncertainties in the calculations were so large that it was not possible definitely to rule out the hypothesis.

13 Barrow (2002: 107).

14 Letter from Dirac to Gamow, 10 January 1961, Gamow archive LC.

15 Quoted in Barrow (2002: 108).

16 Private papers of Mary Dirac. Dirac wrote the notes on 17 January 1933.

17 Letter to Dirac from Gamow, 26 October 1957, Dirac Papers, 2/5/4 (FSU).

18 John Douglas Cockcroft, *Biographical Memoirs of Fellows of the Royal Society* (1968): 139–88; see p. 185.

19 Mitton (2005: 127–9).

20 Overbye (1991: 39).

21 Letter from Gamow to Dirac, June 1965 (undated), Dirac Papers, 2/5/13 (FSU).

22 Letter from Heisenberg to Dirac, 2 March 1967, Dirac Papers, 2/14/1 (FSU). Letter from Dirac to Heisenberg, 6 March 1967, quoted in Brown and Rechenberg (1987: 148).

23 Letter from Geoffrey Harrison, HM Ambassador in Moscow, to Sir John Cockcroft, 19 April 1966, Cockcroft archive, CKFT 20/17 (CHURCHILL).

24 Kapitza gave the lecture at 5 p.m. on Monday, 16 May. Source: *Cambridge University Reporter*, 27 April 1966, p. 1,649.

25 Letter from Manci to Barbara Gamow, 12 May 1966, LC (Gamow archive). Other information from an interview with Mary Dirac, 21 February 2003.

26 Letter from Manci to Rudolf Peierls, 8 July 1986, Peierls archive, additional papers, D23 (BOD).

27 Boag et al. (1990: 43–4).

28 Batelaan, H. (2007) *Reviews of Modern Physics*, 79, pp. 929–42.

29 Dirac greatly admired Gell-Mann's skills as a physicist but went out of his way to avoid him on social occasions. Source: interview with Leopold Halpern, 26 February 2006.

30 Gell-Mann (1967: 699). For more examples of Gell-Mann's initial scepticism about the reality of quarks, see Johnson (2000: Chapter 11).

31 Gell-Mann (1967: 693).

32 'Methods in Theoretical Physics', 12 April 1967, Dirac Papers, 2/28/5 (FSU).

33 Tkachenko was handed back to the Soviet Embassy on 18 September. The British authorities' story was that Tkachenko had 'freely expressed' his wish to return to Russia, but privately they were fearful that he was going to die in their custody. See *The Times*, 18 June 1967, p. 1; *New York Times*, 16 September 1967, p. 1. See also the obituary of John Cockcroft by Kenneth McQuillen, former Vice-Master of Churchill College. I thank Mark Goldie, a Fellow of the college, for providing me with this anecdote.

34 E-mail from Chris Cockcroft, 17 May 2007. See also Oakes (2000: 82). The anecdotes were confirmed by Mary and Monica Dirac.

35 Letter from Wigner to Office of International Affairs, 1 September 1965, PRINCETON, Wigner archive.

36 See, for example, letter from Wigner to Manci, 2 September 1965 (FSU, Wigner letters, annex to Dirac Papers).

37 Letters from the Wigners, 6 and 13 May, and 14 September 1968 (FSU, Wigner letters, annex to Dirac Papers).

38 Letter from Manci to Wigner, 10 February 1968, Wigner archive (Margit Dirac file) PRINCETON.

39 Telegram 17 September 1968 (FSU, Wigner letters, annex to Dirac Papers); interview with Mary Dirac, 26 February 2006.

40 Interview with Mary Dirac, 26 February 2006.

41 Letter from Mary Wigner to the Diracs, 7 October 1968, Dirac Papers, 2/6/6 (FSU).

42 Letters from the Wigners to the Diracs, 20 and 25 September and 9 October 1968 (FSU, Wigner letters, annex to Dirac Papers). Interview with Mary Dirac, 26 February 2006 and e-mail 7 June 2006.

43 Interview with Mary Dirac, 26 February 2006 and e-mail 7 June 2006.

44 Interview with Helaine and Yorrick Blumenfeld, 10 January 2004.

45 Interview with Philip Mannheim, 8 June 2006. See also the article on Kurşunoğlu, 'The Launching of La Belle Epoque of High Energy Physics and Cosmology' in Curtright et al. (2004: 427–46).

46 An account of Dirac's time at the University of Miami is given by Kurşunoğlu's wife in Kurşunoğlu and Wigner (1987: 9–28).

47 Manci wrote to Gamow's wife on 4 February 1969 to complain that Dirac had not accepted the offer made by the University of Miami: 'It makes me feel awful' (LC, Gamow archive, Manci Dirac file).

48 The reaction of Rabbit and Janice Angstrom to *2001* are in *Rabbit Redux*, 1971, Chapter 1 (in the Fawcett Crest Book paperback edition, pp. 58 and 74).

49 LoBrutto (1997: 277).

50 I am grateful to Tony Colleraine, then Mary's husband, for his recollections of Dirac's first visits to see *2001: A Space Odyssey*, interview 15 July 2004 and e-mails on 26 September and 22 October 2004.

51 Interview with Monica Dirac, 7 February 2003.

52 Letter from Manci to Barbara Gamow, 16 March 1971, Gamow archive LC.

53 Letter from Manci to Wigner, 10 February 1968, PRINCETON, Wigner archive.

54 These FBI documents were declassified in 1986. I thank Bob Ketchum for obtaining a copy of these documents under Freedom of Information/Privacy Acts.

55 Letter from Dirac to Alfred Shild, 29 August 1966 (copy held by Lane Hughston).

56 See, for example, the letter from the Senior Secretary at the University of Texas at Austin to the Immigration and Naturalization Service, 8 December 1967, part of the CIA file on Dirac in the 1960s and 1970s. I am grateful to Robert Ketchum for obtaining these documents.

57 Tebeau (1976: 151–71 and 219–35). Stanford (1987: 54–5). Interview with Henry King Stanford, 3 July 2006.

58 Wicker (1990).

59 Letter from Wigner to Manci Dirac, 9 October 1968 (FSU, Wigner letters, annex to Dirac Papers).

60 *Miami Herald*, 7 May 1970, p. 1.

61 According to Morris (1972), the population of Tallahassee in 1970 was 72,000. The

total population of Miami in the same year was 335,000.

62 The Physics Department at Florida State University had recently obtained a Center of Excellence grant from the National Science Foundation to assist in its aspiration to become such a centre.

63 Letter from Colleraine to Dirac, 2 February 1970, Dirac Papers, 2/6/9 (FSU).

64 *Tallahassee Democrat*, 29 November 1970.

65 Interview with Peter Tilley, 2 August 2005; interview with Leopold Halpern, 26 February 2006.

66 Letter from Norman Heydenburg (Chair of the FSU physics department) to Dirac, 4 January 1971, Dirac Papers, 2/6/11 (FSU).

67 Interview with Helaine and Yorrick Blumenthal, 10 January 2004.

Chapter twenty-eight

1 Press release from Dorothy Turner Holcomb, 'Barbara Walters . . . I needed you!', 9 March 1971, Dirac Papers, 2/6/11 (FSU).

2 Walters (1970: 173).

3 Notes on 'The Evolution of our Understanding of Nature', 8 March 1971, in Dirac Papers, 2/28/21 (FSU).

4 Between 1969 and 1983, Dirac gave about a hundred and forty talks, an average of ten talks a year. He gave about eighty-eight talks in the USA, and fifty-two talks over-seas, mainly in Europe but occasionally further afield, notably in Australia and New Zealand in 1975. See Dirac Papers, 2/52/8 (FSU).

5 Interview with Kurt Hofer, 21 February 2004.

6 Interview with Pam Houmère, 25 February 2003.

7 E-mail from Hans Plendl, 5 March 2008, and another from Bill Moulton, 5 March 2008.

8 Interview with Kurt Hofer, 21 February 2004. Hofer recalls that Dirac would melt when he realised that the person he had dismissed was a friend.

9 Interview with Hofer. Leopold Halpern independently confirmed this description of Dirac's telephone manner.

10 Pais (1997: 211). Many of Dirac's colleagues at Florida State University, including Steve Edwards (interview, 27 February 2004) and Michael Kasha (interview, 18 February 2003), attest to the enjoyment he took in telling this joke.

11 M. Dirac (2003: 39).

12 Interview with Barbara Dirac-Svejstrup, 5 May 2003.

13 Letter from Manci to Dirac, undated, August 1972, Dirac Papers, 2/7/2 (FSU).

14 Letter from Manci to Dirac, 18 August 1972, Dirac Papers, 2/7/2 (FSU).

15 Interview with Ken van Assenderp, 25 February 2003.

16 Interview with Helaine and Yorrick Blumenfeld, 10 January 2004. Helaine Blumenfeld recalls: 'When I was pregnant with my second son, Manci called me all the time to check on things.' Shortly before one of Mrs Blumenfeld's appointments up at Addenbrooke's Hospital, Manci advised her, 'Well, you know they have a lot of black doctors there. Don't let them touch you, they're all dirty.' Monica Dirac recalls that her mother was 'the most anti-Semitic person I've ever met', quite surprising as Manci her-self was Jewish. Monica learned of her Jewish ancestry when she was twenty-one years old. Interviews with Monica Dirac, 7 February 2003 and 3 May 2006.

17 Interview with Yorrick and Helaine Blumenfeld, 10 January 2004.

18 Interview with Lily Harish-Chandra, 12 July 2007.

19 Quoted in Chandrasekhar (1987: 65).

20 The clearest account of Dirac's research agenda during his later years is in the summary he wrote for Joe Lannutti in November 1974, Dirac Papers, 2/7/9 (FSU).

21 Halpern (2003: 25). Interview with Leopold Halpern, 18 February 2003.

22 Halpern (2003: 24–5).

23 Leopold Halpern took me on this same trip on Sunday 26 February 2006. During this trip, and in earlier interviews, he described their trips down the river and their reception at home by Manci. In a separate interview, on 27 February 2004, Steve Edwards described the infamous incident in which Dirac dumped Kurşunoğlu in the Wakulla River

24 Weinberg (2002).

25 The special type of gauge theory, was first written down by Yang and his collaborator Robert Mills in 1954. Yang has described the theory as 'a rather straightforward generalization of Maxwell's equation' (quoted in Woolf 1980: 502).

26 Crease and Mann (1986: Chapter 16).

27 In the late 1970s Dirac erroneously analysed the opacity of the universe and his error involved a misunderstanding of the Kapitza–Dirac effect (e-mail from Martin Rees, 27 November 2006). Another error is noted in Dalitz and Peierls (1986: 175).

28 Interview with Leopold Halpern, 18 February 2002. Halpern recalled that Dirac took the discovery seriously and wanted to understand it. 'How can you explain this portrait of Jesus? How can this happen?' Dirac said several times. (The shroud was later proved to be a fake.)

29 There is no record of Dirac's taking any interest at all in the modern theory of renormalisation. He did, however, acknowledge the brilliance of physicists who worked on the theory, including Abdus Salam, Gerhard 't Hooft and Edward Witten, whom he nominated for awards. Evidence of these nominations is in the Tallahassee archive.

30 Interview with Rechenberg, 3 June 2003.

31 Dirac (1977).

32 Brown and Hoddeson (1983: 266–8).

33 Interview with Lederman, 18 June 2002.

34 Interview with Lederman, 18 June 2002. See Farmelo (2002b: 48). Einstein came close to predicting the existence of the positron in his 1925 paper 'Electron and General Relativity', see Fölsing (1997: 563–5).

35 Many female acquaintances attest to Dirac's behaviour in this respect, notably Lily Harish-Chandra, Rae Roeder, Helaine Blumenfeld and Colleen Taylor Sen.

36 Kurşunoğlu and Wigner (1987: 26). See Mill (1869), especially Chapter 3, 'Of Individuality, as One of the Elements of Well-Being'.

37 Interview with Kurt Hofer, 21 February 2004.

38 E-mail from Kurt Hofer, 6 March 2004.

39 Letter from Manci to Rudolf Peierls, 23 December 1985, Peierls archive, additional papers, D23 (BOD).

40 Interview with Christine Teszler, 22 January 2004, and an e-mail, 27 March 2004.

41 This incident occurred in 1978 as Dirac and Hofer passed the Mormon church on Stadium Drive, Tallahassee. Interview with Hofer, 21 February 2004.

42 Talk on 'Fundamental Problems of Physics', 29 June 1971 (audio recording from LINDAU). See Dirac Papers, 2/28/23 (FSU).

43 In the talk, Dirac suggested a probability for the formation of life that he considered would make it overwhelmingly unlikely without the presence of a God: a chance of one in 10^{100} (a power of ten also known as a googol).

44 E-mail from Kurt Hofer, 28 August 2006.

45 Halpern (1988: 466 n.). See also Dirac's notes on his lecture 'A Scientist's Attitude to

Religion', c. 1975, Dirac Papers, 2/32/11A (FSU).

46 Isenstein contacted Dirac after meeting him at Bohr's home: letter from Isenstein to Dirac, 29 June 1939, Dirac Papers, 2/3/9 (FSU). Isenstein renewed contact with Dirac in 1969, see letter from Isenstein to Dirac, 29 June 1969, Dirac Papers, 2/6/7 (FSU).

47 For correspondence concerning the bust, see the correspondence in the summer of 1971, Dirac Papers, 2/6/11 (FSU).

48 I thank Michael Noakes for his comments on Dirac's sitting for this portrait (interview, 3 July 2006). Noakes points out that Frank Sinatra did not sit for his portrait, though he much liked the result, which he hung on a wall of his study.

49 Dirac liked the picture, though he grumbled slightly: 'It makes me look a bit old.' Dirac was sensitive about the mark on the left side of his nose, the remains of a precancerous cyst, removed in the summer of 1977. For this reason, Noakes's portrait of Dirac shows only the right side of his face. Dirac looked rather more resolute in the two chalk drawings by Howard Morgan in 1980, commissioned by the National Portrait Gallery.

50 Feynman's drawing is reproduced in the frontispiece of Kurşunoğlu and Wigner (1987). An example of Feynman's 'I'm no Dirac' is in interview by Charles Weiner of Richard Feynman, 28 June 1966, p. 187 (CALTECH).

51 Lord Waldegrave points out that 'the award was largely the result of the intervention of Victor Rothschild, the late Lord Rothschild, who was well placed at that time as a Permanent Secretary in the Cabinet Office as Head of the Central Policy Review Staff of Prime Minister Edward Heath' (interview with Lord Waldegrave, 2 June 2004).

52 Letter from Manci to Barbara Gamow, 1 May 1973, LC.

53 Salaman and Salaman (1986: 70). Dirac raised this issue in the context of the experience of his daughter Monica, who 'had studied geology but had given it up to look after her baby'.

54 Interview with Mary Dirac, 21 February 2003.

55 Interview with Leopold Halpern, 18 February 2003.

56 The British part of the project was eventually delivered by the British Aircraft Corporation in collaboration with the French company Sud Aviation, following an agreement signed in 1962. The British Aircraft Corporation had been formed in 1960 from the Bristol Aeroplane Company and other aeronautical firms. I thank Andrew Nahum for advice on this.

57 The Diracs flew from Dulles to Paris on 5 May 1979 (DDOCS). Letters to Dirac from Abdul-Razzak Kaddoura, Assistant Director-General for Science at UNESCO, dated 29 March 1979, are in Dirac Papers, 2/9/3 (FSU).

58 *New York Times*, 5 May 1979.

59 A copy of the speech is in Dirac Papers, 1/3/8 (FSU).

60 Kapitza wrote to Dirac on 18 February 1982, 'Knowing of your going will certainly stimulate my travelling,' Dirac Papers, 2/10/6 (FSU).

61 A recording of Dirac's 1982 talk to the Lindau meeting, 'The Requirements of a Basic Physical Theory' (1 July 1982), and other details are available at LINDAU.

62 Details of the accommodation are in Dirac Papers, 2/10/7 (FSU).

63 Interview with Kurt Hofer, 21 February 2004; interview with Leopold Halpern, 26 February 2006.

64 Dirac gave this lecture on 15 August 1981, Dirac Papers, 2/29/45 (FSU).

65 The Erice Statement is readily available on the internet.

66 On 7 December 1982, Dirac wrote to the Master of St John's to apologise for not being able to attend a gathering at college on 27 December to toast Dirac's health in his eightieth year: 'For 59 years, the College has been the central point of my life and

a home to me' (STJOHN).

67 Interview with Peter Goddard, 7 June 2006.

Chapter twenty-nine

1 The account of Ramond's encounter with Dirac is taken from an interview with Ramond on 18 February 2006 and from subsequent e-mails. Note that the date of the encounter given here is later than the one given in an earlier version of the story (Pais 1998: 36–7); Ramond confirmed the date quoted here, after checking his departmental records. It is not possible to give the precise date of the meeting.

2 E-mail from Pierre Ramond, 22 December 2003.

3 *Tallahasse Democrat*, 15 May 1983, page G1.

4 Letter to Dirac and Manci from Dirac's mother, 8 April 1940, Dirac Papers, 1/4/10 (FSU).

5 Interview with Dr Watt on the telephone, 19 July 2004.

6 Dirac's last talk, 'The Future of Atomic Physics', was in New Orleans on 26 May 1983: Dirac Papers, 2/29/52 (FSU).

7 Dirac's surgeon was Dr David Miles. I thank Dr Hank Watt for providing me with a copy of the post-operation report.

8 Solnit (2001: 104).

9 Halpern (1985). Interview with Halpern, 24 February 2006.

10 The essences Halpern used were echinacea, milk thistle and ginseng: interview with Halpern, 24 February 2006.

11 Dirac (1987: 194–8).

12 Letter from Manci Dirac to Lily Harish-Chandra, 30 September 1984 (property of Mrs Harish-Chandra).

13 Letter from Manci Dirac to Lily Harish-Chandra, 16 March 1984 (property of Mrs Harish-Chandra).

14 Interview with Barbara Dirac-Svejstrup, 5 May 2003.

15 Interview with Barbara Dirac-Svejstrup, 5 May 2003.

16 Interview with Peter Tilley, 2 August 2005.

17 Dirac's death certificate says that he died of respiratory arrest. The coroner found that the final cause of his death was not kidney failure but clogged arteries. See Dirac Papers, 1/9/17 (FSU).

18 Telephone call with Hansell Watt, 19 July 2004.

19 Manci chose an Episcopalian service because the American Episcopal Church is the Anglican Church in America and is a province of the Anglican Communion under the Archbishop of Canterbury. Information from Steve Edwards, interview, 16 February 2006.

20 E-mail from Pierre Ramond, 23 February 2006.

21 I am grateful to Mary Dirac, Steve Edwards, Ridi Hofer and Pierre Ramond for their recollections of the funeral.

22 The details of Judy's case are from Mercer County Surrogate's Office. The papers that closed the case of Judith Thompson are dated 29 October 1984.

23 Letter from Dick Dalitz to Peter Goddard, 3 November 1986 (STJOHN; permission to quote this letter from Dalitz during interview with him 9 April 2003).

24 Letter from Peter Goddard to the Master of St John's College, 26 May 1990, STJOHN.

25 Interview with Richard Dalitz, 9 April 2003.

26 Letter from Michael Mayne to Richard Dalitz, 20 May 1990, STJOHN.

27 The memorial stone was designed and cut by the Cardozo Kindersley workshop in Cambridge, see Goddard (1998: xii).

28 Letter from Dalitz to Gisela Dirac, 30 November 1995, property of Gisela Dirac.

29 Goddard (1998: xiii).

30 Interview with Richard Dalitz, 9 April 2003.

31 Letter from Dalitz to Gisela Dirac, 30 November 1995, property of Gisela Dirac.

32 Letter from Manci to Gisela Dirac, 4 July 1992, property of Gisela Dirac. Manci was wrong about Byron's burial. When his remains were brought back to England, burial in the Abbey was refused, and he was interred at Hucknall. Three subsequent unsuccessful attempts were made to insert a memorial to him in the Abbey, the last being in 1924, when the supporting letter was signed by Hardy, Kipling and three former prime ministers (Balfour, Asquith and Lloyd George). Permission for a plaque in Poets' Corner was finally given only in 1969.

33 See, for example, the letter from Manci to the editor of *Scientific American*, August 1993, p. 6.

34 Letter from Manci to Abraham Pais, 25 November 1995, in Goddard (1998: 29).

35 The Ledermans had become friendly with the Diracs since May 1980, when Dirac attended the conference on the history of particle physics. Lily Harish-Chandra was married to the mathematician Harish-Chandra, Dirac's colleague; Erika Zimmerman was the daughter of Wigner from a relationship he had in Göttingen in the late 1920s.

36 Interview with Peggy Lannuti, 25 February 2004.

37 Manci did arrange for his Nobel Medal and certificate to be returned to St John's College (letter from Manci to 'Anna', 15 October 1986, Wigner archive PRINCE-TON). Manci's version of the story of Elizabeth Cockcroft's alleged ejection from Churchill College is told in Oakes (2000: 82).

38 Letter from Manci to 'Anna', 15 October 1986, Wigner archive PRINCETON.

39 Interview with Kurt Hofer, 21 February 2004; interview with Leopold Halpern, 26 February 2006.

40 Interview with the Ledermans, 30 October 2003.

41 Letter to Manci from Hillary Rodham Clinton, 12 February 1996 (DDOCS). Ms Rodham Clinton wrote: 'It is a pleasure to hear from individuals who share a vision of a better life for all Americans. It is particularly rewarding to hear from people who realize that achieving that vision will not always be easy.' Interview with Monica Dirac, 1 May 2006.

Chapter thirty

1 The prize was funded by Rolls Royce and British Aerospace. William Waldegrave recalls that Dirac supported this prize and asked him to send photographs of the Bishop Road School, where his formal education began.

2 I am grateful to Laura Thorne, of Brunel 200, for details about the programme.

3 These details and others in this paragraph were confirmed in a telephone conversation with John Bendall, 18 October 2007.

4 Interview with Mary Dirac, 10 August 2006.

5 This visit took place on 22 June 2004. Don Carleton, a historian of Bristol, kindly arranged it.

6 Letter from Manci to 'Anna', 15 October 1986, in PRINCETON, Wigner archive (Margit Dirac file).

7 These three statements are based on the more rigorous ones given by the autism

expert Uta Frith in her definitive introduction to the condition (2003: 8–9). Her statements are consistent with the most detailed and most recent scheme described in the *Diagnostic and Statistical Manual* of the American Psychiatric Association (2000), 4th edition, Washington DC, and a similar scheme issued by the World Health Organization, 'The ICD-10 Classification of Mental and Behavioural Disorders: Clinical Descriptions and Diagnostic Guidelines' (1992).

8 *Stockholms Dagblad*, 10 December 1933.

9 Walenski et al. (2006: 175); for the data on depression see p. 9.

10 Wing (1996: 47, 65 and 123).

11 Anon. (2007) 'Autism Speaks: The United States Pays Up', *Nature*, 448: 628–9; see p. 628.

12 Frith (2003: Chapter 4).

13 Unlike people with autism, people with Asperger's Syndrome show a delay neither in acquiring language when they are young nor in other aspects of intellectual development. But people with Asperger's Syndrome, when they are older, have similar social impairments to people with autism. See Frith (2003: 11).

14 Frith (2003: 182).

15 Interview with Simon Baron-Cohen, 9 July 2003; Baron-Cohen (2003: Chapters 3 and 5).

16 Fitzgerald (2004: Chapter 1).

17 Frith (2003: 112).

18 E-mail from Simon Baron-Cohen 25 December 2006.

19 Grandin (1995: 137).

20 Park (1992: 250–9); Temple Grandin's quote is from *Morning Edition*, US National Public Radio, 14 August 2006. See http://www.npr.org/templates/story/story.php?storyId=5628476 (accessed 16 August 2006).

21 Dirac (1977: 140).

22 Letter to Dirac from Manci, 2 September 1936, DDOCS.

23 'Many patients with tuberculosis present with general symptoms, such as tiredness, malaise, loss of appetite, weakness or loss of weight': Seaton et al. (2000: 516).

24 There are insights into the childhood of autistic children in the memoir of Gunilla Gerland (translated by Joan Tate), *A Real Person: Life on the Outside*. Gerland writes powerfully of her perception of the misunderstandings in her early relationship with her parents, notably with her father. 'He had no respect for anyone's needs [. . .] The effect of my father's actions was one of pure sadism, although he was not really a sadist. He didn't enjoy my humiliation in itself – he couldn't even imagine it' (Gerland 1996). See also Grandin (1984).

Chapter thirty-one

1 Weinberg wrote these words for me to read aloud at the Centenary meeting. Text checked by Weinberg, 22 July 2007 (e-mail).

2 Interview with Freeman Dyson, 27 June 2005.

3 Quoted in Charap (1972: 332).

4 E-mail from Sir Michael Atiyah, 15 July 2007.

5 Woolf (1980: 502).

6 Letter from Dirac to Abdus Salam, 11 November 1981, reproduced in Craigie et al. (1983: iii).

7 't Hooft (1997: Chapter 14).

8 Stephen Hawking appeared in an episode of *Star Trek* first broadcast on 21 June

1993, and in episodes of *The Simpsons* first broadcast on 9 May 1999 and 1 May 2005.

9 Letter from Nicolas Kurti to *New Scientist*, 65 (1975), p. 533; letter from E. C. Stern (1975) to *Science*, 189, p. 251. See also the comments by Dalitz in 'Another Side to Paul Dirac', in Kurşunoğlu and Wigner (1987: 87–8).

10 Freimund et al. (2001). The Kaptiza–Dirac effect had been observed for atoms, but not for electrons, in 1986 (Gould et al. 1986). I thank Herman Betelaan for his advice on modern experiments on the effect.

11 Deser (2003: 102).

12 Interview with Nathan Seiberg, 26 July 2007, and e-mail, 20 August 2007.

13 In his interviews, Leopold Halpern often stressed the importance to Dirac of the large numbers hypothesis (interview with Halpern, 26 February 2006).

14 By conventional measure, the gravitational force is a millionth of a billionth of a billionth of a billionth the strength of the next strongest fundamental force, the weak interaction.

15 Rees (2003). I thank Martin Rees for his advice on the status of Dirac's large numbers hypothesis.

16 E-mails from James Overduin, 20–2 July 2006.

17 Overduin and Plendl (2007).

18 I thank Rolf Landua of CERN for his expert help on the current state of experimental research into anti-matter.

19 See Yang (1980: 39).

20 These words, written on 27 November 1975, seem to have been special to Dirac. He wrote them on a single sheet of paper and filed them among his lecture notes: Dirac Papers 2/29/17 (FSU). The words replaced by [this happened] are 'I have felt the mathematics lead me by the hand.'

21 The first reference to beauty in Dirac's papers appears to be in the paper he co-wrote with Kapitza in 1933, 'The Reflection of Electrons from Standing Light Waves', where they refer to the beauty of the colour photography introduced by Gabriel Lippmann.

22 Green and Schwarz's paper was received on 10 September 1984 by the academic journal *Physics Letters B*, which published it on 13 December.

23 For a popular account of modern string theory, see Greene (1999).

24 Dirac told his student Harish-Chandra, 'I am not interested in proofs but only in what nature does': Dalitz and Peierls (1986: 156).

25 Dirac's notes commend Witten's 'brilliant solutions to a number of problems in mathematical physics', Dirac Papers, 2/14/9 (FSU).

26 Interview with Edward Witten, 8 July 2005, and e-mail, 30 August 2006.

27 E-mail from Veltman, 20 January 2008. For a sceptical assessment of string theory, see Woit (2006), especially Chapters 13–19.

Bibliography

Genealogy

The genealogy of the Dirac family is presented at http://www.dirac.ch. The website is maintained by Gisela Dirac-Wahrenburg.

References

In the text, I do not normally give a reference for Dirac's technical papers. They are listed in full in Dalitz (1995) and in Kragh (1990).

Aaserud, F. (1990) *Redirecting Science: Niels Bohr, Philanthropy, and the Rise of Nuclear Physics*, Cambridge: Cambridge University Press.
Annan, N. (1992) *Our Age: Portrait of a Generation*, London: Weidenfeld and Nicolson.
Anon. (1935) *The Frustration of Science*, foreword by F. Soddy, New York: W. W. Norton.
Anon. (1993) *Operation Epsilon: The Farm Hall Transcripts*, Bristol, Institute of Physics Publishing.
Anon. (2001) *The Cuban Missile Crisis: Selected Foreign Policy Documents from the Administration of John F. Kennedy, January 1961–November 1962*, London: The Stationery Office, pp. 109–34.
Anon. (2007) 'Autism Speaks: The United States Pays Up', *Nature*, 448: 628–9.
Badash, L. (1985) *Kaptiza, Rutherford and the Kremlin*, New Haven, Conn.: Yale University Press.
Baer, H. and Belyaev, A. (eds) (2003) *Proceedings of the Dirac Centennial Symposium*, London: World Scientific.
Baldwin, T. (1990) *G. E. Moore*, London and New York: Routledge.
Barham, J. (1977) *Cambridgeshire at War*, Cambridge: Bird's Farm.
Baron-Cohen, S. (2003) *The Essential Difference*, New York: Basic Books.
Barrow, J. (2002) *The Constants of Nature*, New York: Pantheon Books.
Batterson, S. (2006) *Pursuit of Genius*, Wellesley, Mass.: A. K. Peters Ltd.
– (2007) 'The Vision, Insight, and Influence of Oswald Veblen', *Notices of the AMS*, 54 (5): 606–18.
Beller, M. (1999) *Quantum Dialogue: The Making of a Revolution*, Chicago, Ill.: University of Chicago Press.
Bernstein, J. (2004) *Oppenheimer: Portrait of an Enigma*, London: Duckworth.
Billington Harper, S. (2000) *In the Shadow of the Mahatma*, Richmond: Curzon.
Bioley, H. (1903) *Les Poètes du Valais Romand*, Lausanne: Imprimerie J. Couchoud.
Bird, K. and Sherwin, M. J. (2005) *American Prometheus: The Triumph and Tragedy of J. Robert Oppenheimer*, New York: Vintage.
Blackett, P. M. S. (1955) 'Rutherford Memorial Lecture 1954', *Physical Society Yearbook 1955*.
– (1969) 'The Old Days of the Cavendish', *Rivista del Cimento*, 1 (special edition): xxxvii.
Blackwood, J. R. (1997) 'Einstein in a Rear-View Mirror', *Princeton History*, 14: 9–25.

Blokhintsev D. I. and Gal'perin F. M. (1934) 'Gipoteza neutrino I zakon sokhraneniya energii', *Pod znamenem marxisma*, 6: 147–57.

Boag, J. W., Rubinin, P. E. and Shoenberg, D. (eds) (1990) *Kapitza in Cambridge and Moscow*, Amsterdam: North Holland.

Board of Education (1905) *Suggestions for the Consideration of Teachers and Others Concerned with Public Elementary Schools*, London: Her Majesty's Stationery Office.

Bohr, N. (1950) *Open Letter to the United Nations*, Copenhagen: J. H. Schultz Forlag.

– (1972) *The Collected Works of Niels Bohr*, Amsterdam: North Holland.

Bokulich, A. (2004) 'Open or Closed? Dirac, Heisenberg, and the Relation Between Classical and Quantum Mechanics', *Studies in the History and Philosophy of Modern Physics*, 35: 377–96.

Born, M. (1978) *My Life: Recollections of a Nobel Laureate*, London: Taylor & Francis.

– (2005) *The Born–Einstein Letters 1916–55*, Basingstoke: Macmillan. (First published 1971.)

Bowyer, M. J. F. (1986) *Air Raid! The Enemy Air Offensive Against East Anglia*, Wellingborough: Patrick Stephens.

Boys Smith, J. S. (1983) *Memories of St John's College 1919–69*, Cambridge: St John's College.

Brendon, P. (2000) *The Dark Valley: A Panorama of the 1930s*, New York: Alfred A. Knopf.

Broad, C. D. (1923) *Scientific Thought*, Bristol: Routledge. Reprinted in 1993 by Thoemmes Press, Bristol.

Brown, A. (1997) *The Neutron and the Bomb*, Oxford: Oxford University Press.

– (2005) *J. D. Bernal: The Sage of Science*, Oxford: Oxford University Press.

Brown, L. M. (1978) 'The Idea of the Neutrino', *Physics Today*, September, pp. 23–8.

Brown, L. M. and Rechenberg, H. (1987) 'Paul Dirac and Werner Heisenberg: A Partnership in Science', in B. M. Kurşunoğlu and E. P. Wigner (eds), *Reminiscences about a Great Physicist: Paul Adrien Maurice Dirac*, Cambridge: Cambridge University Press, pp. 117–162.

Brown, L. M. and Hoddeson, L. (1983) *The Birth of Particle Physics*, Cambridge: Cambridge University Press.

Bryder, L. (1988) *Below the Magic Mountain: A Social History of Tuberculosis in Twentieth Century Britain*, Oxford: Clarendon Press.

– (1992) 'Wonderlands of Buttercup, Clover and Daisies', in R. Cooter (ed.), *In the Name of the Child: Health and Welfare 1880–1940*, London and New York: Routledge, pp. 72–95.

Budd, M. (2002) *The Aesthetic Appreciation of Nature*, Oxford: Clarendon Press.

Bukharin, N. (1931) 'Theory and Practice from the Standpoint of Dialectical Materialism', available at http://www.marxists.org/archive/bukharin/works/1931/dia-mat (accessed 22 May 2008).

Cahan, D. (1989) *An Institute for an Empire*, Cambridge: Cambridge University Press.

Cannadine, D. (1994) *Aspects of Aristocracy*, New Haven, Conn.: Yale University Press.

Cassidy, D. C. (1992) *Uncertainty: The Life and Science of Werner Heisenberg*, New York: W. H. Freeman and Co.

Cathcart, B. (2004) *The Fly in the Cathedral*, London: Penguin Books.

Chadwick, J. (1984) 'Some Personal Notes on the Discovery of the Neutron', in J. Hendry (ed.), *Cambridge Physics in the Thirties*, Bristol: Adam Hilger, pp. 42–5.

Chandrasekhar, S. (1987) *Truth and Beauty: Aesthetic Motivations in Science*, Chicago, Ill.: University of Chicago Press.

Charap, J. (1972) 'In Praise of Paul Dirac', *The Listener*, 14 September, pp. 331–2.

Christianson, G. E. (1995) *Edwin Hubble: Mariner of the Nebulae*, Bristol: Institute of Physics Publishing.

Churchill, Randolph (1965) *Twenty-One Years*, London: Weidenfeld and Nicholson.

Conquest, R. (1986) *Harvest of Sorrow: Soviet Collectivization and the Terror-Famine*, New York: Oxford University Press.

Copeland, B. J. (2006) *Colossus: The Secrets of Bletchley Park's Codebreaking Computers*, Oxford: Oxford University Press.

Cornwell, J. (2003) *Hitler's Scientists*, London: Penguin Books.

Craigie, N. S., Goddard, P. and Nahm, W. (eds) (1983) *Monopoles in Quantum Field Theory*, Singapore: World Scientific.

Crease, R. P. and Mann, C. C. (1986) *The Second Creation: Makers of the Revolution of Twentieth Century Physics*, New Brunswick, NJ: Rutgers University Press.

Crowther, J. G. (1970) *Fifty Years with Science*, London: Barrie & Jenkins.

Cunningham, E. (1970) 'Ebenezer: Recollections of Ebenezer Cunningham', unpublished, archive of St John's College, Cambridge.

Cunningham, E. (1956) 'Obituary of Henry Baker', *The Eagle*, 57, p. 81.

Curtright, T., Mintz, S. and Perlmutter, A. (eds) (2004) *Proceedings of the 32nd Coral Gables Conference*, London: World Scientific.

Da Costa Andrade, E. N. (1964) *Rutherford and the Nature of the Atom*, New York: Anchor Books.

Dalitz, R. H. (1987a) 'Another Side to Paul Dirac' in B. M. Kurşunoğlu and E. P. Wigner (eds), *Reminiscences about a Great Physicist: Paul Adrien Maurice Dirac*, Cambridge: Cambridge University Press, pp. 69–92.

– (1987b) 'A Biographical Sketch of the Life of P. A. M. Dirac', in J. G. Taylor (ed.), *Tributes to Paul Dirac*, Bristol: Adam Hilger, pp. 3–28.

– (ed.) (1995) *The Collected Works of P. A. M. Dirac 1924–1948*, Cambridge: Cambridge University Press.

Dalitz, R. H. and Peierls, R. (1986) 'Paul Adrien Maurice Dirac', *Biographical Memoirs of the Royal Society*, 32: 138–85.

Daniel, G. (1986) *Some Small Harvest*, London: Thames and Hudson.

Darrigol, O. (1992) *From c-Numbers to q-Numbers*, Berkeley, Calif.: University of California Press.

Darrow, K. K. (1934) 'Discovery and Early History of the Positive Electron', *Scientific Monthly*, 38 (1): 5–14.

De Maria, M. and Russo, A. (1985) 'The Discovery of the Positron', *Rivista di Storia della Scienza*, 2 (2): 237–86.

Delbrück, M. (1972) 'Out of this World', in F. Reines (ed.), *Cosmology, Fusion and Other Matters: George Gamow Memorial Volume*, Boulder, Col.: Colorado Associated University Press, pp. 280–8.

Deser, D. (2003) 'P. A. M. Dirac and the Development of Modern General Relativity', in H. Baer and A. Belyaev (eds) (2003) *Proceedings of the Dirac Centennial Symposium*, London: World Scientific, pp. 99–105.

Dingle, H. (1937a) 'Modern Aristotlelianism', *Nature*, 8 May.

– (1937b) 'Deductive and Inductive Methods in Science: A Reply', supplement to *Nature*, 12 June, pp. 1,001–2.

Dirac, M. (1987) 'Thinking of My Darling Paul', in B. M. Kurşunoğlu and E. P. Wigner (eds), *Reminiscences about a Great Physicist: Paul Adrien Maurice Dirac*, Cambridge: Cambridge University Press, p. 3–8.

Dirac, M. (2003) 'My Father', in H. Baer and A. Belyaev (eds) (2003) *Proceedings of the Dirac Centennial Symposium*, London: World Scientific, pp. 39–42.

497

Dirac, P. A. M. (1936) 'Does Conservation of Energy Hold in Atomic Processes?' *Nature*, 22 February, pp. 803–4.

– (1954) 'Quantum Mechanics and the Ether', *Scientific Monthly*, 78: 142–6.

– (1963) *The Evolution of the Physicist's Picture of Nature*, Scientific American, May, vol. 208, no. 5, pp. 45–53.

– (1966) *Lectures on Quantum Field Theory*, New York, Belfer Graduate School of Science, Yeshiva University.

– (1970) 'Can Equations of Motion Be Used in High-Energy Physics?', *Physics Today*, April, pp. 29–31.

– (1977) 'Recollections of an Exciting Era', in C. Weiner (ed.), *History of Twentieth Century Physics*, New York: Academic Press, pp. 109–46.

– (1978) 'The Monopole Concept', *International Journal of Physics*, 17 (4): 235–47.

– (1982) 'Pretty Mathematics', *International Journal of Physics*, 21: 603–5.

– (1982a) 'The early years of relativity' in G. Holton and Y. Elkana (eds) *Albert Einstein: Historical and Cultural Perspectives*, Princeton: Princeton University Press, pp. 79–90.

– (1987) 'The Inadequacies of Quantum Field Theory', in B. M. Kurşunoğlu and E. P. Wigner (eds), *Reminiscences about a Great Physicist: Paul Adrien Maurice Dirac*, Cambridge: Cambridge University Press, pp. 194–8.

Dorozynski, A. (1965) *The Man They Wouldn't Let Die*, New York: Macmillan.

Dyson, F. (1992) 'From Eros to Gaia', Pantheon Books, New York.

– (2005) 'Hans Bethe and Quantum Electrodynamics', *Physics Today*, October, pp. 48–50.

– (2006) *The Scientist as Rebel*, New York: New York Review of Books.

Earman, J. and Glymour, C. (1980) 'Relativity and Eclipses: the British Eclipse Expeditions of 1919 and Their Predecessors', *Historical Studies in the Physical Sciences*, 11: 49–85.

Eddington, A. (1918) 'Report on the Meeting of the Association Held on Wednesday November 27 1918 at Sion College, Victoria Embankment, E.C.', *Journal of the British Astronomical Association*, 29: 35–9.

– (1928) *The Nature of the Physical World*, Cambridge: Cambridge University Press.

Einstein, A. (1931) in *James Clerk Maxwell: A Commemorative Volume 1831–1931*, Cambridge: Cambridge University Press.

Elsasser, W. (1978) *Memoirs of a Physicist in the Atomic Age*, London: Adam Hilger.

Enz, C. P. (2002) *No Time to Be Brief: A Scientific Biography of Wolfgang Pauli*, Oxford: Oxford University Press.

Farmelo, G. (ed.) (2002a) *It Must Be Beautiful: Great Equations of Modern Science*, London: Granta.

– (2002b) 'Pipped to the Positron', *New Scientist*, 10 August, pp. 48–9.

– (2005) 'Dirac's Hidden Geometry', *Nature*, 437, p. 323.

Feinberg, E. L. (ed.) (1987) *Reminiscences about I. E. Tamm*, Moscow: Nauka.

Fellows, F. H. (1985) 'J. H. Van Vleck: The Early Life and Work of a Mathematical Physicist', unpublished PhD thesis, University of Minnesota.

Fen, E. (1976) *A Russian's England*, Warwick: Paul Gordon Books.

Fermi, L. (1968) 'Illustrious Immigrants: The Intellectual Migration from Europe 1930–41', Chicago, Ill.: University of Chicago Press.

Feynman, R. P. (1985) *QED: The Strange Theory of Light and Matter*, London, Penguin Books.

Fitzgerald, M. (2004) *Autism and Creativity*, New York: Brunner-Routledge.

Fitzpatrick, S. (1999) *Everyday Stalinism*, Oxford: Oxford University Press.

– (2005) *Tear Off the Masks!*, Princeton, NJ: Princeton University Press.

Fölsing, A. (1997) *Albert Einstein: A Biography*, New York: Viking.

Freeman, J. (1991) *A Passion for Physics*, London: Institute of Physics Publishing.

Freimund, D. L., Aflatooni, K., and Batelaan, H. (2001) 'Observation of the Kaptiza-Dirac Effect', *Nature*, 413: 142–3.

French, A. P, and Kennedy, P. J. (eds) (1985) *Niels Bohr: A Centenary Volume*, Cambridge, Mass.: Harvard University Press.

Frenkel, V. Y. (1966) 'Yakov Ilich Frenkel: His Life, Work and Letters', Boston, Mass.: Birkhäuser Verlag.

Frith, U. (2003) *Autism: Explaining the Enigma*, 2nd edn, Oxford: Blackwell.

Galison, P. (1987) *How Experiments End*, Chicago, Ill.: University of Chicago.

– (2000) 'The Suppressed Drawing: Paul Dirac's Hidden Geometry', *Representations*, autumn issue, pp. 145–66.

– (2003) *Einstein's Clocks, Poincaré's Maps*, London: Sceptre.

Gamow, G. (1966) *Thirty Years that Shook Physics*, Doubleday & Co, New York.

– (1967) 'History of the Universe', *Science*, 158 (3802): 766–9.

– (1970) *My World Line: An Informal Autobiography*, New York: Viking Press.

Gardiner, M. (1988) *A Scatter of Memories*, London: Free Association Books.

Gardner, M. (2004) *The Colossal Book of Mathematics*, W. W. Norton & Co, New York.

Garff, J. (2000) *Søren Kierkegaard*, trans. B. H. Kirmmse, Princeton, NJ: Princeton University Press.

– (2005) *Søren Kierkegaard: A Biography*, trans. B. H. Kirmmse, Princeton, NJ: Princeton University Press.

Gaunt, W. (1945) *The Aesthetic Adventure*, London: Jonathan Cape.

Gell-Mann, M. (1967) 'Present Status of the Fundamental Interactions', in A. Zichichi (ed.), *Hadrons and Their Interactions: Current and Field Algebra, Soft Pions, Supermultiplets, and Related Topics*, New York: Academic Press.

– (1994) *The Quark and the Jaguar*, London: Little, Brown & Co.

Gerland, G. (1996) *A Real Person: Life on the Outside*, trans. Joan Tate, London: Souvenir Press.

Gleick, J. (1992) *Richard Feynman and Modern Physics*, London: Little, Brown.

Goddard, P. (ed.) (1998) *Paul Dirac: The Man and His Work*, Cambridge: Cambridge University Press.

Goodchild, P. (1985) *J. Robert Oppenheimer: Shatterer of Worlds*, New York: Fromm International.

Gorelik G. E. and Frenkel, V. Y. (1994) *Matvei Petrovich Bronstein and Soviet Theoretical Physics in the Thirties*, Boston, Mass.: Birkhäuser Verlag.

Gottfried, K. (2002) 'Matter All in the Mind', *Nature*, 419, p. 117.

Gould, P. et al. (1986) *Physical Review Letters*, 56: 827–30.

Gowing, M. (1964) *Britain and Atomic Energy 1939–45*, Basingstoke: Macmillan.

Grandin, T. (1984) 'My Experiences as an Autistic Child and Review of Selected Literature', *Journal of Orthomolecular Psychiatry*, 13: 144–74.

– (1995) 'How People with Autism Think', in E. Schopler and G. B. Mesibov (eds), *Learning and Cognition in Autism*, New York: Plenum Press: 137–56.

Gray, A. (1925) *The Town of Cambridge*, Cambridge: W. Heffers & Sons Ltd.

Greene, B. (1999) *The Elegant Universe*, New York: W.W. Norton & Co.

Greenspan, N. T. (2005) *The End of the Uncertain World: The Life and Science of Max Born*, Chichester: John Wiley & Sons Ltd.

Halpern L. (1988) 'Observations of Two of Our Brightest Stars', in K. Bleuler and M.

Werner (eds), *Proceedings of the NATO Advanced Research Workshop and the 16th International Conference on Differential Geometrical Methods in Theoretical Physics*, Boston, Mass.: Kluwer, pp. 463–70.

– (2003) 'From Reminiscences to Outlook', in H. Baer and A. Belyaev (eds), *Proceedings of the Dirac Centennial Symposium*, London: World Scientific, pp. 23–37.

Harap, J. (1972) 'In Praise of Paul Dirac', *The Listener*, 14 September, pp. 331–2.

Hardy, G. H. (1940) *A Mathematician's Apology*, Cambridge: Cambridge University Press.

Hayward, F. H. (ed.) (1909) *The Primary Curriculum*, London: Ralph, Holland & Co.

Hearn, L. (1896) *Kokoro: Hints and Echoes of Japanese Inner Life*, London: Osgood & Co.

Heaviside, O. (1899) *Electromagnetic Theory*, Vol. II, London: Office of 'The Electrician'.

Heilbron, J. (1979) *Electricity in the 17th and 18th Centuries*, Berkeley, Calif.: University of California Press.

Heisenberg, W. (1967) 'Quantum Theory and its Interpretation', in S. Rozental (ed.), *Niels Bohr: His Life and Work As Seen by His Friends and Colleagues*, New York: Wiley, pp. 94–108.

– (1971) *Physics and Beyond*, London: George Allen & Unwin.

Hendry, J. (ed.) (1984) *Cambridge Physics in the Thirties*, Bristol: Adam Hilger Ltd.

Hennessey, P. (2006) *Having It So Good*, London: Allen Lane.

– (2007) *Cabinets and the Bomb*, Oxford: Oxford University Press.

Hermann, A., v. Meyenn, K. and Weisskopf, V. F. (eds) (1979) *Wolfgang Pauli: Scientific Correspondence with Bohr, Einstein, Heisenberg*, 3 vols, Berlin: Springer.

Hodge, W. V. D. (1956) 'Henry Frederick Baker', *Biographical Memoirs of Fellows of the Royal Society*, Vol. II, November, pp. 49–68.

Holloway, D. (1994) *Stalin and the Bomb*, New Haven, Yale University Press.

Holroyd, M. (1988) *Bernard Shaw, Vol. I: 1856–98*, New York: Random House.

't Hooft, G. (1997) *In Search of the Ultimate Building Blocks*, Cambridge: Cambridge University Press.

Howarth, T. E. B. (1978) *Cambridge Between Two Wars*, London: Collins.

Hoyle, F. (1987) 'The Achievement of Dirac', *Notes and Records of the Royal Society of London*, 43 (1): 183–7.

– (1994) *Home is Where the Wind Blows*, Mill Valley, Calif.: University Science Books.

Hughes, J. (2003) *Thinker, Toiler, Scientist, Spy? Peter Kapitza and the British Security State*, Manchester: University of Manchester.

Huxley, A. (1928) *Point Counterpoint*, New York, Random House.

Huxley, T. H. (1894) *Biogenesis and Abiogenesis: Collected Essays, 1893–1894: Discourses, Biological and Geological*, vol. 8, Basingstoke: Macmillan.

Infeld, I. (1941) *Quest: The Evolution of a Scientist*, London: The Scientific Book Club.

Ito D. (1995) 'The Birthplace of Renormalization Theory', in M. Matsui (ed.), *Sin-itiro Tomonaga: Life of a Japanese Physicist'*, Tokyo: MYU, pp. 171–82.

Jeffreys, B. (1987) 'Reminiscences at the Dinner held at St John's College', in J. G. Taylor (ed.), *Tributes to Paul Dirac*, Bristol: Adam Hilger, pp. 38–9.

Jerome, F. (2002) *The Einstein File*, New York: St Martin's Griffin.

Jerome, F. and Taylor, R. (2005) *Einstein on Race and Racism*, New Brunswick, NJ: Rutgers University Press.

Johnson, G. (2000) *Strange Beauty*, London, Jonathan Cape.

Jones, D. (2000) *Bristol Past*, Chichester: Phillimore.

Jones, R. (1998) *Gender and the Formation of Taste in Eighteenth-Century Britain*, Cambridge: Cambridge University Press.

Kedrov, F. B. (1984) *Kapitza: Life and Discoveries*, Moscow: Mir Publishers.

Kevles, D. J. (1971) *The Physicists: The History of a Scientific Community in Modern America*, New York: Alfred A. Knopf.

Khalatnikov, I. M. (ed.) (1989) *Landau: the Physicist and the Man*, trans. B. J. Sykes, Oxford: Pergamon Press.

Knight, A. (1993) *Beria: Stalin's First Lieutenant*, Princeton, NJ: Princeton University Press.

Kojevnikov, A. (1993) *Paul Dirac and Igor Tamm Correspondence Part 1: 1928–1933*, Munich, Max Planck Institute for Physics.

– (1996) *Paul Dirac and Igor Tamm Correspondence Part II: 1933–36*, Munich, Max Planck Institute for Physics.

– (2004) *Stalin's Great Science: The Times and Adventures of Soviet Physicists*, London: Imperial College Press.

Kragh, H. (1990) *Dirac: A Scientific Biography*, Cambridge: Cambridge University Press.

– (1996) *Cosmology and Controversy*, Princeton, NJ: Princeton University Press.

Kronig, R. and Weisskopf, V. F. (eds) (1964) *Collected Scientific Papers by Wolfgang Pauli*, Vol. II, New York: Interscience Publishers.

Kurşunoğlu, B. N. and Wigner, E. P. (eds) (1987) *Reminiscences About a Great Physicist: Paul Adrien Maurice Dirac*, Cambridge: Cambridge University Press.

Kurşunoğlu, S. A. (1987) 'Dirac in Coral Gables', in B. M. Kurşunoğlu and E. P. Wigner (eds), *Reminiscences about a Great Physicist: Paul Adrien Maurice Dirac*, Cambridge: Cambridge University Press, pp. 9–28.

Lamb, W. (1983) in 'The Fine Structure of Hydrogen' in L. M. Brown and L. Hoddeson (eds) (1983), *The Birth of Particle Physics*, Cambridge: Cambridge University Press, pp. 311–28.

Lambourne, L. (1996) *The Aesthetic Movement*, London: Phaidon Press.

Lanouette, W. (1992) *Genius in the Shadows: A Biography of Leo Szilard*, New York: Scribner's.

Lee, S. (ed.) (2007a) *Sir Rudolf Peierls: Selected Private and Scientific Correspondence, Volume 1*, London: World Scientific.

– (ed.) (2007b) *The Bethe–Peierls Correspondence*, London: World Scientific.

LoBrutto, V. (1997) *Stanley Kubrick: A Biography*, London: Faber & Faber.

Lützen, J. (2003) 'The Concept of the Function in Mathematical Analysis', in M. J. Nye (ed.), *The Cambridge History of Science, Vol. V: The Modern Physical and Mathematical Sciences*, Cambridge: Cambridge University Press, pp. 468–87.

Lyes, J. (n.d.) 'Bristol 1914–19', Bristol Branch of the Historical Association (undated but apparently c. 1920).

McCrea, W. H. (1987) 'Eamon de Valera, Erwin Schrödinger and the Dublin Institute', in C. W. Kilmister (ed.), *Schrödinger: Centenary Celebration of a Polymath*, Cambridge: Cambridge University Press, pp. 119–34.

McGucken, W. (1984) *Scientists, Society and State*, Columbus, Ohio: Ohio State University Press, pp. 40–1.

McMillan, P. J. (2005) *The Ruin of J. Robert Oppenheimer and the Birth of the Modern Arms Race*, New York: Penguin.

Matthews, G. B. (1914) *Projective Geometry*, London: Longmans, Green & Co.

Matsui, M. (1995) *Sin-Itiro Tomonaga: Life of a Japanese Physicist*, Tokyo: MYU.

Mehra, J. (ed.) (1973) *The Physicist's Conception of Nature*, Boston, Mass.: D. Reidel.

Mehra, J. and Rechenberg, H. (1982) *The Historical Development of Quantum Theory*, Vol. IV, New York: Springer-Verlag.

Michalka, W. and Niedhart, G. (eds) (1980) *Die ungeliebte Republik*, Munich: DTV.

Michelet, H. (1988) in *Les Echos de Saint-Maurice*, Saint-Maurice, Editions Saint-Augustin, pp. 91–100.

Mill, J. S. (1869) *On Liberty*, London: Penguin Books.

– (1873) *Autobiography*.

– (1892) *A System of Logic*, London: George Routledge and Son.

Miller, A. I. (2005) *Empire of the Stars*, London: Little, Brown.

Mitton, S. (2005) *Fred Hoyle: A Life in Science*, London: Aurum Press.

Moldin, S. O. and Rubenstein, J. L. R. (eds) (2006) *Understanding Autism: from Basic Neuroscience to Treatment*, New York: Taylor & Francis.

Møller, C. (1963) 'Nogle erindringer fra livet på Bohrs institute I sidste halvdel af tyverne [Some memories from life at Bohr's institute in the late 1920s]', in *Niels Bohr, et Mindeskrift [Niels Bohr, a Memorial Volume]*, Copenhagen: Gjellerup, pp. 54–64.

Moore, G. E. (1903) *Principia Ethica*, Cambridge: Cambridge University Press.

Moore, W. (1989) *Schrödinger: Life and Thought*, Cambridge: Cambridge University Press.

Morgan, K., Cohen, G. and Flin, A. (2007) *Communists and British Society 1920–91*, London: Rivers Oram Press.

Morley, A. (1919) *Strength of Materials*, London: Longmans, Green and Co.

Morrell, G. W. (1990) 'Britain Confronts the Stalin Revolution: The Metro-Vickers Trial and Anglo-Soviet Relations, 1933', Ph.D. thesis, Michigan State University.

Morris, A. (1972) *The Florida Handbook 1971–72*, Tallahassee, Fla.: Peninsular Publishing Company.

Mott, N. F. (1986) *A Life in Science*, London: Taylor & Francis.

Nahin, P. J. (1987) *Oliver Heaviside: Sage in Solitude*, New York: IEEE Press.

Needham, J. (1976) *Moulds of Understanding*, London: George, Allen & Unwin.

Newhouse, J. (1989) *War and Peace in the Nuclear Age*, New York: Knopf.

Nye, M. J. (ed.) (2003) *The Cambridge History of Science, Vol. V: The Modern Physical and Mathematical Sciences*, Cambridge: Cambridge University Press.

– (2004) *Blackett: Physics, War, and Politics in the Twentieth Century*, Cambridge, Mass.: Harvard University Press.

Oakes, B. B. (2000) 'The Personal Papers of Paul A.M. Dirac: Their History and Preservation at the Florida State University', Unpublished PhD thesis, Florida State University.

Oliphant, M. (1972) *Rutherford: Recollections of the Cambridge Days*, Amsterdam: Elsevier Publishing Company.

Orwell, G. (2004) *Why I Write*, London: Penguin.

Osgood C. (1951) *Lights in Nassau Hall. A Book of the Bicentennial 1746–1946*, Princeton, NJ: Princeton University Press.

Overbye, D. (1991) *Lonely Hearts of the Cosmos*, New York: Harper Collins.

Overduin, J. M. and Plendl, H. S. (2007) 'Leopold Ernst Halpern and the Generalization of General Relativity', in H. Kleinert, R. T. Jantzen and R. Ruffini (eds), *The Proceedings of the Eleventh Marcel Grossmann Meeting on General Relativity*, Singapore: World Scientific.

Pais, A. (1967) 'Reminiscences from the Post-War Years', in S. Rozental (ed.), *Niels Bohr: His Life and Work as Seen by His Friends and Colleagues*, New York: Wiley, pp. 215–26.

– (1982) *Subtle is the Lord*, Oxford: Oxford University Press.

– (1986) *Inward Bound*, Oxford: Oxford University Press.

– (1991) *Niels Bohr's Times, in Physics, Philosophy and Polity*, Oxford: Clarendon Press.

– (1997) *A Tale of Two Continents: A Physicist's Life in a Turbulent World*, Princeton, NJ: Princeton University Press.

– (1998) 'Paul Dirac: Aspects of His Life and Work', in P. Goddard (ed.), *Paul Dirac: The Man and His Work*, Cambridge: Cambridge University Press, pp. 1–45.

– (2000) *The Genius of Science*, Oxford: Oxford University Press.

Park, C. (1992) 'Autism into Art: a Handicap Transfigured', in E. Schopler and G. B. Mesibov (eds), *High-Functioning Individuals with Autism*, New York: Plenum Press, pp. 250–9.

Parry, A. (1968) *Peter Kapitza on Life and Science*, Basingtoke: Macmillan.

Peierls, R. (1985) *Bird of Passage*, Princeton, NJ: Princeton University Press.

– (1987) 'Address to Dirac Memorial Meeting, Cambridge', in J. G. Taylor (ed.), *Tributes to Paul Dirac*, Bristol: Adam Hilger, pp. 35–7.

Penny, J. (2006) 'Bristol During World War Two: the Attackers and Defenders', unpublished.

Perkovich, G. (1999) *India's Nuclear Bomb*, Berkeley, Calif.: University of California Press.

Pincher, C. (1948) *Into the Atomic Age*, London: Hutchinson and Co.

Polkinghorne, J. C. (1987) 'At the Feet of Dirac', in B. M. Kurşunoğlu and E. P. Wigner (eds), *Reminiscences about a Great Physicist: Paul Adrien Maurice Dirac*, Cambridge: Cambridge University Press, pp. 227–9.

Popplewell, W. C. (1907) *Strength of Materials*, London: Oliver and Boyd.

Pratten, D. G. (1991) *Tradition and Change: The Story of Cotham School*, Bristol: Burleigh Press Ltd.

Raymond, J. (ed.) (1960) *The Baldwin Age*, London: Eyre and Spottiswoode.

Rees, M. (2003) 'Numerical Coincidences and "Tuning" in Cosmology', *Astrophysics and Space Science*, 285 (2): 375–88.

Reines, F. (1972) (ed.) *Cosmology, Fusion and Other Matters: George Gamow Memorial Volume*, Boulder, Col.: Colorado Associated University Press.

Rhodes, R. (1986) *The Making of the Atomic Bomb*, London: Simon and Schuster.

Robertson, M. (1985) 'Recollections of Princeton: The Princeton Mathematics Community in the 1930s', available at http://www.princeton.edu/~mudd/finding_aids/mathoral/pmo2.htm (accessed 22 May 2008).

Roqué, X. (1997) 'The Manufacture of the Positron', *Studies in the Philosophy and History of Modern Physics*, 28 (1): 73–129.

Ross, S. (1962) 'Scientist: The Story of the Word', *Annals of Science*, 18 (June): 65–85.

Rowlands, P. and Wilson, J. P. (1994) *Oliver Lodge and the Invention of Radio*, Liverpool: PD Publications.

Rozental, S. (ed.) (1967) *Niels Bohr: His Life and Work as Seen by His Friends and Colleagues*, New York: Wiley.

Russell, B. (1972) *The Collected Stories*, London, George Allen & Unwin.

Sachs R. G. (ed.) (1984) *The Nuclear Chain Reaction: Forty Years Later*, Chicago, Ill.: University of Chicago Press.

Salam, A. (1987) 'Dirac and Finite Field Theories', in J. G. Taylor (ed.), *Tributes to Paul Dirac*, Bristol: Adam Hilger, pp. 84–95.

Salam, A., and Wigner, E. P. (eds) (1972) *Aspects of Quantum Theory*, Cambridge: Cambridge University Press.

Salaman, E. and M. (1986) 'Remembering Paul Dirac', *Encounter*, 66 (5): 66–70.

Schilpp, P. A. (1959) (ed.) *The Philosophy of C. D. Broad*, New York: Tudor Publishing Company.

– (1970), *Albert Einstein: Philosopher-Scientist*, Library of Living Philosophers, Volume VII, La Salle, Ill.: Open Court Publishing Company.

Schücking, E. (1999) 'Jordan, Pauli, Politics, Brecht, and a Variable Gravitational Constant', *Physics Today*, October, pp. 26–36.

Schultz, B. (2003) *Gravity from the Ground Up*, Cambridge: Cambridge University Press.

Schuster, A. (1898a) 'Potential Matter: A Holiday Dream', *Nature*, 18 August, p. 367.
– (1898b) 'Potential Matter', *Nature*, 27 October, pp. 618–19.
Schweber, S. S. (1994) *QED and the Men Who Made It: Dyson, Feynman, Schwinger and Tomonaga*, Princeton, NJ: Princeton University Press.
Scott Fitzgerald, F. (1931) 'Echoes of the Jazz Age', *Scribner's Magazine*, November, pp. 459–65.
Seaton, A., Seaton, D. and Leitch, A. G. (2000) *Crofton and Douglas's Respiratory Diseases*, Vol. II, Oxford: Blackwell.
Serber, R. (1998) *Peace and War*, New York: Columbia University Press.
Service, R. (2003) *A History of Modern Russia*, Cambridge, Mass.: Harvard University Press.
Shanmugadhasan, S. (1987) 'Dirac as Research Supervisor and Other Remembrances', in J. G. Taylor (ed.), *Tributes to Paul Dirac*, Bristol: Adam Hilger, pp. 48–57.
Shoenberg, D. (1985) 'Piotr Leonidovich Kapitza', *Biographical Memoirs of Fellow of the Royal Society*, 31: 326–74.
Sinclair, A. (1986) *The Red and the Blue: Intelligence, Treason and the Universities*, London: Weidenfeld and Nicolson.
Skorupski, J. (1988) 'John Stuart Mill', in E. Craig (ed.), *Routledge Encyclopaedia of Philosophy*, London: Routledge.
Slater, J. (1975) *Solid-State and Molecular Theory: A Scientific Biography*, New York: John Wiley and Sons.
Smith, A. K. and Weiner, C. (eds) (1980) *Robert Oppenheimer: Letters and Recollections*, Stanford, Calif.: Stanford University Press.
Snow, C. P. (1931) 'A Use for Popular Scientists', *Cambridge Review*, 10 June, p. 492–3.
– (1934) *The Search*, London: Victor Gollancz.
– (1960) 'Rutherford in the Cavendish', in J. Raymond (ed.), *The Baldwin Age*, London: Eyre & Spottiswoode, pp. 235–48.
Solnit, R. (2001) *Wanderlust: A History of Walking*, New York: Penguin Books.
Sponsel, A. (2002) 'Constructing a "Revolution in Science": The Campaign to Promote a Favourable Reception for the 1919 Solar Eclipse Experiments', *British Journal of the History of Science*, 35 (4): 439–67.
Spruch, G. M. (1979) 'Pyotr Kapitza, Octogenarian Dissident', *Physics Today*, September, pp. 34–41.
Stanford, H. K. (1987) 'Dirac at the University of Miami', in B. M. Kurşunoğlu and E. P. Wigner (eds), *Reminiscences about a Great Physicist: Paul Adrien Maurice Dirac*, Cambridge: Cambridge University Press, pp. 53–6.
Stanley, M. (2007) 'Practical Mystic: Religion, Science and A. S. Eddington', Chicago, Ill.: University of Chicago Press.
Stansky, P. and Abrahams, W. (1966) *Journey to the Frontier: Julian Bell and John Cornford: Their Lives and the 1930s*, London: Constable.
Stoke, H. and Green, V. (2005) *A Dictionary of Bristle*, 2nd edn, Bristol: Broadcast Books.
Stone, G. F. and Wells, C. (eds) (1920) *Bristol and the Great War 1914–19*, Bristol: J. W. Arrowsmith Ltd.
Szasz, F. M. (1992) *British Scientists and the Manhattan Project*, London: Macmillan.
Tamm, I. E. (1933) 'On the Work of Marxist Philosophers in the Field of Physics', *Pod znamenem marxizma (Under the Banner of Marxism)*, 2: 220–31.
Taubman, W. (2003) *Khrushchev*, London: Free Press.
Taylor, J. G. (ed.) (1987) *Tributes to Paul Dirac*, Bristol: Adam Hilger.
Taylor Sen, C. (1986) 'Remembering Paul Dirac', *Encounter*, 67 (2): 80.
Tebeau, C. W. (1976) *The University of Miami – a Golden Anniversary History*

1926–1976, Coral Gables, Fla.: University of Miami Press.

Thomälen, A. (1907), trans. George How, London: Edward Arnold & Co.

Thorp, C. and Shapin, S. (2000) 'Who was J. Robert Oppenheimer?', *Social Studies of Science*, August.

Van der Waerden, B. L. (1960) 'Exclusion Principle and Spin', in M. Fierz and V. F. Weiskopff (eds), *Theoretical Physics in the Twentieth Century*, London: Interscience Publishers Ltd, pp. 199–244.

– (ed.) (1967) *Sources of Quantum Mechanics*, New York: Dover.

Van Vleck, J. (1972) 'Travels with Dirac in the Rockies', in A. Salam and E. P. Wigner (eds), *Aspects of Quantum Theory*, Cambridge: Cambridge University Press, pp. 7–16.

Vint, J (1956) 'Henry Ronald Hassé', *Journal of the London Mathematical Society*, 31: 252–5.

Von Kármán, T. (with Edson, L.) (1967) *The Wind and Beyond*, Boston, Mass.: Little, Brown and Company.

von Meyenn, K. (1985) 'Die Faustparodie', in K. von Meyenn, K. Soltzenburg and R. U. Sexl (eds), *Niels Bohr 1885–1962: Der Kopenhagener Geist in der Physik*, Brunswick: Vieweg, pp. 308–42.

von Meyenn, K. and Schücking E. (2001), 'Wolfgang Pauli', *Physics Today*, February, p. 46.

Walenski, M., Tager-Flusberg, H. and Ullman, M. T. (2006) 'Language and Autism', in S. O. Moldin and J. L. R. Rubenstein (eds), *Understanding Autism: From Basic Neuroscience to Treatment*, New York: Taylor & Francis, pp. 175–204.

Wali, K. C. (1991) *Chandra: A Biography of S. Chandrasekhar*, Chicago, Ill.: University of Chicago Press.

Walters, B. (1970) *How to Talk with Practically Anybody About Practically Anything*, New York: Doubleday & Co., Inc.

Warwick, A. (2003) *Masters of Theory: Cambridge and the Rise of Mathematical Physics*, Chicago, Ill.: University of Chicago Press.

Watson, J. D. (1980) *The Double Helix*, ed. G. S. Stent, New York: W. W. Norton & Co.

Wattenberg, A. (1984) 'December 2, 1942: The Event and the People', in R. G. Sachs (ed.), *The Nuclear Chain Reaction: Forty Years Later*, Chicago, Ill.: University of Chicago Press, pp. 43–53.

Weart, S. and Weiss Szilard, G. (eds) (1978) *Leo Szilard: His Version of the Facts*, Cambridge, Mass.: MIT Press.

Weinberg, S. (2002) 'How Great Equations Survive', in G. Farmelo (ed.), *It Must Be Beautiful: Great Equations of Modern Science*, London: Granta, pp. 253–7.

Weiner, C. (1977) *History of Twentieth Century Physics*, New York: Academic Press.

Weisskopf, V. (1990) *The Joy of Insight*, New York: Basic Books.

Wells, H. G. (2005) *The Time Machine*, London: Penguin.

Wells, J. C. (1982) *Accents of English 2*, Cambridge: Cambridge University Press.

Werskey, G. (1978) *The Visible College*, London: Allen Lane.

Westfall, R. S. (1993) *The Life of Isaac Newton*, Cambridge: Cambridge University Press.

Wheeler, J. A. (1985) 'Physics in Copenhagen in 1934 and 1935', in A. P. French and P. G. Kennedy (eds), *Niels Bohr: A Centenary Volume*, Cambridge, Mass.: Harvard University Press, pp. 221–6.

– (1998) *Geons, Black Holes, and Quantum Foam*, New York: W. W. Norton & Co.

Wicker, W. K. (1990) 'Of Time and Place: The Presidential Odyssey of Dr Henry King Stanford', Doctor of Education thesis, University of Georgia.

Wigner, E. P. (1992) *The Recollections of Eugene P. Wigner as Told to Andrew Szanton*, New York: Plenum Press.

Wilson, A. N. (2002) *The Victorians*, London: Hutchinson.

Wilson, D. (1983) *Rutherford: Simple Genius*, London: Hodder and Stoughton.

Wing, L. (1996) *The Autistic Spectrum*, London: Robinson.

Winstone, R. (1972) *Bristol as It Was 1914–1900*, Bristol: published by the author.

Woit, P. (2006) *Not Even Wrong: The Failure of String Theory and the Continuing Challenge to Unify the Laws of Physics*, London: Jonathan Cape.

Woodhead, M. (1989) 'School Starts at Five . . . or Four Years Old', *Journal of Education Policy*, 4: 1–21.

Woolf, H. (ed.) (1980) *Some Strangeness in the Proportion: A Centennial Symposium to Celebrate the Achievements of Albert Einstein*, Reading, Mass.: Addison-Wesley.

Yang, C. N. (1980) 'Beauty and Theoretical Physics', in D. W. Curtin (ed.), *The Aesthetic Dimension of Science*, New York: Philosophical Library, pp. 25–40.

List of Plates

1 Dirac family, 3 September 1907 (courtesy Monica Dirac).

2 Paul Dirac, 17 August 1907 (courtesy Monica Dirac).

3 Felix, Betty and Paul Dirac c.1909 (courtesy Monica Dirac).

4 Technical drawing by Paul Dirac (FSU, Dirac archive, 1/10/F5).

5 Bristol University Engineering Society's visit to Messrs Douglas's Works (FSU, Dirac archive, 1/10/F128).

6 Charles Dirac, c.1933 (FSU, Dirac archive, 1/15/F1D).

7 Felix Dirac, 1921 (FSU, Dirac archive, 1/15/F1J).

8 6 Julius Road, Bristol.

9 Max Born entertaining his younger colleagues at his home in Göttingen, spring 1926 (FSU, Dirac archive, 1/14/F6).

10 Some members of the Kapitza Club, after a meeting c. 1925 (courtesy Giovanna Blackett).

11 Patrick Blackett and Paul Ehrenfest, c.1925 (courtesy Giovanna Blackett).

12 Isabel Whitehead and her husband Henry, with their son Henry, 1922 (courtesy Archives, The United Theological, Bangalore, India).

13 Dirac at a meeting in Kazan, Russia, 12 October 1928 (FSU, Dirac archive, 1/14/F12).

14 Heisenberg's mother, Schrödinger's wife, Flo Dirac, Dirac, Heisenberg and Schrödinger (AIP Emilio Segrè Visual Archives).

15 Extract from a letter from Dirac to Manci Balazs, 9 May 1935 (courtesy Monica Dirac).

16 Dirac and Manci on their honeymoon, Brighton, January 1937 (courtesy Monica Dirac).

17 The Dirac family in the garden of their Cambridge home, c.1946 (courtesy Monica Dirac).

18 Dirac and Manci with a party during a crossing of the Atlantic on the SS America, 2 April 1963 (FSU, Dirac archive, 1/14/F63).

19 Dirac and Richard Feynman at a conference on relativity, Warsaw, July 1962 (photograph by A. John Coleman, courtesy AIP Emilio Segrè Visual Archives, Physics Today collection).

20 Dirac at the Institute for Advanced Study, Princeton, *c.*1958 (courtesy Monica Dirac).
21 The Diracs' home in Tallahasse, 223 Chapel Drive.
22 Kapitza and Dirac at the Hotel Bad Schachen, Lindau, summer 1982 (FSU, Dirac archive, 1/14/F98).
23 One of the last photographs taken of Dirac, Tallahassee, *c.*1983 (courtesy Monica Dirac).

Acknowledgements

Art is I, science is we.
CLAUDE BERNARD (1865) 'Introduction' to *L'Étude de la médecine experimental*

Claude Bernard was right. Biographies of scientists, too, are 'we', not 'I', in the sense that none could be written satisfactorily without a good deal of help. So I'd like to begin by acknowledging the huge contribution of the scientists, historians, archivists and writers who have preserved memories and other information about Paul Dirac. My gratitude extends to Dirac himself, who evidently took care to preserve documents about many crucial events in life, right down to the row about his Cambridge parking permit.

But let me be more specific. First I should like to thank Dirac's closest family. His daughter Monica has been unfailingly helpful, welcoming my enquiries and going out of her way to make available family documents to me. Her friend John Amy has been immensely accommodating to me throughout the project, and I am duly grateful to him. No less kind than Monica was Dirac's other daughter, Mary, who died in Tallahassee on 20 January 2007. Her guardian, Marshall Knight, has been extremely generous and obliging to me, especially during my visits to Florida.

Other family members who have given generously of their time in helping me: Gisela and Christian Dirac, Leo Dirac, Vicky Dirac, Barbara Dirac-Svejstrup, Christine Teszler, Pat Wigner, Charles and Mary Upton, Peter Lantos and Erika Zimmermann. Past family members who provided valuable testimonies are Tony Colleraine and Peter Tilley. Gisela Dirac, the family genealogist, has been indefatigable in helping to clarify the French and Swiss provenance of the Dirac family.

Four institutions to which I owe special gratitude are St John's College, Cambridge, the Institute for Advanced Study in Princeton, Florida State University in Tallahassee and Bristol University.

St John's invited me to stay in the college several times, enabling me to experience day-to-day life there, to use its superb facilities and to talk at length with several of Dirac's former colleagues and acquaintances. I am grateful to the Master and Fellows of the college for this hospitality and for making available the facilities of the college to me, notably the library. For enlightening conversations, I thank the late John Crooke, Duncan Dormor, Clifford Evans, Jane Heale, John Leake, Nick Manton, George Watson and Sir Maurice Wilkes. I have received a huge amount of support from the college library, especially from Mark Nicholls, Malcolm Underwood and the special collections librarian Jonathan Harrison, whose industry has enormously benefited the book. The university library has been most helpful, and I would like to thank Elisabeth Leedham-Green and Jackie Cox for taking so much trouble to answer my queries. Also in Cambridge, I should like to thank Yorrick and Helaine Blumenfeld, Richard Eden, Peter Landshoff, Sir Brian Pippard, the Reverend Sir John Polkinghorne and Lord (Martin) Rees.

At the Institute for Advanced Study, I was fortunate enough to spend four productive and very happy summers researching the book and writing it. I benefited considerably from conversations there with Yve-Alain Bois, Freeman Dyson, Peter Goddard, Juan Maldacena, Nathan Seiberg, Morton White and Edward Witten. The library facilities at the institute are peerless, and I should like to thank all the staff there who were unstinting in their support: Karen Downing, Momota Ganguli, Gabriella Hoskin, Erica Mosner, Marcia Tucker, Kirstie Venanzi and Judy Wilson-Smith. Among the other colleagues who made my stays there so rewarding: Linda Arntzenius, Alan Cheng, Karen Cuozzo, Jennifer Hansen, Beatrice Jessen, Kevin Kelly, Camille Merger, Nadine Thompson, Sharon Tozzi-Goff and Sarah Zantua. Also in Princeton, I should like to thank Gillett Griffin, Lily Harish-Chandra, Louise Morse (*mère et soeur*) and Terri Nelson.

I should like to give my special thanks to Peter Goddard, formerly Master of St John's, now Director of the Institute for Advanced Study. No one has been more supportive of the project or shown more interest in its progress. I owe him an enormous debt.

At Florida State University, I have benefited from the excellent library facilities and from invaluable help from the staff responsible for the Dirac archive. Sharon Schwerzel, Head of the Paul A. M. Dirac Science Library, could not have been more helpful to me – her understanding of the challenges faced by a biographer working thousands of miles from the primary archive has been hugely beneficial.

It has also been a delight to work with Chuck McCann, Paul Vermeron, with Lucy Patrick and all the librarians in Special Collections: Burt Altman, Garnett Avant, Denise Gianniano, Ginger Harkey, Alice Motes, Michael Matos and Chad Underwood. On the past and present faculty of the university, I should like to thank Howie Baer, Steve Edwards, the late Leopold Halpern, Kurt Hofer, Harry Kroto, Robley Light, Bill Moulton and Hans Plendl. Through colleagues at Florida State, I also met many other people in Tallahassee who shared their memories of Dirac with me: Ken van Assenderp, Pamela Houmère, Peggy Lannutti, Jeanne Light, Pat Ritchie, Rae Roeder and Hansell Watt.

At Bristol University, I have been supported by Debra Avent-Gibson, Sir Michael Berry, Chris Harries, Michael Richardson, Margaret and Vincent Smith and Leslie Warne. Many others in Bristol have also done much to shed light on Dirac's early life, especially Karen and Chris Benson, Dick Clements, Alan Elkan, Andrew Lang, John Penny and John Steeds. I was fortunate to be introduced to Don Carleton, a local historian, who has done an inordinate amount of work to illuminate the history of Bristol in the early twentieth century.

I should like to thank the following institutions for granting permission to quote from their archives: American Philosophical Society; Bodleian Library, University of Oxford; University of Bristol Library; Bristol Record Office; British Broadcasting Corporation; Masters and Fellows of Christ's College, Cambridge; The Syndics of Cambridge University Library; Council for the Lindau Nobel Laureate Meetings; Institute for Advanced Study, Princeton; Master, Fellows and scholars of St John's College Cambridge; Archives for the History of Quantum Physics, College Park, MD, USA; Archives for the Society of Merchant Venturers, held at the Bristol Records Office, UK; Provost and Scholars of King's College, Cambridge; Niels Bohr Archive, Copenhagen; Princeton University Library; Royal Commission for the Exhibition of 1851; International Solvay Institutes, Brussels; Special Collections at the University of Sussex; Archives at the United Theological College, Bangalore, India.

During my research, many friends and colleagues at archives and other institutions have given me valuable support. At the California Institute of Technology archives: Shelley Erwin and Bonnie Ludt. At the Center for History of Physics of the American Institute of

Physics, Maryland: Melanie Brown, Julie Gass, Spencer Weart and Stephanie Jankowski. At CERN, Geneva: John Ellis, Rolf Landua, Esthel Laperrière. In the Archives Centre: Anita Hollier. At Christ's College, Cambridge: Candace Guite. At the archive of the Royal Commission for the Exhibition of 1851: Angela Kenny and Valerie Phillips. At the Archives in the College of Aeronautics, Cranfield University: John Harrington. At the Archives of Imperial College, London: Anne Barrett. At Lambeth Palace Library, London: Naomi Ward. At the Royal Society, London: Martin Carr and Ross MacFarlane. At the Max Planck Institute, Munich: Helmut Rechenberg. At the Niels Bohr Archive, Copenhagen: Finn Asserud and Felicity Pors. At the University of Madison, Wisconsin: Vernon Barger, Tom Butler, Kerry Kresse, Ron Larson, David Null and Bill Robbins. At Firestone Library, Princeton University: AnnaLee Pauls and Meg Sherry Rich. At the Solvay archive in the Free University of Brussels: Carole Masson, Dominique Bogaerts and Isabelle Juif. At the Science Museum, London: Heather Mayfield, Doug Millard, Andrew Nahum, Matthew Pudney and Jon Tucker. It is a special pleasure to thank past and present staff at the Science Museum Library: Ian Carter, Allison Pollard, Prabha Shah, Valerie Scott, Robert Sharp, Joanna Shrimpton, Jim Singleton, Mandy Taylor, Peter Tajasque, John Underwood and Nick Wyatt. Thanks also to Ben Whelehan at Imperial College Library. At the Tata Institute in Bombay: Indira Chowdhury. At the National Media Museum, York: Colin Harding and John Trenouth. At Special Collections, University of Sussex: Dorothy Sheridan and Karen Watson. For their help with determining the detailed weather conditions in towns and cities in the UK and USA, it is a pleasure to thank Steve Jebson at the Met Office and Melissa Griffin at Florida State University.

Others who have been extremely helpful in responding to my enquiries: Sir Michael Atiyah, Tom Baldwin, John Barnes, Herman Batelaan, Steve Batterson, John Bendall, Giovanna Blackett, Margaret Booth (née Hartree), Gustav Born, Olaf Breidbach, Andrew Brown, Nicholas Capaldi, David Cassidy, Brian Cathcart, Martin Clark, Paul Clark, Chris Cockcroft, Thea Cockcroft, Flurin Condrau, Beverley Cook, Peter Cooper, Tam Dalyell, Dick Dalitz, Olivier Darrigol, Richard Davies, Stanley Deser, David Edgerton, John Ellis, Joyce Farmelo, Michael Frayn, Igor Gamow, Joshua Goldman, Jeffrey Goldstone, Jeremy Gray, Karl Hall, Richard

Hartree, Peter Harvey, Steve Henderson, Chris Hicks, John Holt, Jeff Hughes, Lane Hughston, Bob Jaffe, Edgar Jenkins, Allan Jones, Bob Ketchum, Anne Kox, Charles Kuper, Peter Lamarque, Willis Lamb, Dominique Lambert, Ellen and Leon Lederman, Sabine Lee, John Maddox, Philip Mannheim, Robin Marshall, Dennis McCormick, Arthur I. Miller, Andrew Nahum, Michael Noakes, Mary Jo Nye, Susan Oakes, James Overduin, Bob Parkinson, John Partington, Sir Roger Penrose, Trevor Powell, Roger Philips, Chris Redmond, Tony Scarr, Robert Schulmann, Bernard Shultz, Simon Singh, John Skorupski, Ulrica Söderlind, Alistair Sponsel, Henry King Stanford, Simon Stevens, George Sudashan, Colleen Taylor-Sen, Laura Thorne, Claire Tomalin, Martin Veltman, Andrew Warwick, John Watson, Russell Webb, Nina Wedderburn, John Wheeler, the late David Whitehead, Oliver Whitehead, Frank Wilczek, Michael Worboys, Nigel Wrench, Sir Denys Wilkinson and Abe Yoffe. Special thanks to Alexei Kojevnikov, who has been unstinting in the guidance and help he has given to me concerning the development of Russian physics in the past century.

For their help with primary research, my sincere thanks to Anna Cain, Martin Clark, Ruth Horry, Anna Menzies, James Jackson, Joshua Goldman, Katie Kiekhaefer, Tadas Krupovnickas and Jimmy Sebastian.

For technical support, thanks to: Paul Chen at Biblioscape (the marvellous bibliographic software) and Ian Hart.

For translating documents, I am indebted to Paul Clark, Gisela Dirac, Karl Grandin, Asger Høeg, Anna Menzies, Dora Bobory and Eszter Molnar-Mills.

For reading parts of the manuscript and for their constructive comments, thanks to: Simon Baron-Cohen, Paul Clark, Olivier Darrigol, Uta Frith, Freeman Dyson, Roger Highfield, Kurt Hofer, Bob Jaffe, Ramamurti Rajaraman, Martin Rees and Jon Tucker. And for reading the entire manuscript and for dozens of helpful comments, thanks to: Don Carleton, Stanley Deser, Alexei Kojevnikov, Peter Rowlands, Chuck Schwager, Marty Schwager and David Ucko. I am especially grateful to my friends David Johnson and David Sumner for reading several drafts of the book, each time providing extremely insightful and constructive feedback.

Finally, I should like to acknowledge the huge contribution of my publisher, Faber and Faber, to the book. Kate Ward supervised the production of the book with great attention to detail, and Kate Murray-Brown read the book with a keen and sensitive eye on content and style and provided many valuable suggestions and comments. Liz O'Donnell has been a dream of an editor – meticulous, sensitive, questioning and collegiate. I am indebted most of all to Neil Belton, who has supported the project from its inception, given me no end of wise advice and kept the bar high.

The concept of 'we' extends only so far: I take responsibility for any remaining inaccuracies in the book and for its portrayal of Paul Dirac's work and personality. In that sense, the book is 'I'.

Graham Farmelo
June 2008

Index

'PD' indicates Paul Dirac

2001: A Space Odyssey (film) 386, 391
Aarhus, Denmark 407
Abstein, Dr W. Robert 414
action principle 215–16, 229, 333, 430
Adcock, Frank 320
Adler, Dorothy and Sol 352
Adrian, Edgar 269
Aesthetic Movement 16, 443*n*42
algebra 48, 49, 71
  Grassmann 73, 85
  non-commutative 188
American Physical Society, annual meeting of
  (New York, 1959) 367–8
American Science News Service 178
Amsterdam
  Ehrenfest's suicide in 232, 347
  Betty and Joe Teszler live in 291, 311,
    383–4, 401
  Betty and Joe flee from their home 325
Anderson, Carl 196–7, 211–13, 218, 223
  chooses the name positron 224
  'The Apparent Existence of Easily
    Deflectable Positives' 212, 218, 224
Anglo-French Society of Sciences 330
anti-electrons 187, 195, 196, 203, 204, 211,
    212, 218, 247, 464*n*29
  Blackett and Occhialini's discovery
    214–15, 223
  see also positrons and anti-matter
anti-matter
  anti-quarks 433
  the Big Bang 2, 433
  PD predicts its existence 2, 187, 195, 226
  surplus of matter over anti-matter 433–4
  a universe made from equal amounts of
    matter and anti-matter 243, 433
anti-Semitism 131, 151, 153, 180, 217, 219,
    297, 395, 487*n*16
Apollo space programme 386–7, 391
Aquitania (liner) 159
Arbuckle, Fatty 30, 446*n*4
Armstrong, Neil 386
Arts School, Cambridge 72
Asperger's Syndrome 423, 492*n*13
Asuma Bura, MS 267
Atiyah, Sir Michael 415, 430

atomic bomb *see* nuclear weapons
Atomic Energy Commission 351
atomic physics
  and classical laws 171
  PD attends Tyndall's lectures 51
  PD writes on 159, 210
  see also quantum theory, quantum physics
atoms
  atom visualized as a mechanical device 129
  Balmer's formula for hydrogen spectrum
    70–71, 91
  Bohr's work on atomic structure 69–72,
    83, 88, 90, 95, 348
  electrons as a constituent of 51
  energy levels 71
  heavy 158, 159
  Rutherford's discovery of the nucleus 61–2,
    69
Austria, Hitler's invasion of (1938) 297, 396
autism 421–6, 491–2*n*7, 492*n*13, 492*n*24
aviation industry 19, 24, 47
Avon Gorge 11
Aydelotte, Frank 307, 308
Ayer, A. J. 313

Baker, Henry 72, 73
  his tea parties 72–3, 85, 326
  appearance 72
  personality 73
  and the Greeks' love of beauty 73
Balázs, Nandor 448*n*18, 484*n*45
Balázs, Richard 256, 469*n*33
Baldwin, Stanley 226
Balmer, Johannes: formula for hydrogen spec-
    trum 70–71, 91, 94
Baltimore Dairy Lunch, Princeton (the Balt)
    252, 253
bare electron 356, 357
bare energy 336–7
Barnes, Julian: *Flaubert's Parrot* 374
Baron-Cohen, Simon 423–4
Batchelor, George 365, 367, 374, 387, 484*n*3
Battle of Britain 309–10, 312
BBC (British Broadcasting Corporation (later
    Company)) 67, 328, 329, 379
  Home Service 384
  PD declines numerous interviews 377, 485*n*10
  Start the Week (Radio 4 programme) 420

Beatles, The 372, 374, 389
Beaufort, Lady Margaret 56, 57, 140
beauty
  Baker's fascination with the Greeks' love of beauty 73
  concept of 16, 443n41, 443n42
  discussion between PD and Heisenberg 377–8
  of a fundamental theory in physics 74
  Kant and 74, 450n54
  in mathematics 45, 73, 301, 380, 402
  Moore on 74
  PD's first recorded mention of 73
  in vogue as a concept at Cambridge 74, 156
Beeston Hall School, West Runton, Norfolk 346, 480n2
Belgium, Queen of (Elisabeth of Bavaria) 180
Bell, James 190, 297, 462n16, 473n54
Bell Laboratories, New Jersey 379
Bendall, John 420
Beria, Lavrentiy 341, 358
Berlin
  global capital of theoretical physics 96
  Oppenheimer in 123
  Einstein in 152, 180
  anti-Semitism 180
  nuclear fission discovered in 299
  Debye in 307–8
Berlin Wall, fall of (1989) 417
Bernal, Desmond 98–9, 146, 189, 191, 228, 241, 260, 313, 320, 327, 330, 350, 362
Berne, Switzerland 11
Bethe, Hans 471n26
Bhabha, Homi 353, 354, 481n36
Bialobrzeski, Czeslaw 468n22
Big Bang 2, 379, 403, 431, 433
Birge, Raymond 124
Birmingham 77, 305, 306, 311, 321
Bishop Road Junior School, Bristol 13–17, 23, 26, 27, 48, 104, 184, 419, 443n31, 491n1
Bishopston, Bristol 10, 269, 278
Bismarck, Prince Otto von 123
black holes 432
blackbody radiation 52, 81, 104, 105
Blackett, Patrick 133, 182, 270, 330, 332, 341, 403
  serves in World War I 63
  personality 63
  influences PD 63
  appearance 63
  resents Kapitza 63
  experimental physics 63
  attempted poisoning by Oppenheimer 97
  and cosmic rays 196, 214, 218–19
  anger at Rutherford's despotic style 213–14
  discovery of the anti-electron 214–15, 218
  revelations at the Royal Society 222–4
  supports the Labour Party 228
  family 284
  and nuclear fission 300
  a wartime Government scientific adviser 319
  and manufacture of a nuclear weapon 319–20
  and the Manhattan Project 325
  refused a visa for the Soviet Union (spring 1945) 327
  Nobel Prize 405, 479n36
Bletchley Park, Buckinghamshire 320
Bloomsbury Group 74
Blumenfeld, Helaine 385, 390, 395, 487n16, 488n35
Blumenfeld, Yorrick 385, 395
Blunt, Anthony 276
Boer War 10, 20
Bohr, Margrethe 109, 232, 244, 245, 280, 371
Bohr, Niels 76, 136, 139, 161, 231, 280, 300, 308, 338, 349, 350, 407, 428, 453n1
  theory of atomic structure 69–72, 83, 88, 90, 95, 348
  PD's mastery of his atomic theory 75
  Nobel Prize for physics 81
  visits Cambridge 81
  appearance 81, 108
  personality 81, 108, 120, 244
  and Rutherford 81
  gloomy about the state of quantum physics 82
  and Heisenberg's theory of 1925 83, 85
  PD's visit to the Institute 107–11, 113–15, 120
  concern with words 111, 114
  PD on 120, 371
  on PD 120
  complementarity principle 128
  coat of arms 128
  defends Heisenberg's uncertainty principle 137
  and a relativistic equation of the electron 139, 143
  response to PD's hole theory 168–9, 175, 206
  and PD's Bristol lecture 177, 178
  at the 1930 Solvay Conference 180
  and the neutrino 195
  represented in a special version of *Faust* 204
  and Hitler's appointment as Chancellor 219–20

and philosophy 220
and the positron 224
and the bas-relief of Rutherford 226–7
his mansion 230, 244, 466*n*47
congratulates PD on his Nobel Prize 235
party to honour the Nobel Prize winners
 243–5
and Shankland's results 274
death of his eldest son 280
at Rutherford's memorial service 294
and nuclear fission 300, 306
meeting with Heisenberg (1941) 316–17
escapes from occupied Denmark 476*n*53
and genetics 426
death 371
Bohr orbits 70
Bollobas, Gabriella 417
Bolshevik Party 190
Bolshevik Revolution (1917) 191, 293, 363
Bolshevism 138, 149, 152, 153
Bolshevo, near Moscow 268, 292–3
Bombay (Mumbai) 352, 353
Bordeaux, France 4, 9
Born, Gustav 310
Born, Max 137, 217, 338, 341, 406, 428,
 454*n*10
quantum mechanics named by 88
and PD's first paper on quantum mechanics
 91
works with Heisenberg and Jordan at
 Göttingen 92, 96
and Heisenberg's quantum theory 96
and Jordan's work on Fermi-Dirac statistics
 105
interpretation of Schrödinger's waves
 109–10
quantum probabilities 110
appearance 122
personality 122
and Oppenheimer's behaviour 124
surprised at PD's knowledgeability 124
and field theory 126
and the rise of anti-Semitism in Göttingen
 131, 151
and the Dirac equation 143–4
nervous breakdown 151
considers emigration 200
appointment at Cambridge 219
resents PD's Nobel Prize 235
message from the Nazi Government 250
professorship in Edinburgh 272
in the Lake District with PD 310
PD asks him to support Heisenberg 340–41
Nobel Prize 405
death 400

Bose, Satyendra 331, 453*n*20
bosons 331
Boston, Massachusetts 172
Boston University: PD's lecture (1972) 130
Boulton, Edmund 31
Boys Smith, John 132
bra 326
Bradman, Sir Donald 175, 176
Bragg, Sir Lawrence 294–5, 467*n*63
Bragg, William 245
Bridges, Robert: *A Testament of Beauty* 149,
 156–7
Brighton, PD's honeymoon in 284, 286
Bristol
 Charles Dirac settles in 5, 6
 described 8, 11, 18–19
 and Catholicism 8–9
 aviation industry 19, 20, 24, 404
 First World War 20, 23–4
 protestors baton-charged by police (1932)
  217
 Second World War 311, 312, 317
Bristol Aeroplane Company 404–5, 489*n*56
Bristol Central Library 8
Bristol Citizens' Recruiting Committee 20
Bristol Downs 11, 78, 155, 271, 401
Bristol Evening News 24
Bristol Records Office 419
Bristol Shiplovers' Society 271
British Aeroplane Company 309
British Aerospace 491*n*1
British Aircraft Corporation 489*n*56
British and Colonial Aeroplane Company 19
British Association for the Advancement of
 Science
 meeting (Bristol, 1930) 174, 176–8, 180
 meeting (Leicester, 1933) 229, 299
British Thomson-Houston Company, Rugby
 35, 37–8
Broad, Charlie 74, 387, 395
 Professor of Philosophy at Bristol 39
 as a lecturer 39, 42, 447*n*40
 treatment of relativity 39–40
 and PD's interest in philosophy 43
 moves to Cambridge 65
Brookhaven National Laboratory 360
Brown, Dan: *Angels and Demons* 434
Brown, Miss Josephine 57, 449*n*5, 449*n*6
Brunel, Isambard Kingdom 11, 165, 419–20,
 429
Budapest 152, 256, 261, 268, 269, 275, 280,
 284, 285, 286, 292, 297, 303, 325, 330,
 394
Bukharin, Nikolai 190–91
Bulletin of the Soviet Academy of Sciences 293

Bullock, W.H. 446*n*19
Bunin, Ivan 241
Bush, George, Snr. 417
Butler, Samuel: *The Way of all Flesh* 1
Byron, Lord 174, 416, 491*n*32

Cadet Corps 24
California Institute of Technology (Caltech)
  163, 196, 198, 211, 212, 213, 224, 226,
  436
Cambodia, US invasion of 388
Cambridge
  described 55, 101, 122, 140
  Manci's dislike of 296–7, 339–40, 347,
  367, 383, 411
  Socialist Society march (1933) 236
  wartime 303–4, 309, 310, 315, 323, 325,
  477*n*32
  VE-Day celebrations 327
  celebration of Japan's surrender 329
Cambridge Borough Cemetery (now
  Cambridge City Cemetery), Bristol
  476*n*59
Cambridge Review 156, 227, 228, 276
Cambridge Union 207, 227–8
Cambridge University
  mathematics as its largest department 59
  social life 59, 175
  opposition to the General Strike 98
  Marxist scientists' efforts to establish radical politics 98–9
  applications from refugee scientists 219
  in the Second World War 304
  women in 340, 479*n*28
  offers a professorship to Oppenheimer
  352
  Department of Applied Mathematics and
  Theoretical Physics 365
  PD moves to Florida State 3, 390
Canadian Rockies 163
Canford Cemetery, Westbury on Trym,
  Bristol 279
canonically conjugate variables 455*n*27
Cardoza Kindersley workshop, Cambridge
  490*n*27
Cario family 121–2, 125, 406
Carpenter, Edward, Dean of Westminster
  414, 415
Carroll, Lewis: *Alice through the Looking
  Glass* 107–8
Carter, Jimmy 3
Carus, Paul: *Reflections on Magic Squares* 284
Casimir, Hendrik 119
Caucasus 155, 176, 267, 276, 279
Cavendish Avenue, Cambridge (No.7) 287–8,

296, 315, 317, 322, 324, 330, 350, 351,
  352, 363
Cavendish Laboratory, Cambridge 62, 146,
  183, 218, 226, 233, 280, 380, 381
  Rutherford succeeds J. J. Thomson 62
  seminars 62–3, 152
  PD talks on quantum discoveries 97
  ode to the electron 171
  Millikan's presentation on cosmic rays
  196, 197
  Chadwick's work on the neutron 203
  splitting of the atom 206–7
  discovery of the anti-electron 213–15
  and the Nobel Prize (1933) 235
  Bragg succeeds Rutherford 294–5
  Second World War 319
Cavendish Physical Society: annual dinner
  181–2
Central Intelligence Agency (CIA) 349
centrifugal jet stream method 307, 311, 314
CERN (European Organization for Nuclear
  Research) 404, 430
Chadwick, James 270, 319, 341, 399, 405
  and cosmic rays 196
  'Possible Existence of the Neutron' 202–3,
  212
Chamberlain, Neville 298, 300, 302, 308
Chandrasekhar, Subrahmanyan 180, 183
Channel Islands 279
Chaplin, Charlie 277
Charlesworth, Martin 189, 197, 198, 463*n*46
Cher 3, 394, 412
Chicago 157, 161, 323, 400
Chopin, Fryderyk 3, 260
Christie, Agatha 204
Christ's College, Cambridge 313
Chukovsky, Korney: *Crocodile* 64, 449*n*29
Churchill, Sir Winston 38, 98, 227, 308–9,
  319, 320, 324, 326, 327, 328, 379, 403
Churchill College, Cambridge 379, 382, 383,
  417, 491*n*37
Civil Defence offices, St Regis 312
Clark, Sir Kenneth (later Lord) 313
classical mechanics 88, 102, 114, 152, 156,
  215, 321
classical physics 71, 84, 96, 145, 215, 296,
  343
Cleese, John and Chapman, Graham: *Monty
  Python's Flying Circus* script 107
Clifton Suspension Bridge 11
Clinton, Bill 417
Clinton, Hillary Rodham 417, 491*n*41
cloud chamber 197, 211, 224
Cockcroft, Lady Elizabeth 367, 383, 417,
  491*n*37

Cockcroft, Sir John 189, 206, 207, 222, 313, 319, 324, 328, 330, 341, 367, 379–83, 403, 404, 405, 485*n*33
Cold War 349, 382
Cole, Sir Henry 16
Coliseum ice-rink, Bristol 17, 20, 444*n*48
Colleraine, Tony 386, 390
Columbia Radiation Laboratory 335
Columbia University, New York 307, 335, 361
Communism 153, 388
Communist Academy 198
Communist Party 65, 98, 99, 146, 249, 322, 362, 457*n*17
complementarity principle 128, 455*n*30
Compton, Arthur 264
  electromagnetic radiation behaving as discrete particles 69
  PD declines his offer of a post in Chicago 157
Compton, Karl 264, 265
Comte, Auguste 43, 222, 465*n*13
Conan Doyle, Sir Arthur 39
Concorde 404–5
Congress of Russian Physicists (1928) 154–5, 168
conservation of energy, law of 169, 273, 274
Copenhagen 145
  PD in 106, 107–20, 125, 127, 148, 243–4, 265, 371
Coral Gables conferences 385
Cornwall 346, 421
correspondence principle 71, 87
cosmic rays 213, 348
  Millikan's investigations 196, 197, 225
  Blackett's interest in 196, 197, 213, 214
  Anderson's use of a cloud chamber 196–7, 211
  Blackett and Occhialini's work 218–19, 223
  Anderson identifies the muon 369
cosmology 261, 280–81, 288, 300, 378, 432
Coughlin, Joseph ('Roundy') 162
Council on Foreign Relations 349
counter-current centrifuge 321
Coward, Noël 277
Crimea, the 210, 211, 249, 250, 264
Crowther, Jim 145–6, 176, 204, 206, 216, 222, 228, 271, 313, 315, 327, 330, 453*n*1, 457*n*17
  Soviet Science 276
Cuban crisis (1962) 370–71
Cunningham, Ebenezer 69, 82, 118, 448*n*14
  Hassé's letter supporting PD 52–3

PD asks to study relativity with him 53
  on PD 76
Curie, Marie 137, 202, 428
Czechoslovakia 298, 300

Daily Express 329, 351
Daily Herald 223
Daily Mail 235
Daily Mirror 206, 277
Daily Telegraph 175
Daladier, Édouard 298
Dali, Salvador 347
Dalitz, Dick 21, 414, 415, 416
Dalyell, Tam 363, 483*n*11
Daniel, Glyn 287
Darwin, Charles 190, 294, 415
  bottom-up thinking 2
  compared with Dirac 3
  theory of evolution 68
Darwin, Charles (grandson of the naturalist) 143, 145, 150, 182, 205, 328
Davisson, Clinton 160, 161, 227
de Broglie, Louis: wave theory of matter 81, 99
de Sitter, Wilhelm 255
de Valera, Éamon 323, 476–7*n*20
Debye, Peter 307–8
V²V Club 66
Delbrück, Max 125, 205
Delhi 354
Delta function 113, 454*n*20
Dent, Beryl 49–50
Department of Scientific and Industrial Research 53
Depression 198, 200, 257
Descartes, René 188
Deutsches Volkstum ('German Heritage') 151
dialectical materialism 199
Dicke, Robert 378
Dickens, Charles 66
differential geometry 463*n*56
Dingle, Herbert 224, 290
Dirac, Betty (PD's sister) 38, 115, 116, 146, 201, 245, 404
  birth 11
  names 442*n*16
  childhood 5, 17, 193
  education 18, 33
  her father's favourite child 27, 49, 165
  personality 49, 134, 286
  and Felix's death 80
  attends PD's Ph.D. ceremony 101
  lack of employment 115, 134, 155, 366
  chauffeurs her father to and from work 183

forced to sell her car 191
and her parents' marriage crisis 192
degree studies 209, 217, 229, 272, 291
goes to Lourdes with her father 272
supports her parents 276–7
moves to London to become a secretary
    277, 279
in Budapest 284, 285
possible reason for her parents' failed
    marriage 285, 472n11
marries Joe Teszler 286
lives in Amsterdam 291
birth of son 291
in the Second World War 303, 325, 328
stays in Cambridge 330
her suffering in Budapest 330
birth of daughter 330
relationship with Manci 330, 394
in Alicante 394
stroke 401
Dirac, Charente, France 9
Dirac, Charles (PD's father) 66, 67, 82–3,
    420
birth (in Monthey, Switzerland) 4
childhood 7
education 7
in London 7
teaches at Merchant Venturers' Secondary
    School 7–8, 13, 19, 21–2, 33–4
settles in Bristol 5, 9–10
appearance 8, 12, 49, 134, 209
personality 6, 8, 18, 21, 426
meets Florence Holten 8, 426
and religion 8–9, 209, 221, 272, 278
marries Flo 5, 9
insistence on his children speaking French
    5, 10, 13, 27, 287
champions Esperanto in Bristol 11
relationship with PD 5, 6, 7, 12, 13, 54,
    427
careful with money 18, 76, 101, 116, 134,
    201–2
work ethic 18, 26, 102, 134
effects of his rigorous educational regime
    at home 26
tyranny of 5, 6, 27, 239, 256
his favourite child 27, 49, 165
forces Felix to study engineering instead of
    medicine 28, 372
deceptions by 33–4, 291
acquires British nationality 34, 46
efforts to send PD to Cambridge 45, 46,
    52
helps PD financially 54, 101, 278, 291
interest in PD's career 65, 115, 146, 271

family radio 67
deeply affected by the death of Felix 79,
    90, 98, 115, 372
death of his mother 98
attends PD's Ph.D. ceremony 101
letters to his 'only son' 115, 210
vegetarianism 134, 201
PD continues to feel intimidated by 169
and PD's FRS election 173, 174
retirement 183, 191
infidelity 184, 192, 193
marriage crisis 192–3, 201, 217, 271
loses his grip on his family 209
continues to teach from home 210,
    421
plans to visit Geneva 210
rediscovery of his childhood Catholicism
    221
visits Geneva with Betty 229
Flo attacks in the Swedish press 239
tries to understand PD's work 271, 272
goes to Lourdes 272
ill with pleurisy 276
serial tax evader 291
PD blames him for Felix's suicide 372
'loathed' by PD 372, 484n45
death and funeral 277–8
his estate 278–9
gravestone 279
Dirac, Felix (PD's brother) 94, 202
birth 10
names 442n16
appearance 10, 14, 22, 67, 80, 443n32
education 10, 14, 16, 18, 19, 22, 27, 28,
    34, 272
childhood in Bristol 5, 10–12, 17, 193,
    401
bullied by his father 6, 28
personality 14, 77, 80
rift with PD 27, 34, 38, 49, 77, 78
forced to study engineering instead of
    medicine 28
student apprenticeship in Rugby 34–5,
    37
based near Wolverhampton 49, 67
a draughtsman 67
Buddhism and astrology 67–8
acquires a girlfriend 68
settles in Birmingham 77
volunteers for the Ambulance Corps 77
leaves his job at a machine-testing labora-
    tory 77–8
personality 14, 77, 80
suicide 78–80, 82, 256, 318, 345, 372,
    401, 421, 451n7

the family's response to his death 79–80, 90, 98, 105, 115, 221–2
memorial service and inquest 80
gravestone 279
Dirac, Florence (née Holten; PD's mother) 33, 159, 413, 426–7
first meets Charles 8, 426
appearance 8, 134
personality 8, 156, 318
absent-minded 8, 238
and religion 8
correspondence with Charles 9, 11–12
marries Charles 9
birth of Felix 10
birth of Paul 10
poem about PD 14, 444n49
Paul as her favourite child 27, 116
and Charles's deception 34
correspondence with PD 38, 66–7, 76, 80, 90, 146, 183, 317, 454n29
fears competition for PD's affections 68, 239–40, 242, 281, 318
asks PD for money 76–7, 116, 209
and the death of Felix 79–80
poetry 80, 296, 442n16, 443n34, 444n49, 451n15, 473n49
interest in politics 90, 165
attends PD's Ph.D. ceremony 101
worried about PD's emaciated appearance 101–2
evening classes 115, 134, 155, 456n17
admits her unhappiness 115–16, 133, 134
housework, dislike of 116, 192
PD pays for a diamond ring 116, 133–4
PD's visits home 135, 169
visits PD in Cambridge 147
and PD's visits to Russia 153–4, 176
opposes the idea of a woman prime minister 155–6
fussing over PD 169
and PD's FRS election 173–4
dreads Charles's retirement 183
the charade of her marriage 184, 318
affinity with the sea 192, 271; see also Richard Holten (her father)
marriage crisis 192–3, 201, 217, 271
Mediterranean cruises 229, 272, 296
at PD's Nobel Prize ceremony 237–40, 242
at Bohr's party in Copenhagen 244, 245
and Charles's pleurisy 276
meets Manci 283
disputes with Manci 305, 318
in the Second World War 309, 311, 312, 317–18, 421
death and funeral 318

Dirac, Gabriel (PD's step-son) 256, 288, 310, 311, 322, 339, 349, 366, 373, 384, 407, 412, 483n26
Dirac, Gisela 426
Dirac, Judy (PD's step-daughter) 256, 288, 310, 311, 315, 322, 339, 355, 366, 373, 384–5, 414
Dirac, Louis (PD's paternal grandfather) 9, 441n10
Dirac, Margit (Manci; née Wigner; PD's wife) 4, 6, 265–6, 278
meets PD 253
personality 253, 256–7, 269, 281, 282, 297, 342, 346, 395, 417
and PD's talk of his unhappy childhood 5, 256
on her first marriage and divorce 256, 413
and religion 257, 413–4, 416
a keen follower of the arts 257, 315, 346
pursuit of PD 261–3, 270–71
PD visits her in Budapest 269–70, 275
Isabel Whitehead's assessment 281–2
PD's proposal of marriage 282–3
marriage and honeymoon 284
relationship with Betty 285–6, 328, 330, 366–7, 394, 417
'Wigner's sister' appellation 286, 472n17
settles in Cambridge 286–8
in the Soviet Union 292–3
pregnancies 302, 305, 320, 474n10
as an alien in wartime England 303
and air raids on Cambridge 310
Flo helps with housework 317–18
orders Judy out of the house 322
and the Nazi concentration camps 327
complains about the exodus from Cambridge 338
scorns Heisenberg 340
in Princeton 342
and politics 362, 417
marriage under strain 365–6, 394, 395
a better wife than mother 366
and disappearance of Judy 384, 385
worsening arthritis 390, 417
and PD's decision to move to Florida State University 390
at Florida State 392, 393, 394
Jewish and occasionally anti-Semitic 395, 487n16
as a hostess 3, 400
fraught relationship with Halpern 411, 413, 417
PD's death and funeral 413–14
lively and active for ten years after PD's

death 416–17
letter from Hillary Rodham Clinton 417
death 418
Dirac, Mary (PD's daughter; later Colleraine, then Tilley) 318, 339, 366, 381, 390, 416, 418, 421
  birth 305
  childhood 325, 339, 352
  personality 345–6
  education 346
  emigration to the USA 367, 372
Dirac, Monica (PD's daughter) 418, 421
  birth 323
  childhood 339, 346, 347, 352–3
  personality 346
  at PD's commemoration 416
Dirac, Paul Adrien Maurice
  LIFE STORY
  birth (8 August 1902) 10
  appearance and dress sense 1, 3, 4, 10, 14, 22, 26, 47, 56, 60, 125, 164, 174, 179, 183, 194, 208–9, 240, 253, 258, 365, 377, 378, 393, 395, 408, 410, 443n32
  digestive problems 1, 5, 57, 410, 427
  foresees the existence of the positron 2, 187, 195, 224–6, 229, 243
  childhood in Bristol 4, 5, 6, 10–13, 17–18, 193, 426–7
  relationship with his father 5, 6, 7, 12, 13, 54, 67, 68, 79, 80, 82–3, 101, 115, 135, 169, 177, 210, 237, 239, 256, 276–8, 280, 281, 287, 291, 426–7
  nicknamed 'Tiny' 10, 12
  school education 10, 13–19, 22–7
  visits Switzerland 11, 404
  Bristol accent 14–15, 56, 177, 375, 443n36
  and technical drawing 15, 16, 23, 48, 130, 443n44
  handwriting 16, 37, 45–6
  his mother's favourite 27, 116
  rift with Felix 27, 34, 38, 49, 77, 78
  engineering degree 28, 29–32, 34, 37, 38, 39, 44, 45, 46, 49
  public impact of relativity theory 35–6
  trainee engineer in Rugby 37–8, 45
  applied maths degree studies 47–51
  and projective geometry 48–9, 73, 85, 94, 130, 181, 436
  wins scholarships to St John's College, Cambridge 53
  supervision by Fowler 53, 59, 60, 75, 76, 83, 85, 97
  Charles helps him financially 54
  arrives at Cambridge 55

  manner at the dinner table 58, 345
  attends Eddington's lectures 61
  Blackett and Kapitza become his closest friends 63
  and Soviet ideology 65, 99, 138
  and his mother's possessiveness 68
  first academic papers 75, 76
  Felix's death 78–80, 105, 221–2, 345, 372, 401
  first great epiphany 86
  first paper on quantum mechanics 87–9, 91–2
  Ph.D. thesis 99–101
  combines logic and intuition 114, 141–2
  as 'the strangest man' (Bohr) 120
  successful period in Copenhagen 106, 107–20, 125
  in Göttingen 120–32, 148, 151–3
  friendship with Oppenheimer 121, 133, 135, 267, 342, 350, 352, 355–6
  his visits home 135, 169–70, 191–2, 275
  elected Fellow of St John's College 140
  his rooms in college 140–41
  makes his most famous contribution to science 141–5
  relationship with Isabel Whitehead 147–8
  first visit to Russia 153–5
  reductionism 158
  first visit to US 159, 160–63
  elected Fellow of the Royal Society 173–4, 460n10
  buys his first car 174
  represented in a special version of *Faust* 204, 205
  Lucasian Chair 207–8, 284, 374, 387
  Wittgenstein, opinion of 220
  and moral philosophy 220–22
  works with Kapitza in his laboratory 226–7, 229
  last meeting with Ehrenfest 232
  Nobel Prize for physics 234–5, 237–46, 468n20
  first public comment on social and economic affairs 241–2
  smitten with Rho Gamow 250–51, 253
  first meets Manci 253
  campaign for Kapitza's release 259, 264–5, 266, 424
  sends the Gamows a baby alligator 259, 469n43
  guardian of Kapitza's sons 270, 274
  graduate supervisor 272–3
  proposes to Manci 282–3
  marriage and honeymoon 284
  first love letter 285

wants his own children 288, 472*n*23
refuses Princeton's job offer 296, 473*n*50
Scott lecture 300–1
offered war work 312–13
Baker Medal 316
and the death of his mother 318
refused a visa for the Soviet Union 327,
  477*n*46
declines honours 347–8, 403
refused a US visa 351, 352
visits India 352–4
jaundice 354–5, 482*n*37
marriage under strain 365–6, 394–5
marginalised in Cambridge 365, 374, 375
emigration to US 375, 383, 388
Scientific American article (1963) 376–7,
  485*n*10
Horizon interview (1965) 377–8, 485*n*11
quarks, likes concept of 382
decision to move to Florida State
  University 390
routine at Florida State 393–4
busts and paintings of PD 402–3
accepts the Order of Merit 403–4
visits CERN 404
flies on Concorde 404–5
sees his life as a failure 409
surgery on tubercular kidney 410–11
death (20 October 1984) 413, 490*n*17
funeral 414, 490*n*19
commemoration in Westminster Abbey
  414–16
centenary of his birth 420, 428
possible autism 421–6
names 442*n*16
memorial stone 490*n*27
PERSONALITY
– aloofness 112, 339, 422
– confident 4, 32, 75, 99, 208, 218
– defensiveness 153, 256
– determination 4, 75, 99, 208, 218, 425
– diffidence 58, 269
– equability 5
– frugality 208–9
– inhibition 37
– lack of social sensitivity 27, 34, 89,
  107–8, 109, 111–2, 120, 145, 422
– literal-mindedness 6, 120, 346, 422, 477*n*44
– modesty 27, 157, 235, 366, 402
– narrow-mindedness 37
– objectivity 253
– obsession with taking long walks 22, 60,
  109, 111, 112, 123, 128, 132, 154, 161,
  163, 267, 327, 355, 390, 411, 455*n*28,
  477*n*46

– otherworldiness 37, 75, 120, 343
– passivity 422
– physical ineptitude 26, 422
– private enthusiasms 3, 346–7
– reticence 3, 5, 60, 126, 189, 253, 305, 422
– rigid pattern of activities 80, 109, 135,
  260, 270, 287, 312, 422
– self-centredness 125, 422, 425
– shyness 208, 235, 241
– stubborness 151, 169, 412
– taciturnity 1, 89, 345, 421, 427
– top-down thinker 2, 94, 322
– verbal economy 58, 107, 120, 130, 162,
  211, 256, 267, 287, 326–7, 342–3, 345,
  391, 397, 408–9, 427
– work ethic 18, 26, 102, 132, 174–5
INTERESTS, APTITUDES AND OPINIONS
– beauty, mathematical, fascination with
  42, 45, 73, 153, 255, 300–1, 357, 376,
  377–8, 380, 396, 402, 428, 435, 436,
  438, 493
– board games and mathematical
  puzzles, enjoyment of 97, 98, 115,
  172–3, 208, 353–4, 366, 411, 460*n*6
– driver, skills as a 172, 174, 190, 251,
  283, 346
– fondness for Mickey Mouse films 3, 172,
  274, 460*n*5
– food, tastes and appetite 1, 5, 10, 13, 57,
  106, 140, 323, 345, 410, 413, 423, 427,
  449*n*6,
– gardening 27, 298, 305, 322, 374–5,
  421, 445*n*33
– Hamiltonian approach to mechanics,
  strong belief in 88, 114, 332, 343, 367,
  368, 405, 432
– jokes, appreciation of 22, 138, 259, 292,
  381, 393, 487*n*10
– lecturer, skills as a 97, 150, 169, 179,
  208, 241, 243, 277, 300, 320–1, 330,
  331, 353, 392, 405, 476*n*11
– mountain-climbing 153, 190, 279–80,
  310
– philosophy, opinion of 16, 43, 44, 74,
  111, 137, 220, 300
– relativity, fascination with 35–6, 38, 41, 45,
  50, 53, 74, 85, 300, 301, 338, 355, 396
– religion, opinions about 137, 138, 220–2,
  257, 325, 401–2, 415, 416, 488*n*45,
– renormalisation, distaste for and dislike
  of 336–7, 356, 398, 399, 405, 409, 412,
  425, 437
– swimming 174, 390, 397
– team games and teams, aversion to participation in 26, 93, 312–3

- technology of space flight, interest in 364, 386–7, 391
- top-down thinking 2, 94, 322
- tree-climbing 174, 190, 460n17

CONTRIBUTIONS TO PHYSICS AND MATHEMATICS

- action principle in quantum mechanics 215–6, 229, 333, 430
- antimatter, foresees, see also positron and antiproton 2, 226, 243, 400, 433–4
- anti-electron predicts, see positron
- anti-proton, predicts 187
- blackbody radiation spectrum derived 105
- bra and ket notation 326
- classical theories of the electron 296, 343
- cosmology, thoughts on 280–1, 300–1, 378, 432–3
- density matrix 156
- delta function 113, 454n20
- Dirac equation 2, 132, 139, 141–145, 149, 150, 156, 163, 166, 181, 255, 334, 398, 420, 430, 436, 458–9n30,
- Dirac sea 167, 213
- dispersion theory 126
- ether, post-Einstein view of 344
- Fermi-Dirac statistics 102–3, 105, 331
- general relativity, Hamiltonian formulation of 343, 367–8, 432
- gravity, weakening of – postulates, see also large numbers hypothesis 290
- high-spin theory 396
- hole theory 156, 166–9, 171–2, 175, 177–8, 180, 181, 187,195, 205, 206, 211, 215, 218, 223, 224, 225, 230, 247, 362, 463n7
- indefinite metric 316
- jet-stream method of isotope separation 248–9, 307, 311, 313–4, 321, 431
- Kapitza-Dirac effect 227, 381, 431, 432, 488n27
- large numbers hypothesis 289–291, 378, 396, 433, 493n13
- magnetic monopole 185–9, 194, 204, 205, 213, 343, 430–31, 462n5, 480n48
- many-times formulation of quantum electrodynamics by PD, Fock and Podolsky 210, 464n25
- neutron diffusion in matter, theory of 321, 324
- non-commutation in quantum mechanics 84–5, 86, 117, 128, 188, 448n11
- parity violation, foresees possibility of 361–2
- philosophy of physical science 300–1, 359, 432

- Poisson bracket in quantum mechanics 86–87
- positron, prediction of 2, 187, 195, 224–6, 229, 230, 233, 235, 243, 246, 248, 400, 488n34
- principle of mathematical beauty 301, 359
- quantum electrodynamics 117–9, 123, 126, 132, 138, 200, 201, 205, 210, 229, 233–4, 247, 248, 296, 336, 337–8, 398, 405, 409, 410, 464n25
- quantum field theory, co-discovery see quantum electrodynamics 126
- quantum mechanics of heavy atoms 158
- quantum mechanics, later contributions 370, 376, 410
- Schrödinger equation (time dependent), independent discovery by PD 102
- sphere, quantum-relativistic treatment of 370, 410
- spinors 430
- string concept in quantum electrodynamics 356, 357
- transformation theory 112–4
- vacuum polarisation 233–4
- virtual states 126

THE ARTS, TASTE AND APPRECIATION OF

- art (visual) 4, 120, 347, 374,
- cinema 261, 386, 391
- comics and comic characters 3, 22, 78, 172, 274, 346, 391, 395, 401, 444n6, 460n5
- music 23, 238, 257, 261, 346, 347, 479n15
- novels 25, 26, 172, 257, 274, 339, 356, 393
- poetry 18, 26, 121
- radio and television, appreciation of 3, 67, 197, 297, 309, 310, 328, 347, 362, 364, 371, 386, 388, 394, 411
- theatre and opera 65, 151, 154, 238, 287, 346

BOOKS

- *General Theory of Relativity* 392–3
- *Principles of Quantum Mechanics* 145, 156, 158, 178–9, 254, 391, 417, 420, 428, 482n49

Dirac, unofficial unit of frequency of speech 89
Dirac, Walla (PD's paternal grandmother) 9, 11, 98, 134, 442n16
'Dirac stories' 120, 161–2, 163, 422
DNA, double-helix structure of 295
Dneproges hydroelectric power station 267
Dobb, Maurice 65
Dostoevsky, Fyodor: *Crime and Punishment* 339
Douglas' Works, Kingswood 446n16
Dublin 328

Dublin conference (1942) 323
Durango, Colorado 252
Duranty, Walter 198
Dutton, S. T.: *Social Phases of Education in the School and the Home* 47
Dyson, Freeman 320, 336, 337–8, 342, 428, 461*n*31, 476*n*11, 478*n*6, 479*n*13, 480*n*45

Eddington, Sir Arthur 53, 75, 88, 94, 157, 178, 204, 205, 261, 271, 282, 403
  mathematician and astronomer 36, 60
  understanding of relativity theory 36
  solar-eclipse experiments 36, 41, 61, 223
  on Einstein's $E = mc^2$ equation 36–7
  introduces PD to relativity 60
  appearance 60
  personality 60, 61
  mathematical approach to science 61
  and Rutherford 61
  congratulates PD on his Ph.D. thesis 101
  and the splitting of the atom 207
  media savvy 224
  pilloried by his younger colleagues 295
  and nuclear energy 299, 474*n*60
  disagreement with PD 316
  Dublin conference (1942) 323
  death 338
  The Mathematical Theory of Relativity 454*n*20
  The Nature of the Physical World 128–9
  Space, Time and Gravitation 42
Edward VII, King 10
Ehrenfest, Paul 119, 125, 132–3, 137, 149, 150, 230–33, 347, 403
Ehrenhaft, Felix 480*n*48
Einstein, Albert 76, 127, 133, 175, 194, 197, 232, 245, 288, 300, 331, 338, 339, 347, 348, 402, 428, 453*n*20
  personality 355
  most successful spurt of creativity 11
  appearance 35, 254
  studies Mill's *System of Logic* 43
  $E = mc^2$ equation 36–7, 40, 41, 207, 219
  and Planck's blackbody radiation spectrum formula 52
  and solar eclipse results 61
  light quanta idea 69
  and Bohr 71
  and Heisenberg's theory of 1925 85
  suspicious of the new quantum mechanics 94, 136
  top-down approach to physics 94
  on PD 114, 218
  stimulated emission process and the laser 118
  attacks Heisenberg's uncertainty principle 137
  differs from PD in his approach to science 137
  praises PD's textbook 179
  at the 1930 Solvay Conference 180
  despises Hitler 180–81
  Nazis' view of his 'Jewish physics' 200
  and the photon 203
  and the splitting of the atom 207
  flees from Germany to the USA 217, 219
  at Princeton 253–4, 342, 344, 375
  and Kapitza's detention 259, 265
  dislike of quantum electrodynamics 273–4
  treats Heisenberg with contempt 340
  Hoover's campaign against 344
  suggests the existence of a positive electron 400
  in search of generalisations 435
  death 355
  centenary of his birth 404, 405
  'Electron and General Relativity' 488*n*34
  see also relativity
Eisenhower, Dwight D. 363
electrical charge 185–6, 233–4
electromagnetic interaction 398
electromagnetism 255
  laws of 71, 361
  Maxwell's theory 51, 68–9, 116, 117, 185, 296
  PD's magnetic monopole theory 185–9
electron-positron pairs 224, 225
electrons
  bare 356, 357
  behaving as discrete particles 69
  Cavendish annual dinner, toast to 182
  describing behaviour of a single, isolated electron 136
  diffraction by light 381, 431
  Dirac equation 2, 142–5, 369
  discovered by J. J. Thomson 51, 243
  extended 370
  moving in a straight line 83
  negative-energy 150, 165–8, 171, 175, 177, 213, 233
  orbiting the nucleus 70, 71, 72, 81, 83, 84, 399
  particle-like 431, 432
  Pauli's exclusion principle 103–4
  positive-energy 167, 171, 234
  scattering 95, 96, 110
  self-energy of 295–6, 336
  spin of 90, 132, 141, 142, 143, 195, 248, 376, 382, 429

wave nature, 81, 161, 227, 431, 432
see also Fermi-Dirac statistics
Eliot, T. S. 114, 157, 228
Elizabeth, Queen (later the Queen Mother)
    340, 479n28
Elizabeth II, Queen 347, 348, 404
Elsasser, Walter 125
empiricism 43
energy quanta 52, 81
'Erice Statement' (1982) 407
Erice summer school, Sicily (1982) 406–7
Esperanto 11, 192, 210, 278
ether
    belief in 40, 199
    PD's ether 344
Euclid 36
    Euclidean geometry 17, 42, 152
evolution, theory of 68
exclusion principle 103–4, 166–7, 247

Falklands War (1982) 406
Faraday, Michael 176, 294
Farm Hall, Godmanchester, Cambridgeshire
    329
Farmelo, Amelia (née Jones) 420
Faust, performance of special version (1932)
    204–5
Federal Bureau of Investigation (FBI) 344,
    350, 352, 388, 390, 486n54
Fen, Elisaveta *see* Jackson, Lydia
Fermi, Enrico 105, 331, 332
    radioactive decay of nuclei 233
    quantum field theory of beta decay 255
    and nuclear fission 300
    builds first nuclear reactor 323
    weak interaction 348
    Nobel Prize 405
Fermi-Dirac statistics 102–3, 105, 331
Fermi National Accelerator Laboratory
    (Fermilab) 400
Fermilab Symposium (1980) 400, 483n8
fermions 331
Feynman, Richard 335, 368–9, 412
    at 'The Future of Nuclear Science' confer-
        ence 332, 333–4
    personality 332
    new version of quantum mechanics 333
    analyst and intuitionist 334
    Wigner on 334
    portrait of PD 403
    says he is 'no Dirac' 403
Filton, Bristol 19, 309, 405
First World War 20–21, 23–4, 27, 29–30, 36,
    53
Fisher family 191–2, 235

Fitzgerald, F. Scott 121, 160, 166
Flexner, Abraham 197, 217, 218, 259, 265,
    342
Florida State University 433
    Physics Department 389–90, 486–7n62
    PD moves from Cambridge to 3, 390
    treatment of PD 392
    Keen Building 393
    and PD's funeral 414
    Dirac Science Library 417
fluid mechanics 365
Fock, Vladimir 210, 359, 464n25
Folies Bergère 347, 480n5
Ford, Henry 277
Foreign Office 258, 260, 320
Fourier, Joseph 113
Fowler, Ralph 82, 83, 88, 143, 145, 158,
    183, 215, 272
    PD's supervisor at Cambridge 53, 59, 60,
        75, 76, 85, 97
    and Rutherford 62
    lectures on Bohr's theory 71
    works with Bohr 80
    and PD's paper 'The Fundamental
        Equations of Quantum Mechanics' 89
    elected a Fellow of the Royal Society 89
    PD's visits to Copenhagen and Göttingen
        106
    co-edits the 'International Series of
        Monographs on Physics' 145
    failing health 295
    death 338
Franck, James 133
Franco, General 291, 292
Frank, Anne 325
Frank, Sir Charles 446n9
Fraser, Peter 48, 49
Frayn, Michael: *Copenhagen* 317
French Circle 278
French Riviera 302
Frenkel, Yakov 168, 403
Friedmann, Alexander 261
Frisch, Otto 306, 319, 324
Frisch, Otto and Peierls, Rudolf:
    'Memorandum on the Properties of a
        Radioactive "Super-Bomb"' 306–7
Frith, Uta 491–2n7
Frost, Robert 354
Fuchs, Klaus 324, 350
fundamental interactions, unified theory of
    385
fundamental particles 102–3, 396, 405–6
'Future of Nuclear Science' conference
    (Graduate College, Princeton, 1946)
    332–4

Gabor, Dennis 480*n*50
Galileo Galilei 56
Galsworthy, John: *Forsyte Saga* 285
Gamow, Barbara (née Perkins) 379, 463*n*2
Gamow, George 151–2, 172, 177, 178, 198,
    199, 250, 251, 258, 259, 264, 288, 379,
    458*n*13
Gamow, Lyubov Vokhminzeva ('Rho')
    250–51, 253, 259, 264, 379
Gandhi, Mahatma M.K. 106, 148
Gardiner, Margaret 260
Gaspra, Crimea 210
gauge invariance 356
gauge theory 398, 488*n*25
Gautier, Théophile 16, 443*n*42
Gebhard's Hotel, Göttingen 406
Geiger counters 214
Gell-Mann, Murray 381–2, 399, 485*n*29,
    485*n*30
General Strike (1926) 98, 99, 207
genetics 280, 426
Geneva, Switzerland 11, 210, 229
geometry
    differential 463*n*56
    Euclidean 17, 42, 152
    non-Euclidean 188
    PD immersed in at Cambridge 72
    projective 48–9, 73, 85, 94, 130, 181,
        436
    Riemannian 26, 42, 278
George V, King 34, 35, 98
Gercke, Achim 131
Germany
    Youth Movement 123
    the Depression in 200
    new militarism in 201
    Einstein flees to the US 217, 219
    Hitler becomes Chancellor 219
    book-burning ceremonies 219
    annexes Austria 297
    invades Czechoslovakia 298, 300
    invades Poland 302
    Britain declares war on 302
    overwhelms Norway and Denmark 308
    blitzkrieg on Belgium, Luxembourg and
        the Netherlands 308
    U-boat fleet 320
Germer, Lester 160, 161
Gill, Eric 226
Glacier National Park 193
gluons 399
Go (a.k.a. Wei Chi) 256, 411
Goddard, Peter 407
Goethe, Johann Wolfgang von: *Faust* 204–5
Gog Magog Hills 174

Goldfinger (film) 381
Goldhaber, Maurice 467*n*67
Goldschmidt, Victor 326
Gonville and Caius College, Cambridge
    381
Gottfried, Kurt 226
Göttingen 82, 92, 96, 105, 128, 145, 194,
    340
    PD visits 120–33, 148, 151–3, 265
    Mathematics Institute 125
    anti-Semitism in 131, 180, 200, 219
    Nazism in 151, 180, 217, 219, 406
    seething with political tensions 200
    Heisenberg returns to 340
    the Diracs and Kapitzas in 406
gramophone 169
Grand Canyon 163, 165
Grandin, Temple 425
Grant, Cary 14, 15, 21, 257, 419, 420
Granta 301–2
Grassmann algebra 73, 85
Graves, Robert 30
gravitational waves 432
graviton 368, 484*n*33
gravity 255, 378, 433
    and the general theory 41, 68, 73, 116
    laws of 361
    and Newton's falling apple 42
    and Riemann's geometric ideas 26
    and string theory 410
Great Britain, SS 420
Great Exhibition (1851) 16, 82
Greater Soviet Encyclopedia 199
Green, Michael 436
group theory 463*n*56
Groves, General Leslie 322

Haddon, Mark: *The Curious Incident of the
    Dog in the Night Time* 419
Hahn, Otto 298–9, 329
Halpern, Leopold 404, 448*n*18, 477*n*46,
    484*n*45
    born and raised in Austria 396
    personality 396–7
    friendship with PD 397–8
    opposes surgery on PD's tubercular kidney
        410–11
    fraught relationship with Manci 411, 413,
        417
    homoeopathic treatment of PD 411, 413
    and PD's funeral 414
    satellite-based experimental programme 433
    death 433
Hamilton, William 50–51, 73, 75, 86, 432
Hamiltonian 50, 332–3, 367, 368, 405

Harding, Gardner L. 160
Hardy, G. H. 74, 175, 176, 312, 366
Harish-Chandra (Harish Chandra Mehrotra) 491n35
Harish-Chandra, Lily 273, 395, 416, 469n34, 488n35, 491n35
Harvard University 90, 123, 144
Harz Mountains 153
Hassé, Ronald 45, 47, 51, 52–3, 59, 72, 174
Hawking, Stephen 375, 415–16, 431, 492–3n8
Hayward, F. H. 16
Heath, Edward 387, 489n51
Heaviside, Oliver 44, 113
Hebblethwaite, Cyril 21–2
Heckmann, Otto 456n5
Hegel, Georg Wilhelm Friedrich 450n53
Heisenberg, Werner 87, 104, 205, 210, 273, 300, 338, 371, 428, 485n8
    personality 82, 109, 242, 377
    addresses the Kapitza Club 82, 452n28
    quantum theory (1925) 83–5, 88, 90, 94–6, 99, 101, 112, 136, 198, 316, 333, 421
    non-commuting quantities 84–5, 86
    and PD's first paper on quantum mechanics 91–2
    works with Born and Jordan at Göttingen 92, 96
    and Schrödinger's work on wave mechanics 100
    uncertainty principle 113, 127–9, 137, 160, 199, 220
    pianistic skills 122, 244, 340
    and PD's attack on religion 138
    appointed full professor in Leipzig 144
    and the Dirac equation 144, 150
    visits Japan with PD 163–5
    Soviet government's attitude to his work 198–9
    pleased at Hitler's coming to power 219, 220
    and the positron 224
    atomic nucleus structure 233
    Nobel Prize for physics 234, 237, 238, 240, 242, 245, 246
    celebrations in Copenhagen 243–4, 245
    message to Born from the Nazi Government 250
    a 'White Jew' 308
    meeting with Bohr (1941) 316–17
    attests to Betty's non-Jewish status 325, 477n34
    interned near Cambridge 329
    explanation of his wartime conduct 340
    PD supports 340–41, 424
    quarrels with Pauli 360
    at Lindau 377
    interviewed with PD 377
    appearance 377
    death 399–400
Heisenberg-Pauli theory 200–201
Hellman, Bruce 412
Henri Poincaré Institute, Paris 169
Hess, Rudolf 235
Hessen, Boris 190, 199, 292
high-dimensional field theories 408, 409
high-energy particle accelerators 233, 348
high-energy physics 334, 363
Highgate Cemetery, London 191
Hilbert, David 122
Hippodrome theatre, Bristol 21, 146, 420
Hiroshima, bombing of (1945) 328, 329
Histon, Cambridgeshire 340
Hitler, Adolf 131, 180–81, 216–17, 219–20, 227, 228, 231, 235, 254, 270, 291, 297, 298, 300, 302, 305, 311, 317, 323, 325, 340, 341, 396
Hofer, Kurt 1, 2, 3–6, 12, 291, 372, 392, 401, 402, 427, 448n18
Hoffman, Dustin 422
Holborn Registry Office, central London 284
Holcomb, Dorothy 391
hole theory 156, 166–9, 171–2, 175, 177–8, 180, 181, 187, 195, 205, 206, 211, 215, 218, 223, 224, 225, 230, 247, 362, 464n7
Holiday Inn, Tallahassee 390
Holmes, Sherlock 187, 346, 462n7, see also 39
Holten, Beatrice (Flo's sister) 442n16
Holten, Fred (Flo's brother) 192
Holten, Nell (Flo's sister) 78, 79, 192
Holten, Richard (PD's maternal grandfather) 8, 10
Hong Kong 354
Hoover, Herbert 198
Hoover, J. Edgar 344, 350
Horizon (BBC programme) 'Lindau' 377–8, 452n28, 452n49, 485n11
Horthy, Admiral 152
Hotel Britannique, Brussels 137
Hotel Metropole, Moscow 258
Houmère, Pam 392
Housman, A.E. 65, 80
Hoyle, Fred 295, 365, 374, 379, 393
Hubble, Edwin 281
Hubble's law 281
Hungary 152, 328, 350, 362–3, 417
Huntingdon Road, Cambridge 135, 286
Huxley, Aldous: Point Counterpoint 274
Huxley, Thomas 143

hydrogen atom
  Bohr theory and 70, 71
  Dirac equation and 143, 150
  lamb-shift of 376
  quantum mechanics and 91
hydrogen bomb 359

Immigration Act (1952) 350
Imperial Hotel, Bloomsbury, London 283
indefinite metric 316
India
  PD visits (1954) 352–4, 353, 354
  becomes a nuclear power (1974) 481n36
Infeld, Leopold 211
Institut de Physiology Solvay au Parc
  Léopold, Brussels 456n22
Institute for Advanced Studies, Dublin 297
Institute for Advanced Study, Princeton 197,
  217, 252, 298, 307, 308, 342, 343, 344,
  350–1, 355, 367, 370, 375, 383, 396,
  484n6
Institute for Physical Problems (Soviet Union)
  263, 269, 270, 280, 293
Institute for Theoretical Physics (Niels Bohr
  Institute), University of Copenhagen
  106–9, 113, 132, 145, 204, 230, 280,
  288, 426
International Congress on the History of
  Science and Technology, second (Science
  Museum, London, 1932) 190
International Esperanto Congress (Trinity
  College, Cambridge, 1907) 11
'International Series of Monographs on
  Physics' 145–6
Inyom, Revd. Sapasvee Anagami 67–8
Isenstein, Harald 402, 488–9n46
isotope separation 248–9, 311, 313–15, 321,
  431
Israel 375
Ivanenko, Dmitry 'Dimus' 199, 200, 269

Jackson, Lydia (previously Elisaveta Fen) 251–2
Japan
  PD and Heisenberg visit 163–5
  new militarism in 201
  bombing of Hiroshima 328, 329
  bombing of Nagasaki 329
  surrender of 329
'Jazz Band' (informal group of Soviet theo-
  rists) 199, 305
Jeans, Sir James 82, 205, 377
  The Mysterious Universe 182
Jeffreys, Harold 158
'Jewish physics' 200, 308
John Paul II, Pope 401

Joliot-Curie, Frédéric 202, 332
Joliot-Curie, Irène 202, 300, 332
Jones, Norman 49, 445n34, 448n8
Jordan, Pascual 151, 300, 341, 428
  works with Born and Heisenberg at
    Göttingen 92, 96
  and groups of electrons 105
  personality 124
  appearance 124
  and field theory 126, 139
  and the Dirac equation 144
  Nazi past 219, 405
Joyce, James 228
  Finnegans Wake 382
  A Portrait of the Artist as a Young Man 140
Julius Road, Bristol (No.6) 14, 18, 31, 33,
  66, 67, 83, 134–5, 191, 192, 193, 201,
  208, 210, 229, 237, 275, 277, 311, 312,
  364, 421, 426–7, 445n33, 462n24

Kant, Immanuel
  and beauty 74, 450n54
  and truth 43
Kapitza, Anna ('Rat') 135, 208, 249,
  257–60, 263, 264, 265, 267, 270, 286,
  297, 381
Kapitza, Peter 133, 154, 174, 176, 189, 327,
  328, 371
  settles in the UK 63–4
  personality 63, 64, 65, 358
  influences PD 63
  resented by Blackett 63
  obsession with the crocodile 64, 226, 259,
    449n29
  Russia's industrialisation and electrification
    64
  relationship with Rutherford 64, 97, 182,
    226, 249
  compared with PD 65
  supports Communist goals 65, 99
  under surveillance 65, 191, 249, 250, 258,
    292
  sets up the Kapitza Club 66
  attitude to experimental physics 74
  marries Anna Krylova 135
  co-edits the 'International Series of
    Monographs on Physics' 145
  at the Cavendish Physical Society annual
    dinner 181–2
  the Bukharin visit to Cambridge 191
  MI5 monitors him 191, 249–50
  vacation with PD in the Crimea 210
  and the anti-electron 215
  PD works with him in his laboratory
    226–7, 229, 248, 249

detained by the Soviet Government
257–60, 263–6, 292
and Rutherford's death 294
seeks Landau's release 358–9, 473*n*40
wartime telegram to PD 317
nominated by PD for a Nobel Prize 341,
479*n*37
invents method of liquefying oxygen 341
'Hero of Socialist Labour' 341
and Beria 341
in disgrace 342
letters to Stalin 358
and PD's passion for beauty 380
visits Cambridge in 1966 380–81
Nobel Prize in Physics 405
death 412
PD spends his last hours talking about him
413
'The Training of the Young Scientist in the
USSR' 380
Kapitza-Dirac effect 227, 381, 431, 432,
488*n*27
Kapitza Club 66, 81, 82, 94, 185, 193, 203,
273, 381, 432, 452*n*28, 469*n*52
Keats, John 156
Kennedy, John F. 370
Kent State University 388
ket 326
Keynes, John Maynard 64, 74, 82, 242
Kharkhov 210
Khrushchev, Nikita 357, 362, 370
Kierkegaard, Søren 111–12
Kitchener, Lord 20
Klampenborg Forest, Denmark 108
Klein, Oskar 139, 141, 458–9*n*30
Koh-i-Noor restaurant, St John's Street,
Cambridge 347, 480*n*7
Kronborg castle, Denmark 110
Kubrick, Stanley 386–7
Kuhn, Thomas 371, 372
Kun, Béla 152, 153
Kurşunoğlu, Behram 385, 386, 397, 488*n*23
Kyoto 164

Labour government 67
Labour Party 165, 228, 363
Lagerlöf, Selma 467*n*10
Lagrange, Jopseph Louis 215
Lagrangian 215
Lake District 297, 310, 415
Lake Elmore 384
Lamb, Charles: 'The Old Familiar Faces'
360
Lamb, Willis 334–5, 376
Landau, Lev 155, 177, 178, 199, 293, 358,

359, 362, 405, 473*n*40
Landshoff, Peter 483*n*18
Langer, Rudolph 213
Lannutti, Joe 389–90
Large Hadron Collider 430
large numbers hypothesis 289–91, 378, 396,
433, 493*n*13
Larmor, Sir Joseph 157, 207, 452*n*44
lasers 118, 381, 431, 432
Lawrence, Ernest 233
Lawrence, T.E.: *Seven Pillars of Wisdom* 274
Lederman, Ellen 416, 417, 491*n*35
Lederman, Leon 361, 370, 400, 416, 417,
491*n*35
Lee, T. D. 361, 362, 467*n*63
left-right symmetry 361, 482*n*4
Leiden, Netherlands, PD visits 132–3, 148,
149–50, 458*n*5
Leipzig
Heisenberg appointed full professor 144
PD in 150–51
Lemaître, Abbé Georges 260–61, 401,
469*n*52
Lenin, Vladimir 64, 65, 175, 190, 265, 341,
359
Leningrad, PD in 154, 210, 230
leptons 370, 398
Liberal Party 165, 420
Life magazine 363
light
viewed as photons 69, 82, 451*n*20
in a continuous wave 51, 52, 68–9, 82
emitted and absorbed by atoms 82
energy of light tranferrable to atoms only
in quanta (Planck) 52
as particles 82
see also radiation, electromagnetic
Lindau, Germany
1965 meeting 377
1971 meeting 401–2
1982 meeting 405, 489*n*61
Lindemann, Frederick 219, 313, 319, 320
Lindstrom, Andy 410
Lippmann, Gabriel 493*n*21
Liverpool 311
Lloyd George, David 29–30, 165
Locarno, Treaty of (1925) 91, 207
logical positivists 220
London
Charles Dirac in 7
in Second World War 309, 311, 316
Dirac family stays in 346
London Mathematical Society 465*n*16
Los Alamos headquarters, New Mexico 324,
360

Lost Lake, near Tallahassee 397
Lourdes 272
Lucasian Professorship of Mathematics 157, 207–8, 374, 387, 431
Luftwaffe 309, 325
Lyons, Eugene 268

McCarthy, Joseph 350
MacDonald, Ramsay 67, 154, 165, 226, 294
magnetic monopole 185–9, 194, 204, 205, 213, 343, 430–31, 462n5, 480n48
Manchester 81
Manchester Guardian 146, 166, 175, 176, 216, 223
Manchester University 157, 243
Manhattan, New York 160, 227, 257
Manhattan Project 321, 324, 325, 332, 342, 350, 351, 352
Martineau, Harriet 456n29
Marx, Karl 191
Marxism 98–9, 189, 198–9, 222, 228
mathematics
    aesthetic view of 74
    applied 48, 49, 50, 448n7
    beauty of 45, 73, 382, 435
    Bohr's attitude to 111
    God as a mathematician 2, 376–7
    mathematical rigour 49
    PD's sometimes cavalier attitude to 95
    pragmatic approach to the mathematics of engineering 44
    pure 48, 200, 437, 448n6
    game in which people invent the rules, PD's view as 300
matrices
    and electron spin 142
    Heisenberg's quantum theory 84, 96, 100
MAUD committee 319, 476n3
Maugham, W. Somerset
    A Writer's Notebook 237
    Of Human Bondage 315
    Then and Now 356
Mauthausen-Gusen concentration camp, Austria 328
Maxwell, James Clerk
    electromagnetic theory 51, 52, 68–9, 71, 74, 116, 117–18, 185, 296, 368
    the universe as a giant mechanism 129
Maysky, Ivan 250
mechanics, laws of (Newton) 49, 50
Meitner, Lise 306
Merchant Venturers' Secondary School, Bristol (later Cotham Road School) 98, 278
    Charles teaches at 7–8, 13, 19, 21, 33–4, 442n19

PD's education 20, 22–8, 42, 445n29
    relocates to Cotham Lawn Road 33
    celebration of PD's success 174, 217
Merchant Venturers' Society 8
Merchant Venturers' Technical College, Bristol 83, 278, 444n56
    Felix studies at the Faculty of Engineering 28, 29–32
PD's studies 28, 29, 337, 446n10
    wartime bombing of 312
mesons 368, 369, 483n22
Metropolitan Police Special Branch 65
MI5 65, 191, 249, 250, 350, 383
Miami 385, 389
Miami Museum of Science 391
Mickey Mouse 3, 172, 274, 460n5
microelectronics 2, 429
Mill, John Stuart 43, 126, 137, 138, 143, 176, 188, 222, 300, 401, 436, 450n53
    On Liberty ix
    A System of Logic 43–4, 137, 300
Miller, Arthur 334, 345
Millikan, Robert 198, 225
    electrical charge 185
    cosmic rays 196, 197, 225
    and Anderson's evidence for a positive electron 212
    and electron-positron pairs 224
    efforts to get Kapitza released 264, 265
Mills, Robert 488n25
Milne, Edward 80, 205, 290
Miners' Union 98
Minkowski, Hermann 40
Molière 260
Møller, Christian 112
Mond Laboratory, Cambridge 226, 248, 264, 276, 293
Monge, Gaspard 48
Monk Road, Bishopston, Bristol (No.15) 10, 18, 442n15
monopole problem 431
Monthey, Switzerland 4, 9, 10
Moore, George 450n53
    Principia ethica 74
Morgan, Howard 489n49
Morrisville, Vermont 385
Morse, Louise 343
Moscow 260
    PD in 154, 165, 268–9, 278, 357–9
    Kapitza detained in 257, 258, 260
    science community 263
Moscow News 341
Moscow Polytechnic 199
Moscow University 359
Moseley, Sir Oswald 227

Mott, Nevill 106, 145, 157–8, 174, 478n46
Mount Brocken 153
Mount Elbrus 279
Mount Wilson Observatory, near Pasadena 163
Mountfield Nursing Home, London 474n10
Much Wenlock, Shropshire 78
Munich 7, 101
Munich agreement 298, 300
muon 369, 370, 382, 410
Mussolini, Benito 227, 270, 298, 312

Nagasaki, bombing of 329
nanotechnology 429
NASA 433
nature
  fundamental equations of Nature as only
    approximations 45
  laws of 44, 171, 301, 402
  metaphor of a colossal clockwork mecha-
    nism 130
  unity and beauty of 300–301
Nature journal 36, 178, 202–3, 243, 273,
    288, 289, 290, 294, 297, 371, 423
Nazis (National Socialists)/Nazism 131, 151,
  180, 200, 217, 219, 235, 250, 297, 298,
  308, 311, 317, 325, 327, 328, 329, 340,
  371, 406
negative energy states 166–7
Nehru, Jawaharlal 354
neutrinos 194, 195, 203, 204, 205, 233, 255,
  273, 360, 367, 369, 382, 483–4n32
neutron stars 432
neutrons
  atomic nuclei 62
  Chadwick's discovery 202–3, 212, 399
  Rutherford proposes 186, 202
  strongly interacting 382
New Statesman 175–6, 267, 276, 362
New York 160, 257
New York Times 160, 178, 195, 196, 198,
  223, 335, 351, 352, 361, 368, 405
Newlin's restaurant, Princeton 260
Newman, Max 189, 320
Newnham College, Cambridge 479n28
News Chronicle 263
Newton, Sir Isaac 65, 72, 97, 99, 116, 147,
  157, 190, 294
  childhood 17
  Einstein's theory refutes his ideas 35, 36, 61
  theory of gravity 41, 42
  mechanics 49, 50, 88
  burial in Westminster Abbey 414, 415
  and autism 424
Nightingale, Florence 8, 403
Nixon, Richard 389

Noakes, Michael 403, 489n48, 489n49
Nobel, Alfred 235, 241, 245
'Nobel disease' 236
Nobel Foundation 237, 238, 245–6
non-commuting quantities 84–5, 86, 128
non-interacting quantum particles 156
Norway, PD in 230
nuclear fission 299, 303, 306
nuclear industry 431
nuclear weapons 380, 406
  destruction of incoming 358
  Second World War 299, 300, 303, 306, 307,
    311, 316, 319, 324, 326, 328, 334, 341

Occhialini, Giuseppe 214–15, 218–19
Oklahoma! (film soundtrack) 366
Old Faithful geyser 163
Oppenheimer, Frank 349
Oppenheimer, J. Robert ix, 111, 281, 332,
    355, 356, 375
  personality 90, 121, 124, 133, 352
  dislike of Cambridge life 90
  clinical depression 90, 124
  tries to poison Blackett 97
  works with PD 97
  friendship with PD 121, 133, 135, 247,
    267, 342, 351, 352, 355, 356, 375
  at Born's Department of Theoretical
    Physics 121
  poetry 121
  Ph.D. on the quantum mechanics of mole-
    cules 123, 133
  and the rise of anti-Semitism 131
  disappointed with PD's work in Göttingen
    132
  at University of California at Berkeley 163,
    267
  and PD's hole theory 172, 181, 187, 212,
    247
  on the Heisenberg-Pauli theory 201
  and Anderson's positive electron 212–13
  quantum electrodynamics 248
  Scientific Director of the Manhattan
    Project 321–2, 324, 351, 352
  celebrated as a hero in the USA 329
  director of the Institute for Advanced
    Study 342
  former Communist sympathies 344
  adviser on nuclear policy 349
  US withdraws his security clearance 349–52
  appearance 370
  retirement and death 383
  and black holes 432
Orpington, south-east London 420
Orwell, George 75, 251, 302

Coming Up for Air 209
The Lion and the Unicorn 303
Oseen, Carl 246
Ottawa 355–6
Oxford 314

Pais, Abraham 416, 451*n*24
Subtle is the Lord 468*n*20
Palais de la Découverte, La, Paris 330–31, 353
pantheism 402
Papal Academy 401, 415, 437
parity violation 361, 362
particle accelerators 430, 434
particle physics 381, 396, 398, 399, 405
Pauli, Wolfgang 91–2, 100, 127, 138, 155,
    210, 218, 234, 308, 338, 341, 415, 428,
    479*n*33
  an analytical conservative analyst 362
  exclusion principle 103–4, 166–7
  personality 109
  and electron spin 142
  PD's harshest critic 166, 175, 247–8, 343,
    362, 400, 415
  Second Principle 172
  praises PD's textbook 179
  and PD's hole theory 180, 181, 187, 206,
    225, 230, 362
  co-presents seminar with PD at Princeton
    194–5
  appearance 194–5
  the neutrino 194, 195, 203, 205, 233, 360
  problems in his personal life 195
  second marriage 298
  quarrels with Heisenberg 360
  Nobel Prize 405
  death 360, 362
Pavlov, Ivan 269
'Peanuts' 391
Pearl Harbor, bombing of 318
Peierls, Genia 305–5
Peierls, Rudolf 151, 176, 194, 305–7, 311,
    314, 316, 319, 321, 324, 340, 350, 429,
    457*n*12, 458*n*10
Penrose, Roger 130
Peterhouse College, Cambridge 236
Phillips, Leslie Roy 21, 444*n*4, 445*n*26, 446*n*31
philosophy 43, 44, 74, 111, 137, 377, 391
Phoney War 304, 308
photography, amateur 12
photons 368
  and Einstein 203
  and Langer 213
  light consisting of 69, 82, 451*n*20
  scattering by a single electron 95
  stimulated emission process and the laser 118

Picasso, Pablo 157, 228
Pickering, Arthur 25–6, 42, 278
Pincher, Chapman 351, 477*n*44
Pippard, Brian 359
Planck, Max
  quantum hypothesis 51–2, 81,
    104–5
  blackbody radiation spectrum 52, 105
Planck's constant 52, 87
Plato 74
Podolsky, Boris 210, 254, 464*n*25
Poisson bracket 86, 87, 452*n*40
Poland
  Hitler's invasion of 302
  collapse of 304
  Manci's view of Poles 327
Polkinghorne, John 365, 374
Poncelet, Jean-Victor 48
Portishead, Bristol 11, 192, 401
Portland Street Chapel, Bristol 9
position and momentum symbols 87, 91
positive energy states 166
positivism 43, 222
positron emission tomography (PET) 423, 434
positrons 2, 187, 195, 224–6, 229, 230, 233,
    235, 243, 246, 434, 488*n*34
  see also anti-electrons
Pottier family 442*n*16
Pravda 190, 412
'primitive atom' theory 261, 281
Princeton 161, 193–4, 268, 286, 335
Princeton University 191, 198, 252, 259–62,
    265, 281, 296, 473*n*50
  bicentennial celebrations 332
  Fine Hall (later Jones Hall) 194, 197, 200,
    201, 252, 253, 254, 256, 257, 260
  Fuld Hall 342, 343
  Graduate College 332
Proceedings of the Royal Society 216
projective geometry 48–9, 85, 94, 130, 181,
    436, 448*n*5
protons
  atomic nuclei 62
  negative 243
  strongly interacting 382
Pryce, Gritli (née Born) 310
Pryce, Maurice 310, 311, 316
pulsars 432
Punch magazine 36
Pythagoras's theorem 42

quanta 51
  energy 52, 81
  Schrödinger's wave theory 100
  see also photons

quantum chromodynamics 398
quantum electrodynamics 117–9, 123, 126,
    132, 138, 139, 233–4, 247, 248, 250,
    270, 273, 274, 296, 335–8, 343, 348,
    356, 376, 380, 398, 410
quantum field theory 116–19, 123, 126, 132,
    138–9, 200–201, 204, 205, 210, 216,
    226, 229, 234, 247, 267, 295, 336, 343,
    344, 369, 382, 437
    see also quantum electrodynamics and
    quantum chromodynamics
quantum jumps 110, 455n24
quantum mechanics 1–2, 300, 348, 371, 405
    named by Born 88
    birthplace of 406
    building of the complete theory 92–3
    first prediction of 96
    mathematical symbols in 96, 109
    central role of probability 119
    relationship with classical mechanics 156,
        215–16
    relativistic 316
    and miniaturisation 429
quantum numbers 70, 71
quantum theory
    discovered by Planck 51–2
    Einstein lays its foundations 11
    laws of 51
    PD introduces the mathematics of creation
        and annihilation 118–19
    the universe as fundamentally granular
        116–17
quarks 382, 398, 399, 433
quaternions 50, 73, 85
Queen Mary (liner) 281

Rabi, Isidor 335
radiation
    electromagnetic 52, 68–9, 81
    gravitational 432
radio 67, 165, 311
radioactive decay 152, 168, 195, 233, 255
Rain Man (film) 422
Ramond, Pierre 408–9, 414, 490n1
Reagan, Ronald 411
Redlands Girls' School, Bristol 33
reductionism 158
relativity 11, 25, 52, 300
    as PD's passion 35–6, 38–9
    Broad's teaching of 39–40
    Einstein's general theory 40, 41, 42, 45,
        60, 68, 74, 75, 131, 142, 145, 157, 188,
        199, 255, 261, 280–81, 367, 368, 379,
        392, 396, 432, 433, 436, 483–4n32,
        484n35

Einstein's special theory 40, 41, 42, 50, 53,
    72, 74, 75, 81, 85, 95, 136, 141, 142,
    144, 201, 210, 316, 343, 448n10
    Hassé speaks on the subject at Cambridge
        45
Rembrandt van Rijn 347
renormalisation 336–7, 356, 398, 399, 405,
    409, 412, 425, 437, 488n29
Retherford, Robert 335
Reynolds's Illustrated News 206
Richards, Sir Gordon 348
Riemann, Bernhard 26, 42
Rijksmuseum, Amsterdam 347
Robertson, Andrew 173–4
Robertson, David 31–2, 38, 45, 47
Robertson, Howard 131–2
Robertson, Malcolm 193–4
Robeson, Paul 254
Roentgen Institute, Leningrad 154
Rolls-Royce 388, 491n1
Röntgen, Wilhelm 235
Roosevelt, Eleanor 323
Roosevelt, Franklin D. 217, 264, 307, 324
Roselawn cemetery, Tallahassee 414
Rosen, Nathan 254
Rosenfeld, Léon 219–20
Rothschild, Victor, Lord 489n51
'Roundy' (Joseph Coughlin) 162
Rousseau, Jean-Jacques 11, 404
Royal Air Force 304, 310, 326
Royal Astronomical Society, Burlington
        House, London 465n16
Royal Commission for the Exhibition of
        1851 82, 105, 132
Royal Navy 63, 304
Royal Society 89, 91, 105, 118, 143
    PD elected a Fellow 173–4
    funds the Mond Laboratory 226
    Baker Medal 316
    and Heisenberg 340–41, 479n33
    and Schrödinger 341
Royal Society of Scotland 300
Rugby, Warwickshire 37–8, 45
Russell, Bertrand 347
    'Zahatopolk' 480n6
Rutherford, Ernest, Baron Rutherford of
        Nelson 25, 74, 89, 152, 157, 223, 249,
        380, 403, 414, 430
    and Eddington 61, 289
    personality 61, 64, 81, 182
    appearance 61
    discovery of the atomic nucleus 61–2, 69
    proposes the neutron 186, 202
    director of the Cavendish Laboratory 62
    Kapitza's nickname for him ('the

Crocodile') 64
and Kapitza's support of Communism 65
and Bohr 81, 108
relationship with Kapitza 97, 182
loathes Bernal 99, 146
ennobled 183
death of his daughter 183
and Chadwick's discovery of the neutron 203
and the Cockcroft-Walton splitting of the atom 206
leadership of Cambridge experimental physicists 207
Blackett's anger at his despotic style 213–14
bottom-up approach to physics 224
bas-relief in the Mond Laboratory 226–7
stays in Bohr's mansion 466n47
and Kapitza's detention 258, 260, 269, 270, 276, 280
and PD's marriage 286
on Eddington 288–9
death 294
memorial service at Westminster Abbey 293–4, 415

St John's College, Cambridge 78, 135, 247, 285, 288
PD unable to take up a place at (1921) 45–6, 53
PD wins two scholarships (1923) 53
PD arrives at 55
described 55–8, 59–60, 77
encourages PD to apply for a Fellowship 132
awards PD a special lectureship 158, 459n36
Tamm's visit (1931) 188–90
Born's honorary position 219
Combination Room 140, 326, 327, 407
Fellowship extended for life 387
Isenstein bust of PD 402
PD's last visit to 407
first women undergraduates 407
nurturing environment for PD 424
PD apologises for absence at eightieth birthday celebrations 489n66
PD's Nobel Medal and certificate returned 491n37
St Maurice, Switzerland 9, 442n16
Sakharov, Andrei 433–4
Salam, Abdus 273, 369, 395, 398
Salaman, Esther and Myer 339, 404, 442n24
scattering matrix 369, 399
Schnabel, Artur 261

Schönberg, Arnold 157, 337, 479n15
Schrödinger, Annemarie (Anny) 242, 305, 328
Schrödinger, Erwin 274, 308, 328, 338, 396, 428, 485n8
his quantum theory 99–101, 112, 136, 160, 316, 333
reputation as a polymath 99
wave mechanics 100–101, 104, 109, 112, 142, 160
visits the Bohr Institute 110
Nobel Prize for physics 234, 237, 238, 242, 245, 246
personality 242
a refugee in Oxford 244
affirms his loyalty to the Nazi regime 297, 298, 371
accepts Dublin post 297–8
Dublin conference (1942) 323
elected to the Royal Society 341
death 371
PD's obituary 371, 474n56
Schrödinger's equation 100, 102, 112, 376, 377
Schuster, Arthur 243
Schwarz, John 436
Schwinger, Julian 335
Science journal 212, 213
Science Museum, London 190, 346
Scientific American 376, 485n10
Scott lecture 300–301, 432
Second Physics Institute, Göttingen 122
Second World War
    Chamberlain declares war on Germany 302
    Cambridge 303–4, 309, 310, 315, 323, 325
    blitzkrieg of Belgium, Luxembourg and the Netherlands 308
    end of the war in Europe 327
Seiberg, Nathan 433
Sen, Colleen Taylor 488n35
Shakespeare, William 3, 110–11
    Hamlet 109, 110, 115
    Love's Labour's Lost 247
    Richard II 76
Shankland, Robert 273
Shaw, George Bernard 176
    Getting Married 271, 285
    – Preface 7
    The Irrational Knot 28
Shelter Island Conference, Long Island, New York (1947) 334–5
Shinyo Maru (steamer) 163
Sidgwick, Henry 450n53
Silver Lake, near Tallahassee 397
Simon, Sir Francis 314

Simpsons, The 431, 493*n*8
Sinatra, Frank 403, 489*n*48
Skye, Isle of 190
Slater, John 96, 144, 453*n*8
Sliger, Bernie 392
Snow, C. P. 183
    The Search 183, 461*n*51
Social-Democratic Workers' Party 149
Socialist Society 236
sociology 43
solar-eclipse experiments (1919) 36, 41, 61, 223
Solvay Conferences
    1927 136–9, 143, 180, 221, 334
    1930 180
    1933 233–4
    1961 368–9
Sommerfeld, Arnold: *Atomic Structure and Spectral Lines* 72
Sound of Music, The (film) 1
Soviet Academy of Sciences 188, 327, 388
Soviet Conference on Nuclear Physics (Leningrad, 1933) 230
Soviet Embassy, Washington 265
Soviet Union
    PD's first visit 148, 153–5
    PD's second visit 175–6
    and the British press 175–6, 190
    the Soviet experiment 181, 189, 190, 198, 210, 228, 231–2
    the Jazz Band 199, 305
    PD falls foul of the censors 199–200
    PD's support for Soviet physics 216
    PD attends Leningrad conference (1933) 230
    PD unaware of the cost of the collectivisation programme 231–2
    PD in Bolshevo 268, 292–3
    Great Purge 292
    trials in 297
    Nazi invasion of 317
    PD and colleagues refused visas by Churchill 327
    Fuchs passes secrets to 350
    early detonation of the Soviet nuclear weapon 350
    Sputnik missions 363
    space programme 364
    Cuban crisis 370–71
space-time
    curved 41, 42, 131
    and de Sitter 255
    more than four dimensions of 437
    special theory of relativity 42
    unified 40

Spanish Civil War (1936-9) 291–2
Spender, Stephen
    Journals 484*n*38
    World Within World 121
Spielberg, Steven 386
spinors 430
spintronics 429
Spinoza, Baruch 402
Sputnik missions 363
SS 328
Stalin, Joseph 181, 227, 251, 260, 263, 269
    rise to absolute power 154
    industrialisation policy 64, 175
    collective farming programme 175
    interviewed in *New Statesman* 175–6
    and the intelligentsia 190
    attitude to science 198, 231
    Cambridge students favour over Hitler 227–8
    his government becomes more repressive 258
    non-aggression pact with Hitler 302, 317
    and Kapitza 341, 358
    death 358
    Khrushchev denounces 357
Stalingrad 155
Standard Model 398, 399, 405, 431, 436, 438
Stanford, Henry King 388, 389
Stanford University 433
Star Trek 431, 434, 492–3*n*8
Start the Week (Radio 4 programme) 420
steady-state theory 379, 403
Stockholm, Sweden 237–43, 245, 246
Stockman, Gertrude 244
Stokes, Sir George 365
Stony Brook, New York 484*n*6
Stoppard, Tom: *Arcadia* 94
Strassman, Fritz 298–9
Strategic Defence ('Star Wars') Initiative 358
stress diagrams 44, 447*n*54
string theory 409–10, 436, 437–8, 493*n*27
strings 356, 357
strong interaction 348, 369, 398
subatomic particle accelerators 363
Sudarshan, George 481*n*31
Suez crisis (1956) 363
Sunday Dispatch 235
Svenska Dagbladet 237, 239
Swift, Jonathan: *Gulliver's Travels* 171
Swirles, Bertha 465*n*15
Switzerland, PD visits 11, 404
Szilárd, Leó 152, 264, 299, 300, 307, 308

't Hooft, Gerard 398
Tallahassee, Florida 389, 390, 393, 411

Tallahassee Democrat 413
Tallahassee Memorial Hospital, Florida 1, 411
Tamm, Igor 151, 176, 229, 267, 359, 433
  at Leiden 149–50
  personality 149
  politics 149
  first Soviet theoretician to use quantum mechanics 149
  friendship with PD 149–50
  meets up with PD in Moscow 165
  and PD's hole theory 172, 180
  on 'brigade education' 181
  and the positron's detection 224–5
  in Bolshevo 268
  climbing vacation in the USSR with PD 276, 279
  secret project to build the hydrogen bomb 359
  Nobel Prize 405
Tata Institute, Bombay 353, 481*n*30
technical drawing 15–16, 23, 48, 130, 443*n*40, 443*n*44
Teller, Edward 152, 300, 307, 351, 355, 363, 378, 407, 485*n*12
Tennyson, Alfred, Lord 146
Teszler, Betty (PD's sister) *see* Dirac, Beatrice
Teszler, Christine (PD's niece) 330, 445*n*29
Teszler, Joe 286, 291, 325, 328, 330, 477*n*34
Teszler, Roger (PD's nephew) 291, 325, 328
Thatcher, Margaret, Baroness 3, 405, 406
theoretical physics 45, 118
  Berlin as its global capital 96
  PD introduces a new approach to 187–8
  Weyl's approach 396
Thomson, J. J. 51, 62, 157, 182, 243, 403
Tilley, Peter 412–13
Times, The 35, 146, 235, 299, 383
Tkachenko, Vladimir 382–3, 485*n*33
Todd, Horace 448*n*2
Tollast, Robert 403
Tolstoy, Count Leo 238
  Anna Karenina 154
  War and Peace 339
Tomonaga, Sin-Itiro 336
Tots and Quots dining club 313
Trans-Siberian Railway 165
transformation theory 127, 128, 141
transistors 429
Trieste symposium (1971) 395
Trinity College, Cambridge 11, 63, 65, 66, 74, 135, 176, 203, 249, 280, 292, 380, 424
Trotsky, Leon 64
Troyanovsky, Aleksandr 264, 265
Truman, Harry S. 328, 329, 342
'Tube Alloys' project 319, 321, 326

tuberculosis 22, 332, 410, 427, 492*n*23
Turin Shroud 399, 488*n*28
Turing, Alan 320
Tyndall, Arthur 51, 52, 104, 271–2

uncertainty principle 113, 127–9, 198, 220, 317
Under the Banner of Marxism journal 368
UNESCO 405
United States of America
  development of quantum mechanics 96
  PD's first visit (1929) 159, 160–63
  PD's 1931 visit 191, 193–201
  depression in 198
  Einstein emigrates to 217
  prominent role in the Second World War 323
  American-led experiments to build a nuclear bomb 323, 324
  funding of theoretical physics 334
  anti-Communist paranoia (1950s) 349
  space programme 364
  Judy settles in 373
  PD's regular visits 375
universe
  expanding 301, 378
  'primitive atom' theory 261, 281
University of Aarhus, Denmark 384
University of Bristol 68, 83, 136, 349, 419
  University Engineering Society 32
  Dirac Centenary Meeting (2002) 428
  Faculty of Engineering 28
  mathematics department 45
  PD declines an honorary degree 347
  PD takes the qualifying examinations early 445*n*40
  PD's FRS election 173–4
University of British Columbia 355
University of California at Berkeley 163, 267, 322
University of Cambridge *see* Cambridge University
University of Florida, Gainesville 408
University of Geneva 7
University of Leiden, Netherlands 119
University of Liverpool 310
University of London 436
University of Madison, Wisconsin 161–3, 281
University of Manchester 61
University of Miami 385, 388, 389, 484*n*6, 486*n*47
University of Minnesota 349
University of Nebraska 432
University of Swansea 384

University of Texas at Austin 388, 484n6
Updike, John 386
uranium 298–300
    235 isotope 306, 307, 313, 321
    238 isotope 306, 307, 313
Urey, Harold 321
utilitarianism 43, 137, 450n53

vacuum concept 225, 233–4, 344
vacuum cleaner 134
Valais canton, Switzerland 4
Van Vleck, John 144, 159, 161, 163, 193,
    252, 332, 371, 405, 460n5
Vancouver 354, 355
VE-Day celebrations 327
Veblen, Oswald 194, 197, 218, 253, 254,
    307, 308, 342
Veltman, Martin 398, 438
Vermont 384–5, 414
Victoria, Queen 10
Vienna 349
Vietnam war 388, 389
Vieux, Annette (née Giroud; PD's paternal
    great-grandmother) 442n10
Viktor Frankl Institute, Vienna 349
virtual states 126
Vladikavkas 155
Vladivostock 165
von Neumann, John 152, 156, 261, 307, 371
VSO (MI5 informant) 249, 250

Wakulla river 397
Waldegrave, William, Baron Waldegrave of
    North Hill 419, 489n51, 491n1
Wall Street crash (1929) 166
Waller, Ivar 166, 246, 483n5
Walters, Barbara: How to Talk to Practically
    Anybody about Practically Anything 391
Walton, Ernest 206, 207, 222, 270, 341, 404,
    405
Walton, Sir William 403
Washington, D.C. 178, 258, 259, 264–5, 350
Watt, Dr Hansell 410, 411, 413
Wattenberg, Al 323–4
Waugh, Evelyn: Brideshead Revisited 58
wavicle 161
weak interaction 348, 360, 361, 398, 483n7
Wei Chi (a.k.a. Go) 256, 411
Weimar Republic 121–2
Weinberg, Steven 398, 428
Weinberg-Salam theory 398
Weisskopf, Vicki 247, 248, 400
Wells, H. G. 28, 32, 299, 313, 328–9
    The Time Machine 25
    The World Set Free 40

Westminster Abbey, London 142, 293–4,
    414–16, 491n32
Weyl, Hermann 167, 181, 187, 396
Wheeler, John 204, 306, 352
Whewell, William 176
Whiston, William 207–8
White, Sir George 19
Whitehead, Henry 147, 159
Whitehead, Right Reverend Henry 147–8, 284
Whitehead, Isabel 147–8, 159, 220, 240,
    281–2, 284, 353, 457n22
Whittaker, Edmund 86–7
Wigner, Amelia (née Frank) 281, 297, 384–5,
    473n53
Wigner, Jenő (later Eugene) 125, 259, 260,
    293, 340, 362, 388, 389, 407, 491n35
    childhood 152
    field theory of the electron 139
    personality 153, 383
    and PD's impoliteness 249
    aims to bring modern quantum mechanics
        to Princeton 253
    at the University of Madison 281
    marriage to Amelia 297, 473n53
    and nuclear fission 300
    marriage to Mary 303
    and nuclear weapons 307
    organises 'The Future of Nuclear Science'
        conference 332
    on Feynman and PD 334
    and Kuhn's interviewing of PD 371–2
    an elder statesman of American science
        383
    and Judy's disappearance 384–5
    Manci's response to his death 417
    The Recollections of Eugene P. Wigner 391
Wigner, Mary 303
Wilczek, Frank 142
Wilde, Oscar 16, 121
    The American Invasion 332
    The Importance of Being Earnest 282
Wilhelm II, Kaiser 10, 29
Wilkinson, Sir Denys 476n11
Williams, Edith 26
Willis, D. C. 135
Wiltshire, Herbert Charles (Charlie) 31, 36, 37
Wisconsin Journal 162
Witten, Edward 437, 493n25
Wittgenstein, Ludwig 220
Wolverhampton 49, 67, 78
Woolf, Virginia 74
Wordie, James 132, 448n17
Wordsworth, William 50, 57, 310, 411
    The Prelude 55
Wu, Chien-Shiung 361

X-rays 69, 168, 235

Yang, C. N. 361, 362, 430, 488*n*25
Yeats, W. B.: *The Living Beauty* 408
Yeshiva University, New York 376

Yosemite National Park 163

Zimmerman, Erika 416, 491*n*35
Zurich 7, 360
Zweig, George 382